1000714665

GRANT MACEWAN
COMMUNITY COLLEGE

DATE DUE			
NOV 22 1996			
OCT 20 1997			
MAR 31 2000			
SEP 08 2000			
NOV 10 2003			
FEB 02 2004			

Addison-Wesley

Biology
SI EDITION

Addison-Wesley
Biology
SI EDITION

Edward J. Kormondy
Bernice E. Essenfeld

Addison-Wesley Publishers
Don Mills, Ontario • Reading, Massachusetts
Menlo Park, California • London • Amsterdam • Sydney

Authors

Edward J. Kormondy is Vice President for Academic Affairs at California State University—Los Angeles. Dr. Kormondy received his Ph.D. degree in zoology at the University of Michigan and has taught college biology and ecology for a number of years. He is the author of many articles and the popular college text, *Concepts of Ecology*. He has served as consultant to many organizations, including the National Science Foundation, and has contributed to a variety of educational life science programs. Dr. Kormondy is past president of the National Association of Biology Teachers.

Bernice E. Essenfeld received her Ph.D. degree in science education at Columbia University. For many years, she taught at Irvington High School in Irvington, New York, where she also served as Science Department Head. Dr. Essenfeld has taught courses at Columbia University on teaching biological sciences, and she has written articles for professional educational magazines.

Consultants

Mary Alice Fryar
Science Department Head, Sandalwood Junior
   High School
Jacksonville, Florida

Maren McDonald
Teacher, College Park High School
Pleasant Hill, California

Karin Rhines
Science Consultant, Bedford Hills, New York

George Zahrobsky
Science Department Head, Glenbard West
   High School
Glen Ellyn, Illinois

Cover photo Seven to ten days after spawning, threespine stickleback fish eggs begin hatching. The young fish have large yolk sacs and generally remain in the nest made by the male parent for two days after they hatch.

Copyright © 1984 Addison-Wesley Publishers Limited. All rights reserved. No part of this publication may be reproduced, stored in a retrieval system, or transmitted, in any form or by any means, electronic, mechanical, photocopying, recording, or otherwise, without the prior written permission of the publisher. Printed in the United States of America.

ISBN 0-201-16816-2

ABCDEFGHIJKL-VH-8987654

Acknowledgments

Illustrations

Edith Allgood: Figs. 8–27, 8–28, 8–30, 10–24, 11–30, 12–4, 12–5, 12–10, 12–11, 12–13, 12–21, 12–26, 13–6, 13–8, 13–10, 13–28, 14–12, 14–24, 14–26, 14–30, 20–19, 20–20, 26–9, 26–18, 26–22
Sherry Balestra: Figs. 4–30, 4–32, 5–26, 7–2, 13–23, Perspective 14 (p. 419), 25–9
Kathy Barbour: Figs. 14–11, 22–3, 22–4, 22–5
Ellen Blonder: Perspective 23 (p. 659)
Dale Bogaski: Fig. 7–22
Shirley Bortoli: Figs. 6–3, 7–30, Art-in-text (p. 233), 8–31, 8–32, 10–25, 11–23b, Art-in-text (p. 345), Art-in-text (p. 375), 13–16b, 13–30, 14–17, 15–11, 16–6, 16–10, 16–11, 16–12, 17–3, Table 17–1 (p. 480), Art-in-text (pp. 500-501), 20–10, 21–7, Art-in-text (p. 638), 25–16, 26–17b
Cyndie Clark-Huegel: Figs. 2–32, 2–34, 2–40, 2–41, 2–54, 2–60, 3–14, 8–10, 8–19, 8–20, 9–28, 9–34, 10–6, 10–22, 15–10c, 15–15a, 17–4, 18–6, 18–7, 18–8, 18–12, 18–13
John Dawson: Figs. 1–15, 2–3
Jennifer Dewey: Figs. 1–25, 1–30, 9–22, 19–11, 25–2, 25–12, 25–30
Michael Easker: Fig. 11–22
Pam Ford: Fig. 2–33
Lisa French: Figs. 10–2, 12–17, 12–18, 12–19, 12–22, 12–23, 16–9, 16–14, 18–2b, 18–3b, 22–8, 22–9, 22–10, 26–14, 26–16
George Gershinowitz: Fig. 2–26
Barbara Hack: Figs. 2–22, 2–24, 3–22, 3–25, 4–2a, 4–3, 4–4, 4–5, 4–7c, 4–8c, 4–9, 4–13
Gunter Hagenow: Figs. 6–17, 7–24, 8–7, 8–23, 9–26, 13–4, 13–21, 14–5
Linda Harris-Sweezy: Fig. 10–23
Jean Helmer: Figs. 3–23b, 3–24b, 3–29, 5–5, 5–6a, 6–6, 9–13b, 9–16, 9–17, 10–20, 11–5, 11–24, 11–25, 11–31, 13–9b, 13–17, 13–25, 14–15, 14–19, 14–27a, 17–6, 17–7, 17–8c, 17–10, 17–22, 17–23, 17–24, 17–25b, 17–26b
Pamela Joyce Hettinger: Figs. 15–3, 20–5, 20–6, 20–9, 20–15, 23–2, Art-in-text (p. 660), 25–11
Barbara Hoopes: Figs. 10–18, 12–14, 12–15, 14–8, 14–10, 14–14, 14–32, 14–33b, 16–8, 21–17, 21–19, 21–23, 25–7, 25–15, 25–24, 25–25a
Julia Iltis: Figs. 8–15, 8–17, 8–18, 9–21, 9–31, 10–15, 10–19, 13–14, 13–19, 13–20, 13–22
Susan Jaekel: Perspective 24 (p. 693)
Tony Kenyon: Perspective 16 (p. 463), Perspective 20 (p. 575)
Heather King: Figs. 17–11d, 18–11
James LaMarche: Fig. 2–2
Dennis Leatherman: Figs. 1–16, 2–5, 3–26
Jeff Leedy: Perspective 4 (p. 145)
Judith Lopez: Figs. 1–24, 5–4, 5–18, 7–4b, 7–15, 7–16, 8–12, 8–14, 9–19, 9–36, 10–14, 10–21, 11–6, 11–15, 11–17, 12–12, 13–13, 13–15, 14–21b, 14–22b, 18–17, 19–13, 20–2, 20–4, 20–7, 20–13, 21–16, 22–7
Marlene May: Figs. 1–9, 2–9, 2–38, 2–44, 2–47, 2–59, 2–61, 7–19, 10–17, 11–8, 11–19, 11–20, 12–3, 13–16a, 13–26, 14–3, 17–9, 17–16, 17–15, 18–10, 22–12, 22–19, 24–8, 24–28
Jane McCreary: Table 2–1 (p. 43), Figs. 4–29, 5–3
Virginia Mickelson: Figs. 1–19, 4–34, 7–12, 7–13, 7–20, 8–6, 8–22, 9–5, 9–6, 9–7, 9–20, 9–30, 9–39, 11–14, 11–16, 11–18, 11–26, 12–7, 12–24, 12–25, 13–12, 14–34
Debby Morse: Fig. 3–11, Perspective 5 (p. 177), Perspective 9 (p. 277), 14–4, 14–6b, 20–22, 21–3, 21–4, 21–8, 21–11
Dennis Nolan: Perspective 6 (p. 205), Perspective 8 (p. 259)
Sharron O'Neil: Figs. 19–10, 19–15, 19–18, 21–18, 21–21, 22–6, 24–7, 25–4
Betsy Palay: Fig. 2–57
Vida Pavesich: Figs. 10–10, 10–11
Heather Preston: Figs. 2–30, 16–3, 16–4, 19–3, 19–4, 19–5, 19–6, 19–12
Doug Roy: Perspective 19 (p. 555)
Margaret Sanfilippo: Figs. 8–9b, 8–24, 9–29, 9–33
Alan Say: Perspective 2 (p. 45)
Carla Simmons: Figs. 16–13, 17–12a, 17–13, 17–14, 17–19, 21–20, 24–5, 25–17, 25–19, 25–27
Blanche Sims: Perspective 1 (p. 19)
Doug Smith: Perspective 11 (p. 343)
Robert Steele: Fig. 1–3
Sara Lee Steigerwald: Figs. 5–24, 7–3, 7–9, 7–25, 8–8, 9–8, 9–9, 10–4, 10–12, 10–16, 13–3a, 13–7, 13–11, 13–18, 14–18, 14–25
Cynthia Swann-Brodie: Figs. 1–18, 2–12, 7–21, 11–11, 11–28, 14–13, 14–23, 21–2, Perspective 22 (p. 637), 24–3, 24–9, 26–11
Ed Tabor: Perspective 17 (p. 499)
James Teason: Fig. 21–15
Tom Wilson: Perspective 10 (p. 315), Perspective 13 (p. 401), Figs. 15–4, 15–5a, 15–6 right, 15–16 right, 15–17, 15–18, 15–19, 20–25
Stephen Zinkus: Figs. 1–4, 2–6, 2–7, 3–2, 19–2, 21–22

Foreword

The common thread that connects human beings with all other creatures is simply this: we are alive. Because we are alive we are similar in many ways. All living things need food for building their bodies and providing energy for life's activities. All living things react to their environment and interact with each other.

Addison-Wesley Biology emphasizes the fundamental similarities of all living things. At the same time, it compares the varied ways that different types of living things perform the activities basic to their survival. Because biology is so visually rich, learning biology must be a visual experience. This book is illustrated with numerous photographs, diagrams, and drawings. Together, the text and illustrations capture the many dimensions of biology, from the structure of minute viruses to the dynamic interactions among plants, animals, and other living things.

The text is made up of eight units and 26 chapters. Each chapter is divided into three sections. At the end of each section is a list of the vocabulary words contained in the section and several questions that ask you to recall information you read in the section. Each chapter also contains a one-page essay entitled *Perspective*. These essays highlight careers in biology, biotechnology, the work of important biologists, environmental issues, and other topics of interest.

Each chapter concludes with a summary of the major concepts presented in the chapter and two types of questions. The first asks you to apply what you have learned. The second asks you to relate the material in the present chapter to other chapters or to go beyond the text in evaluating a situation relevant to the material covered. Suggestions for individual research projects appear at the end of the chapter, along with descriptions of two or three biology-related careers. Each unit ends with an annotated list of suggested readings.

New science words are carefully introduced. When an unfamiliar word is used for the first time, it is set off in **boldface** type. For some of these, a simple phonetic pronunciation is given in parentheses. A pronunciation key appears at the back of the book. The word origins of many new science terms appear in the margins of the text.

Several resources to facilitate using the text are included at the end of the book. There are appendixes on the metric system and classification of living things. The glossary, which includes the pronunciation key, gives definitions of many science words introduced in the text. Finally, there is a complete index, a valuable reference tool in learning about living things.

Contents

Unit I Introduction to Living Things — 1

1 The World of Life — 3
1.1 An Introduction to Biology — 3
1.2 An Overview of Structure and Function in Living Things — 9
  Perspective Everything's Connected to Everything Else — 19
1.3 An Overview of Interactions in the Environment — 20
  Chapter Summary and Review — 28

2 Classification of Living Things — 31
2.1 Scientific Names and Classification — 31
2.2 The Basis for Classification — 38
  Perspective Same But Different — 45
2.3 Kingdoms of Living Things — 46
  Chapter Summary and Review — 84
  Unit I Bibliography — 86

Unit II Cell Structure and Function — 87

3 The Cell: Basic Unit of Life — 89
3.1 Early Observations of the Cell — 89
3.2 Cell Investigation — 95

3.3	A Closer Look at the Cell	100
	Perspective So You Like the Water?	113
	Chapter Summary and Review	114

4 The Molecular Machinery of the Cell — 117

4.1	The Chemical Components of Matter	117
4.2	The Chemical Components of Cells	126
4.3	Chemical Reactions in Cells: Energy and Enzymes	136
	Perspective Hazardous Environments	145
	Chapter Summary and Review	146

5 Food Production and Nutrition — 149

5.1	Food Production	149
5.2	The Mechanism of Photosynthesis	157
5.3	Food and Human Nutrition	165
	Perspective Is It Real Food?	177
	Chapter Summary and Review	178

6 Obtaining Energy from Food — 181

6.1	The Process of Obtaining Energy from Food	181
6.2	The Mechanism of Cellular Respiration	189
6.3	Cellular Respiration and Metabolism	197
	Perspective Petroleum from Plants	205
	Chapter Summary and Review	206
	Unit II Bibliography	208

Unit III Basic Body Processes — 209

7 Gas Exchange — 211

7.1	An Overview of Gas Exchange	211
7.2	Respiration in the Representative Organisms	215
7.3	Breathing and Respiration in Humans	223
	Perspective So You'd Like to Work Outdoors?	231
	Chapter Summary and Review	232

8 Food Processing — 235

8.1	Getting and Using Food: A General Plan	235
8.2	Food Processing in the Representative Organisms	243
8.3	Food Processing in Humans	248
	Perspective The Evil Spirit in Milk	259
	Chapter Summary and Review	260

9	**Transport**	263
9.1	Characteristics of Transport	263
9.2	Transport in the Representative Organisms	270
	Perspective Butterflies and Beetles	277
9.3	Transport in Humans	278
	Chapter Summary and Review	292
10	**Excretion**	295
10.1	An Overview of Excretion	295
10.2	Excretion in the Representative Organisms	303
10.3	Excretion in Humans	307
	Perspective A Growing Parts Catalog	315
	Chapter Summary and Review	316
11	**Movement and Locomotion**	319
11.1	Movement in Living Things	319
11.2	Movement and Locomotion in the Representative Organisms	326
11.3	Movement and Locomotion in Humans	331
	Perspective Biology as Art	343
	Chapter Summary and Review	344
	Unit III Bibliography	346

Unit IV Body Regulation and the Senses 347

12	**Chemical Control**	349
12.1	The Role of Hormones in Organisms	349
12.2	Hormone Function in Animals	355
12.3	Hormonal Control in Humans	361
	Perspective The ABCs of Teaching Biology	373
	Chapter Summary and Review	374
13	**Nervous Control**	377
13.1	Characteristics of Nervous Systems	377
13.2	An Overview of Invertebrate and Vertebrate Nervous Control	385
13.3	Nervous Control in Humans	391
	Perspective Tomorrow's Dinner	401
	Chapter Summary and Review	402

14 The Senses — 405

- 14.1 Receptors Gather Information — 405
- 14.2 Perception of Sound, Pressure, Pain, Temperature, and Body Position — 413
- *Perspective* Call of the Wild — 419
- 14.3 Perception of Sight, Smell, and Taste — 420
- Chapter Summary and Review — 430
- Unit IV Bibliography — 432

Unit V Reproduction — 433

15 Reproduction of Molecules and Cells — 435

- 15.1 The Molecular Basis of Reproduction — 435
- 15.2 The Cellular Basis of Reproduction — 440
- 15.3 Protein Synthesis — 446
- *Perspective* Competition and Cooperation in Science — 453
- Chapter Summary and Review — 454

16 Reproduction of Organisms — 457

- 16.1 Asexual Reproduction — 457
- *Perspective* Myself All Over Again — 463
- 16.2 The Basis of Sexual Reproduction — 464
- 16.3 Sexual Reproduction — 469
- Chapter Summary and Review — 474

17 Reproduction, Growth, and Development of Plants — 477

- 17.1 Asexual Reproduction in Plants — 477
- 17.2 Sexual Reproduction in Plants — 482
- 17.3 Development of Flowering Plants — 490
- *Perspective* What Good's a Plant? — 499
- Chapter Summary and Review — 500

18 Reproduction and Development of Vertebrates — 503

- 18.1 Vertebrate Reproduction — 503
- 18.2 Human Development — 508
- 18.3 Adaptations in Vertebrate Reproduction and Development — 516
- *Perspective* Agriculture—A Misunderstood Science — 525
- Chapter Summary and Review — 526
- Unit V Bibliography — 528

Unit VI Continuance and Change 529

19 Genetics: Mendel's Laws of Heredity 531
19.1 Gregor Mendel: First Geneticist 531
19.2 Mendel's Laws and Probability 539
19.3 Going Further with Mendel's Laws 545
Perspective Designing an Organism 555
Chapter Summary and Review 556

20 Heredity: Chromosomes and Genes 559
20.1 Research in Chromosomal Genetics 559
20.2 Chromosomes Contain Genes 567
Perspective Simultaneous Discovery 575
20.3 The Nature of Genes 576
Chapter Summary and Review 586

21 Change Over Time 589
21.1 Theories About the Origin of Life 589
21.2 Information About the Past 599
21.3 Theories About How Living Things Change 608
Perspective Nature's Way: Adapt or Become Extinct 615
Chapter Summary and Review 616
Unit VI Bibliography 618

Unit VII The Environment 619

22 Interactions in the Ecosystem 621
22.1 Food and Energy in the Ecosystem 621
22.2 Chemical Cycles in the Ecosystem 627
22.3 Interactions Among Organisms 632
Perspective Can There Be Too Much Carbon Dioxide? 637
Chapter Summary and Review 638

23 Populations in Ecosystems 641
23.1 Characteristics of Population Growth 641
23.2 Population Size in a Balanced Ecosystem 649
23.3 Human Population Growth 655
Perspective Population is Everybody's Problem 659
Chapter Summary and Review 660

24	**The Geography of Ecosystems**	663
24.1	Dispersal of Organisms	663
24.2	Biomes	673
24.3	Aquatic Ecosystems	685
	Perspective What Good Is a Wetland?	693
	Chapter Summary and Review	694
	Unit VII Bibliography	696

Unit VIII Reacting to Each Other and the Environment   697

25	**Behavior**	699
25.1	Overview of Behavior	699
25.2	Cyclic Behavior and an Overview of Social Behavior	709
25.3	Social Behavior in Insects and Vertebrates	718
	Perspective A Woman in the Wild	727
	Chapter Summary and Review	728

26	**Health and Disease**	731
26.1	Causes of Disease	731
26.2	Infectious Disease and Body Defense	738
	Perspective The Research Laboratory: Who Does What?	747
26.3	Conquering Infectious Diseases	748
	Chapter Summary and Review	754
	Unit VIII Bibliography	756

Appendixes   757

The International System of Units and the Metric System	757
Five-Kingdom Classification of Living Things	758
Glossary	762
Index	772
Acknowledgments	786

Unit I
Introduction to Living Things

Think of the world around you. Everywhere there are living things: squirrels, maple trees, wildflowers. What makes them different from rocks and other nonliving things? You can probably immediately think of many differences. Squirrels move. They eat. They interact with other animals. Maple trees grow tall and lose their leaves in the fall. Wildflowers bloom. Rocks merely stay in one place until they weather away or are moved by wind or water or an earthquake.

People observe living things to better understand what makes them different from nonliving things. They look for similarities and differences among the millions of kinds of living things on earth. They seek to understand the numerous fascinating relationships between all living things and their environment.

The idea of studying millions of kinds of living things is overwhelming. Only by organizing them into groups can one begin to make sense of the amazing diversity of living forms. Animals, plants, and all other living things are classified according to similarities in the ways they look and live. Classification makes order out of chaos. It enables scientists to better see how similar all living things are in fundamental ways.

Eucalyptus trees dominate a tropical rain forest in Hawaii. The largest living things on earth are trees.

Chapter 1

The World of Life

Focus *Living things are abundant on earth. They exist in many different forms, sizes, and colors. You can find living things nearly everywhere, even in a hot spring, on the rim of a volcano, or in the frozen Arctic. In spite of the diversity of living things, they are similar in fundamental ways.*

1.1 AN INTRODUCTION TO BIOLOGY

Elizabeth and Carol are identical twins. If you know twins or have twins in your family, you know how amazing the resemblances between them can be. Many twins can fool teachers, friends, and even their parents by switching identities. Some speak a special language with each other that no one else knows. A few twins are able to sense what their twin is thinking or feeling. Some even seem to have invisible ties that link them together when they are apart.

In 1979 a pair of male twins who had been separated soon after they were born were reunited at the age of 39. The researchers who brought them together found that they had many things in common. Both married and later divorced women named Linda; both remarried women named Betty. One twin named his first son James Allen. The other twin also named his first son James Alan. Both named their dogs Toy. They both

1–1 The American white ibis commonly roosts in trees and on the ground in swamps and marshes. Ibises have been valued since prehistoric times for controlling animal pests in irrigated crops.

1–2 These twins share many of the same attitudes and have the same hobbies and abilities. When they were younger it was almost impossible to tell them apart. Now you can see a few differences between them.

liked woodworking, had the same kind of headaches, worked part time as deputy sheriffs, and drove the same kind of car. These similarities are not unusual. Many pairs of separated twins have been studied and remarkable similarities show up in almost every case.

Many researchers feel that twins hold the answers to some important questions that affect all of us. For example, health researchers are interested in how food influences health. Since twins are so similar physically, they will usually react similarly to the same diet. Suppose that twins were separated and ate different foods. Would they both stay in the same state of health? Would some food factor make a difference? Twins might also help us understand how stress affects the length of a person's life. Suppose one member of a separated pair of twins had a stressful life, while the other one had a peaceful life. Would one live longer than the other? Questions such as these involve the nature of living things—the topic of this book.

Not only do many twins have similar appearances and activities, some even get the same illnesses.

What Is Biology?

bios = life
-logy = study of

The branch of science that deals with living things is called **biology**. Living things are part of your everyday life. You encounter people, dogs, cats, trees, insects, grass, and worms. Some living things around you are not as familiar or noticeable as these things. For example, you cannot see bacteria, yet they are found practically everywhere.

zōion = an animal
botanē = a plant, herb
-logy = study of

Biology is divided into specialized areas. This is necessary because there are many kinds of living things and different ways of studying them. Two of the many specialized areas of biology are **zoology** (zō-AHL-uh-jee) and **botany** (BAHT-en-ee). Zoology is the study of animals; botany is the study of plants.

1–3 With careful observation, you can observe the changes a caterpillar undergoes in becoming a butterfly.

Biology is considered a branch of science because its information can be tested. Someone may have told you that caterpillars develop into moths or butterflies. You can test this idea by watching caterpillars over a period of time and looking for changes. Scientists constantly test the ideas and information of biology. The results of these tests may show that an idea or fact is accurate, or they may show that it is in error.

Aristotle, a Greek philosopher who lived in the fourth century B.C., was one of the first people to study nature in a scientific way. Aristotle observed plants and animals and recorded his findings. Much of Aristotle's work provided important information about living things. Nevertheless, some of his ideas were later disproved. For example, Aristotle wrongly believed that blood is formed in the liver, flows to the heart, and then flows to all parts of the body through veins. This explanation was believed to be correct until the seventeenth century, when two scientists, William Harvey (1578-1657) and Marcello Malpighi (1628-1694), correctly explained how blood circulates.

1–4 Aristotle (likeness at right) was fascinated by living things. On the left is an artist's conception of one of his illustrations of the human circulatory system.

Section 1.1 An Introduction to Biology

Most of our present information about biology has been discovered within the last century.

Scientists discover more about living things every year. There is so much information that it must be organized carefully and updated constantly. It might seem that we know a tremendous amount about living things, yet the truth is that our knowledge is like the visible tip of an iceberg. The vast majority of what there is to know in biology is hidden. There are countless things still to be discovered.

What Biologists Do

The scientists who study biology are called **biologists**. As part of their work, biologists observe living things. Most biologists also perform scientific tests or experiments to learn more about the forms of life around them. Perhaps the earliest biologists were people who lived ten thousand years ago in the Middle East and used the grain of a wild-growing form of wheat for food. These people observed that when they broke the wheat heads apart, some grains fell on the soil and later began to grow. So they performed a test: they planted the wheat grains in areas of their own choosing and observed where they grew best. The results of the test had widespread effects—they led to the beginning of farming!

Biologists have different areas of interest and different ways of working. Some biologists are interested in individual living things. They may study the body parts of an organism and try to find out how the parts function. Other biologists study the behavior of living things and how they interact with each other. For example, a twentieth century biologist named Karl von Frisch wanted to know how honeybees "tell" each other where food is. Von Frisch's careful observations of honeybees in their natural surroundings and in experimental beehives helped him figure out the "language" of the bees.

Other biologists also study living things in their environment. Rachel Carson, another twentieth century biologist, wanted to know the effects on living things of chemicals that are released into the environment. The results of her studies alerted the public to the problems of pollution and inspired others to continue her work.

Many biologists begin their work in natural surroundings. They may collect living things for closer study. Then they continue their research in a laboratory where they use microscopes and other equipment to study their collections from the field. Other biologists spend most of their time in the laboratory. They study the chemical substances that make up organisms or search for information about health and disease.

1–5 Ecologist samples water from a stream. Back in the laboratory the types of chemicals and organisms present in the water will be determined.

Why Study Biology?

Picture a summer evening. A loud chorus of crickets breaks the silence. A child asks, "What is making that noise?" After getting an answer, the child asks more questions. "How do crickets make that noise? Why do they make the noise?" Many people study biology to satisfy their desire to understand the living world around them. Others study biology to find solutions to human problems. Many improvements in medical technology and food production have been the work of dedicated biologists.

Science is often in the news. Many science news reports deal with advances in health, medicine, and other aspects of biology. When you study biology, these news reports take on more meaning. Studying biology helps you keep up with ideas, discoveries, and opinions about living things that occur in the news.

Studying biology can also help you prepare for a career in several different fields. Understanding biology is important in health care. Medical doctors, dentists, nurses, laboratory technicians, medical secretaries, and dieticians all use biology in their work. Veterinarians, farmers, and foresters must also understand biology, because their careers involve plants and animals. Biology helps industries find safeguards that protect living things and their surroundings from harmful chemicals. Biology even helps artists, photographers, and writers record their impressions of the natural world.

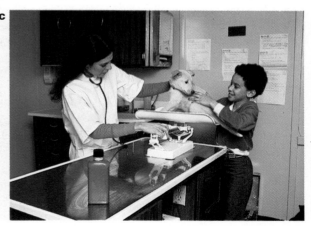

1–6 Biology-related careers. **a.** A dietician supervises food preparation. **b.** A nature photographer spots an empty beehive. **c.** A veterinarian weighs a puppy, with the help of its owner.

1–7 Waste water treatment facilities for a paper mill. The proper handling of chemicals used in many industries requires understanding the effects of the chemicals on the environment.

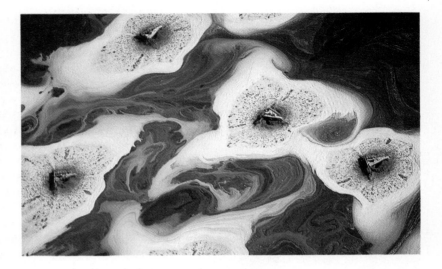

No matter what career you choose, biology relates to your personal life and your life as a community member. It deals with everyday problems: How can I lose weight? What is the best way to train my dog? How does my body fight disease? Biology also deals with far-reaching problems: What does pollution do to the environment? Can more food be produced for starving populations? Why is there an energy crisis? How can an epidemic be prevented? By understanding biology you will be able to make better decisions for yourself and for your community.

Building a Science Vocabulary

biology   botany
zoology   biologist

Questions for Review

1. Why is biology considered a branch of science?
2. How do scientists change our understanding of biology?
3. What method did Karl von Frisch use to study bees?
4. Which of the reasons for studying biology that are given in this chapter do you consider most important?

1.2 AN OVERVIEW OF STRUCTURE AND FUNCTION IN LIVING THINGS

A good place to begin studying biology is with individual living things. Each individual living thing is an **organism**. All plants and animals are organisms; so are bacteria and other living things too small for us to see.

Biologists have named most of the kinds of organisms on earth. A name makes it easier to talk about an organism. Instead of saying that a small, rounded, orange "bug" with 13 black spots was found in the garden, you can simply say that it was a ladybug. A biologist would be more precise in naming and call it *Hippodamia convergens*. This name distinguishes it from more than 250 000 kinds of beetles in the world. To some people, it might seem unnecessary to name each kind of beetle, butterfly, or worm. Yet names make it easier for biologists to talk about the organisms they are studying.

Each kind of organism is called a **species** (SPEE-sheez). Members of a species are very much alike and are able to mate with each other. One species of bird is the mallard duck. You may have observed mallard ducks swimming in freshwater ponds, lakes, or marshes. The male has a green head, white neck ring, grayish body and is usually about 50 cm long. Females have tan heads and mottled brown bodies. Both male and female mallards have webbed feet, long necks, and short tails. Mallard ducks mate and produce offspring like themselves. Contrast the mallard duck with another species of duck: the shoveler. The male shoveler looks similar to the male mallard. It has a green head, white neck, and black and white tail. Female shovelers also look very much like female mallards; they are mottled brown with a small blue patch on their wings. Shovelers are a bit smaller than mallards and they have a larger bill. In spite of all their similarities, however, shovelers do not mate with mallards. They are two separate species of ducks.

1–8 Biologists refer to this common ladybug, which has 13 black spots on its back, as *Hippodamia convergens*.

Though members of a species may look alike, there are differences among them. For example, adult male mallard ducks range in length from 46 cm to 68 cm.

1–9 The shoveler's large bill and the male mallard's white neck ring help to identify these two species of ducks.

1–10 Various species of cacti are found in deserts. Their ability to store water enables them to thrive in hot, dry conditions.

The Classification of Living Things

You would need a book larger than a big dictionary just to list the names of all the species in the world. It would be an impossible task to learn about each of them. Fortunately there are many similarities between species. We put those that are similar together in the same group.

Biologists place organisms in large and small groups according to the similarities and differences among them. For example, there are many different species of birds, including hummingbirds, robins, pheasants, mallard ducks, and bald eagles. Although these species vary greatly in size, color, and other characteristics, they are all classified as birds because they have feathers and wings.

Some other familiar biological groups include insects, ferns, mammals, fungi, and reptiles.

Characteristics of Living Things

You can find living things almost anywhere on earth: in a cornfield, at the bottom of the ocean, or on the rim of a volcano. Many of the organisms that inhabit these places do not appear very similar to each other. There are similarities between them, however. By studying living things from all over the world, biologists have discovered that all organisms possess certain characteristics. Some characteristics are structures—living things are all made of similar basic units and contain many similar chemicals. Other characteristics are actions or processes—all living things perform a few basic life functions.

1–11 A young fern emerges from a lava crack in an area of recent volcanic activity in Hawaii.

Basic Life Structures

It is hard to imagine that pine trees and dogs have anything in common. Their colors, sizes, and shapes are very different. Yet pine trees and dogs are made up of similar small structures. These structures are **cells**, the smallest units of life. Most cells are too small to be seen without a microscope, but they make up the bodies of all living things.

Cellular Organization Although many organisms look very different from each other, their cells are similar in many ways. Most cells are too small to be seen by the naked eye. On the average, one hundred cells lined up end-to-end would make a line only 1 mm long. All cells are surrounded by a boundary called the **cell membrane**. The cell membrane controls which substances enter and leave the cell. Most cells also have a **nucleus** (NOO-klee-us), which directs the cell's activities.

Cells contain many different chemicals. Some of the chemicals are simple and familiar to everyone. Water, for example, makes up over half of most cells. Table salt is a chemical that is also found in all cells. Other chemical substances in cells are quite complex, including protein and fats.

The structure of a cell is related to its function in an organism's body. The trunk of a tree, for example, supports the tree and carries water from the roots to the needles or leaves. As you might expect, the cells in the tree trunk are strong and tubelike. Similarly, an animal's muscles move its body parts by contracting. Muscle cells are long and narrow, a shape that allows them to contract.

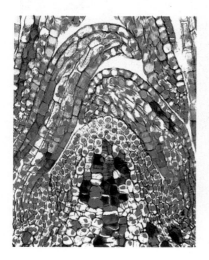

1–12 Many cells are visible in the growing tip of a white pine tree stem (magnified 100 times). Nuclei appear as dark spots.

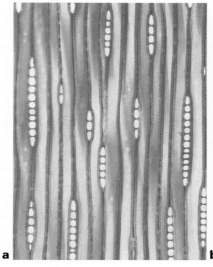 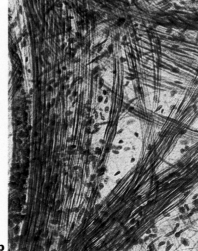

1–13 The structure of a cell is related to its function. **a.** Tubelike stem cells in a tree carry water and provide support. **b.** Elongated strands of muscle cells are able to contract.

Section 1.2 An Overview of Structure and Function

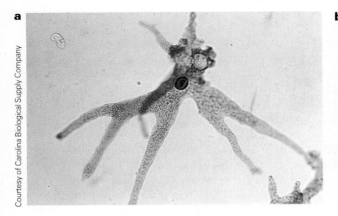 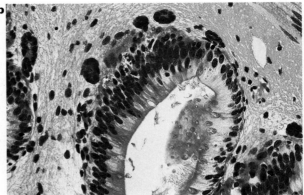

1–14 Unicellular and multicellular organisms: **a.** ameba; **b.** tissue, made up of many cells, from a multicellular animal.

Some organisms are **unicellular**: they are composed of only a single cell. Yeast and bacteria are unicellular. So are the organisms that form a green scum on trees and aquariums. Other organisms are **multicellular**: they are composed of more than one cell. Seaweed, potatoes, dogs, and pine trees are all multicellular organisms. There are advantages to being multicellular. One is that a multicellular organism continues to live even if a few of its cells die.

Multicellular Organization Multicellular organisms are more than a mass of individual cells. The cells are connected together in groups that are organized to perform specific jobs. A **tissue** is a group of similar cells that work together. Bone is one of the tissues in your body. The cells in bone tissue work together to support and protect body parts. Muscles and nerves are other animal tissues. Some plant tissues store food; others carry water and support the stem.

In many organisms tissues are organized into **organs**. The tissues in an organ work together to perform a certain function. Roots, stems, and leaves are organs in plants. A plant's roots absorb water for the plant. The tissue on the outside of the root takes in water from the soil. Water-carrying tissue inside the root carries the water up to the stem. Your lungs, brain, and stomach are some of the organs in your body. The function of the stomach is to digest food. The action of muscle tissue in the stomach wall mashes the food. Gland tissue produces a fluid that helps digest food. Muscle tissue and gland tissue work together in carrying out the job of digestion.

The organs in your body are organized into **organ systems**. An organ system is a group of organs that works together in performing a function. Organ systems carry out functions such as digestion, movement, and reproduction. The digestive system contains many organs, which work together in digesting food.

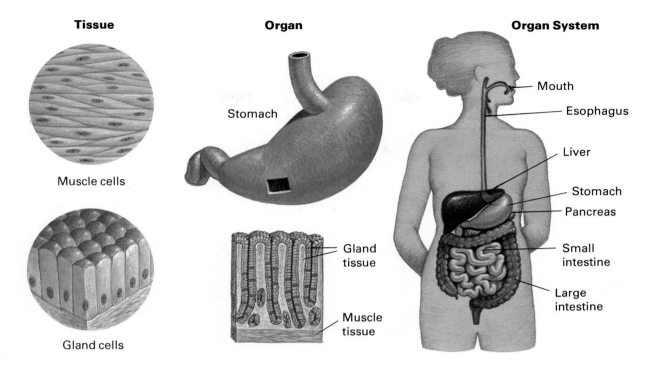

1–15 Groups of muscle cells and gland cells form tissues of the stomach, one of the organs of the human digestive system.

Basic Life Functions

In addition to being made up of similar basic units, organisms also perform similar functions. All organisms perform three basic life functions. They are:

1. Obtaining and using food
2. Controlling life processes
3. Responding to the environment

These functions are important for the survival of individual organisms. The survival of a species, however, depends on a fourth function: producing offspring.

Obtaining and Using Food Both you and a plant such as the dandelion obtain and use food, though you do it in somewhat different ways. Dandelions make their own food in a process called **photosynthesis** (FŌT-ō-SIN-thuh-sis). To do this they must have light, water, and air. Without all of these the dandelion will eventually die. Like most other organisms, dandelions can store some food. Food stored in the dandelion root helps it survive when it cannot make food.

phōtos = light
syn- = together
tithenai = to place

Section 1.2 An Overview of Structure and Function

Humans cannot make food. We find, eat, and digest food in order to survive. As food is digested, it is broken down into several basic kinds of chemicals, which are small enough to enter cells. Blood picks up these chemicals from the intestine and transports them to all the cells in the body. Cells use food for energy, growth, repair, or storage, depending on their needs. Most cells obtain energy by combining the chemicals from food with oxygen. Cells also use some of the chemicals from digested food to replace worn out parts and to increase in size.

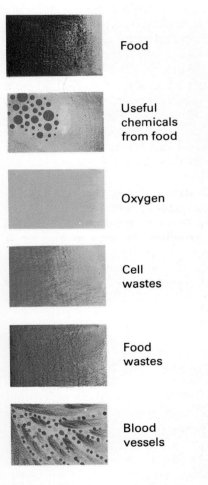
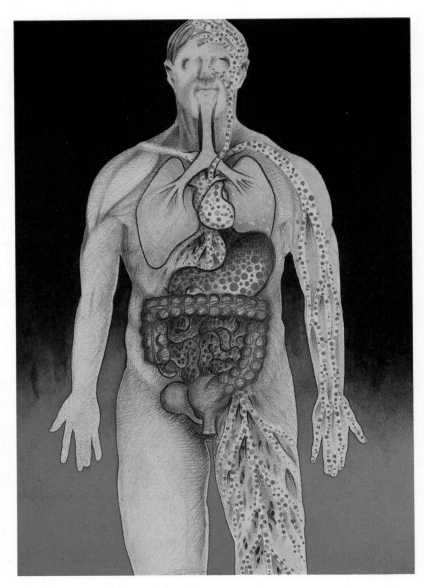

1-16 As it passes through the human body, food is first broken down, then transported through blood vessels to cells, which use it for energy, growth, repair, or storage.

Most organisms cannot digest all the food they eat. Some wastes are left at the end of the digestion process. In addition, body cells produce chemical wastes. All organisms rid their bodies of their food and chemical wastes.

The processes discussed above—making food; eating and digesting food; transporting food chemicals to all parts of the body; moving, repairing, and growing cells; and getting rid of food wastes and cell wastes—are all part of obtaining and using food.

Controlling Life Processes A rock rolling down a steep hill will roll to the bottom unless it is blocked in some way. A mouse running down the same hill, however, might stop anywhere on the hillside. Living things control their actions and life processes in ways that nonliving things cannot.

Most of the processes in our bodies are controlled without our thinking about them. Normally you do not think about breathing. Your body automatically controls how often and how deeply you breathe. You breathe faster and deeper when you exercise. Your heart also beats more quickly. Later, at rest, your heart rate and breathing rate slow down. Your body makes this adjustment automatically.

All organisms use chemical signals to control and coordinate their life processes. For example, chemicals control the growth of tree branches. The tip of a tree produces a chemical that keeps nearby twigs from growing. As long as the tip is intact, few twigs form and the tree will not have many branches. If the tip is broken, however, the chemical signal is no longer produced. Without the chemical signal, nearby twigs grow and branches develop on the tree.

1–17 When the tip of a tree is broken, side branches take over as growth leaders.

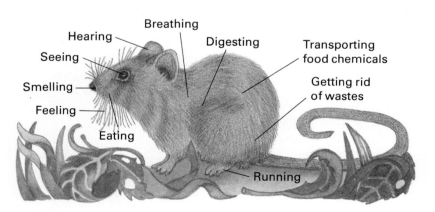

1–18 Even the smallest animals are complex organisms that carry on all of life's processes.

Section 1.2 An Overview of Structure and Function

1–19 The octopus has a highly developed nervous system. Nerves controlling the octopus' movement radiate from the central brain to the tips of the tentacles.

In addition to chemical signals, animals use nerve signals to control their life processes. Unlike chemical signals, which usually control continuing processes such as growth, nerve signals control adjustments that must be made quickly and accurately. Nerve signals are carried along nerve tissue between the brain and all the parts of the body. Touch, sight, hearing, and our other senses are all the result of nerve signals to the brain. Nerve signals from the brain adjust the heart and breathing rates during exercise. They also control other movements, such as walking and running.

Responding to the Environment Living things are sensitive to conditions in their environment. They respond to changes in temperature and the presence of other organisms. These are two types of **stimuli** (STIM-yuh-lī). Light, rain, sound, and the words on this page are also stimuli. The smell of food is a stimulus and so is a fire alarm. You detect and respond to thousands of stimuli every day. These responses are part of your behavior.

1–20 Responses to stimuli. **a.** A prairie dog stands alert at the edge of its burrow ready to hide if danger approaches. **b.** Sunflowers respond to sunlight by bending toward the sun.

An organism's response to stimuli helps it to survive. Plants turn their leaves toward light, which enables them to make more food. People quickly remove their fingers from a hot object, which prevents them from being burned. The ability to respond to stimuli is a basic function of all living things.

Producing Offspring Most organisms have a definite life span. For example, after a giraffe is born, it grows and matures and eventually becomes old and dies. All giraffes would die out if they did not reproduce. Species survive over long periods of time only by producing offspring. In most species, all organisms are capable of producing offspring. In some, however, such as honeybees, only a few individuals carry out this life function.

An individual cell reproduces by making copies of cell parts and dividing into two smaller cells. The two offspring cells are identical to the original parent cell. In this way, multicellular organisms grow and replace dead cells, and unicellular organisms produce offspring.

In some species, one parent can give rise to offspring. This type of reproduction is called **asexual** (ay-SEK-shoo-wul) **reproduction**. Offspring that are produced by asexual reproduction look almost exactly like the parent. The ameba undergoes asexual reproduction; so do many plants, such as the geranium, that can be grown from a branch cut off from the parent plant.

1–21 Asexual reproduction. **a.** Paramecia reproduce by splitting in half. **b.** At certain times of the year, female aphids reproduce asexually. The young female aphids are exact copies of their mother. **c.** Plantlets of a piggyback plant will each grow into mature plants.

Section 1.2 An Overview of Structure and Function

1–22 Sexual reproduction: **a.** strawberries, showing mature fruit and flowers; **b.** a female millipede surrounding her brood of eggs; **c.** a female lion nursing her cubs.

The most usual type of reproduction, **sexual reproduction**, involves two parents. Each parent contributes a cell to the formation of the offspring. Offspring produced by sexual reproduction sometimes look different from either parent. The production of offspring in fish, frogs, snakes, robins, humans, and many other organisms is accomplished by sexual reproduction.

Building a Science Vocabulary

organism	unicellular	photosynthesis
species	multicellular	stimulus
cell	tissue	asexual reproduction
cell membrane	organ	sexual reproduction
nucleus	organ system	

Questions for Review

1. Which of the following terms includes the other three? organ, tissue, cell, organ system
2. How does the structure of bone tissue help it perform its function?
3. How might a biologist tell if the individuals in a certain group of mice belong to the same species?
4. Why are cells called the building blocks of organisms?

Perspective: A New View
Everything's Connected to Everything Else

There's an old camp song that goes, "The thigh bone's connected to the knee bone, the knee bone's connected to the shinbone, the shinbone's connected to the ankle bone . . ." The song describes a very important aspect of nature. Everything is connected to everything else. This idea will be brought out many times in this book, and now is a good time to start thinking about it.

Everything in your body is connected to everything else in your body by blood vessels and nerves. The organs of your body are connected to each other by tubes and passages, such as the tube between your mouth and your stomach.

Organisms are connected by links such as those between parent and child. People are linked together as members of societies, such as tribes or nations. Nations may be linked together for the exchange of goods or for protection.

Organisms are connected to one another through what they eat. These links begin with a connection between the sun and plants. Sunshine strikes plants and provides the energy they need to make food. The plants may be eaten by cattle. Then we eat the cattle in the form of hamburger or steak. Likewise, field mice eat seeds and in turn may be eaten by foxes or hawks.

Organisms are connected to their environment in many ways. For example, all organisms exchange materials with their environment. Most take oxygen out of the atmosphere and put carbon dioxide in. Plants take carbon dioxide out of the atmosphere and put oxygen in.

One way humans connect with their environment is by discharging dust, soot, and other particles into the atmosphere. These particles reduce the amount of sunlight that reaches the earth. Less sunlight means less food production. Less food production means less food for the world's population. One step links to another step.

Indeed, everything is connected to everything else.

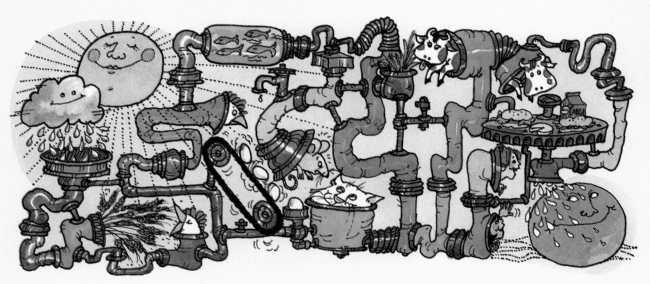

1.3 AN OVERVIEW OF INTERACTIONS IN THE ENVIRONMENT

Most organisms live together in groups along with many other kinds of organisms. For example, mice, grass, earthworms, ground squirrels, grasshoppers, and meadowlarks are commonly found living in a field. These organisms constantly interact with each other and with the air, soil, rocks, and water in the field. Earthworms burrow through the soil, while the grasshoppers eat many plant parts. All these animals breathe the air and add their wastes to the surrounding soil.

Biologists study the interactions between organisms and their environment to better understand the effects organisms and their surroundings have on one another. The study of these interactions is called **ecology** (ee-KOL-uh-jee). The word *ecology* comes from the Greek word *oikos*, which means "house." This word is a good name, because it implies that organisms and their environment are all part of a natural household.

Interactions Between Organisms

The different kinds of organisms living and interacting in an area make up a **community**. For example, you can find woodpeckers, squirrels, and wood-eating insects living in an oak tree. These four types of living things make up part of a community. The animals use the tree for shelter. The woodpecker eats the insects, while the squirrels and insects get their food from the tree. The woodpeckers and squirrels may also have fleas, which feed on them. These are but a few of the interactions that go on between community members.

1–23 Aquatic and land communities: **a.** fish swimming among sponge coral; **b.** a deer browsing in a forest; **c.** bison herd grazing on a grassland.

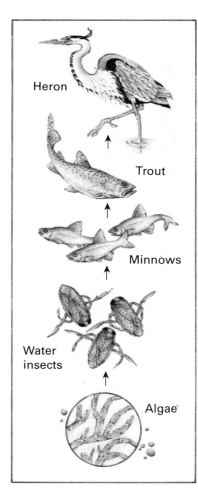

1-24 Aquatic food chain. Each organism is eaten by the organism above it. Algae are the producers; great blue heron is the final consumer.

1-25 Forest food web. Even in this simplified food web, numerous feeding relationships are possible.

Many kinds of communities exist. A pond is a community that may include fish, frogs, dragonflies, cattails, and mallard ducks. You are part of a community too. People, dogs, trees, grass, and houseflies are some of the organisms that interact in a human community.

The most obvious interactions between organisms in a community have to do with obtaining food. Organisms feed upon other organisms and are then fed upon themselves. For example, caterpillars eat grass, leaves, and stems. The caterpillars, in turn, are eaten by quail and quail are eaten by foxes. This is one example of a **food chain**. A food chain is a sequence of organisms in a community in which each member of the chain feeds on the member below it.

Every community has many possible food chains. For example, one species of insect may eat many different kinds of plants; and many species of birds, lizards, and other animals may eat the same species of insect. The many different food chains in an area are collectively referred to as a **food web**. Food webs can be fairly simple (such as the one in Figure 1-25), or incredibly complex.

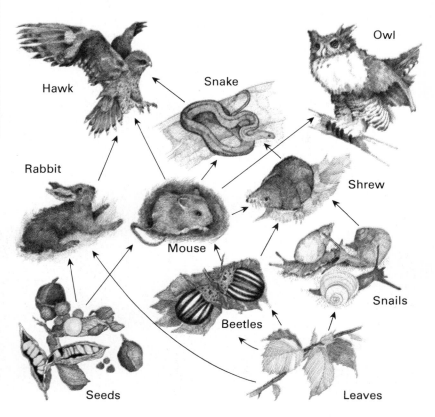

Section 1.3 An Overview of Interactions in the Environment

1–26 A consumer (the mouse) has found an ample supply of food manufactured by a producer (the corn).

Communities vary in many ways, but they all have organisms that fill certain roles. The food chains in any community begin with **producers**, which make their own food. Producers include plants such as elm trees, grass, daffodils, and grape vines. They also include plantlike forms of life, called algae (AL-jee), which grow in lakes, streams, and oceans. Producers make food from air, water, and light by photosynthesis. Since they are the original source of all food, every living thing depends on producers directly or indirectly.

All organisms that do not make their own food eat other organisms and are called **consumers**. Some consumers eat producers; for example, caterpillars feed on green leaves. Other consumers eat consumers. For example, hawks, foxes, and snakes eat live consumers; vultures, bears, and some insects eat dead consumers.

Many consumers in the food web are **decomposers**, including bacteria, mushrooms, and molds. They play an important role in the environment by breaking down the bodies of dead organisms and the waste products of living organisms into simple chemical substances.

1–27 Mushrooms growing on a fallen log absorb nutrients as they decompose the wood.

syn- = together
bios = life

Close Interactions The relationship between two kinds of organisms that interact closely and repeatedly with each other is called **symbiosis** (SIM-bī-Ō-sis). In some types of symbiosis, both members in the relationship benefit. Such is the case in the symbiosis between bees and many kinds of flowers. The bees obtain food (nectar and pollen) from the flowers. The flowers get their pollen spread to other flowers, a process that aids reproduction.

1–28 In this symbiotic relationship, the anemone fish attracts fish that serve as food for the sea anemone. In return the fish receives scraps of food missed by the anemone and the protective cover of the anemone's tentacles.

In other types of symbiosis, only one member benefits. Organisms called **parasites** obtain food directly from the live organisms in or on which they live (their **hosts**). In this case, only the parasite benefits; the host is harmed in the process.

Another type of symbiosis is **competition**. Competition occurs when something is in short supply. Organisms compete for food, space, shelter, and other resources. In certain cases of competition the competitors fight and the winner gets the resource. In most cases of competition no fighting occurs. For instance, forest trees compete for sunlight by growing taller.

1–29 Parasitism. **a.** A tick resides on the back of its python host. **b.** An oak gall houses parasitic wasp larvae.

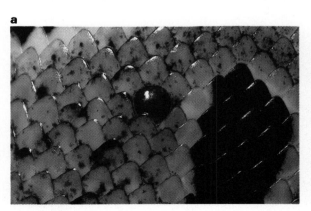

Section 1.3 An Overview of Interactions in the Environment

Interactions with the Environment

Whereas a community is made up of all the organisms in an area, an **ecosystem** (EE-kō-sɪs-tum) includes all the living and nonliving things in an area. The nonliving aspects of the environment include light, temperature, space, water, air, soil, rocks, and minerals.

Living and nonliving things interact in an ecosystem. Many organisms find shelter in the nonliving environment. Moles and earthworms burrow into the ground, snakes live between crevices in rocks, and large animals find shelter in caves.

Living things also obtain chemical substances for their bodies from the nonliving environment. For example, producers obtain the water and other materials for making food from the soil and the air. They also use a variety of other chemicals in the environment, including nitrogen and phosphorus. These chemicals are essential for a producer's growth. Chemical substances from the nonliving environment are passed from the producers to the consumers in the food chain. Consumers use these substances in their bodies. When organisms die, these substances are put back into the environment by the action of decomposers. The cycling of chemicals in this way is a dynamic and important interaction between organisms and their environment.

1–30 Water cycles continuously in an ecosystem.

Adaptations of Organisms

If you study the organisms living in any ecosystem, you will find that they seem to be suited for their environment. Organisms have characteristics that help them avoid enemies, find food, find mates, or withstand high or low temperatures. For example, certain insects look like twisted leaves. Their appearance on plants conceals them from hungry birds. Characteristics that help organisms survive are called **adaptations**. An adaptation might be an organism's physical appearance, the way it behaves, or the way that its body functions.

Each organism pictured in Figure 1–32 is well suited for its environment. The peregrine falcon dives down from high in the sky at 290 km/h to capture a bird for food. Its excellent eyesight helps it locate its prey. Its streamlined body gives it speed. The deep-growing root of the carrot plant helps it obtain water far below the surface of the soil. The bright color of the poisonous "painted frog" of Panama is a signal that it is poisonous. If a bird tastes a painted frog and survives, it will not try to eat another. The Arctic fox has fur that is brown and thin in summer. In winter its fur grows white and thick. The coat color of the fox helps it hide in its different surroundings each season. The thickness helps the animal keep its proper body temperature as the weather changes.

1–31 This species of praying mantis is hidden by its coloration.

1–32 Adaptations: **a.** long roots of carrots; **b.** brightly colored painted frogs; **c.** peregrine falcon, with keen eyesight and streamlined body; **d.** arctic fox in its summer coat.

Section 1.3 An Overview of Interactions in the Environment

Adaptations occur inside the body as well as on the surface. Tissues are suited for their functions. The long extensions of nerve cells are an adaptation that helps them carry messages. Lightweight bone tissue is an adaptation inside a bird's body that helps it fly.

Changes over Time

Organisms continually interact with other organisms and with their nonliving environment. As a result, species change over time. Some species die out; new ones develop.

The changes that species undergo may be affected by differences between individuals. Look around you. People do not all look alike. Dogs do not all look alike either. They vary in height, weight, and body coloration. The differences that exist between members of a species are called **variations**. Some variations are adaptations that help organisms survive in their environment.

Consider the Arctic fox as an example of how species may change. Its thick winter coat helps it survive the bitter cold of winter. Suppose that some time in the past, not all Arctic foxes had thick coats. Instead there may have been variations in the thickness of their coats. Some may have had thick ones and others thin ones. It is possible that one winter the weather was so severe that only the thick-coated foxes survived. In the spring,

1-33 All dogs belong to the species *Canis familiaris*. Many years of breeding have produced numerous variations.

1-34 Arctic fox in winter coat. The thickness and color of the fur are adaptations for winter.

the surviving foxes reproduced and passed on the thick-coat trait to their offspring. Because the thin-coated foxes died of the cold and did not reproduce, most Arctic foxes then had thick coats. This is a possible explanation for how changes occur in species over time.

Building a Science Vocabulary

ecology	consumer	competition
community	decomposer	ecosystem
food chain	symbiosis	adaptation
food web	parasite	variation
producer	host	

Questions for Review

1. How are the members of a species different from the members of a community?
2. Give an example of a producer and a consumer. Explain the differences between them.
3. What causes competition in an ecosystem? Give an example of competition between plants.
4. Explain why the Arctic fox's coat color can be considered an adaptation to changing seasons.

Chapter 1
Summary and Review

Summary

1. Biology is the scientific study of living things. Our understanding of biology is subject to change based on the results of scientific observations and tests.

2. There are many different specialized areas and many different approaches to research in biology.

3. The study of biology can relate to a person's personal life, social life, and career.

4. Each individual organism is a member of a species. The members of a species look similar and are able to breed.

5. Classifying living things makes it easier to study them.

6. The cell is the basic unit of structure of all living things. All cells have similar features, but differ in size and shape, depending on their function.

7. Organisms may be unicellular or multicellular. The cells of multicellular organisms can be organized into tissues, organs, and organ systems.

8. All living things perform three basic life functions: obtaining and using food, controlling life processes, and responding to the environment. In addition, all species perform a fourth function: producing offspring.

9. Organisms interact with other organisms and with the nonliving things in an ecosystem.

10. Many of the interactions between community members involve food, giving rise to food chains and food webs. Food webs have producers, consumers, and decomposers.

11. Chemical substances cycle between living things and the nonliving environment through producers, consumers, and decomposers.

12. Adaptations help suit organisms for their environment and help aid their survival.

13. Variation, adaptation, survival, and reproduction are four factors that may play a role in the way species change.

Review Questions
Application

1. Suppose you wanted to test the idea that symbiosis exists between a bee and a flower. What could you do to find out if bees are really needed for a plant's reproduction?

2. Certain small animals called ticks live in folds of the skin of the rhinoceros and get food from its body. Small birds called tick birds walk all over the body of a rhinoceros and eat the ticks. The rhinoceros allows the birds to do this.
   a. What is the relationship between the rhinoceros and the tick bird?
   b. How is each member of the pair affected?

3. Why is good eyesight considered an adaptation in organisms that hunt for their food?

Interpretation

1. Do all organisms have tissues? Explain your answer.
2. Biologists believe that almost all organisms on earth are made of cells. How could they come to such a conclusion?
3. Why is reproduction *not* considered to be a life function of an individual?
4. Many minute organisms live on the skin of an animal and get food from the animal. Therefore, some people refer to the skin as an ecosystem. Is ecosystem the correct term to use to describe the skin and its inhabitants? Explain.

Extra Research

1. Begin a scrapbook of news items about biology that interest you. Add news stories about these areas of interest as the year progresses.
2. Observe a plant and an animal carefully. Make a list of the similarities and differences you see between them.
3. Make a collection of pictures of plants and animals. For each organism in your collection, state the features that help adapt it to its environment.
4. Set up an aquarium, terrarium, or window box as a model ecosystem. Observe how the living and nonliving factors interact.
5. Take a nature walk or visit a zoo or open area in your neighborhood. Record your observations with photographs or pencil sketches. Include examples of adaptations of plants and animals to their natural homes and examples of symbiosis.

Career Opportunities

Life scientists study living things and their environment to understand the living world and to solve biological problems. Most have four or more years of university training. Many life scientists do basic research for the government, private industry, or universities. Others teach, work in management, or work as consultants. Contact the Canadian Federation of Biological Sciences, Box 498, Saskatoon, Saskatchewan S7N 0W0.

Science writers include both journalists and technical writers. They must write well and be knowledgeable in a scientific field. Some of their work includes writing research proposals, technical reports, instruction manuals, and magazine articles. Science writers play an important role in helping the public understand and use new scientific and technological developments. Write the Canadian Science Writers Association, 160 Wellesley St. E., Toronto, Ontario M4Y 1J3.

Animal caretakers feed and exercise animals in zoos, laboratories, and veterinary hospitals. They also clean cages and keep health reports. On-the-job training is usually provided. Contact the Canadian Association for Laboratory Animal Science, 2627 Morley Trail N.W., Calgary, Alberta T2M 4G6, the Canadian Parks/Recreation Association, 333 River Rd., Tower B, Vanier, Ontario K1L 8B9, or the Canadian Veterinary Medical Association, 339 Booth St., Ottawa, Ontario K1R 7K1.

Chapter 2

Classification of Living Things

Focus *There are millions of kinds of living things on earth. Since early times, people have given them names and organized them into groups with similar characteristics. Grouping makes the study of living things easier. Today, scientific systems of naming and grouping organisms are used by biologists all over the world.*

2.1 SCIENTIFIC NAMES AND CLASSIFICATION

With such a large number of living things in the world, it might seem that identifying different species would be very difficult. Biologists, however, have developed a way for different people in different places to identify organisms. Consider the following example.

One day, a small bird flew against a window at a high school and fell to the ground. Two students rescued the stunned bird, took it into the biology classroom, and put it into a cage, where it soon revived. Before they released the bird, the students wanted to know what kind of bird it was. To help them identify it, the biology teacher gave them a **dichotomous** (dī-KOT-uh-mus) **key**. A dichotomous key contains questions about the appearance of the thing being identified. Each question in the key involves a choice between two characteristics. The students

2–1 Tulips are cultivated in many parts of the world for their beauty.

2–2 The students are observing a bird they found on the ground. The cloth covering the top of the cage helps to calm the bird.

2–3 The bird found by the students is shown next to a centimetre ruler. In identifying birds, the length of the body is measured from the tip of the beak to the tip of the tail.

looked carefully at the bird in the cage as they answered the questions.

1. Are the toes separate?
   a. If yes, go to 2.
   b. If no, go to 19.
2. Is the bill small and pointed?
   a. If yes, go to 3.
   b. If no, go to 20.
3. Is the length of the body more than 16 cm?
   a. If yes, go to 4.
   b. If no, go to 21.
21. Does the bird have a cap and bib?
    a. If yes, go to 31.
    b. If no, go to 35.
31. Is the cap brown?
    a. If yes, brown-capped chickadee.
    b. If no, go to 36.

The dichotomous key indicated that the bird was a brown-capped chickadee.

    A teacher whose hobby was bird watching saw the bird and was sure that it was a Hudsonian chickadee. To settle the question, the students consulted a bird book. In the book, the bird was called the boreal chickadee. At this point, everyone was confused. Who was right?

Chapter 2 Classification of Living Things

Common Names and Scientific Names

Actually, the teacher, the students, and the books were all right. All were using **common names**. Common names vary from place to place and from person to person. For example, there is a small, blue flower that grows in the spring in open, grassy places. Depending upon where you live, the flower is called Bluet, Quaker Lady, Innocence, Eyebright, Little Washerwoman, or Quaker Bonnet.

Plants and animals also have different common names in different languages. A bird is called *pajaro* in Spanish, *oiseau* in French, and *Vogel* in German. Since common names are so different, biologists use **scientific names** that are the same everywhere. Thus, scientists all over the world can use the same name for a particular organism.

If the students had continued to read the key, they would have found that the scientific name for the brown-capped chickadee is *Parus hudsonicus* (PAYR-us hud-SŌN-ih-kus). No other organism has the same scientific name.

Origin of Scientific Names Imagine the task of giving every living thing a scientific name! Over five million different kinds of living organisms inhabit the earth. They range in size from organisms that are too small to be seen without a microscope to giant redwood trees and blue whales. They range in complexity from simple viruses and bacteria to humans. Naming all these living things involves three steps. First, they must be found and

2–4 Bluet is one of several common names for these flowers. Their scientific name is *Houstonia caerulea*.

2–5 There are many kinds of living things.

2–6 Carolus Linnaeus, born in Sweden in 1707, studied medicine and later devoted himself to teaching biology and classifying organisms.

bi- = two
nomen = name
calator = caller

Scientific names are often descriptive and are either in Latin or are Latinized. Names like Hudsonicus *or* Houstonia *are Latinized.*

2–7 No two species have the same scientific name.

identified. Then, they must be studied and compared. Finally, they must be sorted into categories, or classified, according to their similarities. The science of naming and classifying living things is called **taxonomy** (tax-ON-uh-mee).

Many different taxonomic systems have been designed. The most important system was created during the 1700s by a Swedish biologist named Carolus Linnaeus. As part of his taxonomic system, Linnaeus designed a naming system called **binomial nomenclature** (bī-NŌ-mee-ul NŌ-men-CLAY-chur). Stated simply, binomial nomenclature assigns every living organism a two-word scientific name. These names are written in Latin, which was the language of science when Linnaeus lived.

Several common organisms are shown with their scientific names in Figure 2–7. *Homo sapiens* (HŌ-mō SAY-pee-unz) is the scientific name for humans; *Felis domesticus* (Fee-lis dō-MES-tih-kus) is the name for the house cat; *Canis familiaris* (KAY-nis fa-MIL-ee-AYR-is) is the name for dog; and *Quercus rubra* (QUAYR-kus ROO-bruh) is the name for the red oak tree.

Selection of Scientific Names In the past, taxonomists named many organisms after people. For example, *Chlamydomonas steini* (KLA-mee-duh-MŌ-nus STĪN-ee), a one-celled pond organism, was named after someone named Stein. Unfortunately, this name does not describe any of the characteristics of

Chapter 2 Classification of Living Things

longi- = long
cauda = tail

Chlamydomonas. Today, scientists try to name organisms by selecting Latin words that describe the organism. For example, the name *Apis longicaudatis* (AY-pis LON-jih-kah-DA-tis) indicates that this kind of bee (*Apis*) has a long tail.

Taxonomic Categories

You use a classification system every time you go grocery shopping. Food in a grocery store is arranged in groups. Fruits and vegetables are in one area; meats are in another; dairy foods are in still another area. These general categories include a variety of foods. For example, the dairy section includes milk, cream, butter, cheese, and yogurt. These food items are also categories of their own. The milk category includes several specific kinds of milk: nonfat, low fat, whole, and buttermilk.

Taxonomists categorize living things in a similar way to how grocers arrange food in their stores. The categories range from general to specific. The most general category includes many different kinds of organisms, while the most specific category has only one kind of organism. The major categories in biological classification are **kingdom, phylum** (FĪ-lum), **class, order, family, genus** (JEE-nus), and **species**. Figure 2–8 shows how these categories are arranged.

The most general category of biological classification is the kingdom. All organisms within a kingdom have the same general characteristics. Kingdoms are divided into one or more

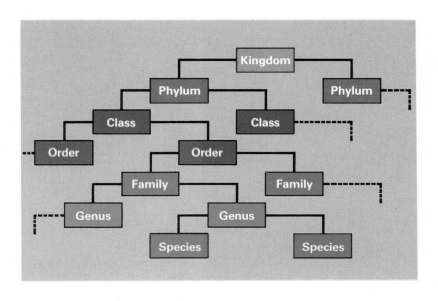

2–8 Categories of classification from kingdom to species. Each category may be subdivided, resulting in many branches.

Section 2.1 Scientific Names and Classification 35

Members of a species may differ, but not enough to be considered separate species. Instead, they are classified in categories called varieties. For example, a poodle is a variety of the species Canis familiaris *(dogs).*

subgroups called phyla. All organisms in a phylum share characteristics that are more specific than those shared by members of a kingdom. Phyla are divided into subgroups called classes. The members of a class are alike in more ways than the members of a phylum or a kingdom. Classes are made up of smaller groups called orders. Orders are further subdivided into families, families are subdivided into genera, and genera are subdivided into species. Each subgroup is made up of organisms that are more alike. The organisms in an individual species have thousands of characteristics in common.

Scientific Names: Genus and Species The system of binomial nomenclature gives a two-word scientific name to each species. These two words are the genus name and the species name. For example, *Quercus rubra* is the scientific name for the red oak tree. The genus name *Quercus* indicates that the tree is an oak. But there are other species of oak. For example, *Quercus alba* is the white oak, and *Quercus macrocarpa* is the bur oak. The species name *rubra* means red. Many plants and animals are identified by the name *rubra*. *Piranga rubra* is a red bird, the summer tanager. Only when the two names *Quercus* and *rubra* are put together can you identify the species as the red oak, *Quercus rubra*.

Scientific names are always written with the genus name before the species name. The genus name is always capitalized, but the species name is not. Often the genus name is abbreviated. For example, *Quercus alba* may be written *Q. alba*. Notice that scientific names are shown in *italics*. When they are written in longhand, they are underlined.

2–9 Identifying traits of these three species of the genus *Quercus* are the reddish inner bark of the red oak, the light ash-gray bark of the white oak, and the fringed acorns of the bur oak.

Quercus rubra
(red oak)

Quercus alba
(white oak)

Quercus macrocarpa
(bur oak)

2–10 Classification of human, dog, and red oak. Which two species have more in common?

	Human	Dog	Red Oak
Kingdom	Animalia	Animalia	Plantae
Phylum	Chordata	Chordata	Tracheophyta
Class	Mammalia	Mammalia	Angiospermae
Order	Primates	Carnivora	Fagales
Family	Hominidae	Canidae	Fagaceae
Genus	*Homo*	*Canis*	*Quercus*
Species	*H. sapiens*	*C. familiaris*	*Q. rubra*

Building a Science Vocabulary

dichotomous key   binomial nomenclature   order
common name   kingdom   family
scientific name   phylum   genus
taxonomy   class   species

Questions for Review

1. Which of the following pairs of organisms has more in common? Explain your answer.
   Pair 1: *Rana pipiens*   Pair 2: *Rana pipiens*
   Culex pipiens   *Rana grylio*

2. Which of the following categories is most general?
   species, phylum, genus, order

3. Explain the problem that occurs when common names are used for organisms.

4. Write four questions that can be answered "yes" or "no" to identify a pear.

Section 2.1 Scientific Names and Classification

2.2 THE BASIS FOR CLASSIFICATION

Taxonomists have the problem of placing living things into categories. They realize that related organisms belong in the same groups. But what are related organisms? All the kittens in a litter are related because they have the same mother. Members of human families are related because they have the same parents or grandparents—the same ancestors. In biology, **related organisms** are believed to have the same or common ancestors. Relatedness is the basis for classification.

To determine relatedness, taxonomists first look at the structure of an organism. Related organisms often look very much alike. When organisms with the same ancestors look alike, the taxonomist's job is easy. Related "look-alikes" belong in the same category. If taxonomists find an animal with scales and gills, they can easily classify it as a fish. If they find an animal with feathers, they can classify it as a bird.

Determining relatedness by appearance can be difficult, because close relatives do not always look alike. Related organisms that do not look alike are difficult to classify. In addition, unrelated organisms sometimes do look alike. A whale, for example, looks more like a tuna than it looks like a cow. Yet, whales are more closely related to cows than they are to tunas. A butterfly and a robin both have wings, but insects are not closely

2–11 The Holstein cow (**a**) and the spinner porpoise (**b**) are both mammals. The blue shark (**c**) looks much like the porpoise; however, it belongs to a class of fish.

related to birds. Appearances sometimes make it difficult for taxonomists to properly classify organisms.

Taxonomic Tests

Taxonomists must go beyond outward appearance to find similarities and differences among organisms. They observe internal as well as external features and perform tests to find chemical similarities and differences.

Homologous structures have similar structure and development.

Homologous Structures Related organisms with similar structures develop in the same way. Figure 2–12 shows a human, a bat, and a whale. It also shows the internal structure of the arm and hand of the human, the wing of the bat, and the flipper of the whale. When you look at the internal structures of the arm, wing, and flipper, you find bones that are similar. These bones not only resemble each other, they develop in the same way. The forelimb bones of all three animals develop before birth as part of a skeleton that is composed of bone.

Similar structures in different organisms that develop in the same way are called **homologous** (hō-MOL-uh-gus) **structures**. The forelimbs of animals with bony skeletons are homologous structures. Taxonomists place organisms with homologous structures in the same categories.

2–12 Homologous structures. The whale's flipper, the human's arm, and bat's wing all have similar structures that develop in the same way.

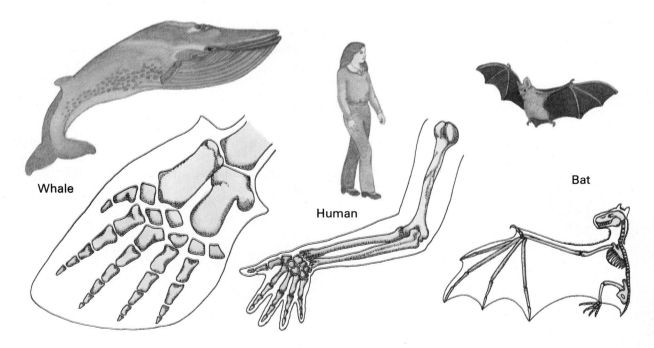

Section 2.2 The Basis for Classification

2–13 Analogous structures. The hummingbird (**a**) and the hummingbird moth (**b**) have wings; however, the structure and development of their wings differ.

Analogous structures have similar functions but different development.

Analogous Structures The butterfly's wing is not homologous to the bird's wing even though they are both used for flight. The bird's wing develops as part of a bony skeleton; however, butterflies do not have bones. (See Fig. 2–13.) Structures in different organisms that have similar functions but develop in different ways are called **analogous** (uh-NAL-uh-gus) **structures**. The wing of a bird and the wing of an insect are analogous structures. Taxonomists place organisms with analogous structures in different categories.

Chemical Tests Taxonomists sometimes analyze the blood of organisms to find chemical similarities or differences. Blood tests show, for example, that whales are more closely related to humans than to fish. Taxonomists compare other chemicals within organisms in addition to blood.

Life History of Organisms Taxonomists also study an organism's life history—the stages of its life. They observe how organisms develop and reproduce. Studies in the life history of plants have been useful in classifying them. The way that a daisy develops and reproduces, for example, is different from the way that a fern does.

The study of life histories also helps taxonomists classify animals. For example, taxonomists obtain further evidence that the whale and the fish are not closely related by observing differences in their reproduction and development. Most fish hatch from eggs laid by the mother. Food for the developing fish is contained within the egg itself. Once the mother fish has laid the eggs, she has little or nothing to do with her offspring. Newly hatched fish catch their food in the water around them. A whale, however, develops within the mother's body and gets food from her during development. After birth, the whale feeds on milk produced by its mother and stays close to her for about six months. These differences in reproduction and development also indicate that the whale and the fish are not closely related.

Studies of the life histories of sharks and bony fish show that they are closely related.

2–14 Life histories. **a.** Once hatched, independent salmon fingerlings will search for their own food. **b.** A humpback whale nurses her calf for up to a year following birth.

Organisms that are similar to the whale in reproductive structure and behavior are related. These include porpoises, bats, foxes, deer, and many others.

Breeding Studies If two organisms breed and produce offspring that also produce the same kind of offspring, they belong to the same species. A species is an interbreeding population. Suppose several pairs of elephants breed and produce a new generation of elephants. This new generation of elephants will breed to produce yet another generation of elephants. These elephants all belong to the same species.

Horses breed successfully with other horses; the same is true of donkeys. But, if a horse and a donkey breed, the offspring is a mule. Mules cannot reproduce their own kind. The horse and the donkey are members of different species; however, the mule is not considered to be a species since it cannot reproduce.

2–15 Breeding donkeys (**a**) with horses (**b**) produces mules (**c**), which have great strength and endurance but cannot reproduce.

The Kingdoms of Living Things

When the Greek philosopher Aristotle first classified living things in 300 B.C., he divided all living things into two kingdoms—plants and animals. An obvious problem of a two-kingdom system is that some organisms do not clearly fit into either category. The microorganism *Euglena* (yoo-GLEE-nuh) is a good example. *Euglena* produces its own food. Because of this, it is considered plantlike; however, *Euglena* also swims around freely like an animal.

To eliminate the uncertain classification of many organisms in the two-kingdom system, taxonomists have created more kingdoms. They do not all agree on how many kingdoms there should be. Different taxonomists use systems based on three, four, and five kingdoms. Figure 2–17 outlines the two-kingdom, three-kingdom, four-kingdom, and five-kingdom classification

2–16 *Euglena* has characteristics of both plants and animals. It makes its own food and swims freely.

2–17 Four possible classification systems.

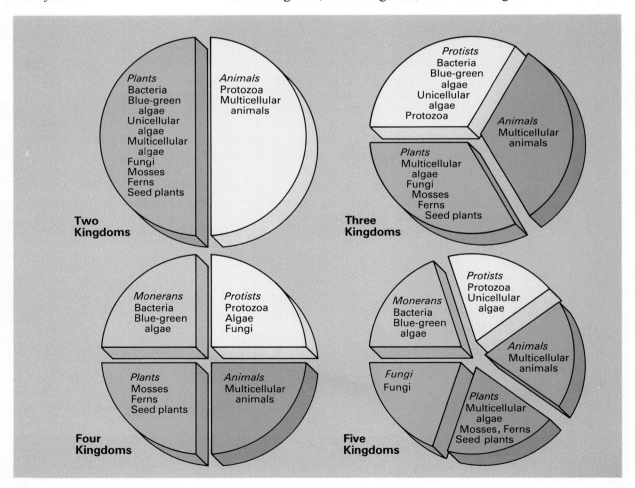

Table 2–1 Kingdoms of Living Things

Kingdom	Characteristics	Members	Examples
Monerans	Simple structure, unicellular prokaryotes (no true nucleus), producers, consumers, decomposers, some move around	Bacteria Blue-green algae	Decay bacteria Tuberculosis bacteria *Nostoc*
Protists	Simple structure, unicellular eukaryotes (true nucleus), producers, consumers, some move around	Protozoa Unicellular algae Slime molds	*Paramecium* *Euglena* Diatoms *Cribraria*
Fungi	Simple structure, most multicellular, eukaryotes, not green, most decomposers, some consumers, do not move around	True fungi	Mushroom *Saccharomyces* Bread mold
Plants	Most have complex structure, multicellular, eukaryotes, producers, do not move around	Multicellular algae Mosses Liverworts Fern relatives Ferns Seed plants	Kelp *Porphyra* *Marchantia* Horsetail Boston fern Pine Beech Oak
Animals	Complex structure, multicellular, all have nerve cells except sponges, consumers, most move around	Invertebrates Simple chordates Vertebrates	*Hydra* Moth Tunicate Flounder Frog Snake Robin Bear

systems. One system is not necessarily better than the others. Biologists in different fields use the systems they think are best.

The five-kingdom classification system is used in this book. The five kingdoms include **monerans** (mō-NEER-ens), **protists** (PRŌT-ists), **fungi** (FUN-jī), **animals**, and **plants**. The major characteristics of each kingdom and examples of organisms in each kingdom are listed in Table 2–1. You may not have seen some words before. They are explained in the next part of this chapter.

You obviously cannot study millions of kinds of living things in just one book or course. The following section describes the major phyla within the five kingdoms. In addition, the section introduces five organisms from different phyla that have been selected for more detailed study. These organisms are the bean (plant), paramecium (protist), hydra (animal), earthworm (animal), and grasshopper (animal). Throughout the book these organisms are referred to as the "representative organisms."

Building a Science Vocabulary

related organism   protist
homologous structure   fungus
analogous structure   animal
moneran   plant

Questions for Review

1. Organisms that look alike are not always related. Give an example of unrelated "look-alikes."
2. Explain why the presence of wings does not make birds and insects relatives.
3. Explain why taxonomists do not consider the mule to be a species.
4. What difficulty in classification arises with a two-kingdom system?
5. What are the names of the kingdoms in the five-kingdom classification system?

Perspective: A New View
Same But Different

In this chapter you learned how to use a dichotomous key. With the right key, you could tell a white pine from a scotch pine. Then you would be able to identify a white pine wherever you saw one, since they all look the same. This is also true for the common earthworm. The one you see in your back yard looks the same as one you might find anywhere in the world. All members of the same species look alike to us.

Now, look around your classroom, school, ball field, playground, or the crowd at a rock concert. Everybody looks different. It is not just the different kinds of clothes. It is things such as skin color, hair color, and eye color; the shape of the head, nose, and mouth; body size and proportions. Even if you lined up a group of people who all had the same hair color and eye color, there would still be differences. Some people would be shorter, others taller; some would have wavy hair, others straight hair; some would have light skin, others dark skin.

There are also many differences you cannot easily see. The big growth spurt in girls begins anytime between the ages of ten-and-a-half and thirteen, while in boys it begins about two years later. Body organs differ in normal adults. For example, one person's thyroid gland can be as much as six times as large as another person's. In most adults, three arteries branch off the main blood vessel leading from the heart. Other people have one, two, four, five, or even six branches.

There are also chemical differences. The amount of pepsin, a chemical in the stomach, can vary a thousandfold, and the optimum amount of vitamin C needed can vary from 250 to 10 000 mg per day in different people.

Yet, we are all members of the same species, *Homo sapiens*. Maybe if we looked closely we would find that although all the earthworms and all the white pines look basically the same, they are different, too.

2.3 KINGDOMS OF LIVING THINGS

The five-kingdom system arranges living things into categories according to their complexity and means of obtaining food. The first division is between unicellular organisms and multicellular organisms. There are two kingdoms of unicellular organisms: monerans and protists. The difference between these two kingdoms is that moneran cells are smaller and much simpler than protist cells.

Multicellular organisms are usually more complex than unicellular ones. There are three kingdoms of multicellular organisms: fungi, plants, and animals. These categories are based on nutritional differences. Most of the organisms within each multicellular kingdom have similar ways of getting food. Plants make their own food, fungi absorb their food, and animals eat their food. You can usually identify a multicellular organism as plant, fungus, or animal by its method of obtaining food.

THE MONERAN KINGDOM

The Moneran Kingdom is made up of two groups of organisms: **bacteria** and **blue-green algae**. The characteristic that unites these two types of unicellular organisms in the same kingdom has to do with their internal structure.

The cells of most living things have a nucleus. The nucleus consists of the material that controls the cell surrounded by a

2–18 Cell types. **a.** A prokaryotic cell (a bacterium) has no nuclear membrane; its nuclear material is found throughout the cell. **b.** Eukaryotic cells have a nucleus and nuclear membranes, as can be seen in the large central nucleus in this lymph cell from a rat. Prokaryotic cells are actually much smaller than eukaryotic cells.

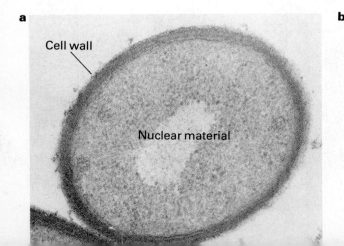
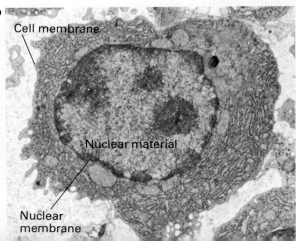

pro- = before
karyon = a nut, kernel

sac, or **nuclear membrane**. A nucleus surrounded by a nuclear membrane is considered a "true" nucleus. In bacteria and blue-green algae, however, there is no nuclear membrane. The material that controls the cell is spread throughout the entire cell. Bacteria and blue-green algae are called **prokaryotes** (prō-KAYR-ee-ōts) because they lack a true nucleus. Almost all other living things have cells with a true nucleus. Because of this, they are called **eukaryotes** (yoo-KAYR-ee-ōts).

Bacteria

As many as 2.5 billion bacteria can be found in one gram of fertile soil.

Bacteria are everywhere. They can be found in water, on land, and inside and on other living things. Bacteria are the most numerous and among the smallest living things on earth. Nearly 1600 species of bacteria have been identified. Most species of bacteria do not produce their own food.

Bacteria are important for many reasons. Some of them benefit the environment by decomposing dead matter and waste products. These decomposers release important chemical substances to the environment during the process of breakdown. Harmful bacteria cause diseases to plants, animals, and humans. These bacteria are parasites, living inside or on the host from which they obtain food. Two bacterial diseases you may be familiar with are strep throat and tooth decay.

2–19 The three basic types of bacteria are: rod-shaped bacilli (**a**), which include types that cause lockjaw and tuberculosis; round cocci, such as *Streptococcus mutans* (**b**), which causes tooth decay; and spirilla (**c**), named for their long, coiled shapes.

The structure of bacteria is simple. A strong cell wall encloses and protects the cell. Bacteria commonly have one of three body shapes: round, rod-shaped, or long and curved. Bacteria of any shape may grow as individual cells or in chains or clusters of cells. Many bacteria form **spores**. A spore is a cell with a special protective coat that can survive severe conditions over long periods of time. Some bacterial spores even survive boiling. When conditions are favorable, spores begin to grow and reproduce normally.

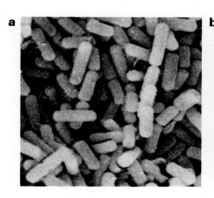
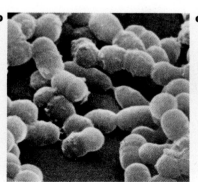
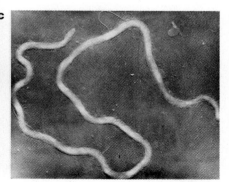

chlōros = green
phyllon = leaf

Blue-Green Algae

Blue-green algae have one major difference from bacteria. Blue-green algae contain a green substance called **chlorophyll** (KLOR-uh-FIL). The presence of chlorophyll enables blue-green algae to make their own food by the process of **photosynthesis**. During photosynthesis, organisms that contain chlorophyll make food (sugar) from carbon dioxide and water in the presence of sunlight. Since blue-green algae make their own food, they are producers.

In addition to chlorophyll, blue-green algae contain blue coloring matter. The two colors combine to give many of them a bluish-green appearance. Blue-green algae also contain other colored substances. As a result, not all of them appear blue-green—some appear red or brown.

Individual cells of blue-green algae are microscopic. About 200 species of blue-green algae have been identified. Some species live as separate cells. In other species, the cells are attached to each other and form large masses. *Nostoc* (NAHS-tahk) is a blue-green alga that forms chains of cells.

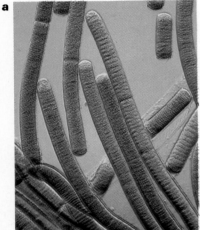

2–20 Blue-green algae. **a.** The disk-shaped cells of *Oscillatoria* form long filaments. **b.** *Anabaena* also forms filaments of cells.

2–21 Under certain conditions, an overgrowth of blue-green algae, or algal bloom, covers the surface of water.

VIRUSES

Viruses (VĪ-rus-uz) are not monerans; in fact, many taxonomists do not place them in the classification system at all. The classification system is largely based on cell structure, and viruses are not made of cells. Some biologists question whether they are even living. Viruses are presented here for two reasons: their structure is very simple, and they have nuclear material that is not surrounded by a cell membrane.

Since viruses are as small as some large molecules, they were not seen until the invention of the electron microscope. Figure 2–23 shows photographs of viruses seen through an electron microscope. The major structure of some viruses is a head with many sides. The head is made of protein and contains nuclear material. Many viruses, including the common cold virus, also have a tail composed of protein.

Viruses are parasites. They enter a host cell and take control of its functioning. Then, they "command" the host cell to produce more viruses.

Viruses cause diseases in humans, plants, and animals. Some virus-caused diseases are common colds, mumps, smallpox, chickenpox, and measles. Some types of cancer are also believed to be caused by viruses.

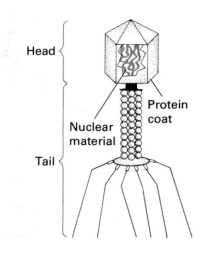

2–22 The structure of a bacteriophage, a common virus that attacks bacteria.

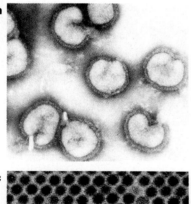
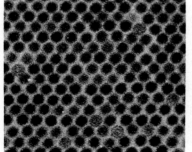
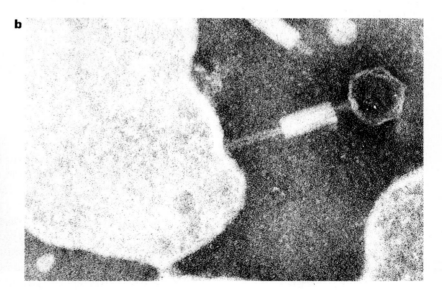

2–23 Viruses: **a.** influenza; **b.** bacteriophage injecting nuclear material into bacterium; **c.** adenovirus, a common cold virus.

Section 2.3 Kingdoms of Living Things

THE PROTIST KINGDOM

In the five-kingdom system, one-celled organisms with a true nucleus are placed in the Protist Kingdom. These one-celled organisms are eukaryotes. Some protists, such as unicellular algae, are producers. Other protists, the **protozoa** (PROT-uh-ZO-uh), are consumers—they eat other living things for food. Still other protists decompose and absorb their food.

prōtos- = first
zōion = an animal

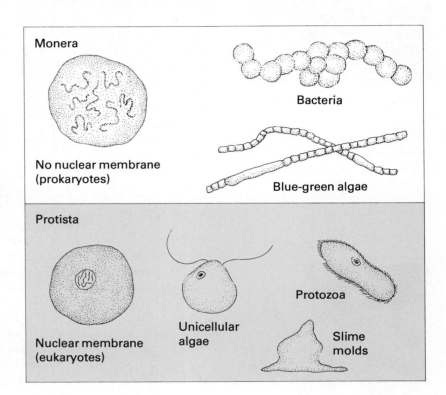

2–24 General structure and representative organisms in the Moneran and Protist Kingdoms, the two major groups of unicellular organisms.

Producer Protists

The producer protists make their own food by photosynthesis. Most producer protists are algae. You may recall *Euglena*. It is difficult to classify because it has characteristics that are both plantlike and animallike. Since *Euglena* produces its own food, it is considered a producer protist. Other unicellular algae differ from *Euglena* because they do not swim around freely.

All producer protists contain chlorophyll. Some, including *Euglena*, appear green. Others contain additional colors. Diatoms (DĪ-uh-TOMZ) are unicellular algae that appear yellow or

A litre of surface sea water commonly contains at least half a million diatoms.

Chapter 2 Classification of Living Things

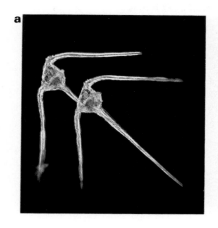 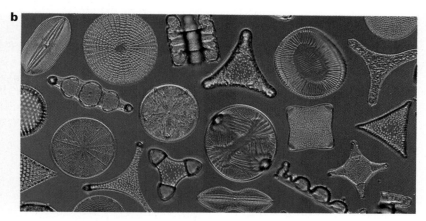

2–25 Producer protists: **a.** a dinoflagellate, one of the "fire algae," which are often red in color; **b.** diatoms, showing their many geometric forms. Both types are marine organisms.

diatomos = cut in two

brown. Diatoms have glasslike walls or shells that fit together like the two parts of a box. Diatoms live in fresh water and salt water and are an essential food for animals in both. When diatoms die, their glasslike shells sink to the bottom of the body of water they inhabit. In time, the shells accumulate and form an abrasive material called **diatomaceous** (DĪ-uh-tuh-MAY-shus) **earth**. Diatomaceous earth is used commercially in many products, including toothpaste, detergents, and insulating materials.

Unicellular algae may live as separate cells or as groups of cells. The only type of producer protist that lives on land is *Protococcus* (PRŌ-tuh-KOK-us). It may be found growing on the bark of trees, or on damp soil or rock. About 8000 species of producer protists have been described.

Protococcus usually grows on moist areas on trees.

Consumer Protists

Protozoa are often much more complex than cells of multicellular organisms. They are common in fresh water and saltwater, in the soil, and in the guts of multicellular organisms.

Most protozoa move about freely. This helps them find food. Protozoa are classified according to how they move. One group of protozoa has tiny, hairlike projections, called **cilia** (SIL-ee-uh), covering the surface of their cells. These protozoa swim by moving their cilia. Other protozoa, such as *Amoeba* (uh-MEE-buh), do not have cilia. They move by **pseudopods** (SOO-duh-PODZ). The word *pseudopod* means "false foot." These protozoa move by extending part of their body—the "foot"—in the direction they are moving. The rest of the organism follows the foot. Still another type of protozoan moves by the whiplike action of a

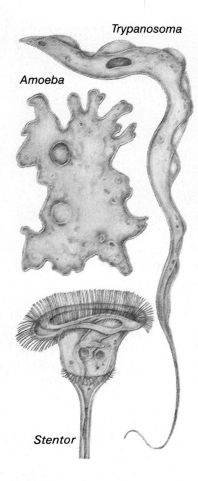

2–26 Protozoa. *Trypanosoma* causes sleeping sickness. *Amoeba* is able to change its form. *Stentor* has a wreath of cilia that beat rhythmically.

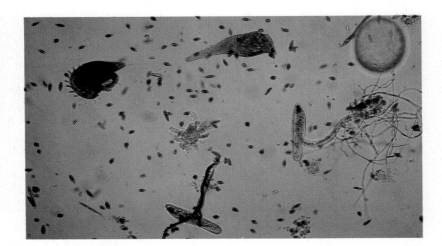

2–27 Drop of water containing protozoa and algae, including *Stentor, Volvox, Paramecium, Euglena,* and *Amoeba.*

long, hairlike structure called a **flagellum** (fluh-JEL-um). Protozoa may have one or more flagella. Some protozoa, such as the parasite that causes malaria, do not move. In all, about 30 000 kinds of protozoa are known.

The Paramecium: A Protozoan The paramecium is a one-celled protozoan that swims by moving its cilia. There are eight species of paramecia. One species, *Paramecium caudatum* (kah-DAY-tum), is a streamlined, slipper-shaped organism between 0.15 mm and 0.30 mm long. Paramecia can be found throughout the world and thrive in quiet, freshwater ponds that have some decaying organic matter. Paramecia usually eat bacteria. Many are used for laboratory research.

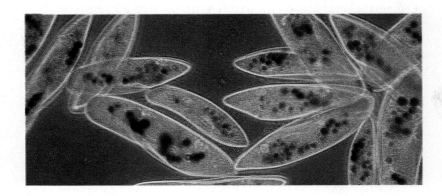

2–28 Paramecia (*Paramecium caudatum*) actively swim in search of food. Food particles inside the paramecia are stained red.

Chapter 2 Classification of Living Things

2–29 A slime mold in this stage is a streaming mass of protoplasm that engulfs and digests small particles of decayed plant matter.

Decomposer Protists

The most important protists that decompose and absorb their food are the **slime molds**. Slime molds grow on damp soil or on rotting logs in moist woods. Some species may even grow on lawns. In one stage of their life cycle, most slime molds resemble large amebas. They move about slowly and feed on decomposed material. Under certain conditions, they stop moving and begin to grow and look like fungi. Many slime molds are white, but some are red or yellow.

THE FUNGAL KINGDOM

2–30 Structure of fungus showing increasing detail of the hyphae, or filaments, that form the stem.

Although some fungi are unicellular, such as yeast, most are multicellular. The cells of many species have more than one nucleus. In general, fungi are composed of threadlike filaments of cells. Even the common mushroom is a bunch of filaments. Fungi live almost everywhere—in water and on land.

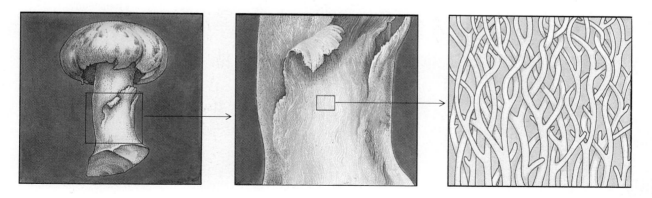

Section 2.3 Kingdoms of Living Things

Without the decomposing activity of bacteria and fungi, the world would soon be cluttered with the bodies of dead plants and animals.

Most fungi are decomposers. Along with bacteria, they break down dead organisms and waste products. By doing this, fungi obtain food and return chemicals to the environment.

Some activities of fungi are particularly beneficial to humans. Some fungi produce chemicals that help fight disease. For example, penicillin is produced by a fungus. Other fungi are important in the food industry. Yeasts are used to produce alcoholic products such as beer and to make bread dough rise.

Some activities of fungi are harmful to humans. Parasitic fungi cause diseases such as ringworm and athlete's foot. They also cause plant diseases that result in extensive crop damage. Wheat rust and corn smut are examples of fungal diseases of crops. Molds decay fruit, bread, and other foods and also decay leather and paper goods. The activity of these fungi causes great loss of food and damage to property.

Fungi include yeasts, bracket fungi, mushrooms, lichens, and molds. Certain fungi are sources of food. Some, like the truffle, are considered to be delicacies; others, including certain types of mushrooms, are highly poisonous. About 100 000 species of fungi have been identified.

2–31 Fungi: **a.** yeast cells; **b.** morels (edible and considered a delicacy); **c.** Western puffball; **d.** chanterelle mushrooms; **e.** bracket fungi, usually found on trees; and **f.** lichen, an association of algae and fungi.

THE PLANT KINGDOM

You are probably more familiar with plants than with monerans, protists, or fungi. Plants are all around us and are important in many ways. They provide food and shelter for us and other living things. They also provide us with clothing.

The most important characteristic of the Plant Kingdom is that plants produce their own food by photosynthesis. Plants, or parts of plants, are usually green in color. All plants are multicellular and do not move from place to place.

Plants are grouped according to their complexity and where they live. One feature that helps determine complexity is the presence or absence of **vascular** (VAS-kyoo-ler) **tissue**. Vascular tissue is made up of hollow tubes, or veins, that support the plant and carry food and water to all parts of the plant. There are five phyla in the Plant Kingdom. Multicellular algae comprise three phyla. Multicellular algae have no vascular tissue and live in salt water or fresh water. The **bryophytes** (BRĪ-uh-FĪTS), which live on land, are another phylum of plants without vascular tissue. The fifth phylum, the **tracheophytes** (TRAY-kee-uh-FĪTS), has vascular tissue. Most tracheophytes live on land. Table 2–2 shows the characteristics of the five phyla in the Plant Kingdom.

The strings in celery are vascular tissue.

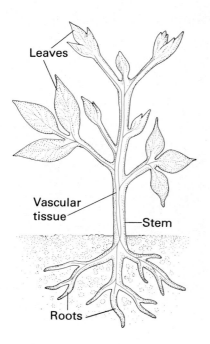

2–32 Vascular tissue in a bean plant. Vascular tissue carries minerals, nutrients, and water to all parts of the plant.

Table 2–2 The Five Phyla in the Plant Kingdom

Phylum	Characteristics	Examples
Green Algae Brown Algae Red Algae	Producers; some have structures that resemble roots, stems, and leaves, but lack vascular tissue; live in water	Freshwater algae Seaweeds
Bryophytes	Producers; have structures that resemble roots, stems, and leaves, but lack vascular tissue; live on land	Liverworts Mosses
Tracheophytes	Producers; have roots, stems, and leaves with vascular tissue; most live on land	Fern relatives Ferns Seed plants

Section 2.3 Kingdoms of Living Things

Multicellular Algae

As you have seen, unicellular algae are grouped with the protists. Multicellular algae are placed in the Plant Kingdom because they are more complex than unicellular algae. About 8000 species of multicellular algae have been identified.

There are three major phyla of multicellular algae: green, brown, and red. All are nonvascular water plants. Most green algae live in fresh water, but some live in the ocean. Most brown and red algae live in the ocean. The body of an alga may be a threadlike filament of cells or a flat plate of cells. Some seaweeds have structures that look something like leaves, stems, and roots. But they are not considered to be leaves, stems, and roots, because they do not contain vascular tissue.

You may be familiar with *Spirogyra* (SPĪ-ruh-JĪ-ruh), a freshwater green alga that consists of filaments of cells. Or, if you go to the seashore, you may recognize *Ulva* (UL-vuh), a light-green leafy alga that is often called sea lettuce. *Fucus* (FYOO-kus), or rockweed, is a branching brown seaweed with air bladders at the tips. Kelp, another type of brown seaweed, may grow to be 30 m long. Red algae, also found in the ocean, do not grow as large as kelps. They usually have a branching form.

Red and brown algae are harvested commercially. They provide thickening agents and emulsifiers for ice cream and many other products.

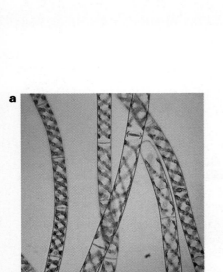
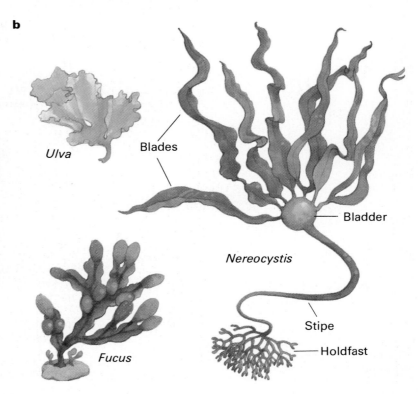

2–33 Multicellular algae: **a.** *Spirogyra*; **b.** *Ulva*, or sea lettuce; *Nereocystis*, or bladder kelp, and *Fucus*, or rockweed. Air bladders act as floats.

Algae (unicellular and multicellular) are the largest producers of food on earth. In addition to providing food, algae also offer protection for animals. Off the coast of California, kelp forms "jungles" that provide protection for sea otters and large fish, such as tuna.

Bryophytes

The bryophytes are divided into three classes: liverworts, hornworts, and mosses. (*Wort* means "herb" and is part of many plant names.) Bryophytes live in warm, moist areas. They are common in tropical forests. A few mosses, however, survive in dry areas and at high altitudes. Bryophytes are small plants ranging in height from 2 cm to 20 cm. They have structures that look like roots, stems, and leaves, but these structures lack vascular tissue. Of the three types of bryophytes, mosses are the most numerous and well-known. There are about 14 000 species of mosses.

2–34 Bryophytes. On the left is the moss *Polytrichum*. On the right is *Marchantia*, a liverwort.

2–35 A moist, temperate forest provides the ideal habitat for bryophytes, which add to the greenness of the forest.

Tracheophytes

Tracheophytes have true roots, stems, and leaves containing vascular tissue. They are divided into plants that produce seeds and plants that do not.

Seedless Vascular Plants Ferns make up the largest group of seedless vascular plants. Fern relatives, or allies, such as horsetails and club moss (*Selaginella*), are also included in this class. Ferns range in size from a small marine fern about 1 cm long to tropical tree ferns over 25 m tall. Ferns are usually found in moist areas. Taxonomists have identified about 11 000 species of ferns.

2–36 Ferns and fern relatives: **a.** *Equisetum*, or horsetails, form spore-bearing cones at the tips of their stems; **b.** *Selaginella*, or spike moss; **c.** maidenhair fern; **d.** ostrich fern; and **e.** tree fern.

angeion = case, capsule
sperma = seed

gymnos = naked
sperma = seed

Seed Plants Most people are familiar with seed-producing plants. Taxonomists divide the seed-producing plants into two categories: **gymnosperms** (JIM-nuh-SPERMZ) and **angiosperms** (AN-jee-uh-SPERMZ).

Gymnosperms generally have leaves shaped like needles. Most gymnosperms stay green all year and do not drop their leaves in autumn. Gymnosperms are so named because their seeds are "naked"; they are not enclosed in a fruit like peaches or apples. The seeds of most gymnosperms are found in cones, such as pine cones. Gymnosperms include ginkgoes, firs, pines, cycads, and redwoods; in all there are about 700 species.

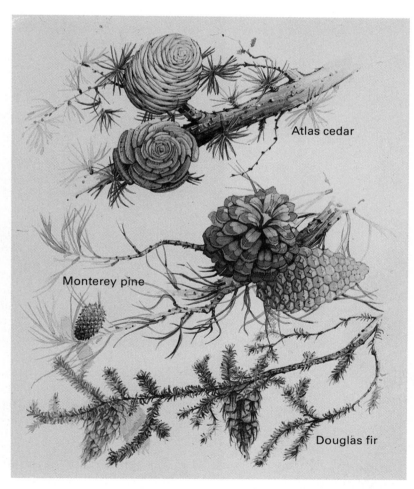

2–37 The redwood trees of California can grow to a height of over 61 m and may live two to three thousand years.

2–38 Branches of three types of gymnosperms, showing immature (green) and mature (brown) female cones. Wind blows pollen grains from the smaller male cones to the female cones, where seeds develop.

Section 2.3 Kingdoms of Living Things

2–39 Angiosperms: **a.** saguaro cactus blooms during a short rainy season; **b.** tree laden with ripening fruit; **c.** young coconut fruits; **d.** clover in bloom; **e.** meadow grasses bear dry fruits, known as grain.

Angiosperms are flowering plants. They have flat leaves that usually fall off in autumn. They produce seeds enclosed in a fruit. Apples, tomatoes, squash, and pears are all fruits. Maple trees, pansies, and nasturtiums are also angiosperms, though their fruits are less obvious. There are 250 000 known species of angiosperms.

Angiosperms can be divided into two groups based on the number of **cotyledons** (кот-ul-EED-unz) in the seed. The cotyledon provides food for the young plant developing within the seed. A plant whose seeds have one cotyledon is a **monocot** (MON-uh-кот). Corn is an example of a monocot. A plant whose seeds have two cotyledons is a **dicot** (DĪ-kot). Lima beans and peanuts are dicots. Monocots and dicots differ in many ways in addition to the number of cotyledons in the seed (see Fig. 2–41).

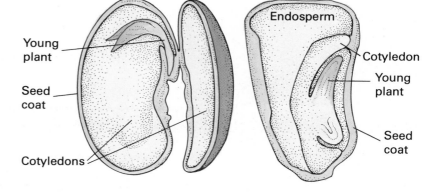

2–40 The two cotyledons of a dicot seed (left) store food for the young plant. The single cotyledon of a monocot seed (right) absorbs stored food from the endosperm.

2–41 Dicot (top) and monocot (bottom) characteristics.

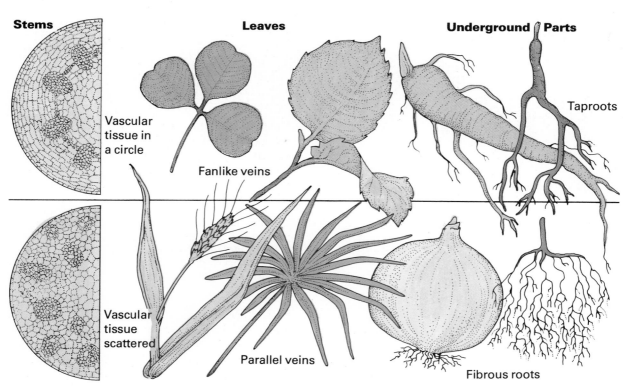

2–42 The bean plant (**a**, pole variety) is a dicot that produces green string beans (**b**), an edible fruit that contains seeds.

The characteristics of the major categories of seed plants are shown in Table 2–3.

The Bean Plant The bean is a dicot angiosperm. You will study the bean *Phaseolus vulgaris* (FAY-zee-ō-lus vul-GAYR-us). Its common names are green bean, snap bean, and string bean. People have grown beans for food since early times. Beans grow as climbing plants or as bushy plants. In North America, farmers cultivate thousands of acres of beans. Heat and frost destroy the plants; as a result, they grow mostly in moderate climates.

Table 2–3 Categories of Seed Plants

Category	Characteristics	Examples
Gymnosperm (seeds uncovered)	Most produce seeds in cones, many with needlelike leaves, many evergreen	Pine, Spruce, Hemlock, Taxus, Juniper, Larch, Cypress
Angiosperm (seeds covered)	Produce flowers, produce seeds in fruit, leaves flat, many lose leaves each year (not evergreen)	
Monocot	One cotyledon in the seed, veins in leaves run parallel, tree stems without annual rings	Grass, Corn, Wheat, Rye, Daffodil, Iris, Lily, Palm
Dicot	Two cotyledons in the seed, veins in leaves resemble a net, tree stems with annual rings	Bean, Oak, Maple, Elm, Rose, Clover, Tomato, Dandelion, Aster

THE ANIMAL KINGDOM

Over one million different kinds of organisms have been identified as members of the Animal Kingdom. Animals are multicellular consumers. Most animals can move around freely. Others, such as sponges, corals, and barnacles, cannot move from place to place. They remain attached to rocks, logs, and other objects, but are able to move parts of their bodies. Many stationary animals are able to move at some time during their development.

If you run your fingers down the center of your back, you can feel your backbone, which is actually a series of bones. These bones, called **vertebrae** (VER-tuh-BREE), enclose your **spinal cord**. Humans and other animals with vertebrae are classified as **vertebrates** (VER-tuh-BRITS). Most members of the Animal Kingdom do not have vertebrae. In fact, many animals do not have a spinal cord. Animals without vertebrae are called **invertebrates**. As you read the following descriptions of the nine major animal phyla, you will see that vertebrates make up only part of one phylum. All other animals are invertebrates.

in- = without
vertebra = a joint

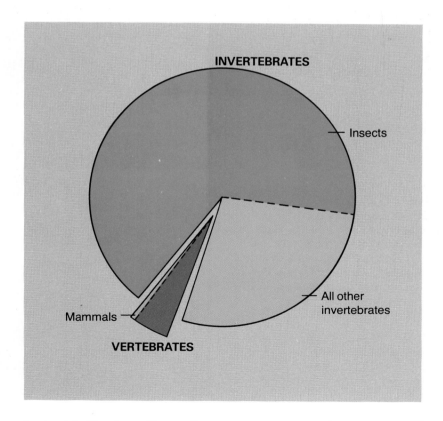

2–43 The animal kingdom consists mostly of insects and other invertebrates, such as worms, snails, spiders and shellfish. Mammals, including dogs, horses, humans, and other familiar species, make up a small portion of the vertebrates.

Section 2.3 Kingdoms of Living Things

poros = entrance, passage
-fer = bearer

Poriferans

The animals in the **poriferan** (puh-RIF-er-un) phylum are sponges. A sponge is a simple organism whose body is made up of groups of cells that are not organized into tissues or organs. Sponges are the only animals that do not have nerve cells. There are nearly 400 species of sponges.

Sponges live in either fresh water or salt water. As adults, they remain in one place, usually attached to rocks, shells, or other objects. The body of a sponge is covered with openings called pores. Water flows into the sponge through the pores and flows out through a larger opening at the end of the sponge. The sponge filters out and eats food carried into it by the water. Sponges have stiff internal skeletons. The cleaned and dried skeletons of sponges were once commonly used for bathing and cleaning. Today, these have been largely replaced by synthetic sponges.

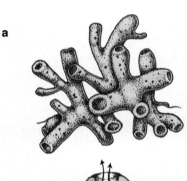

2–44 Sponge structure: **a.** sponge showing tubelike parts with small pores along the sides; **b.** diagram of section of a sponge showing direction of water flow. The opening of the top is called the osculum.

2–45 Sponges: **a.** freshwater sponge covered with algae; **b.** marine sponge colony. Most sponges live in shallow salt-water.

Chapter 2   Classification of Living Things

Coelenterates

koilia = body cavity
enteron = intestine

Most **coelenterates** (sih-LEN-tuh-RAYTZ) live in salt water. Ocean coelenterates include jellyfish, sea anemones (uh-NEM-uh-NEEZ), and corals. Some coelenterates live in colonies, such as the Portuguese man-of-war. Coelenterates have tentacles around the mouth that contain stinging cells that paralyze their prey. The bodies of most coelenterates are made up of two layers of cells: an outer, covering layer and an inner layer surrounding the digestive tract. The digestive tract has only one opening, the mouth, through which food enters and waste products leave the organism.

2–46 Coelenterates: **a.** snowflake coral; **b.** a rare Gorgonian sea whip, a type of coral; **c.** a jellyfish medusa; **d.** strawberry anemones.

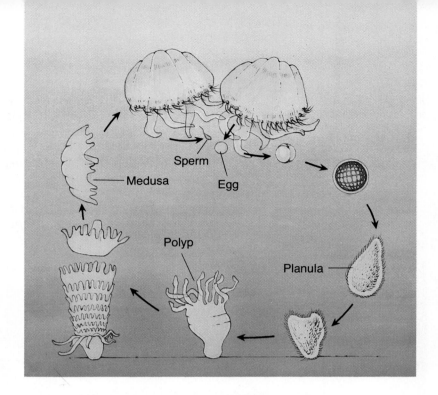

2–47 Life cycle of *Aurelia*. Adult jellyfish release sperm and eggs. After fertilization a larval stage (planula) forms and attaches to the bottom where it becomes a polyp. The polyp grows and produces medusas, which develop into full-sized jellyfish.

Coelenterates occur in two forms. In one form, the tentacles hang down and the organisms resemble jellyfish. In the other form, the tentacles point upright. An example of a coelenterate with upright tentacles is a sea anemone. Sometimes the two forms are parts of the life cycle of a single species (see Fig. 2–47).

The Hydra: A Coelenterate *Hydra* is a freshwater coelenterate. The hydra shown in Figure 2–48 is *Hydra littoralis* (LIH-tor-AL-is). It is a slender animal whose body is about 12 mm long. A hydra has five or six tentacles surrounding its mouth. The tentacles have stinging cells that paralyze prey. They can shoot these cells as far as 30 cm. Hydras live on submerged pond weeds, sticks, dead leaves, and stones, or hang from the surface of a quiet body of water. They eat small pond organisms.

The hydra is often cultured in large numbers and used for laboratory research.

2–48 A brown hydra with two buds. While the buds remain attached to the parent they move and feed independently.

Platyhelminthes

platys = broad, flat
helminth = wormlike animal

The animals in phylum **Platyhelminthes** (PLAT-ee-HEL-men-theez) are flatworms. Taxonomists have identified about 15 000 species of flatworms. The body of a flatworm is more complex than that of a coelenterate. The flatworm has three cell layers and some organ systems; however, flatworms lack systems for circulation and respiration. The digestive system usually has one opening, as in coelenterates.

Phylum Platyhelminthes is divided into three classes. Two of the classes are parasites—the flukes and tapeworms. These organisms cause disease in humans and other animals. Most of them have suckers by which they attach themselves to the host. The other class of Platyhelminthes is nonparasitic. Planaria (pluh-NAYR-ee-uh) are members of this class.

2–49 Flatworms. **a.** A parasitic sheep liver fluke clings to its host with its sucker. **b.** The sheep tapeworm absorbs digested food from the sheep's intestine. **c.** Ingested food particles outline the intestine of this troglobite planarian.

Section 2.3 Kingdoms of Living Things

Nematodes

nēmatos = thread

Nematodes (NEM-uh-TŌDZ) are worms with smooth, round bodies that taper to a point at both ends. They are often called roundworms. The bodies of nematodes are about as complex as the bodies of flatworms; however, nematodes have digestive systems with two openings. Nematodes are extremely abundant; there are nearly 80 000 known species. Some live in soil or in water; others are parasites on plants or animals. Nematode parasites on plants destroy millions of dollars worth of crops each year. Nematode parasites in the human body cause serious diseases, including hookworm and trichinosis.

Nematodes are found almost everywhere on earth but are seldom seen. As many as 7 billion can be found in a hectare of rich topsoil.

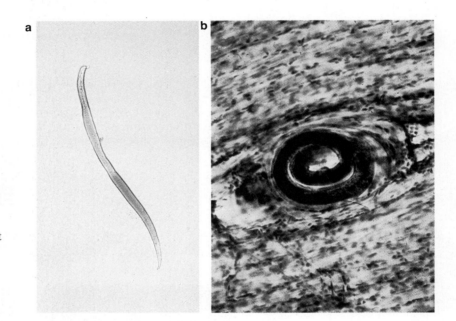

2–50 The roundworm *Trichinella spiralis* (**a**) causes trichinosis in humans. Infection comes by eating raw or incompletely cooked pork. The worms migrate from the intestine to muscle tissue, where they become encysted (**b**). Cooking meat thoroughly kills the worms.

Annelids

anulus = a ring, ringlike part

Annelids (AN-il-idz) are worms whose bodies are divided into segments. Unlike flatworms and roundworms, annelids have a circulatory system and well-developed digestive and nervous systems. Most segmented worms breathe through their skin. There are about 9000 species divided into three classes. Ocean annelids, the sandworms and tubeworms, make up one class. Earthworms and freshwater segmented worms make up a second class. The annelids in both these classes are not parasites. Leeches form the third class. Certain leeches, the bloodsuckers, are parasitic.

2–51 Annelids. **a.** *Nereis* spends most of its life in a burrow underwater; **b.** tubeworm, an ocean annelid, with feathery gills extended; **c.** horse leech swimming.

The Earthworm: An Annelid The earthworm found in gardens and used as bait for fishing is *Lumbricus terrestris* (LUM-brih-kus ter-RES-tris). Hundreds of other species of earthworms live in the soil as well. Earthworms are found almost everywhere. They live in compost heaps, sewage beds, trees, and mud lakes and eat bacteria, small soil organisms, and decaying organic matter. Earthworms, in turn, are eaten by birds and some mammals.

Earthworms range in size from very small (25 mm long and 8 mm wide) to gigantic (120 cm long and 2.2 cm wide). Giant earthworms, such as those found in Australia, can stretch their bodies to a length of 360 cm.

2–52 The earthworm caught by an early bird is one that stayed out too late, since earthworms are nocturnal animals. They burrow underground during the day.

Mollusks

molluscus = soft

Snails, slugs, oysters, clams, squids, and octopuses are a few of the 100 000 species of **mollusks** (MOL-usks). Some mollusks live in salt water, others in fresh water, and still others on land. Most mollusks breathe by means of gills. The soft body may be protected by one shell, as in snails, or two shells, as in oysters. Some mollusks, such as the squid and octopus, have little or no shell.

The mollusk's soft body is usually divided into three main parts. The first part is a head-foot that helps it move. The second part of the body contains the heart and other organs. The third part is the **mantle**, a strong membrane that covers the soft body and often the gills. The mantle also secretes the shell. Except for the squid and octopus, mollusks are very slow-moving.

A pearl results when an oyster coats a sand grain inside its body with a smooth material.

2–53 Mollusks: **a.** octopus; **b.** Philippine nudibranch (sea slug) has no external shell; **c.** tree snail; **d.** white chitons attach to rocks with their strong muscular foot; **e.** Pacific pink scallop, a bivalve.

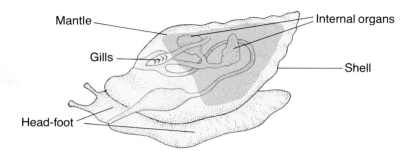

2–54 A snail, showing the basic parts of a mollusk: shell, mantle, head-foot, gills, and internal organs. The snail is a gastropod, all of which have a single external shell.

Arthropods

arthron = a joint
podos = foot

With nearly 1 000 000 species, the **arthropods** (AR-thruh-PODZ) make up the largest phylum of animals on earth. Insects alone make up 700 000 species. Compare these numbers with the other major animal phyla.

Arthropods can be found in water, on land, and in the air. They have a protective skeleton on the outside of the body, and their legs are jointed. It is this last feature that gives the phylum its name.

The arthropod phylum is made up of five important classes. The **crustaceans** (krus-TAY-shunz) include ocean arthropods such as crayfish, lobsters, shrimp, crabs, and water fleas. Sow bugs (or pill bugs) are an example of a land-living crustacean.

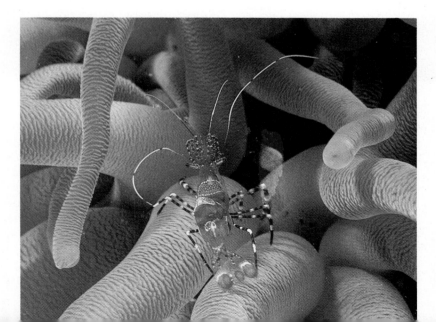

2–55 A shrimp among the tentacles of a sea anemone.

2-56 Arachnids: **a.** jumping spider; **b.** scorpion kills its prey (spiders and insects) with its stinger; **c.** horseshoe crab; **d.** a harvestman, or daddy longlegs; **e.** banded argiope spider; **f.** hard tick; ticks suck blood from mammals, including humans.

Arachnids have two body parts and eight legs. Insects have three body parts and six legs.

The **arachnids** (uh-RAK-nidz) include spiders, mites, ticks, daddy-longlegs, scorpions, and horseshoe crabs. Most adult arachnids have two body regions and four pairs of walking legs. The **insect** class contains the tremendous number of insects on earth. Some common examples are silverfish, dragonflies, grasshoppers, beetles, lice, butterflies, mosquitoes, fleas, and ants. All adult insects have six legs. Their bodies are divided into three regions: head, thorax, and abdomen. The other two important arthropod classes include centipedes and millipedes.

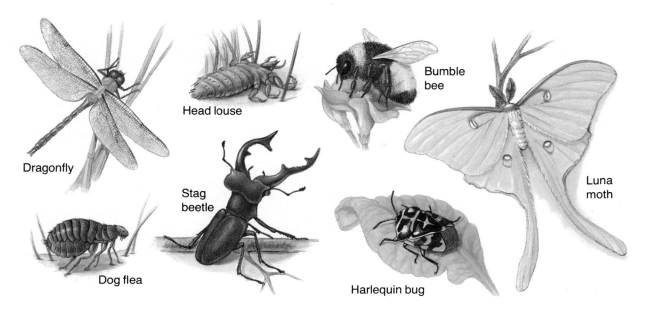

2–57 Insects are found in every land habitat and in all bodies of freshwater. They represent the only group of invertebrates that can fly.

The Grasshopper: An Arthropod Grasshoppers, also called locusts, are insects. Of the nearly 20 000 known species of grasshoppers, 600 live in North America. The grasshopper discussed in this book is *Romalea microptera* (rō-MAYL-yuh mī-KROP-ter-uh).

Grasshoppers live in deserts, grasslands, and forests and eat grass and other plants. Some species are harmful to crops. Birds, frogs, and snakes eat grasshoppers. Some grasshoppers are fairly small, but others can grow to lengths of more than 5 cm.

2–58 A lubber grasshopper from Florida. The third and largest pair of legs is specialized for jumping.

Echinoderms

echino- = spiny
derma = skin

Starfish, sea urchins, sand dollars, sea lilies, and sea cucumbers are examples of **echinoderms** (ee-KĪ-nō-DERMZ). All 6000 species of echinoderms are marine. Most of them live on the ocean bottom. The adult body resembles a wheel, with parts of the body radiating outward from the center. In general, the adult body has five parts (or multiples of five) radiating outward. The most common starfish, for example, has five arms. Echinoderms have internal skeletons with bumps or spines that stick out through the skin. This makes the skin rough and gives the phylum its name. Although they do not even look much like animals, certain of their features cause taxonomists to think that echinoderms are most closely related to the chordate phylum.

2–59 Echinoderms. The center opening of the sand dollar is the mouth. The sea cucumber has rows of tube feet extending along the length of its body. The spines of different species of sea urchins vary greatly. The starfish feeds on clams and oysters.

Chordates

The name chordate *refers to the notochord, which all members of this phylum possess during at least one stage of their lives.*

The most complex and specialized members of the Animal Kingdom are the **chordates** (KOR-dayts). This phylum is made up of about 45 000 species. All chordates have three main characteristics. At some time during their development, all chordates have gill slits. In some animals, fish, for example, the gill slits develop into functioning gills. In other chordates, like cats and dogs, the gill slits disappear before they are born. The second chordate characteristic is the presence of a **notochord** (NŌT-uh-KORD). The notochord is a supporting rod of tissue that runs along the animal's back. The notochord is only present during development in most chordates. The third chordate feature is the presence of a hollow nerve tube that runs along the back. Figure 2-60 shows the placement of the notochord and nerve tube as they are found during development in most chordates.

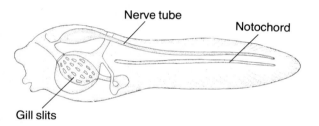

2-60 The tunicate larva has all three chordate characteristics: a supporting notochord, a dorsal nerve tube, and gill slits. After the larva attaches to a rock it loses its tail.

The chordate phylum is divided into three major groups or subphyla. One group of chordates, the **lancelets** (LANS-lits), retains the notochord and gill slits throughout life. Amphioxus (AM-fee-AHK-sus) is one example. Amphioxus is a small, slender animal that lives on tropical and temperate seacoasts. It swims in the ocean and burrows in the sand, burying itself with its tail down.

The second group of chordates has a notochord only during an early stage of development. Gill slits are present in the adult. Like the lancelets, all species of this group live in salt water. Many have saclike bodies. This group is made up of **tunicates** (TOO-nih-kitz), so-named because their body covering resembles a tunic. Tunicates inhabit seas in all parts of the earth. Some are free-swimming; others become attached after a free-swimming stage. The best known tunicates are sea squirts, which squirt water when touched.

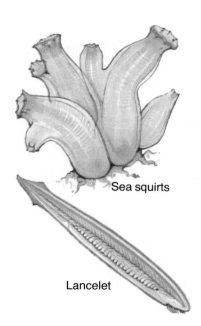

2-61 The sea squirt frequently forms colonies attached to rocks. The lancelet (*Amphioxus*) is a semi-transparent, marine animal that spends most of its time buried up to its mouth in the sand.

Section 2.3 Kingdoms of Living Things

The third group of chordates is the one you are probably most familiar with: the **vertebrates**. Vertebrates have a backbone that develops around the notochord. There are seven living classes of vertebrates: **fish** (three classes), **amphibians**, **reptiles**, **birds**, and **mammals**.

Fish   All fish live in water. Their bodies are covered by scales and they breathe through gills. They are also **coldblooded**. This means that their body temperature changes as the temperature around them changes. Fish with soft skeletons made of **cartilage** (KART-il-ij), the flexible support tissue, make up two classes. One of these classes includes jawless fish, such as lampreys and hagfish. Sharks, skates, and rays make up the other class. The third class of fish have skeletons made of bone. Salmon, tuna, eels, trout, and goldfish are bony fish. Altogether, there are about 23 000 species of fish.

Coldblooded animals do not match the outside temperature degree for degree, but they do vary with it.

2–62   Fish: **a.** spotted eagle ray; **b.** sockeye salmon, a schooling fish; **c.** head of Pacific lamprey, showing jawless mouth with round sucker lined with horny teeth; they feed on fish.

Chapter 2   Classification of Living Things

amphibios = living a double life

Amphibians   The name *amphibian* means "leading a double life." Amphibians spend part of their lives in or near water and part on land. Consider the frog, for example. As a tadpole, the young frog lives in the water and breathes through gills. As an adult, it has lungs and breathes air. Amphibians are cold-blooded and live in a moist environment. Their thin, moist skin would dry out in a dry atmosphere. Frogs, newts, toads, and salamanders are examples of amphibians; in all, this class includes nearly 2000 species.

2–63   Amphibians: **a.** red-eyed frog; **b.** American toad. Toads can live on land; frogs must stay close to water. A toad has poison glands in its skin.

2–64   Red-spotted newt: **a.** green aquatic adult stage; **b.** red eft stage, the juvenile form. The larval form of this species is also aquatic. Newts are small, usually land-dwelling, salamanders.

Section 2.3   Kingdoms of Living Things

Dinosaurs are extinct members of the reptile class.

Reptiles Reptiles are better adapted to life on land than amphibians. Their dry, scaly skin keeps them from drying out. Reptiles are coldblooded and breathe by means of lungs throughout their lives. There are about 5000 species of reptiles, including turtles, alligators, crocodiles, and snakes.

2–65 Northern diamondback terrapins, a freshwater turtle, hatching from their eggs. Reptile eggs are pliable and leathery and retain moisture.

2–66 Red rat, or corn, snake: **a.** beginning to shed its skin; **b.** almost fully emerged from its old skin.

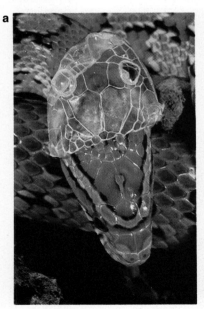

Chapter 2 Classification of Living Things

Birds Birds are the only animals that possess feathers. Feathers help birds stay warm and enable them to fly. Birds breathe through their lungs and are **warmblooded**. Birds differ from fish, amphibians, and reptiles in this latter characteristic. Warmblooded animals maintain a constant body temperature even when the temperature around them changes. Birds live on land and over water. Albatrosses and petrels live over the ocean; gulls and shore birds live on sandy beaches. Ducks inhabit marshes, and sparrows dwell in grasslands. Woodpeckers, hawks, and owls can be found in trees. More than 8000 species of birds have been identified.

2-67 Birds: **a.** mockingbird with its feathers fluffed to keep it warm; **b.** willow ptarmigan in spring plumage; **c.** female mallard duck showing identifying color patches on its wing; **d.** emperor penguin.

Mammalian groups are classified by how their young develop.

Mammals All mammals have hair or fur on their bodies and are warmblooded. Female mammals have **mammary glands** that produce milk, which they feed to their young offspring.

Mammals are divided into three groups. The first group of mammals lays eggs. This group includes the duckbill platypus and the spiny anteater. The second group of mammals has pouches. Offspring of these mammals are not fully developed when they are born. After a short period within the mother's body, they are born and develop further in the mother's pouch. Mammals with pouches include kangaroos, opossums, and koalas. The last, and largest, group of mammals has young that develop inside the mother's body for a comparatively long time. When these mammals give birth, the offspring are very well developed. These mammals include bats, whales, mice, lions, bears, monkeys, apes, and humans. In all, there are about 4000 species of mammals on earth.

The characteristics of the vertebrates are summarized in Table 2–4.

2–68 Mammals: **a.** duck-billed platypus; **b.** impalas grazing on an African plain; **c.** female kangaroo with her "joey," or young, in her pouch.

Chapter 2 Classification of Living Things

2–69 Bats are the only mammals with structures for flight: **a.** Mexican free tail bats searching for food at dusk; **b.** fruit bat in its daytime sleeping position. Most bats eat insects; some eat fruit.

Table 2-4   The Seven Vertebrate Classes

Class	Characteristics
Jawless Fish Cartilaginous Fish Bony Fish	Scales cover body, breathe through gills, most have fins, coldblooded
Amphibians	Moist skin covers body, most adults breathe with lungs, coldblooded
Reptiles	Dry skin with scales covers body, breathe with lungs, coldblooded
Birds	Feathers cover body, breathe with lungs, warmblooded
Mammals	Hair on body, breathe with lungs, female has milk (mammary) glands, warmblooded

Section 2.3   Kingdoms of Living Things

Table 2–5   Major Animal Phyla

Phylum	Characteristics	Examples
Poriferans (sponges)	Do not move around, canals all over body, skeletons of fibers or minerals	Bath sponge Finger sponge
Coelenterates	Soft bodies, tentacles with stinging cells, one body opening for food to enter and wastes to leave	Hydra Sea anemone Jellyfish Coral
Platyhelminthes (flatworms)	Flat bodies, many are parasites	Planaria Tapeworm Liver fluke
Nematodes (roundworms)	Round body, many are parasites	Hookworm Soil nematodes Dog roundworm
Annelids (segmented worms)	Round body, body divided into segments	Earthworm Sandworm Leech
Mollusks	Soft bodies, usually with shell or shells	Oyster Clam Scallop Octopus Squid Snail
Arthropods	External skeleton, jointed appendages	Lobster Grasshopper Spider Centipede Millipede
Echinoderms	Spiny body, radial symmetry	Starfish Sea urchin Sand dollar Brittle star
Chordates	Support structure (notochord) during development, hollow nerve cord along back, gill slits at some time during development	Amphioxus Fish Frog Snake Bird Human

Building a Science Vocabulary

bacterium	gymnosperm	arachnid
blue-green alga	angiosperm	insect
nuclear membrane	cotyledon	echinoderm
prokaryote	monocot	chordate
eukaryote	dicot	notochord
spore	vertebra	lancelet
chlorophyll	spinal cord	tunicate
photosynthesis	vertebrate	fish
virus	invertebrate	amphibian
protozoan	poriferan	reptile
diatomaceous earth	coelenterate	bird
cilium	platyhelminth	mammal
pseudopod	nematode	coldblooded
flagellum	annelid	cartilage
slime mold	mollusk	warmblooded
vascular tissue	mantle	mammary glands
bryophyte	arthropod	
tracheophyte	crustacean	

Questions for Review

1. State one important difference between the terms in each of the following pairs.
    a. monerans and protists
    b. fungi and plants
    c. angiosperms and gymnosperms
    d. monocots and dicots
    e. vertebrates and invertebrates

2. What type of skin covering is found in each of the classes of vertebrates?

3. What characteristics of a euglena cause taxonomists to classify it as a protist?

4. What characteristics of a bean plant cause it to be classified as a plant? as an angiosperm? as a dicot?

Chapter 2
Summary and Review

Summary

1. Taxonomists organize living things into categories and give them scientific names that are used all over the world.

2. Each scientific name is made up of two Latin words (or proper names with Latin endings). This two-word system is known as binomial nomenclature. The first word is the name of the genus. The second word is the name of the species.

3. In order from most general to most specific, the taxonomic categories are: kingdom, phylum, class, order, family, genus, and species.

4. Modern classification is based on relationships between organisms. Related organisms often have similar structures and are placed in the same category.

5. If structure is misleading, chemical and breeding tests are performed to determine relationship.

6. This book recognizes five kingdoms: monerans, protists, fungi, plants, and animals.

7. The five kingdoms are arranged according to complexity and method of nutrition. Monerans include bacteria and blue-green algae, both of which are prokaryotes. Protists are unicellular eukaryotes, including producers, consumers, and decomposers. Fungi, plants, and animals are multicellular eukaryotes. In general, fungi are decomposers, plants are producers, and animals are consumers.

8. Plants include multicellular algae, bryophytes, and tracheophytes. The most dominant groups in the Plant Kingdom are seed producers. The two categories of seed-producing plants are the gymnosperms and angiosperms.

9. The nine major phyla in the Animal Kingdom are: poriferans, coelenterates, platyhelminthes, nematodes, annelids, mollusks, arthropods, echinoderms, and chordates. Vertebrates are classified as chordates and include fish, amphibia, reptiles, birds, and mammals.

10. The five organisms that will be studied in detail in this book are the paramecium, bean plant, hydra, earthworm, and grasshopper.

Review Questions

Application

1. For each of the following groups of organisms, select the one that is least closely related to the other three.
    a. seal, bear, penguin, beaver
    b. frog, snake, lizard, alligator
    c. cactus, fern, daisy, oak tree
    d. robin, bat, sparrow, chickadee
    e. rye, corn, bean, wheat

2. For each of the following groups of four terms, choose the term that includes the other three.
    a. fish, vertebrates, amphibians, animals
    b. arthropods, insects, crustaceans, millipedes
    c. genus, order, phylum, species

Questions 3 and 4 refer to the following breeding experiment:

Scientists find tan mice living on one side of a mountain and gray mice on the other side. The scientists wish to classify these mice. They bring them together into the laboratory to determine whether they will breed with each other.

3. If the mice do not breed with each other, they are members of different (choose one)
   a. kingdoms.
   b. genera.
   c. species.
   c. orders.

4. If the mice breed with each other, but their offspring cannot reproduce, it might mean that the offspring (choose one)
   a. are the same species as the parents.
   b. are a different species than the parents.
   c. are a different genus than the parents.
   d. are not a species.

5. According to the binomial system, the fungus *Acetabularia mediterranea* is most closely related to (choose one)
   a. *Crenulata acetabulara*.
   b. *Acetabularia crenulata*.
   c. *Mediterranean crassa*.
   d. *Mediterranean crenulata*.
   Explain your answer.

6. What problems would a taxonomist have in classifying a whale? Why are whales classified as mammals and not fish?

7. What are the basic chordate characteristics? Which of these characteristics are visible in adult human beings?

8. If you discovered an unclassified multicellular organism, what information would help you classify it?

9. What is the most significant factor in determining the classification of an organism?

Interpretation

1. Suppose you dug up an old bone in your back yard. From your knowledge of taxonomy, what kind of organism did the bone come from? Be as specific as possible in your answer.

2. A German shepherd dog looks more like a wolf than it looks like a bulldog. Yet, German shepherds are placed in the same category as bulldogs and not wolves. Explain the basis for the classification of these animals.

Extra Research

1. Get better acquainted with a single group of plants or animals. Go out in your area and identify birds, trees, flowers, insects, or whatever group interests you. Use a published guide like one of the Peterson series to help you identify the organisms you see. Keep a record of your findings.

2. Harvest *Protococcus* and observe under a microscope.

3. Make up a dichotomous key to identify ten or more late-model cars.

4. Look up the life cycle of a slime mold. Explain the difficulties that might be involved in trying to classify this organism.

Career Opportunities

Naturalists work out-of-doors doing environmental research and helping people enjoy and learn about the natural environment. They must know many plants and animals well. Naturalists often use dichotomous keys and scientific names for organisms in their work. Contact Environment Canada, Environmental Conservation Service, Place Vincent Massey, Ottawa, Ontario K1A 1C7, or the National and Provincial Parks Association of Canada, 47 Colborne St., Toronto, Ontario M5E 1E3.

Horticulturalists plan arrangements of plants in and around buildings, along highways, and in gardens and parks. They use their knowledge of plants, soils, fertilizers, and climate to choose the right plants for each location. For information, write the Canadian Horticultural Council, 1568 Carling Ave., Ottawa, Ontario K1Z 7M5 or the Canadian Society for Horticultural Science, Agriculture Canada Research Stn., 8801 E. Saanich Rd., Sidney, B.C. V8L 1H3.

Unit I
Suggested Readings

Andrews, Michael, *The Life That Lives on Men*, New York: Taplinger Publishing Co., Inc., 1976. *Describes the skin as an ecosystem in which symbiosis occurs. Excellently illustrated and written with humor.*

Batten, Mary, "Earth's Odd Couples," *Science Digest*, November-December, 1980. *Fascinating examples of symbiosis are described in this well-illustrated article.*

Carson, Rachel, *Silent Spring*, New York: Fawcett World Library, 1978. *Well researched account of human's use of poisons to control insect pests and unwanted vegetation and the effect on the balance of nature. Stresses ecology and interrelationships.*

Crockett, Lawrence J., "Plant Science a Varied Field," *Science Digest*, October, 1978. *Describes the many careers possible in the field of botany, including plant taxonomy.*

Gould, Stephen J., "On Heroes and Fools in Science," *Natural History*, September, 1974. *Compares the results of following fact versus superstition in scientific discovery.*

Graham, Frank Jr., "Farewell, Mexican Duck," *Audubon*, September, 1979. *Describes how evidence from breeding helps establish the classification of certain species of ducks.*

Margulis, Lynn, and Karlene V. Schwartz, *Five Kingdoms: An Illustrated Guide to the Phyla of Life on Earth*, San Francisco: W.H. Freeman and Co., 1982. *"A catalog of the world's living diversity" with lively text and beautiful illustrations.*

Marteka, Vincent, "Words of Praise—and Caution—About Fungus Among Us," *Smithsonian*, May, 1980. *Well-illustrated article describes types of mushrooms and their natural history.*

Milne, Lorus, and Margery Milne, *The Audubon Society Field Guide to North American Insects and Spiders*, New York. Alfred A. Knopf, Inc., 1980. *Excellent color photographs complement the text. Includes scientific names.*

Miller, Heather, and Norton Miller, "Mark Ye the Bryophytes," *Horticulture*, January, 1979. *Describes the mosses, liverworts, and hornworts, their reproduction, geography, and uses.*

Mitchell, John, and The Massachusetts Audubon Society, *The Curious Naturalist*, Englewood Cliffs, NJ: Prentice-Hall, Inc., 1980. *Excellent sketches and descriptions of the common phenomena of the four seasons.*

Oleksy, Walter, *Careers in the Animal Kingdom*, New York: Julian Messner, Inc., 1980. *Interviews of 40 men and women who work with many kinds of animals in a variety of settings. Describes education requirements, how to get started, and working conditions.*

Peterson, Roger Tory, Peterson Field Guide Series, Boston: Houghton-Mifflin Co., various copyright dates. *A series of field guides for different organisms: birds, butterflies, mammals, shells, ferns, trees, shrubs, wildflowers, reptiles, amphibians, etc. Sponsored by the National Audubon Society, these guides are clear, well-illustrated, and extremely useful.*

Ruggiero, Michael, Alan Mitchell, and Philip Burton, *Spotter's Handbook: Flowers, Trees and Birds of North America*, New York: Mayflower Books, W. H. Smith Publications, Inc., 1979. *Useful field handbook with many color illustrations to aid identification.*

Shapiro, Stanley Jay, *Exploring Careers in Science*, New York: Richards Rosen Press, Inc., 1981. *Comprehensive guide to careers in the physical, life, and environmental sciences.*

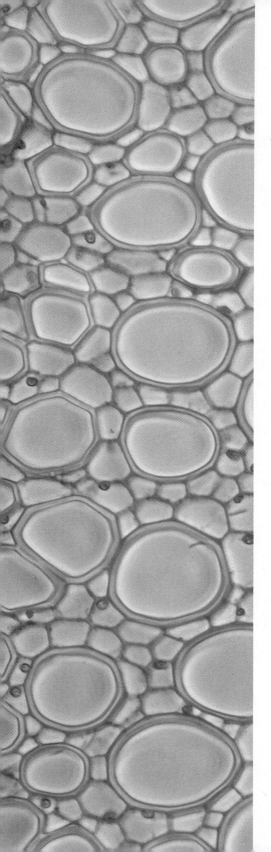

Unit II

Cell Structure and Function

What is the secret of life? Biologists probe the structure and function of organisms in search of the answer. Virtually all living things are made up of cells. Cells vary greatly in size, structure, and function. Yet, all are microscopic workrooms, bustling with the activities of life.

Cells contain numerous smaller structures and are filled with thousands of chemical substances. Like machinery in a factory, cell structures perform life's activities. They harness matter and energy, transforming one chemical substance into another. They produce materials for growth and repair. They get rid of cell wastes and maintain a balanced internal environment.

In a unicellular organism the single cell performs all the functions necessary to sustain its life. The many cells of a multicellular organism work together in performing life's activities. Each has a specialized task and is dependent upon the others. Many similar structures and functions occur in all cells, regardless of where they are found.

Thick-walled cells form tubes that conduct water through a celery stalk, seen here in cross section.

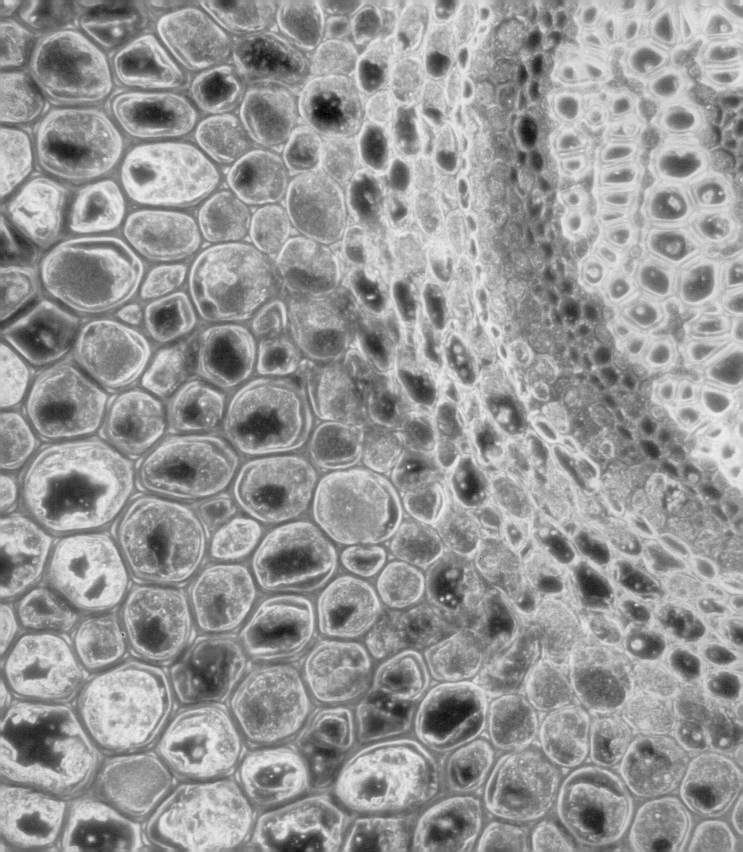

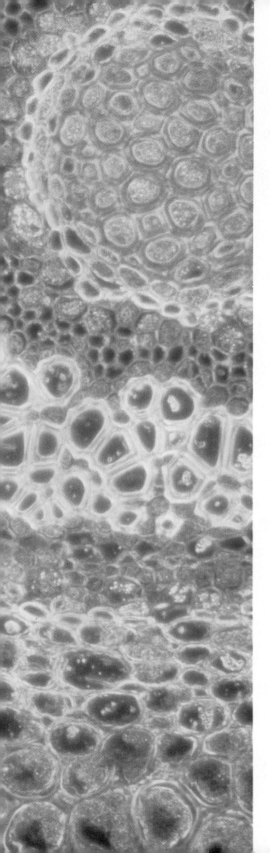

Chapter 3

The Cell: Basic Unit of Life

Focus *All organisms except viruses consist of cells. Most cells are too small to be seen without a microscope, yet they are amazingly complex. Although cells perform different functions and occur in a variety of sizes and shapes, their structure is remarkably similar.*

3.1 OBSERVATION OF THE CELL

People have always been fascinated by the stars and the moon. Yet for much of human history, these distant objects remained a mystery. In 1600 a Dutch spectacle maker created the first telescope by putting an eyeglass at each end of a hollow tube. With this invention, people could see distant objects that they had never seen before. In all fields of science, new instruments and machines have opened the doors to discovery. Just as the telescope revolutionized astronomy, the invention of the microscope has changed our understanding of biology.

Leeuwenhoek's Microscope

The contents of a pot of old rainwater may seem uninteresting compared with the night sky. But to Antony van Leeuwenhoek (LAY-ven-HOOK), a seventeenth century Dutch shopkeeper, the

3–1 Cross section of a plant stem. Dyes (stains) make the cells in different areas of the stem visible.

89

3–2 Antony van Leeuwenhoek and his simple microscope. The specimen is viewed through the small round lens while holding the microscope up to the light.

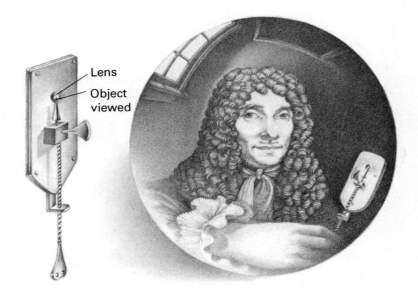

Some of the first microscopes were made in Holland around 1590 by the Janssen brothers. Leeuwenhoek probably did not know of their work.

contents of a pot of rainwater were very exciting. In the 1670s, Leeuwenhoek produced a lens that made an object appear larger than it actually was. Leeuwenhoek made this lens into one of the first microscopes. His microscope is called a **simple microscope** because it has only one lens.

Leeuwenhoek studied rainwater, lake water, soil, and other materials with his microscope. He observed bacteria, yeast, and other unicellular organisms. He called these tiny creatures "animalcules" (little animals).

Even though Leeuwenhoek had no formal science training, he kept very careful records of his observations. In 1674 he compiled these records and submitted them to the Royal Society of England. Through this organization of scientists, others learned about his studies of the microscopic world.

Leeuwenhoek was one of the first to record observations of microscopic life.

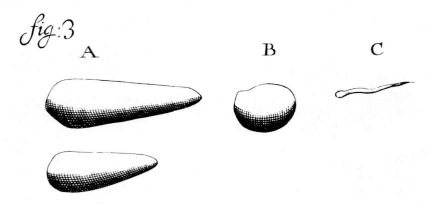

3–3 Drawings of the intestinal protozoa of a frog from an 1863 engraving by Leeuwenhoek.

Chapter 3   The Cell: Basic Unit of Life

Robert Hooke's Microscope

Hooke was also an inventor. He devised the vacuum pump and the balance spring used in watches.

At about the time Leeuwenhoek was making his observations, an English scientist named Robert Hooke built a microscope with two sets of lenses. Such a microscope is called a **compound microscope**. One set of lenses, the **objective**, enlarges or magnifies an object, producing an image. The other set of lenses, the **eyepiece**, further magnifies the image (see Fig. 3–4).

Like Leeuwenhoek, Robert Hooke observed everyday objects. He recorded his findings in detailed pen-and-ink drawings. One of the things he examined through the microscope was an extremely thin sliver of cork. Hooke described the cork as being made up of boxes, pores, and cells. Biologists now use the term "cell" to refer to the smallest unit of living things.

Robert Hooke also submitted the record of his work to the Royal Society of England. In 1665 he published his classic book, *Micrographia* (MĪ-krō-GRAF-ee-uh), which contains descriptions and drawings of his observations.

3–4 Robert Hooke's compound microscope is stationary and has two sets of lenses.

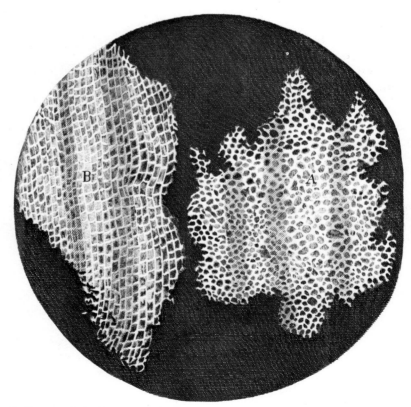

3–5 Hooke's drawings of cork cells: **a.** cross section; **b.** longitudinal section.

Section 3.1 Observation of the Cell

Modern Methods of Observation

For many years after Hooke first described cells, microscopic research proceeded quite slowly. There were two reasons for this. It took many years to develop good microscopes, and it took time for biologists to obtain them. But, by the beginning of the eighteenth century, many biologists were using microscopes to examine plant and animal cells. Today biologists have microscopes that can magnify an object hundreds of thousands of times. This is quite a difference from Hooke's microscope that magnified cork cells about 100 times.

The microscopes used by both Hooke and Leeuwenhoek were **light microscopes**. In a light microscope, light passes through a cell or other thin object to the lens. The lens then magnifies the object. If an object under a light microscope is too thick for light to pass through, the object cannot be seen. Today's compound light microscopes can magnify objects up to 1000 times.

3–6 Modern microscopes: **a.** compound, stereoscopic (two eyepieces) light microscope; **b.** scanning electron microscope with screen and control panel.

During the 1930s, biologists developed a much more powerful microscope called the **electron** (ih-LEK-tron) **microscope**. Light does not pass through the object in an electron microscope. Instead, **electrons**, small parts of atoms, pass through the object. A person using an electron microscope can see the object magnified on a special screen, or it can be viewed on a **micrograph**, which is a photograph taken through the microscope. An electron microscope can magnify biological materials up to 100 000 times. Some nonbiological materials can be magnified up to 1 000 000 times.

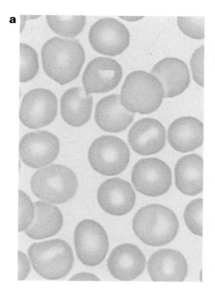 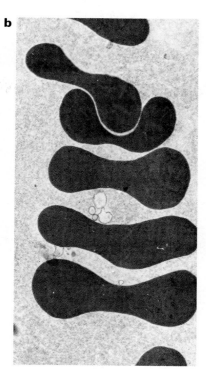 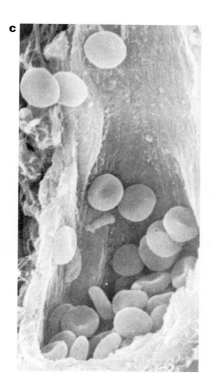

3–7 Red blood cells seen through light microscope (**a**), electron microscope (**b**), and scanning electron microscope (**c**).

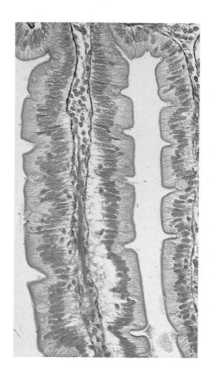

Another type of electron microscope used by biologists is the **scanning electron microscope**. It is similar to the electron microscope in that it uses electrons instead of light. In the scanning electron microscope, however, electrons do not pass through the object. Instead, they bounce off the surface of the object. Thus, biologists are able to see the shape of cells and other objects.

To be seen under an electron microscope, cells must be prepared in a special way. One disadvantage of the electron microscope is that this preparation kills the cells. Light microscopes have the same limitation. Many cell structures are difficult to see under the light microscope. To make these structures more visible, biologists use **stains** to color them. Different structures absorb different kinds and amounts of stain. Unfortunately, stains also kill the cells.

Some new types of microscopes, such as the **phase contrast microscope**, make it possible to view living material without using stains or killing preparations. The phase contrast microscope can magnify cells up to 1000 times, enabling scientists to see structures and substances as they function.

3–8 Stains make cells in this tissue visible.

Section 3.1 Observation of the Cell

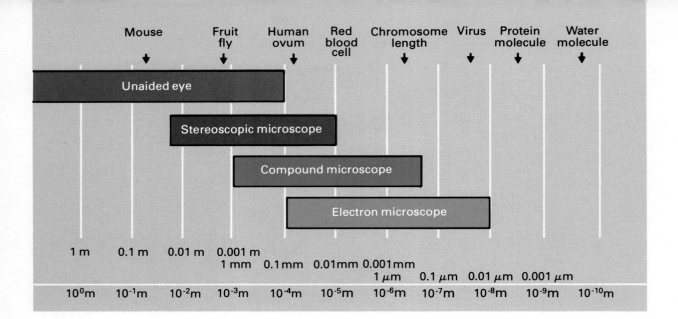

3–9 Relative sizes of biological objects.

1000 mm = 1 m
1000 μm = 1 mm
1 000 000 μm = 1 m

Most cells are too small to see with the unaided eye; in fact, they are generally much less than 1 mm in diameter. To measure objects smaller than 1 mm, a unit called the **micrometre** (μm), is used. There are 1000 μm in 1 mm and 1 000 000 μm in 1 m.

Building a Science Vocabulary

simple microscope	electron
compound microscope	micrograph
objective	scanning electron microscope
eyepiece	stain
light microscope	phase contrast microscope
electron microscope	micrometre (μm)

Questions for Review

1. How did Hooke's microscope differ from Leeuwenhoek's?
2. How did cells get their name?
3. State one advantage and one disadvantage of the electron microscope.
4. The length of one cell is 200 μm. What is the total length in mm of three cells?

3.2 CELL INVESTIGATION

The cork cells Robert Hooke observed were dead and had lost their cytoplasm.

Robert Hooke described cork cells as tiny, empty boxes. But it was not long before scientists observing other cells discovered that they were not empty. An English biologist named Robert Brown found that certain plant cells contain a small, round object, which he called the nucleus. Further observation showed that most plant and animal cells contain a nucleus. Biologists also found a substance between the nucleus and the outside edge of the cell. This substance is called **cytoplasm** (SĪT-uh-plazm).

cyto- = of a cell
-plasm = fluid substance

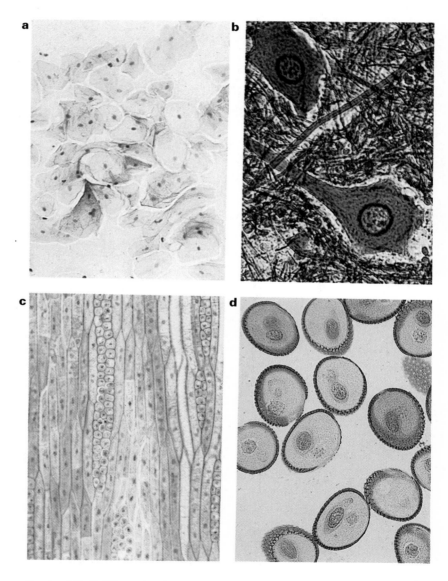

3–10 Cells vary in size, shape, and function: **a.** skin cells; **b.** nerve cells of spinal cord; **c.** plant stem cells; **d.** pollen grains of a lily.

The Cell Theory

As scientists continued to study the microscopic structure of living things, they began to see similarities and differences among them. In 1839 two German scientists, Matthias Schleiden and Theodore Schwann, independently developed the same **hypothesis** (hī-PAHTH-uh-sis). A hypothesis is a possible explanation of observations. Schleiden's and Schwann's hypothesis was that all living things are made of cells. The observations of many scientists supported this hypothesis, and in time, it became known as the **cell theory**. A theory is a general statement that is conditionally accepted as true based on the results of tests and observations. As we know it today, the cell theory has three parts:

A hypothesis may also be a possible answer to a question with no known answer.

1. All organisms are made of cells; some consist of only one cell, others have many cells.

2. The cell carries out the basic functions of living things.

3. All new cells arise from already existing cells.

The Scientific Method

As biologists continued to study cells, they found that cells contain many smaller structures. They called these structures **organelles** (or-guh-NELZ), or little organs. The nucleus was the first of many organelles to be discovered. The compound microscope and the electron microscope made the discovery of most organelles possible. When an organelle is discovered, biologists use microscopes and other equipment to try to determine what it does. In their research, they also use the **scientific method**, which is a logical way of solving problems.

The scientific method starts with an idea or an observation. Assume that, like Robert Brown, you observe a nucleus in all cells. You might wonder, "Can a cell live *without* a nucleus?" Using the scientific method, you would formulate a hypothesis. Your hypothesis might be that a cell cannot live without a nucleus. To test your idea, you would design a **controlled experiment**. The experiment could be to remove the nucleus from a number of cells and watch what happens to them. For comparison, you would leave the nucleus in an equal number of similar cells and watch what happens to them, too. This group is known as the control. The two groups of cells are exactly alike in every way but one: the experimental cells do not have nuclei and the control cells do. This difference between the experiment and the control is called the **experimental variable**. A controlled experiment has only one experimental variable.

Once an experiment is set in motion, your next job is to collect data, keeping careful records of the results. Finally, you study the outcomes and decide if the data support your hypothesis. If all the cells without a nucleus die, while some of the cells with a nucleus continue to live, your hypothesis would be supported. When various experiments or observations support a hypothesis time and time again, the hypothesis becomes a theory. Otherwise, the hypothesis must be rejected or revised. Figure 3–11 summarizes the steps in the scientific method. Scientists do not always follow the scientific method exactly, but they are guided by it.

1. Idea or observation	Cells have nuclei.	
2. Statement of Problem	Can a cell live without a nucleus?	
3. Hypothesis	A cell cannot live without a nucleus.	
4. Controlled Experiment	Experiment: Remove nuclei from group of 20 cells. Grow cells.	Control: Leave nuclei in 20 cells. Grow cells.
5. Collect and Analyze Data	Record number of living cells. 0	18
6. Compare Results with Hypothesis	Hypothesis is supported by data.	
7. Perform Further Experiments for Confirmation		

3–11 Steps in the scientific method.

Specialization and Division of Labor

In a unicellular organism, the cell must perform all life functions. Each cell is an independent unit; if it is destroyed, the organism dies. On the other hand, if some cells of a multicellular organism are destroyed, the organism may still survive. One characteristic of being multicellular is an increased chance of survival.

Another characteristic of multicellular organisms is **specialization**. Specialization is the ability of different parts of an organism to perform specific functions. For example, long fibers of muscle tissue are specialized to move body parts. The structure of each cell in muscle tissue enables the muscle to contract.

Specialization in multicellular organisms results in **division of labor**: different body parts perform different functions. You can compare division of labor in living things to division of labor on an automobile assembly line. Each person on the assembly line does a certain job as the car moves down the line. As a result, production is more efficient than it would be if one person tried to build the whole car alone. In a multicellular organism, division of labor also results in greater efficiency. A specialized cell can perform its limited function better than an individual cell can perform many functions.

A negative aspect of specialization is that cells can no longer carry on all life functions. A muscle cell performs its function of moving, but it depends on blood and cells in blood to deliver food and oxygen to it. If the muscle cell does not get these things, it will stop functioning and eventually die.

Specialization and division of labor lead to greater efficiency in multicellular organisms.

3–12 The parts of this quaking aspen—leaves, branches, trunk, and root—perform specialized functions.

3–13 Each type of blood cell has a different function. Red cells transport oxygen. White cells aid in fighting infections.

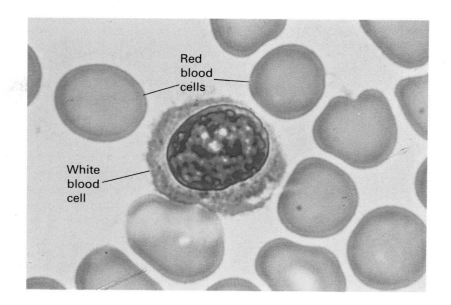

In multicellular organisms, specialized cells perform different functions. Within the cell itself, organelles are also specialized. The functions of organelles are discussed in the next section of this chapter.

Building a Science Vocabulary

cytoplasm	organelle	experimental variable
hypothesis	scientific method	specialization
cell theory	controlled experiment	division of labor

Questions for Review

1. Explain each part of the cell theory in your own words.
2. Why do scientists use the scientific method?
3. State one difference between the terms in each of the following pairs.
   a. hypothesis and theory
   b. experiment and control
   c. cell and organelle
4. State one advantage and one disadvantage of specialization.

3.3 A CLOSER LOOK AT THE CELL

For many years, biologists looking at cells with the compound microscope could not see detail. With the invention of the electron microscope, however, it became possible to see the internal structure of cells more clearly than ever before.

Boundaries of the Cell

The outer boundary of the cell holds it together and controls the passage of substances into and out of the cell. The cell boundary plays an important part in regulating the kinds and amounts of substances in the cell.

The Cell Membrane All cells are enclosed by a **cell membrane**. Through the electron microscope, it can be seen that the cell membrane is made up of two layers that contain large proteins and fatty substances called **lipids**.

The cell membrane has a kind of "memory" that allows it to "recognize" hundreds of substances. If the cell membrane does not recognize a certain substance, it keeps the substance out of the cell. Since the cell membrane allows only certain substances to move into and out of the cell, it is known as a **semipermeable membrane**.

The cell membrane is also known as the plasma membrane.

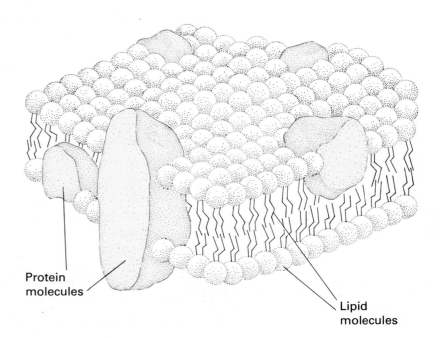

3–14 The cell membrane is composed of lipid molecules and large protein molecules, which function in the passage of materials through the membrane.

Animal cells and some protist cells do not have cell walls.

The Cell Wall In monerans, fungi, and plants, the cell membrane is surrounded by a semi-rigid container, the **cell wall**. Cell walls are not living structures. They provide support and shape to cells and grow as the cell grows. The cell wall is **permeable**; most materials can pass through it freely. Materials that pass through the cell wall do not necessarily pass through the cell membrane inside.

The cell wall of plants contains a fibrous or woody material called **cellulose** (SEL-yoo-lōs). Cellulose fibers are very strong, giving the plant cell wall its rigidity.

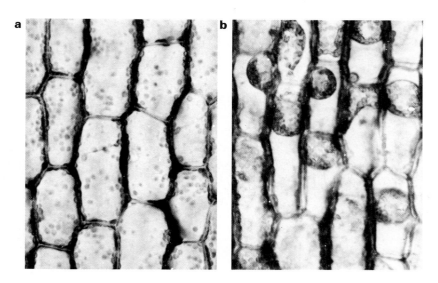

3-15 Cells of *Elodea*, an aquatic plant: **a.** normal cells; **b.** after placing in saltwater. The cell membrane pulls away from the cell wall as water leaves the cytoplasm.

3-16 Cellulose fibers in a plant cell wall. Fibers run parallel to each other in overlapping layers.

Section 3.3 A Closer Look at the Cell

nucleus = a nut, kernel

chrōma = color
sōma = body

The nuclear membrane enables eukaryotic cells to reproduce efficiently.

The Nucleus

The largest and most important structure in a eukaryotic cell is the **nucleus**. The nucleus controls all of the cell's functions. It directs the production of all materials that the cell makes, controls the cell's reproduction, and stores information about the cell's heredity. The important parts of the nucleus are the **nuclear membrane**, the **chromosomes** (KRŌ-muh-sōmz), and the **nucleolus** (noo-KLEE-uh-lus).

The Nuclear Membrane Surrounding the nucleus of all eukaryotic cells is a membrane that controls the passage of materials into and out of the nucleus. Like the cell membrane, the nuclear membrane consists of two layers that are made up of proteins and lipids. There are also holes, called **nuclear pores**, in the nuclear membrane. Materials pass between the nucleus and the cytoplasm through the nuclear pores.

Chromosomes The nucleus contains threadlike bodies called chromosomes that are made of strands of protein and other substances called **nucleic acids**. Chromosomes contain information about heredity, structure, and function. When a cell reproduces, its chromosomes also reproduce. Each species of

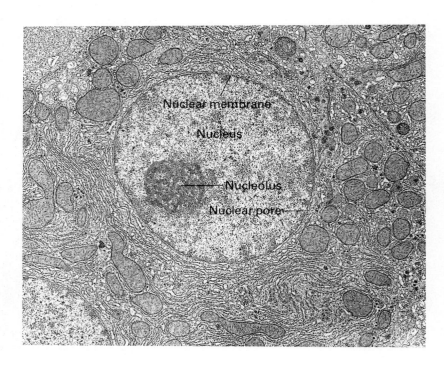

3-17 At the center is the nucleus of a rat liver cell. The dark body to the left is the nucleolus, which functions in the making of ribosomes. Note the double nuclear membrane and the nuclear pore.

Chapter 3 The Cell: Basic Unit of Life

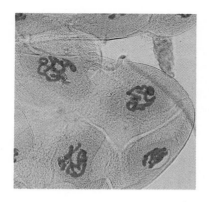

3-18 Coiled chromosomes in the salivary gland cells of a fruit fly.

organism has its own number and kind of chromosomes. All eukaryotic cells have chromosomes contained within the nuclear membrane. Some prokaryotic cells have a single circular chromosome.

The Nucleolus The nucleus contains one or more large organelles called nucleoli. The nucleoli take part in the production of **ribosomes** (RĪ-buh-sōmz), a type of organelle in the cytoplasm. Nucleoli are found only in eukaryotic cells.

Organelles in the Cytoplasm

The cytoplasm within the cell membrane is often called the "living substance of the cell." All cellular organelles except the nucleus and the cell membrane are part of the cytoplasm.

Mitochondria All eukaryotic cells contain **mitochondria** (MĪT-uh-KON-dree-uh). They are fairly large, ranging from 0.5 μm to 10 μm in length. Each mitochondrion is surrounded by a double membrane. The inner membrane is folded, forming shelflike structures inside the mitochondrion.

Mitochondria are often called the "powerhouses" of the cell. Chemical reactions taking place on the shelflike structures within mitochondria produce **ATP**, which is the source of energy for the cell. Cells with high energy requirements, such as muscle cells and heart cells, contain many mitochondria. Cells with low energy requirements have few mitochondria. Mitochondria are not found in prokaryotes.

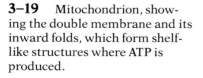

ATP stands for adenosine triphosphate.

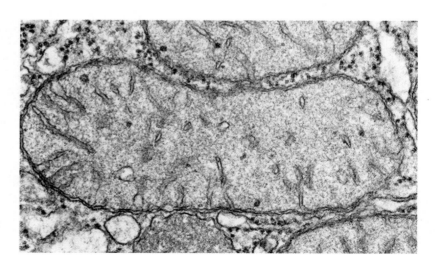

3-19 Mitochondrion, showing the double membrane and its inward folds, which form shelflike structures where ATP is produced.

Chloroplasts Like mitochondria, **chloroplasts** (KLOR-uh-PLASTS) have a double membrane. They are usually larger than mitochondria, ranging from 5 μm to 10 μm in diameter. The inner membrane of the chloroplast is not folded but arranged in layers. Chloroplasts contain chlorophyll. You may recall from Chapter 2 that chlorophyll plays an important role in food production. Chloroplasts are found in the cells of plants and producer protists.

Producer prokaryotes lack chloroplasts, but they do have chlorophyll in their cytoplasm.

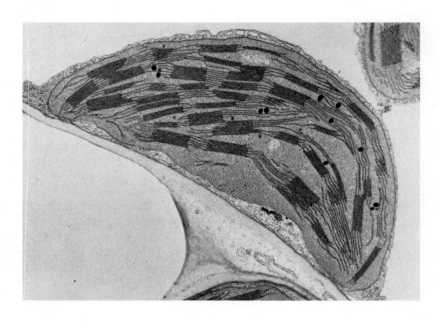

3–20 Chloroplast, showing the double membrane and internal structure of interconnected membranes. Chloroplasts are the site of photosynthesis in eukaryotic producers.

Vacuoles Saclike structures that usually contain water and solutions of salts and food materials are called **vacuoles** (VAK-yoo-ŌLZ). They are surrounded by a single-layer membrane and occur only in eukaryotic cells. Vacuoles are usually larger in plant cells than in animal cells. The vacuoles of animal cells often form around foreign matter such as bacteria. Unicellular protozoa, such as *Amoeba* and *Paramecium*, form vacuoles around the food organisms that they capture. In plants, vacuoles function as storage areas for nutrients and water, and as dump sites for poisonous wastes.

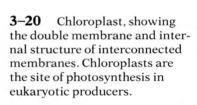

vacuus = empty

Vacuoles in animal cells are often called vesicles.

Lysosomes The "stomachs" of cells are membrane-covered structures about the size of mitochondria called **lysosomes** (LĪ-sō-SŌMZ). They digest food for cells, break down waste products, and destroy worn-out cells and their parts. Lysosomes contain **enzymes** (EN-zīmz) that break down food and other materials into small particles.

lysis = a loosening
sōma = body

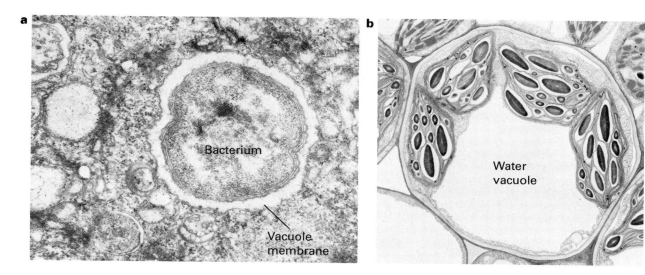

3–21 Vacuoles: **a.** bacterium soon to be digested inside a food vacuole of an ameba; **b.** water vacuole in a corn cell.

In protozoa such as *Paramecium*, lysosomes fuse with food vacuoles. Then the enzymes in the lysosomes digest food that is stored in the vacuoles.

The membrane around the lysosome protects the cell from the powerful enzymes inside. If the lysosome membrane were to break down, the enzymes within the lysosome would leak out and destroy the cell. In a dying cell, the lysosome membrane does break down and the destructive enzymes kill the cell. All eukaryotic cells contain lysosomes.

3–22 Cell digestion: **a.** A food vacuole forms around a food particle. **b.** The vacuole then fuses with a lysosome, which contains digestive enzymes. **c.** The enzymes break down the food into small particles, which pass into the cell. **d.** Undigested material is expelled.

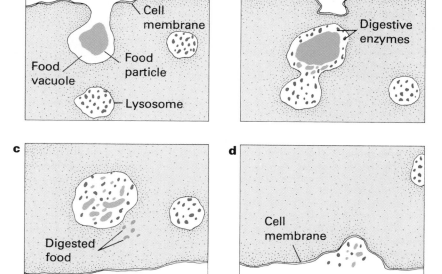

Section 3.3 A Closer Look at the Cell

The Endoplasmic Reticulum The cytoplasm of most cells has an elaborately folded system of membranes running through it. This membrane system is called the **endoplasmic reticulum** (EN-dō-PLAZ-mik ri-TIK-yoo-lum), abbreviated ER. Some parts of the ER are covered with ribosomes which give it a bumpy appearance. For this reason, these parts of the ER are called **rough endoplasmic reticulum**. The parts of the ER without ribosomes are called **smooth endoplasmic reticulum**.

Scientists have determined that the endoplasmic reticulum is involved in transporting substances, such as proteins, throughout the cell. The endoplasmic reticulum is found in all eukaryotic cells.

endo- = within
-plasm = fluid substance
reticulum = a net

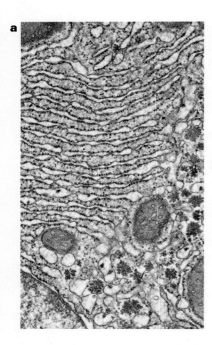

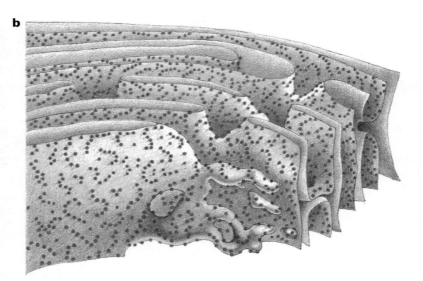

3–23 Endoplasmic reticulum: **a.** cross section of smooth ER; **b.** illustration of rough ER, showing interconnected channels and folded membranes dotted with ribosomes.

Ribosomes found in the cytoplasm of prokaryotic cells differ in structure and are much smaller than those found in eukaryotic cells.

Ribosomes The small, beadlike organelles located on rough endoplasmic reticulum are ribosomes. They assemble proteins, which the cell uses for growth, repair, and control. A cell contains more ribosomes than any other organelle. Ribosomes are found in almost all living cells, prokaryotic and eukaryotic.

The Golgi Apparatus In addition to ribosomes, another organelle is closely associated with the ER: the **Golgi** (GOL-jee) **apparatus**. The Golgi apparatus is a group of flat, saclike structures surrounded by membranes. It surrounds materials in the cell with membrane coverings. These "packages" are then

3–24 Golgi apparatus; **a.** cross section; **b.** interpretation of 3–24a, showing how ends of the flattened membrane sacs pinch off, forming packages.

3–25 Membrane flow in the cell. New membrane may be formed in the nuclear membrane. It then may flow to the rough ER, the smooth ER, the Golgi apparatus, and lysosomes and vacuoles, eventually fusing with the cell membrane.

transported by the ER to the cell membrane, where they are removed from the cell. The Golgi apparatus is found in many eukaryotic cells.

Microfilaments and Microtubules The electron microscope has revealed that cells have internal structures that give strength to the cell and help maintain its shape in much the same way that an animal's skeleton does. These structures are also involved in cell movement. The best known of these structures are **microfilaments**, which are solid and threadlike, and **microtubules**, which are long, thin, hollow tubes. These structures create movement by contracting.

Some unicellular organisms swim by means of hairlike cilia; others swim by beating whiplike flagella. Both cilia and flagella contain microtubules. Microtubules and microfilaments also move certain parts of eukaryotic cells when the cells reproduce.

Section 3.3 A Closer Look at the Cell

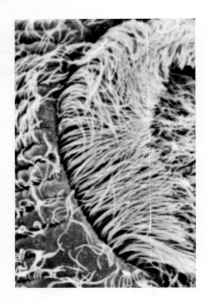

3–27 Cilia on the end of *Stentor*, a protozoan. All cilia have the same internal structure, made up of microtubules.

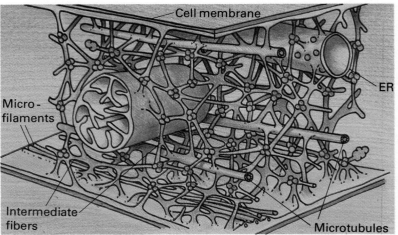

3–26 Illustration of how microtubules, microfilaments and intermediate fibers form an internal skeleton for the cell. Organelles are supported and moved along by this internal structure.

Bacteria and blue-green algae have flagellalike structures, but they are not made up of microtubules. Instead, these prokaryotic cells have special protein threads that enable them to move around.

Centrioles Most cells, except for those of prokaryotes and higher plants, contain small, cylindrical structures made up of microtubules called **centrioles** (SEN-tree-ōlz). Two centrioles are always found near the nucleus of an animal cell, where they function during cell reproduction. Protists with cilia or flagella also have centrioles.

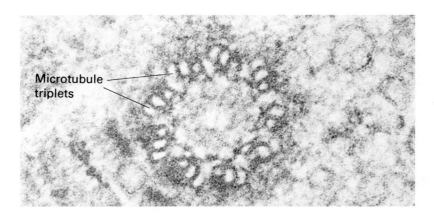

3–28 Centriole in cross section, showing the circular arrangement of nine triplets of microtubules.

Chapter 3 The Cell: Basic Unit of Life

Table 3–1 Structures Found in Cells

Structure	Function
Cell wall	Supports and protects moneran, fungal, and plant cells
Cell membrane	Controls passage of materials into and out of cell
Centriole	Takes part in animal cell reproduction
Chloroplast	Produces food by photosynthesis
Chromosome	Carries information about heredity and control of cell
Cytoplasm	Consists of fluid and organelles in which most cell functions occur; includes parts of cell other than nucleus and cell membrane (and cell wall)
Endoplasmic reticulum	Transports substances within the cell
Golgi apparatus	Transports substances to the surface of the cell
Lysosome	Digests materials in the cell
Microtubule and microfilament	Give strength and help maintain shape of cell; involved in cell movement
Mitochondrion	Produces energy for the cell
Nuclear membrane	Surrounds nucleus and controls passage of materials into and out of nucleus
Nucleolus	Takes part in production of ribosomes
Nucleus	Controls all cell functions
Ribosome	Makes protein for the cell
Vacuole	Stores food, water, and other materials

Differences in Cell Structure

Different cell types can be identified by two characteristics. One characteristic is size. For example, prokaryotic cells are generally much smaller than eukaryotic cells. The typhoid bacterial cell, a prokaryote, is 2.4 μm in size; the ameba, a eukaryote, can be larger than 300 μm. The other identifying characteristic of a cell is the type of organelles it has. If you look at a cell under the microscope and it lacks a nuclear membrane, it is a prokaryote. The prokaryote also lacks all cytoplasmic organelles except ribosomes. Conversely, the presence of a nuclear membrane and cytoplasmic organelles would identify the cell as a eukaryote.

There are also organelle differences that distinguish different types of eukaryotes. Plant cells have a cell wall made of cellulose; animal cells do not. Chloroplasts are found in plant and algal cells, but not in animal cells. A cell without any chloroplasts, however, is not necessarily an animal, protist, or fungal cell. Some plant cells, such as those that make up roots, do not contain chloroplasts. In the case of a root cell, the cellulose cell wall would identify it as a plant cell. Animal cells can be identified by the absence of a cell wall and the presence of centrioles.

Table 3–2 compares structures found in organisms in each of the five kingdoms. In each kingdom, the structures indicated by dots are found in most of the cells that make up the organisms.

Table 3–2 Cellular Structures Found in the Five Kingdoms

Structure	Monerans	Protists	Fungi	Plants	Animals
Cell wall	• (not cellulose)	• (not cellulose)	• (not cellulose)	• (cellulose)	
Cell membrane	•	•	•	•	•
Cytoplasm	•	•	•	•	•
Vacuoles		•	•	•	•
Chloroplasts				•	
Centrioles		• (in cilia and flagella)			•
Nuclear membrane		•	•	•	•
Nucleolus		•	•	•	•
Chromosomes	•	•	•	•	•
Ribosomes	•	•	•	•	•
Mitochondria		•	•	•	•
ER		•	•	•	•
Golgi apparatus		•	•	•	•
Lysosomes		•	•	•	•
Microtubules		•	•	•	•

The Cell: A Functioning Unit

So far, this section has focused on the parts of the cell. It is important to remember, however, that the cell functions as an individual unit in which all of its parts interact. Each part performs its function at the same time as the others. To help you visualize the structure of the cell with its organelles, see Figure 3–29.

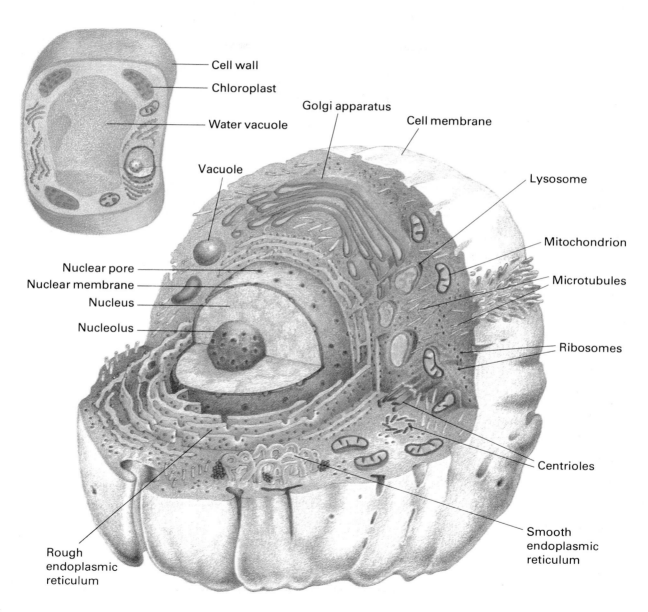

3–29 Cell structure: typical animal cell (below) and plant cell (above).

Building a Science Vocabulary

cell membrane	mitochondrion
lipid	ATP
semipermeable membrane	chloroplast
cell wall	vacuole
permeable	lysosome
cellulose	enzyme
nucleus	endoplasmic reticulum
nuclear membrane	rough endoplasmic reticulum
chromosome	smooth endoplasmic reticulum
nucleolus	Golgi apparatus
nuclear pore	microfilament
nucleic acid	microtubule
ribosome	centriole

Questions for Review

1. What is meant by the term *semipermeable*? What is the function of the semipermeable cell membrane?

2. Name the organelle in a plant cell that produces each of the following.
   a. energy
   b. food
   c. protein

3. Name one feature that would help identify each of the following.
   a. plant cell
   b. animal cell
   c. eukaryote cell

4. How are the organelles in each of the following pairs related?
   a. endoplasmic reticulum and Golgi apparatus
   b. vacuoles and lysosomes
   c. microtubules and flagella

Perspective: Careers
So You Like the Water?

Sea water covers 70 percent of the earth's surface. Although only two percent of the earth's total water supply is tied up in polar ice and glaciers, if it melted the entire earth would be covered with water to a depth of 50 m.

The study of fresh water and sea water is done by physical scientists and biological scientists. Those who work in fresh water are usually called limnologists or aquatic biologists. Those who work in the sea are called oceanographers or marine biologists.

Aquatic and marine biologists are mainly concerned with the study of organisms that live in water. They are interested in their distribution, behavior, ecology, physiology, diseases, pests, and everything else that concerns living things.

Some biologists study small, floating organisms such as protozoa, while others concentrate on large, swimming animals, such as fish. Still others investigate the organisms that inhabit the bottom, such as clams and lobsters.

Some biologists concentrate on the chemistry of marine organisms. They have found that many marine organisms are sources of important food substances, and some can be used for highly sensitive testing of the effects of various drugs.

Aquaculture, or farming from the sea, is a rapidly developing science. Since the time of our early ancestors, some people have obtained much of their food from the sea. Only recently have we begun to scientifically study how to improve the production of food from the sea.

To prepare for a career in marine biology, a person usually needs at least a degree in biology with a good deal of chemistry, physics, mathematics, and computer science. For carrying on advanced research, master's and doctoral degrees are necessary.

Even if you do not want a career in aquatic or marine biology, you may still want to enjoy one of the tools marine scientists use a great deal, SCUBA. This name represents the initials of *S*elf-*C*ontained *U*nderwater *B*reathing *A*pparatus. How about a dive?

Chapter 3
Summary and Review

Summary

1. In the seventeenth century, Antony van Leeuwenhoek observed microorganisms through a simple microscope. In the same century, Robert Hooke developed the compound microscope, which has two sets of lenses. Hooke observed hollow compartments in cork that he called cells.

2. The electron microscope uses a beam of electrons instead of light. It is capable of magnifying biological material up to 100 000 times.

3. Cells are measured in micrometres (μm). A micrometre is equal to one one-millionth (1/1 000 000) of a metre. Cells range in diameter from less than 1 μm to over 300 μm.

4. Schleiden and Schwann helped develop the cell theory. It states that all living things are made of cells, cells carry out all basic life functions, and all new cells come from already existing cells.

5. Scientists use the scientific method to help solve problems. A hypothesis is formed and tested in a controlled experiment. The results either support or reject the hypothesis.

6. Cells in multicellular organisms are specialized to perform different functions. This results in division of labor.

7. All cells are bounded by a semipermeable membrane composed of lipids and proteins. In monerans, fungi, and plants the cell membrane is surrounded by a permeable cell wall that supports and protects the cell.

8. The nucleus controls the cell's functions, heredity, and reproduction. Its major components are the nuclear membrane, the chromosomes, and the nucleolus.

9. The cytoplasm contains structures called organelles that carry out life functions. These include mitochondria, chloroplasts, vacuoles, lysosomes, the endoplasmic reticulum, ribosomes, the Golgi apparatus, microfilaments, microtubules, and centrioles.

10. Prokaryotes can be distinguished from eukaryotes by their small size and lack of organelles. Different types of eukaryotic cells can be identified by their organelles.

Review Questions

Application

1. The diameter of the field seen under a compound microscope is 1.4 mm. What is the diameter in μm? Seven cells of equal size fit across the field. What is the size of each cell in μm? in mm?

2. Scientists hypothesize that the nucleolus takes part in the formation of ribosomes. What must occur before their hypothesis can be considered a theory?

3. As you are observing bacteria under the microscope, you notice that some have a single chromosome. You ask the question, "Is the chromosome necessary for the bacteria to reproduce?" Set up a hypothesis and design a controlled experiment to answer this question. What kind of data would support a correct hypothesis?

4. The following paragraph describes the functions of parts of a cell in a spinach leaf. Name the organelle that performs the function described in each numbered sentence.
   (1) The spinach cell produces its own food. (2) It stores food and water. (3) It breaks down stored food with enzymes. (4) It uses food to produce energy. (5) It builds

up proteins that are used to make cell parts. (6) Raw materials enter and waste products leave through this living structure. (7) The cell's activities are controlled.

5. The electron microscope is used to observe dead cells. Describe the Golgi apparatus as it is seen under the electron microscope. If you could observe live cells at the same magnification, what might you see when you observe the Golgi apparatus?

Interpretation

1. Use the information in each of the following to classify the organism described. Name as many taxonomic categories as you can and explain your reasoning.
   a. one-celled, no nuclear membrane, no chlorophyll
   b. one-celled, nuclear membrane, cilia
   c. multicellular, no chloroplasts, filaments of cells, grows on bread
   d. multicellular, cell walls with cellulose, chloroplast, leaves, stems, roots, fruits containing seeds with two cotyledons
   e. multicellular, chloroplasts, leaves, stems, roots, no seeds
   f. multicellular, centrioles, backbone, gills, scales
   g. multicellular, centrioles, two body layers, digestive tract with one opening, tentacles with stinging cells

2. Assume that you have a group of cells that you know are either plant cells or fungal cells. What organelle would help you identify them as plant cells? Suppose you decided that the cells came either from a plant root or from a fungus. How would you distinguish between the two?

Extra Research

1. Build a three-dimensional model of a cell. Include the organelles.

2. Obtain information in the library about one or more of the following topics.
   a. history of the microscope
   b. types of microscopes not included in the chapter
   c. the microtome
   d. fixing and staining cells
   e. the life of Antony van Leeuwenhoek
   f. the life of Robert Hooke

3. Find pictures of different tissues in your body. Make an "encyclopedia" of tissues. Include:
   a. drawings,
   b. a list of functions,
   c. explanations of how the tissues are adapted for their functions.

4. Find out how microscopes are used in general medicine and in surgery. Write a paragraph describing these uses.

5. Look up the differences and the similarities between cancer cells and normal cells. Write a report about your findings.

Career Opportunities

Medical laboratory workers include technologists, technicians, and assistants who work in clinics and hospitals. They use microscopes and other equipment to run laboratory tests for patients and their doctors. Most medical laboratory workers have two or more years of training. Contact the Canadian Society of Laboratory Technologists, Box 830, Hamilton, Ontario L8N 3N8.

Electron microscopists are usually members of a research team. They need a strong knowledge of chemistry, cell biology, and photography. Electron microscopists prepare cells and tissues for viewing. They also take micrographs that other biologists study and use for reports. Contact the Microscopical Society of Canada, 150 College St., University of Toronto, Toronto, Ontario M5S 1A1.

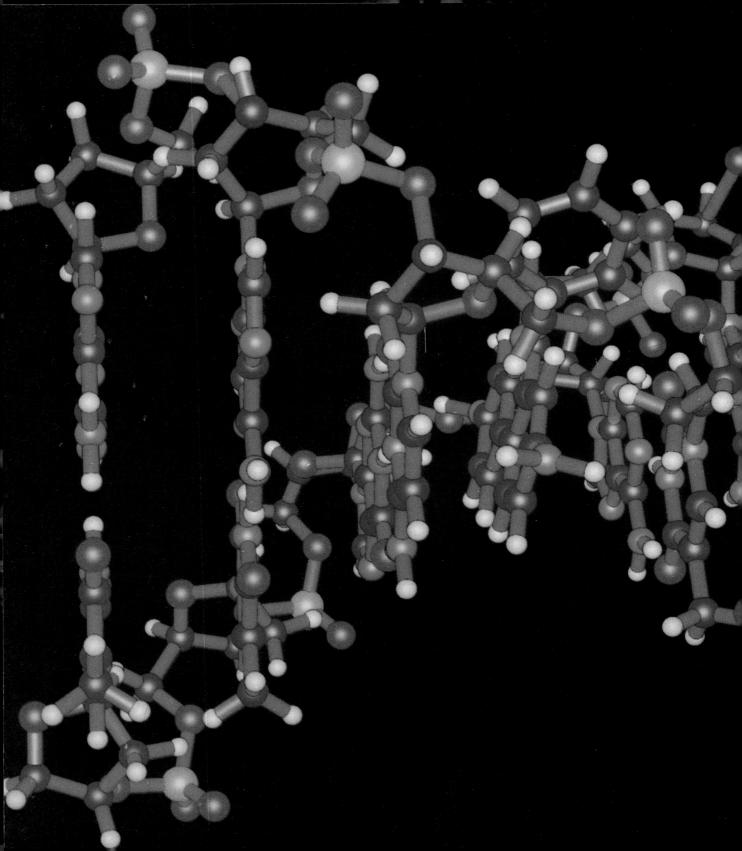

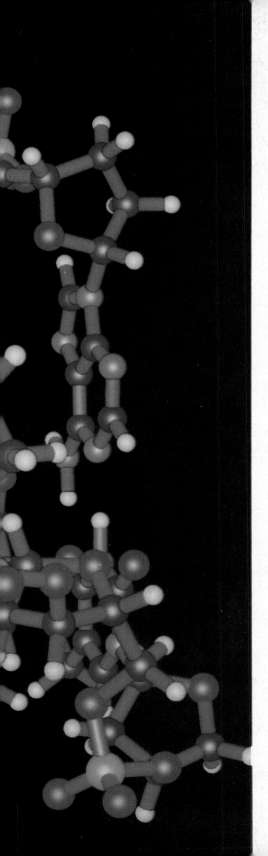

Chapter 4

The Molecular Machinery of the Cell

Focus *How do cells work? This question has puzzled people for more than a century. In trying to find an answer, biologists study the hundreds of substances that make up the cytoplasm, organelles, and membranes of cells.*

4.1 THE CHEMICAL COMPONENTS OF MATTER

To understand the nature of the substances found in cells, it is necessary to be familiar with the substances that make up all matter. These substances combine in different ways to produce the enormous variety of living and nonliving things.

Elements: The Key Substances

Scientists have learned how to separate and break down the chemical substances that make up matter. Their work shows that these chemicals are made from less than 100 simple substances called **elements**. Elements cannot be separated into different kinds of matter. You have probably heard of carbon, hydrogen, oxygen, and nitrogen. These substances occur in all living things.

4–1 Computer-generated model showing part of a DNA molecule that has been stained with ethidium, a cancer-causing substance.

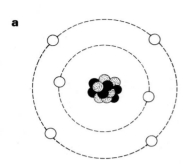

4–2 Carbon occurs in many nonliving and living things: **a.** model of carbon atom; **b.** coal; **c.** diamond; **d.** human; **e.** flower.

All matter is made up of elements and combinations of elements.

An element is the same whether it is found in a tree, in a rock, or in the air. The carbon found in your body is the same as the carbon found in coal, diamonds, and carbon dioxide.

The same elements occur in living and nonliving things, but the amount of a particular element varies. Table 4–1 shows the major elements in the human body and the amounts of these elements in the earth's crust. The most abundant elements in the human body are carbon, oxygen, hydrogen, nitrogen, calcium, and phosphorus. These elements are also found in the earth's crust, but in different amounts. Notice that symbols are used to represent the elements; for example, H stands for hydrogen and C stands for carbon.

Atoms: The Building Blocks of Matter

In addition to the 92 naturally occurring elements there are at least 13 others that are synthetic.

There are 92 different elements in nature. Imagine that you have a small sample of one of these elements—for example a chunk of carbon. Now imagine that you have a way to cut the chunk of carbon into smaller and smaller pieces. After a while, the pieces will be so small that you will not be able to see them even with the most powerful microscope. If you could keep dividing the pieces, you would eventually have the smallest possible par-

Table 4–1 Major Elements in the Human Body: Abundance by Weight in the Body and in the Earth's Crust

Element	Symbol	Approximate Percentage of Human Body	Approximate Percentage of Earth's Crust
Oxygen	O	65.0	47.0
Carbon	C	18.5	0.03
Hydrogen	H	9.5	0.1
Nitrogen	N	3.3	Trace
Calcium	Ca	1.5	3.6
Phosphorus	P	1.0	0.1
Potassium	K	0.4	2.6
Sulfur	S	0.3	0.05
Chlorine	Cl	0.2	0.05
Sodium	Na	0.2	2.9
Magnesium	Mg	0.1	2.1
Iron	Fe	Trace	5.0
Iodine	I	Trace	Trace

ticles of carbon. The smallest particles of an element are called **atoms**. All matter is made of atoms, which are sometimes called the building blocks of matter.

Atoms are made of even smaller particles. In the center of the atom is a dense area called the **nucleus**. The nucleus contains particles that have a positive electrical charge known as **protons** (PRŌ-tonz). In all elements but hydrogen the nucleus also contains **neutrons** (NOO-tronz). Neutrons are about the same size as protons, but they do not have an electrical charge. Surrounding the nucleus are **electrons**. Electrons have a negative charge and are much smaller than protons and neutrons.

Notice that the word nucleus *is used both for the center of the atom and for the organism.*

4–3 Three views of carbon's atomic structure: **a.** charge diagram; **b.** electron cloud; **c.** shorthand Bohr diagram.

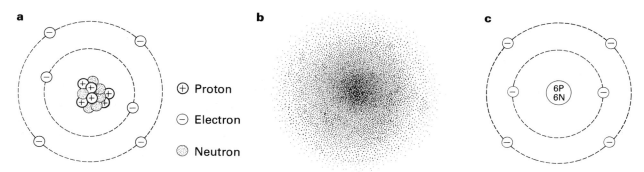

Section 4.1 The Chemical Components of Matter

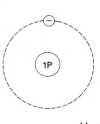 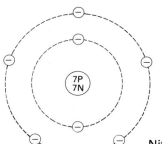 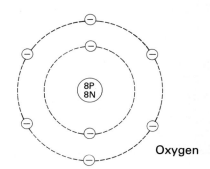

4-4 Atomic structure of three atoms common to all life: hydrogen, nitrogen, and oxygen.

Atoms are mostly space. If you magnified an average atom so that the nucleus were 1 mm in diameter, the electrons would be nearly 100 m away!

Electrons are usually depicted in rings around the nucleus.

The atoms of one element are different from the atoms of another element because they have different numbers of protons. For example, hydrogen, the simplest element, has only one proton. Carbon, on the other hand, has six protons. It is the number of protons in the atom that gives an element its unique characteristics.

The number of electrons in an atom is usually the same as the number of protons, but not always. The number of electrons can vary depending on surrounding conditions. When the numbers of electrons and protons are equal, the numbers of positive and negative charges within the atom are equal, too. These opposite charges cancel each other, making the atom neutral.

The electrons in the atom are not all the same distance from the nucleus. For example, in oxygen, which has eight electrons, some of the electrons are held in close to the nucleus. The others are farther away from the nucleus. The electrons farthest from the nucleus are referred to as the **outer electrons**.

Ions

Under certain conditions, an atom may lose or gain electrons. This results in a different number of electrons than protons. In this case, the atom is not neutral; it has a charge. Atoms that have a charge are called **ions** (Ī-onz). If there are more electrons than protons, an ion is negatively charged. Conversely, if there are fewer electrons than protons, an ion is positively charged. It is important to note that an atom may lose electrons or gain electrons in forming an ion, but the number of protons does not change.

Ions frequently take part in the chemistry of cells. One of the most common ions in living things is the hydrogen ion (H^+).

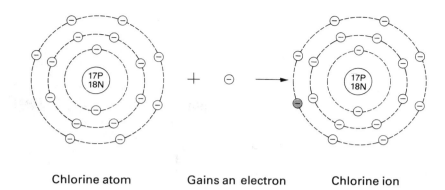

4-5 Formation of ions. A chlorine atom (neutral) becomes a chlorine ion (negatively charged) by gaining an electron.

Chlorine atom   Gains an electron   Chlorine ion

It is a hydrogen atom without its single electron. Therefore the hydrogen ion is simply a proton. Other important ions are the sodium ion (Na^+) and the chlorine ion (Cl^-).

Chemical Compounds

Consider the common white substance known as table salt. Table salt is not an element, because it can be separated into sodium and chlorine, which are elements. Table salt does not resemble these two elements at all: table salt is made up of white grains, whereas sodium is a white metal and chlorine is a greenish-yellow gas. If you heat up sodium metal in a beaker until it is very hot, and then pass chlorine gas into the beaker through a tube, sodium chloride (table salt) will be formed.

Sodium chloride is a **compound**. A compound is a chemical substance that is made up of two or more elements and which has properties different from those of its elements.

Every compound has a specific chemical recipe, or formula. For example, there are always equal numbers of sodium atoms and chlorine atoms in table salt. Chemical formulas can be written using the symbols of the atoms to show the proportions of different atoms in a compound. The formula for table salt is NaCl. Water is a compound made up of hydrogen and oxygen. Hydrogen and oxygen are both gases at room temperature; however, water is a liquid at room temperature. There are always twice as many hydrogen atoms as oxygen atoms in water. Therefore the formula for water is H_2O.

A chemical formula shows the proportion of the elements in a compound.

Organic and Inorganic Compounds   You may recall from Table 4-1 that almost 99 percent of the human body is made up of just six elements. Carbon is one of these important elements.

Section 4.1   The Chemical Components of Matter

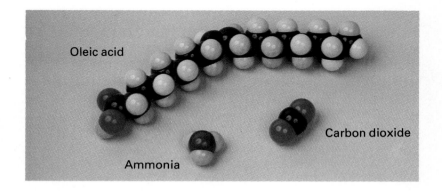

4-6 Space-filling models of oleic acid ($C_{18}H_{34}O_2$), an organic compound found in animal and vegetable fats and oils, and the inorganic compounds carbon dioxide (CO_2) and ammonia (NH_3).

Synthetic fibers and plastics are usually formed from petroleum products, which are organic compounds.

Most of the compounds in living things contain carbon. Almost all compounds that contain carbon and hydrogen belong to a group of chemicals called **organic compounds**. Organic compounds often contain oxygen and other elements in addition to carbon and hydrogen. It was once thought that organic compounds were made only by living things. Then in the nineteenth century, chemists produced organic molecules in the laboratory. Ethylene (EH-thih-leen), (C_2H_4), a gas that aids in ripening fruit, and the sugar glucose (GLOO-kōs), ($C_6H_{12}O_6$), are examples of organic compounds. Most compounds that do not contain carbon and hydrogen, such as carbon dioxide (CO_2) and water (H_2O), are called **inorganic compounds**.

Formation of Compounds

Atoms have a tendency to combine, or react, with other atoms to form compounds. When they combine, the atoms are held together by **chemical bonds**. Chemical bonds are formed by interactions between the outer electrons of different atoms. There are two major types of bonds that form between atoms.

Covalent bonds are formed when different atoms share electrons.

Covalent Bonds In one type of bond, an atom may share some of its outer electrons with one atom, or in many cases, several atoms. When it does, the bond between the atoms is called a **covalent** (kō-VAY-lunt) **bond**. Combinations of atoms held together by covalent bonds are called **molecules** (MOL-eh-KYOOLZ). Some molecules contain atoms of only one element, such as hydrogen gas (H_2) and oxygen gas (O_2). (Since these molecules contain only one element, they are not compounds.) Other molecules contain atoms of two or more different elements, such as carbon dioxide (CO_2), water (H_2O), and calcium carbonate ($CaCO_3$). Compounds made of molecules are known as molecular compounds.

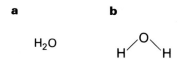

4–7 Representations of the water molecule: **a.** molecular formula; **b.** structural formula; **c.** Bohr diagram; and **d.** space-filling model.

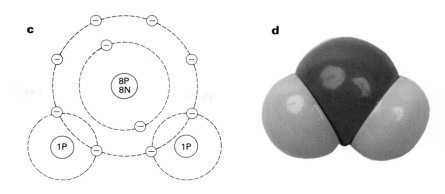

Structural formulas show the arrangement of atoms and bonds in a molecule.

In the water molecule, one oxygen atom forms covalent bonds with two hydrogen atoms. There is a single bond between each hydrogen atom and the oxygen atom. This can be shown in a **structural formula**. Structural formulas show the arrangement of atoms in a molecule using lines between atoms to represent chemical bonds. The structural formula for water is shown in Figure 4–7. Each line represents a pair of electrons being shared between the two atoms.

Sometimes there is more than one bond between the same two atoms. For example, ethylene (C_2H_4) has a double bond between the two carbon atoms. In a structural formula the double bond is shown by two lines, which represent two shared pairs of electrons. Double bonds are much stronger than single bonds; more electrons are shared and the atoms are closer together.

4–8 Covalent bonding in ethylene: **a.** molecular formula; **b.** structural formula; **c.** Bohr diagram, showing electron sharing.

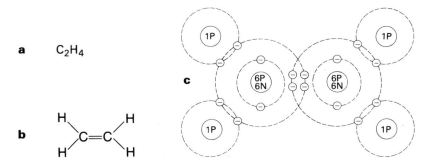

Ionic bonds are formed when atoms transfer electrons to other atoms.

Ionic Bonds In the other type of bond, the **ionic bond**, electrons are not shared. Instead, one atom actually takes electrons from the other atom. For example, chlorine atoms take one electron each from sodium atoms in sodium chloride (Na^+Cl^-). The sodium atoms become positive ions and the chlorine atoms

Section 4.1 The Chemical Components of Matter

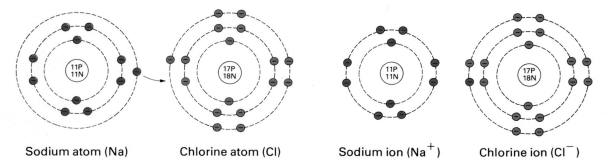

Sodium atom (Na)   Chlorine atom (Cl)   Sodium ion (Na$^+$)   Chlorine ion (Cl$^-$)

4–9 Ionic bonding. Sodium gives up an electron to chlorine in forming the ionic compound sodium chloride. The ions are attracted to each other by their opposite charges.

become negative ions in this compound. They are held together by the force of attraction between oppositely charged particles. Compounds that are held together by ionic bonds are called ionic compounds.

Many ionic bonds break easily in water, leaving the ions separated in the water. In general, covalent bonds do not break in water. Therefore, most ionic compounds are not stable when they are dissolved in water, whereas most molecular compounds are.

Chemical Reactions

Chemical bonds between atoms can be broken, the atoms rearranged, and new bonds formed. In this process, called a **chemical reaction**, new compounds are made that have properties different from those of the original compounds. Chemical reactions can be shown using **chemical equations**. For example, the chemical equation for the formation of water is:

$$2H_2 + O_2 \longrightarrow 2H_2O$$

Stated in words, two molecules of hydrogen react with one molecule of oxygen, yielding two molecules of water. Under certain conditions, it is possible for this reaction to go in the opposite direction; that is, water can be broken up into hydrogen and oxygen molecules. The equation for this reaction is:

$$2H_2O \longrightarrow 2H_2 + O_2$$

4–10 In the formation of water, a chemical reaction occurs between two molecules of hydrogen and one molecule of oxygen.

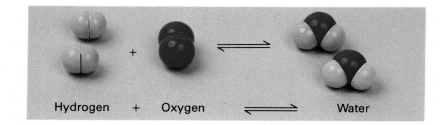

Hydrogen + Oxygen ⇌ Water

The formation and breakdown of water can also be shown with one equation by inserting a double arrow:

$$2H_2 + O_2 \rightleftharpoons 2H_2O$$

Rates of Chemical Reactions The *rate* of a chemical reaction refers to how fast substances come together and react or how fast they break up. Some reactions occur very quickly; others very slowly. One factor that influences the rate of a reaction is temperature. For example, at room temperature, hydrogen and oxygen do not combine very quickly. They combine immediately, however, at a high temperature. In general, increasing the temperature increases the rate of a chemical reaction.

It is also possible to slow down the rate of a reaction by decreasing the temperature. A good example of this involves photographic film. Film contains chemicals that react with each other. When film is exposed in a camera, light speeds up the reactions and produces a picture. Without light, and at room temperature, these reactions still occur, but slowly. Eventually, many of the chemicals in the film will have reacted and the film will no longer be usable. That is why film has an expiration date. By storing film in a refrigerator, as many professional photographers do, it is possible to slow down the rates of the chemical reactions, keeping the film fresh years longer.

A 10°C increase in temperature will double the rate of many chemical reactions.

Building a Science Vocabulary

element	outer electron	covalent bond
atom	ion	molecule
nucleus	compound	structural formula
proton	organic compound	ionic bond
neutron	inorganic compound	chemical reaction
electron	chemical bond	chemical equation

Questions for Review

1. What is the difference between an atom and an ion?
2. How do covalent bonds differ from ionic bonds?
3. Is water an organic or an inorganic compound? Explain.
4. What is a chemical reaction?

4.2 THE CHEMICAL COMPONENTS OF CELLS

The compounds found in organisms can be separated and identified in the laboratory by a variety of techniques. Hundreds of different compounds are now known to exist and more are being discovered every year. Most of these compounds can be classified into a few general groups.

The Separation of Cell Contents

Isolation of Cell Structures In order to separate cell structures, biologists must first break cells apart. This can be done in various ways. Cells can be put into a blender with water or another liquid; water can be made to enter cells until they burst; or cells can be placed into a substance that dissolves the cell membrane. With the membrane broken or gone, the cell contents spill out. They can then be placed into a tube that is put into a machine called a **centrifuge** (SEN-trih-FYOOJ). The centrifuge spins the tube around in a circle.

The basic principle behind the centrifuge is simple. Think of what happens when you hold a bucket of water by the handle and swing it around over your head. The water does not spill out, but instead is forced toward the bottom of the bucket. Similarly, the materials that spin in a centrifuge are forced toward the bottom of the tube.

The centrifuge separates cell structures according to their weight. Heavier structures are forced farther to the bottom of the tube than lighter structures. As a result, the structures sepa-

4–11 A fraction collector is used to separate the components of a solution quickly and automatically. Some types of fraction collectors work on the same principle as chromatography.

centri- = the center
fugere = to flee

4–12 A blood centrifuge (left) is used to separate blood into its component parts. During centrifugation, blood (right) separates into liquid plasma and cells and cell parts.

rate into layers, with the heavier ones toward the bottom and the lighter ones toward the top. When the centrifuge is stopped, it is possible to remove a particular layer for study.

In high speed centrifuges, called **ultracentrifuges**, matter is pushed outward by a force nearly 100 000 times the force of gravity. Such force is necessary to separate some mixtures.

Isolation of Compounds The cytoplasm and its organelles contain many different kinds of complex molecules. One technique for separating a mixture of molecules is **chromatography** (KRŌ-muh-TOG-ruh-fee). In chromatography, a mixture of molecules is dissolved in a solvent (liquid or gas). The solvent mixture moves across a special material. Each kind of molecule in the mixture moves across the material at a different speed. Over time, the molecules spread out across the material and separate. Eventually, groups of molecules appear in different positions on the material.

Picture a horse race. The horses start out together, but separate along the track according to their speeds.

There are different kinds of chromatography depending on the types of solvent and material used. In **paper chromatography**, the mixture to be separated is dissolved in a liquid and placed in a concentrated spot near the bottom of a piece of chromatography paper. The tip of the paper is then placed into a solvent in a jar or test tube. After a while, the solvent begins to creep up the paper. The molecules in the mixture move along the paper with the solvent. Because of their individual properties, each type of molecule moves to a different position on the paper. When the solvent has nearly reached the top of the paper, the different kinds of molecules will be completely separated. The paper with the separated molecules is called a **chromatograph**. The molecules can often be identified according to their color and position on the chromatograph. Sometimes, however, further chemical tests are necessary.

chrōmatos = color
-graphos = record

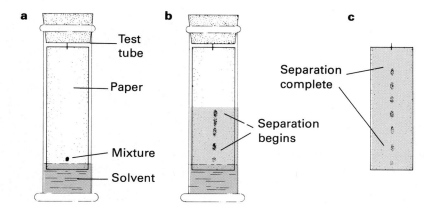

4–13 Paper chromatography: **a.** mixture is placed on paper and paper is put into test tube containing solvent; **b.** separation begins as solvent moves up paper; **c.** separation is complete once solvent reaches top.

Section 4.2 The Chemical Components of Cells

The Major Classes of Biological Compounds

The wide variety of organic compounds found in cells can be classified into four main types: **carbohydrates** (KAR-buh-HĪ-draytz), **lipids**, **proteins** (PRŌ-teenz), and **nucleic acids**.

Carbohydrates We eat many foods that contain carbohydrates, including rice, potatoes, fruits, breads, cakes, and pies. Carbohydrates contain only carbon, hydrogen, and oxygen.

The least complex carbohydrates are **simple sugars**, or **monosaccharides** (MON-uh-SAK-uh-RĪDZ). Simple sugars are the basic structural units of all carbohydrates. The most common simple sugar is **glucose**, also known as grape sugar (see Fig. 4–14). Glucose is the main product of photosynthesis. **Fructose** (FRUK-tōs), or fruit sugar, is another important simple sugar. Notice that fructose has the same molecular formula as glucose, but its structure is different.

mono- = one
saccharum- = sugar

Honey contains a mixture of glucose and fructose.

4–14 Structural formulas of glucose and fructose, two common simple sugars.

Glucose ($C_6H_{12}O_6$)

Fructose ($C_6H_{12}O_6$)

Complex carbohydrates are made by hooking together simple sugars with covalent bonds. Sometimes two sugar units bond, forming a **double sugar**, or **disaccharide**. Malt sugar, **maltose**, is a disaccharide consisting of two glucose units bonded together (see Fig. 4–16). The table sugar that we use in cooking and on cereal is another disaccharide called **sucrose** (SOO-krōs). It is made up of one fructose molecule and one glucose molecule.

di- = two

4–15 Long-chain carbohydrates in grains, especially starch, are the basic ingredient of noodles.

4–16 Maltose, a disaccharide, consists of two glucose molecules bonded together.

poly- = many

Artificial sweeteners, such as saccharine, are synthetic organic molecules that taste sweet. Unlike sugar, artificial sweeteners have little nutritional value.

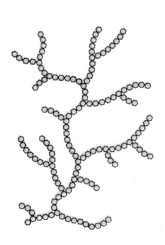

○ = Glucose molecule

4–18 Geometric representation of glycogen. Each hexagonal unit represents a glucose molecule.

Many simple sugars bond together forming long-chain carbohydrates known as **polysaccharides**. **Starch** and **cellulose** are common polysaccharides made up of glucose units. The chemical bonding in cellulose is somewhat different from the bonding in starch. As a result, humans and other mammals can digest starch but cannot digest cellulose. Thus, wood and other plant fibers have little nutritive value to most mammals. Microorganisms living in the intestines of certain animals, such as cows and termites, can break the chemical bonds in the cellulose molecule, enabling these animals to use cellulose for food.

4–17 Structural formulas (simplified) of starch and cellulose. Both are made up of glucose subunits. Starch is highly branched; cellulose is generally a long, straight chain. Notice the difference between the bonding of glucose units in the two polysaccharides.

Plants store carbohydrates in the form of starch. Humans and animals store carbohydrates in the form of **glycogen** (GLĪ-kuh-jun). Glycogen is stored in the liver and in muscle tissue. Like starch and cellulose, glycogen is composed of glucose units bonded together. Glycogen has many more branches than starch.

Section 4.2 The Chemical Components of Cells

4–19 Lipids, such as butter, are a basic part of the diet and are often used to enhance the flavor of other foods.

Lipids When you eat a slice of bread, you consume mostly carbohydrates. When you butter the bread, you add lipids to your diet. Like carbohydrates, lipids contain only carbon, hydrogen, and oxygen. Fats, oils, and waxes are lipids.

Fats, such as butter, are solid at room temperature. Animals use fats to build body parts such as the cell membrane; they also store food in the form of fat. Oils are liquid at room temperature and are produced by both plants and animals. Our skin produces oils, and we obtain corn oil, soybean oil, and other oils from plants. Waxes are also produced by both plants and animals. For example, bees produce wax which they use in building honeycombs, and many plant leaves and stems are coated with wax which prevents them from drying out.

Fats and oils consist of three *fatty acid* molecules bonded to one molecule of *glycerol* (GLIS-ur-ahl). Waxes do not contain glycerol.

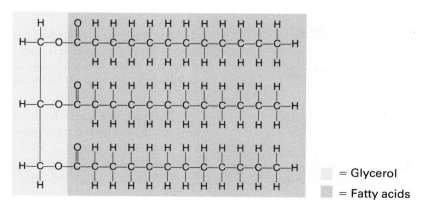

4–20 Structural formula of a fat molecule, showing glycerol bonded to three fatty acid chains.

= Glycerol

= Fatty acids

130   Chapter 4 The Molecular Machinery of the Cell

The three fatty acids in a fat molecule are made of *hydrocarbon chains* (see Fig. 4–20). As the name hydrocarbon implies, the chains contain carbon atoms bonded only to hydrogen atoms. In some fats, only a single bond exists between carbon atoms in the chain:

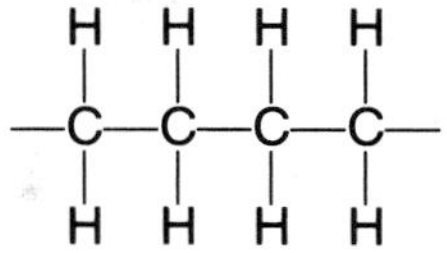

In other fats, double bonds exist between some carbon atoms in the chain:

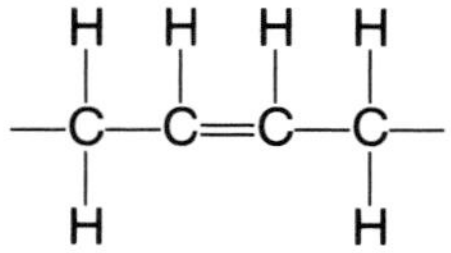

Fats that have fatty acids with only single bonds between carbon atoms are called **saturated fats**. These fatty acids are saturated with hydrogen atoms; all carbon atoms, except the atoms on the ends, have two hydrogen atoms attached. Fats that have fatty acids with double bonds between carbon atoms are called **unsaturated fats**. The carbon atoms are not completely saturated with hydrogen. If a fat has fatty acids with more than one double bond, it is a **polyunsaturated fat**. Animals produce mostly saturated fats, while plants produce mostly unsaturated fats.

There is evidence that eating large amounts of saturated fat can contribute to **arteriosclerosis** (ar-TIR-ee-ō-skluh-RŌ-sis)—a disease in which fatty deposits in the walls of blood vessels interfere with blood flow. Polyunsaturated fats do not appear to have this effect. For this reason, many people believe that vegetable oils are healthier foods than animal fats.

artēria = artery
sklērōsis = a hardening

4-21 Fatty acids: stearic acid is an ingredient of soap; lineolic acid is found in linseed oil.

Stearic acid (saturated) $C_{18}H_{36}O_2$

Linoleic acid (polyunsaturated) $C_{18}H_{32}O_2$

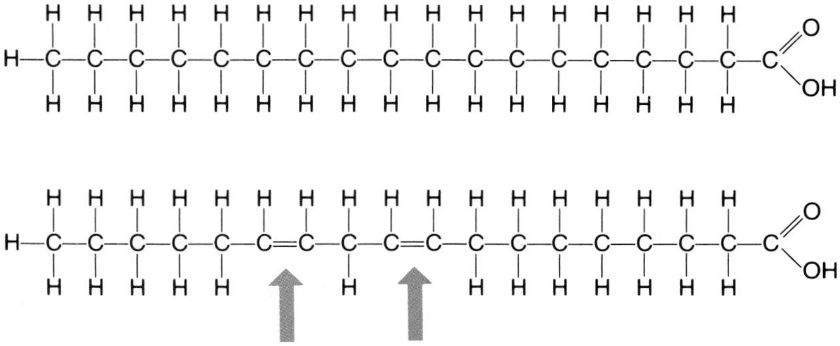

Section 4.2 The Chemical Components of Cells

4–22 Fish, an excellent source of protein, is sometimes eaten raw, as in Japanese cooking.

Proteins When you eat meat, fish, eggs, or dairy products you are adding mostly proteins to your diet. Proteins are large molecules that contain nitrogen, carbon, hydrogen, and oxygen. Many proteins also contain sulfur and other elements. Proteins are important in the growth and maintenance of cells.

In the same way that starch and cellulose consist of glucose units joined together in chains, proteins are made up of units called **amino** (uh-MEE-nō) **acids.** There are 20 different amino acids commonly found in proteins. Each protein has its own number and arrangement of amino acids. Some amino acids appear frequently in the chain, others do not. In general, different proteins contain between 40 and 500 amino acids. Insulin, for example, contains 51 amino acids arranged in two chains (see Fig. 4–23).

Each amino acid has a central carbon atom. Attached to this carbon atom are an *amino group* and an *organic acid group*. A

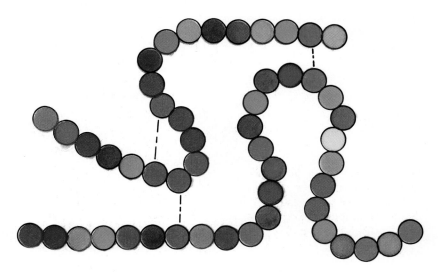

4–23 Diagram of sequence of amino acids in the two linked polypeptide chains of beef insulin. Each circle represents an amino acid; different colors represent different amino acids.

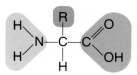
General structure

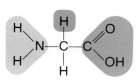
Glycine

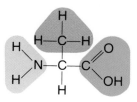
Alanine

4–24 All amino acids have an amino group (blue), an organic acid group (green), and an R group (red) bonded to a carbon atom.

hydrogen atom is also attached to the central carbon. This much of the structure is the same for all amino acids. The difference occurs in the fourth group attached to the central carbon atom. This group is called the *R group* (see Fig. 4–24). Each of the 20 common amino acids has a different R group. For example, the amino acid glycine has a hydrogen atom for its R group, while alanine has CH_3 for its R group.

The different R groups give each amino acid different properties. Some R groups are electrically charged; some cause the protein chain to bend; still other R groups bond with each other in the protein.

In making proteins, a covalent bond forms between the amino group of one amino acid and the organic acid group of another amino acid. Amino acids bonded together are often referred to as **peptides** (PEP-tīdz), and the bonds between amino acids are called **peptide bonds**. **A polypeptide chain** is a series of amino acids bonded together, or simply, it is a protein.

Protein chains are not long and straight; instead, most are coiled and folded into complex globs. The pattern of folding depends on the R groups of the amino acids. Different proteins have distinctive shapes due to their different numbers and arrangements of amino acids.

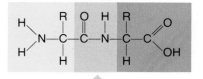
Peptide bond

4–25 Dipeptide: two generalized amino acids connected by a peptide bond.

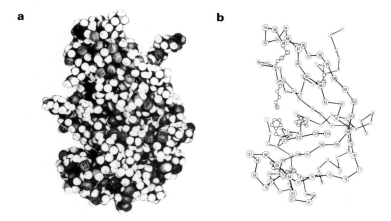

4–26 Twisting and folding of proteins: **a.** space-filling model; **b.** three-dimensional model, showing linkages between amino acids in the chain.

Section 4.2 The Chemical Components of Cells

Nucleic Acids Nucleic acids are the largest and most complicated molecules known in living things. They carry information about the control and reproduction of cells. Nucleic acids are made up of units called **nucleotides** (NOO-klee-uh-TĪDZ), which consist of three parts: a *phosphate group*, a *simple sugar*, and a *nitrogen base*.

All living cells contain an important nucleotide called **adenosine triphosphate** (uh-DEN-uh-seen TRĪ-FAHS-fayt), or **ATP**. It provides energy for life processes. The structural formulas for ATP and its related nucleotides, ADP and AMP, are shown in Figure 4–27.

Nucleic acids are composed of chains of nucleotides bonded together. The bond forms between the phosphate group of one nucleotide and the sugar of the adjacent nucleotide. A portion of a nucleic acid molecule is shown in Figure 4–28. Notice how the sugars and phosphates are bonded together forming a "backbone." The nitrogen bases are attached to the sugars and stick out away from the backbone.

All cells contain two important nucleic acids, **deoxyribonucleic** (dee-AHK-see-RĪ-bō-noo-KLEE-ik) **acid** (DNA) and **ribonucleic acid** (RNA). These nucleic acids play a major role in heredity and in controlling the cell's activities.

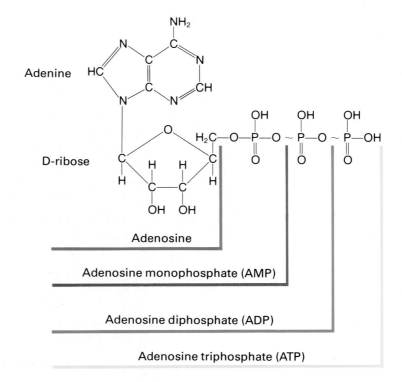

4–27 Adenosine nucleotides. The three parts of each molecule are a nitrogen base (adenine), a simple sugar (d-ribose), and a phosphate group. The addition of phosphates to adenosine results in AMP, ADP, and ATP.

Chapter 4 The Molecular Machinery of the Cell

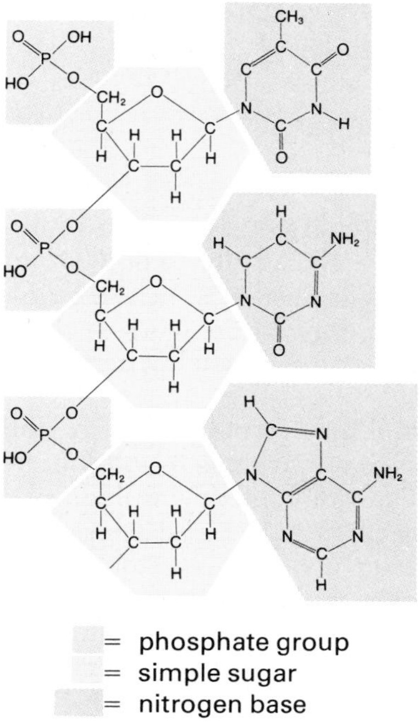

= phosphate group
= simple sugar
= nitrogen base

4-28 Structural formula of part of a nucleic acid molecule.

Building a Science Vocabulary

centrifuge
ultracentrifuge
chromatography
paper chromatography
chromatograph
carbohydrate
lipid
protein
nucleic acid
simple sugar
monosaccharide
glucose
fructose
double sugar
disaccharide
maltose
sucrose

polysaccharide
starch
cellulose
glycogen
saturated fat
unsaturated fat
polyunsaturated fat
arteriosclerosis
amino acid
peptide
peptide bond
polypeptide chain
nucleotide
adenosine triphosphate (ATP)
deoxyribonucleic acid (DNA)
ribonucleic acid (RNA)

Questions for Review

1. Which term in each of the following groups of four terms includes the other three?
   a. sugar, nucleotide, phosphate, nitrogen base
   b. glycogen, starch, glucose, carbohydrate
   c. lipids, fats, waxes, oils
   d. organic acid group, R group, hydrogen, amino acid

2. Assume that you have a solution that contains four different simple sugars. Explain how paper chromatography would separate the sugars.

3. Explain one difference between the terms in each of the following pairs.
   a. saturated and unsaturated fats
   b. starch and glycogen
   c. fats and waxes

Section 4.2 The Chemical Components of Cells

4.3 CHEMICAL REACTIONS IN CELLS: ENERGY AND ENZYMES

Energy

A cell requires a constant supply of **energy** to grow, to reproduce, and even to stay alive. Energy is the ability to do work or to cause change. For example, energy is needed to heat a building or to power a car. You need energy to do work such as cleaning your room, washing dishes, or mowing a lawn.

Forms of Energy There are different forms of energy and they are given different names. When a bicycle is in motion, its wheels have mechanical energy. Water flowing downhill or over a waterfall also has mechanical energy. Electrical energy involves electrons. When you turn on a lamp, electrons move through the lamp cord and into the bulb. Radiant energy is characterized by moving waves of radiation, some of which we see as light. Other forms of energy are heat, nuclear energy, and chemical energy.

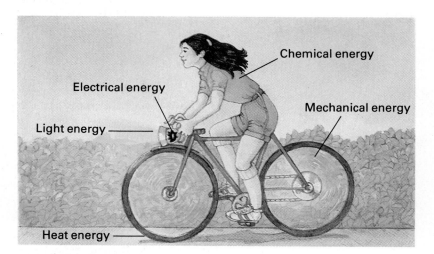

4–29 Even a simple activity may require the use of many different forms of energy.

Kinetic and Potential Energy All the different forms of energy can be divided into two main categories: **kinetic** (kih-NET-ik) **energy** and **potential energy**. Kinetic energy is the energy of motion. Potential energy is inactive or stored energy. For example, if you hold this book in the air, the book has potential energy. If you let go of the book, its potential energy becomes kinetic energy as it falls to the floor.

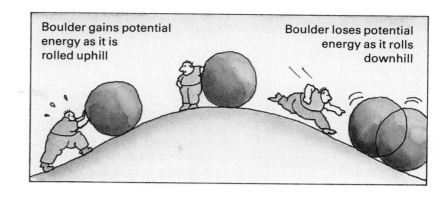

4-30 The energy used to push a boulder to the top of a hill becomes potential energy stored in the boulder as it rests on top of the hill. This potential energy is changed to kinetic energy as the boulder rolls downhill.

Energy Changes Form In the mid-1800s, scientists realized that the total amount of energy in the universe is constant. Energy cannot be created or destroyed; but energy can and does change form. Different forms of energy can be transformed into one another. Mechanical energy becomes electrical energy at hydroelectric power plants as water turns turbines that turn electric generators. Electrical energy is transformed into heat in a toaster. Chemical energy stored in gasoline molecules is changed into mechanical energy and heat when gasoline burns in an automobile engine.

Energy in Chemical Reactions

Chemical reactions involve more than just rearrangements of atoms and molecules; energy plays an important role in almost all chemical processes. When wood (mostly cellulose) and oxygen combine rapidly in a fire, heat and light are given off along with carbon dioxide. The heat represents energy that was stored in the wood. Another example of this is the formation of water. When hydrogen and oxygen combine under the proper conditions to form water, a great deal of energy is given off in a small explosion.

When wood burns, its chemical potential energy is transformed into kinetic energy as light and heat.

4-31 The growth of trees depends on light and heat energy from the sun. Some of the sun's energy is stored as chemical energy in wood. Burning releases this stored energy.

In some chemical reactions, energy is absorbed. A good example is the breakdown of water into hydrogen and oxygen. A great deal of energy must be added to break the bonds in water and re-form hydrogen and oxygen.

Energy to Start Reactions Wood does not burn spontaneously at room temperature. In order to release the great amount of energy stored in wood, it must first be heated. Whether a reaction gives off energy or absorbs it, some energy must be added initially to start the reaction. The energy that is required to start a chemical reaction is called **activation energy**.

Imagine a rock on top of a hill. The rock has potential energy that can be converted into kinetic energy as it rolls down the hill. But to start the rock rolling, someone must push or kick it. The energy it takes to start the rock rolling is similar to activation energy.

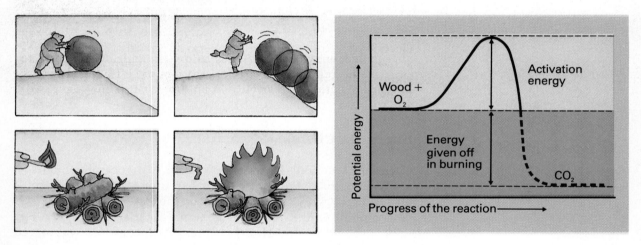

4-32 Activation energy is needed to begin reactions. The graph shows that to obtain the energy in stored wood, some energy must first be added to start the reaction.

Chemical Reactions in Cells

The thousands of chemical reactions that carry out growth, maintenance, and reproduction of cells must occur rapidly to sustain life. It is possible to study many of these reactions in the laboratory. Often this requires a great deal of costly equipment, but sometimes reactions can be observed by putting the compounds in a reaction into a test tube. Scientists have found that most biological reactions occur very slowly in the laboratory—too slowly to carry out life functions.

As you know, some chemical reactions occur faster if the compounds in the reaction are heated. Many biological reactions can be made to occur more quickly by heating, but in general, they will not occur as quickly as they occur in cells. Moreover, cells cannot tolerate too much heat because many biological compounds decompose at high temperatures. Clearly, something other than heat is responsible for fast reaction rates in cells.

Enzymes: Catalysts in Cells

The names of enzymes usually end with the letters -ase. For example, maltase is the name of the enzyme that promotes the formation or breakdown of maltose.

Chemists working in laboratories often use a **catalyst** (KAT-ul-ist) to make a reaction occur quickly. A catalyst is a substance that affects the rate of a chemical reaction. Most catalysts make reactions occur faster. For example, platinum metal acts as a catalyst in the reaction of hydrogen and oxygen to form water. The presence of platinum allows the reaction to occur quickly and at a much lower temperature than without the platinum.

Cells also contain catalysts. The catalysts in cells are proteins called **enzymes**. Enzymes allow biological reactions to occur at the relatively low temperatures of cells. Nearly every chemical reaction that occurs in an organism is made possible by a specific enzyme.

How Enzymes Work

An enzyme increases the rate of a chemical reaction by lowering the activation energy. In other words, an enzyme lowers the amount of energy needed to begin a reaction. This is

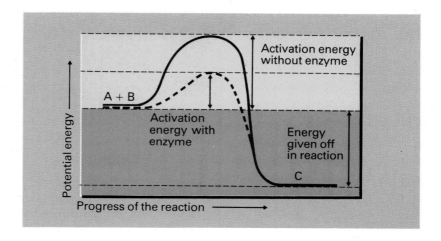

4–33 An enzyme speeds up the rate of a chemical reaction by lowering the activation energy of the reaction (in this case A and B reacting to form C). Almost all chemical reactions that take place in a cell involve enzymes.

accomplished by bringing the molecules in a reaction close together.

In the reaction

$$A + B \xrightleftharpoons{\text{enzyme}} C,$$

the enzyme catalyzes the reaction of A and B to form C, as well as the reverse reaction, the breakdown of C into A and B. Both reactions occur faster when the proper enzyme is present.

Compounds upon which an enzyme acts are known as **substrates**; compounds that are produced during an enzyme reaction are called **products**. In the reaction of A and B forming C, A and B are the substrates and C is the product. In the reaction of C breaking down into A and B, C is the substrate and A and B are the products.

Consider the breakdown of egg white. This can be done without an enzyme by boiling the egg white for 20 hours in strong acid. In the presence of enzyme alone, egg white breaks down in two hours at body temperature without strong acid.

The Lock-and-Key Theory The action of enzymes can be explained by the **lock-and-key theory** (see Fig. 4–34). This theory suggests that an enzyme's function depends on its shape. Recall that proteins are made up of one or more long chains of amino acids that are folded into uniquely shaped globs. An enzyme attaches to its substrate the same way that a key fits a lock. The shape of the enzyme is complementary to the shape of the substrate. Because of this, the substrate molecules fit almost perfectly into the enzyme. This brings them close enough together to react. The place on the enzyme into which the substrate fits is called the **active site**.

All enzymes are proteins, but not all proteins are enzymes. For example, many proteins make up part of the cell membrane and other structures.

An enzyme controls only one or one type of reaction. For example, the enzyme that breaks down starch cannot break down cellulose. This is because the enzymes controlling these reactions have different shapes.

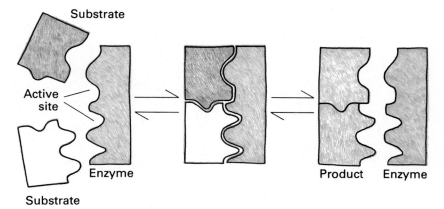

4–34 Lock-and-key theory. Substrates fit into the enzyme and react with each other. The reaction may go in either direction.

Siamese cats have distinctive coloration because the enzyme responsible for their color is very sensitive to temperature. It is more active at lower temperatures than higher temperatures. As a result, the dark fur appears on the cooler areas of the cat.

Factors Affecting Enzyme Activity

Temperature Many experiments have shown that temperature affects enzymes. In one test, researchers placed a solution of substrate and its enzyme into several test tubes. Each test tube contained the same amount of solution. Then they heated groups of these tubes to different temperatures and measured the amount of product formed at each temperature. When the enzyme was not functioning well, little product was formed.

The graph in Figure 4–35 shows the results of this experiment. As the temperature was increased to 33°C, more product was formed. The enzyme worked best at 33°C. The graph shows that the enzyme became less effective above 33°C.

Other experiments show that the enzyme changes shape above 33°C. This information helps explain the data in Figure 4–35. It also supports the lock-and-key theory. According to this theory, the substrates do not fit the enzyme as well after the shape of the enzyme changes. Therefore the rate of the reaction decreases.

Many experiments similar to the one just described show that each enzyme has a temperature range in which it functions best. The optimum, or best, temperature range varies from enzyme to enzyme. Most enzymes in warmblooded animals, however, function best at body temperature.

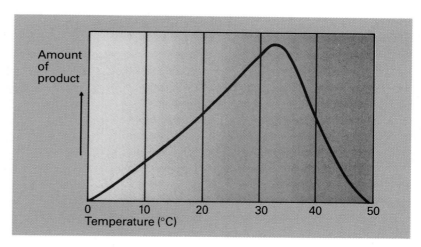

4–35 Effect of temperature on enzyme activity. The enzyme in this study works best (the most product is made) at 33°C.

4–36 Heat affects enzyme shape. Before heating (**a**), protein has many folds and bridges. Heat destroys weak linkages (**b**) and strong ones (**c**). As a result, protein loses its shape.

Section 4.3 Chemical Reactions in Cells: Energy and Enzymes

The pH Value Another factor that affects enzyme action is **pH**. The pH value is a measure of how acidic or basic a solution is. Substances such as vinegar or lemon juice taste sour because they contain acids. Substances such as ammonia or lye, on the other hand, are bases. They are not sour, but tend to feel soapy or slippery to the touch. Strong acids and bases are dangerous because they damage living tissues.

An acid is a substance that gives off, or donates, hydrogen ions (H^+) in solution. The more it gives off, the more acidic it is. You may be familiar with hydrochloric acid (HCl). It is so strong that a small undiluted drop of it will quickly eat a hole in your clothing. HCl is this strong because almost all of it separates into H^+ ions and Cl^- ions in water.

A base combines with, or accepts, H^+ ions. Sodium hydroxide (NaOH) is a strong base. In water, the Na^+ and OH^- separate. The OH^- ions can combine with H^+ ions to form H_2O.

The pH value represents how acidic or basic a solution may be. The pH scale ranges from 0 to 14. A substance with a pH of 7, such as pure water, is neutral. If a substance has a pH lower than 7, there are more H^+ ions than OH^- ions and it is an acid. A substance with a pH of 1 is a strong acid. The acid in a person's stomach varies from pH 1.2 to 3.0; lemons have a pH of about 2.5. If a substance has a pH above 7, there are fewer H^+ ions than OH^- ions and it is a base. A pH of 14 represents the strongest base. Household ammonia has a pH of about 11. Figure 4–37 shows the pH of several other common substances.

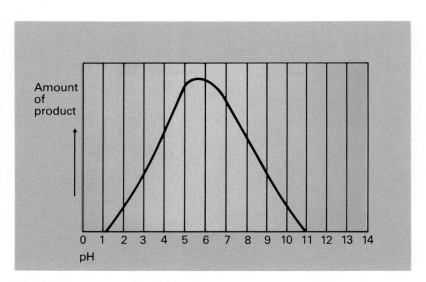

4–37 The pH of an assortment of common substances.

4–38 Effect of pH on enzyme activity. Like all enzymes, the enzyme in this study is most effective within a certain pH range.

Scientists measured the effectiveness of a certain enzyme over the pH range from 1 to 14. They determined enzyme effectiveness by observing how much product was formed at each pH level. The temperature was held constant during these experiments. The results are shown in the graph in Figure 4–38. According to the graph, at which pH does the enzyme work best?

Experiments measuring the effects of pH changes on the effectiveness of many enzymes show that each enzyme has a pH range in which it functions best. Enzymes in the stomach work best in acid conditions. Enzymes in the small intestine, however, work best in basic conditions. The optimum pH range for enzyme function varies from enzyme to enzyme.

Each enzyme has an optimum temperature and pH.

Formation and Breakdown of Chemical Bonds

Several types of chemical reactions occur in cells. Some of these reactions build large molecules from smaller ones. Others break down large molecules into smaller ones.

Dehydration Synthesis In some cellular reactions, covalent bonds form between atoms of different molecules. As this happens, large molecules are formed from small ones. The process of building large molecules from small molecules is called **synthesis**. A common kind of synthesis in cells is **dehydration synthesis**. In dehydration synthesis, a water molecule "splits off" from the reacting molecules at the site where a new bond is formed. Dehydration synthesis forms disaccharides and polysaccharides from simple sugar units, proteins from amino acids, fat from glycerol and three fatty acids, and nucleic acids from nucleotides. There is a specific enzyme for each type of dehydration synthesis.

de- = to reverse
hydrate = combination with water
syn- = together
tithenai = to place

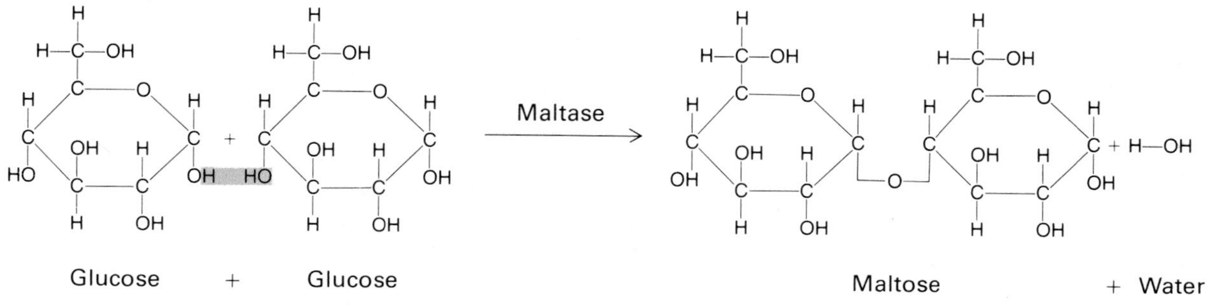

4–39 Dehydration synthesis of maltose. Maltase catalyzes the reaction of two glucose molecules to form maltose and water.

hydro- = (with) water
lysis = a loosening

Hydrolysis The breakdown of a molecule where water is taken up at the broken bond is called **hydrolysis** (hī-DROL-uh-sis). This process is also called digestion. Hydrolysis is the opposite of dehydration synthesis. As you might expect, all four major classes of biological compounds are broken down into their subunits by hydrolysis.

Maltose + Water →(Maltase)→ Glucose + Glucose

4–40 Hydrolysis of maltose. Maltose is broken down into two glucose molecules by the addition of water. Hydrolysis is the opposite of dehydration synthesis.

Building a Science Vocabulary

energy	enzyme	pH
kinetic energy	substrate	synthesis
potential energy	product	dehydration synthesis
activation energy	lock-and-key theory	hydrolysis
catalyst	active site	

Questions for Review

1. What energy changes occur when coal burns?
2. What is activation energy? How does an enzyme affect the activation energy of a reaction?
3. Explain the lock-and-key theory of enzyme action.
4. Why will a boiled enzyme probably not work on its usual substrate?
5. What happens to water during dehydration synthesis? during hydrolysis?

Perspective: Environment
Hazardous Environments

Warning labels appear on many products. Cigarette packages carry a label that reads, "This product may be hazardous to your health." There is convincing evidence that smoking is a cause of lung cancer.

Many products do not carry warning labels. In the 1970s, people in Duluth, Minnesota, were drinking tap water that was not labeled as dangerous. Yet the water was contaminated with cancer-causing asbestos fibers.

Workers in a factory in Virginia were producing a pesticide called Kepone. More than half the employees in the factory suffered from nervous disorders caused by Kepone before anyone realized how harmful it was.

In the mid-1970s, parents clothed their children in garments treated with a flame-retarding substance called TRIS. Later, researchers discovered that TRIS is a cancer-causing agent.

Workers in several chemical plants were producing a pesticide known as DBCP, which farmers used to kill nematodes that attacked the roots of plants. Later, it was found that this chemical affected the ability of the workers to have children.

In 1973, there was an accidental spill in Michigan of PPB, a substance that prevents plastics from catching fire. By 1976, PPB had made its way into the milk of nearly all the mothers in the area. It is now known that PPB causes damage to the liver, kidneys, and skin.

Fluorocarbons were formerly used in aerosol sprays. Fluorocarbons break down the protective layer of ozone in the upper atmosphere, allowing ultraviolet rays from the sun to reach the earth. Ultraviolet radiation is known to cause skin cancer.

There are about 70 000 chemicals in commercial use today. Several hundred new ones are introduced every year. Testing these chemicals to find out if any of them cause cancer or other kinds of diseases is time-consuming and costly. But these tests must be made. It has been estimated that 80 to 90 percent of all cancers are caused by chemicals in our environment. We must learn more about these substances if we want to make our environment less hazardous to our health.

Chapter 4
Summary and Review

Summary

1. All matter is composed of elements. Atoms of different elements bond together forming compounds. Molecular compounds are formed by covalent bonds; ionic compounds are formed by ionic bonds.

2. In chemical reactions, atoms rearrange to form new compounds. The rate of most reactions increases as the temperature increases.

3. Scientists separate the parts of cells with the centrifuge. They use chromatography to separate and identify chemical compounds.

4. Cells are made of inorganic and organic compounds. The four major classes of organic compounds in living things are carbohydrates, lipids, proteins, and nucleic acids.

5. Carbohydrates are sugars. Polysaccharides are chains of simple sugars bonded together. Lipids include fats, oils, and waxes. Fats and oils contain glycerol and fatty acids. Proteins are made up of amino acids. Nucleic acids are chains of nucleotides. A nucleotide consists of a simple sugar, a phosphate, and a nitrogen base.

6. Cells require energy to live and reproduce. Energy exists in different forms; however, all energy can be classified as either energy of motion (kinetic) or stored energy (potential). One form of energy can be transformed into another.

7. Almost all chemical processes involve changes in energy. Some reactions absorb energy; others release energy. Activation energy is the energy required to start a chemical reaction.

8. Enzymes enable chemical reactions in cells to occur quickly at body temperatures. Enzymes are protein catalysts that lower the activation energy of reactions. The lock-and-key theory suggests that a substrate fits into a specific enzyme because its shape complements the shape of the enzyme.

9. Temperature and pH affect enzyme activity. An enzyme is most effective at its optimum temperature range and pH range.

10. Most large molecules in cells are formed by dehydration synthesis and broken down by hydrolysis.

Review Questions

Application

Questions 1–3 refer to the following:

Scientists studied an enzyme called enzymase. They tested the amount of product formed at different temperatures and pH levels. Refer to the graphs in answering the questions.

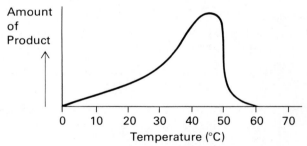

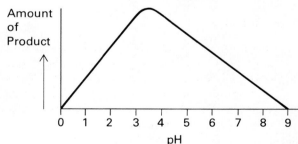

1. The enzyme functions best at (choose one)
   a. 10°C–20°C. c. 30°C–40°C.
   b. 20°C–30°C. d. 40°C–50°C.

2. The enzyme functions best when the environment is (choose one)
   a. acidic.   b. basic.   c. neutral.

3. The enzyme probably does not function at 70°C because (choose one)
   a. the protein in the enzyme changes shape at that temperature.
   b. there is too little substrate at that temperature.
   c. the amount of product increases at a higher temperature.
   d. the pH is not favorable.

4. Egg white is composed mainly of a single kind of protein, albumin. If you boil an egg, the egg white changes and cannot be changed back. What do you think happens to the egg white during boiling?

Questions 5–8 refer to the following chemical equation:

$$\text{amino acid} + \text{amino acid} \xrightleftharpoons{\text{enzyme}} \text{dipeptide} + \text{water}$$

5. During dehydration synthesis, which molecules are considered substrates?

6. During hydrolysis, which molecules are considered products?

7. Why are the arrows shown pointing in opposite directions?

8. According to the lock-and-key theory, what is the relationship between substrate, enzyme, and product in this reaction?

Interpretation

1. Suppose you wanted to classify monerans into groups according to their chemical composition. How could you identify some of the chemical substances in different monerans? What chemical substance distinguishes the two groups of monerans?

2. Bacteria living in the cow's intestine digest the cellulose in the grass cows eat. How is it possible that the bacteria can digest cellulose and the cow cannot?

3. You are asked to identify two cultures of cells based on their chemical composition alone. One culture contains plant cells, the other contains animal cells. Name three substances that will enable you to tell the plant cells from the animal cells.

Extra Research

1. Use toothpicks and gumdrops to make models of glucose and an amino acid. Use different colored gumdrops to represent different atoms. Make a dipeptide and maltose by "bonding" the models together as the molecules would bond in dehydration synthesis. Be sure to account for the water.

2. Using clay or some other construction material, build a model of an enzyme and the substrate that it fits.

3. Separate the colors in blue-black ink on a paper chromatograph. Put a spot of black ink on a strip of paper towel. Dip the edge of the strip of towel in a glass of water. As the water rises up the paper, watch the colors of the ink separate.

Career Opportunities

Food scientists develop new food products, maintain quality control, and devise ways for processing, packaging, and storing foods. A strong background in chemistry is required. Write to the Canadian Institute of Food Science and Technology, 46 Elgin St., Ottawa, Ontario K1P 5K6.

Organic chemists study and work with carbon compounds. They determine the identity and properties of substances and synthesize new ones for specific uses. Some organic chemists work for pharmaceutical companies and other industries. Others perform research in universities. For more information, write the Chemical Institute of Canada, 151 Slater St., Ottawa, Ontario K1P 5H3.

Chapter 5

Food Production and Nutrition

Focus *Food plays a powerful role in the lives of all organisms. Since food requires constant replacement, it is an ever-present need. This chapter concentrates on the production of food by green plants and algae. It also considers our individual food needs and the food needs of people all over the world.*

5.1 FOOD PRODUCTION

It is easy to take food for granted. You walk down the aisles of a supermarket and see the shelves stocked with food. Farm lands produce abundant crops—often surpluses. Nevertheless, reports on television and in newspapers state that people are starving in all parts of the world. A recent international study reported that 1 billion people are hungry and malnourished. Four hundred million of these people live on the edge of starvation. That is more than all the people living in North America. You may be surprised that 12 000 people around the world die of starvation each day.

Reports such as these raise questions about the food we eat. Where does our food come from? What important substances does food supply? How much and what kinds of food do people need to survive? How can we increase the amount of food we produce? What can help the hungry people of the world? These are only a few of the questions that need to be answered about

5–1 Peas, attached to their pod by short stalks, are a renewable food source.

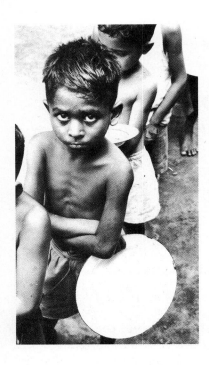

5–2 The present world food supply is not sufficient for the needs of increasing populations.

the world food supply. So far, no simple solutions have been found to the problems of world hunger. Scientists hope, however, that by understanding how food is made, they can find ways to increase food production. They also hope that by understanding the role of food in the body, they can find ways to help people become better nourished.

Photosynthesis: The Food-Making Process

Plants (including multicellular algae) and unicellular algae are the source of all food, which they produce by the process of photosynthesis. In photosynthesis, light energy is used to make glucose ($C_6H_{12}O_6$) from carbon dioxide (CO_2) and water (H_2O). Oxygen gas (O_2) is also released into the atmosphere during photosynthesis.

The Discovery of Photosynthesis

A great deal is known today about the chemical processes in photosynthesis. Before the seventeenth century, however, very little was known about how plants and algae produce food and grow.

One of the earliest investigators of plant growth was Jean-Baptist van Helmont, a seventeenth-century Belgian physician and scientist. He conducted an experiment to determine whether or not the mass of a plant comes from the earth in which it is grown.

Helmont planted a 2.27 kg willow sapling in 90.90 kg of soil. He covered the soil so that nothing could get in or out and added water to the soil from time to time. Five years later, he again determined the mass of the willow tree and the soil. He found

Table 5–1 Helmont's Data

	Original Mass	Mass After Five Years	Change in Mass
Plant	2.27 kg	76.85 kg	+74.58 kg
Soil	90.90 kg	90.85 kg	− 0.05 kg

5–3 Van Helmont's experiment. As the willow grew from sapling to small tree in five years, only water was added and a small amount of soil was lost.

that the willow had increased in mass by 74.58 kg, but the soil had decreased in mass by only 0.05 kg.

Helmont wondered where the additional 74.58 kg of plant matter had come from. It was clear that it could not have come from the soil. He decided that it all came from the water, but he did not perform any experiments to test this idea. Today we realize that his conclusion is not correct, because we know that plants take in carbon dioxide from the air, as well as minerals and water from the soil. Helmont was mistaken in his conclusion, but his work provided important information about plant growth.

Since Helmont's experiment, other scientists have learned more about plant growth and photosynthesis. The accomplishments of several important researchers are shown in Table 5–2.

Table 5–2 Discoveries about Photosynthesis

Year	Scientist	Discovery
1772	Joseph Priestley English clergyman and chemist	Plants change the air by giving off O_2.
1779 & 1796	Jan Ingenhousz Dutch physician and naturalist	The green parts of plants give off O_2 only in the presence of light.
1804	Nicolas de Saussure Swiss chemist and naturalist	Plants take in CO_2 as well as produce O_2 during photosynthesis.
1845	Julius Robert von Mayer German physicist	Sunlight contributes the energy to make food from CO_2 and H_2O.
1960	Melvin Calvin American biochemist	There are many chemical steps in forming $C_6H_{12}O_6$ from CO_2 and H_2O.

Section 5.1 Food Production

An Equation for Photosynthesis

Photosynthesis is the same in all producers, even though plants and algae differ in structure. The bean plant is a good example of a photosynthetic organism. The bean plant obtains the carbon and oxygen for glucose from carbon dioxide in the atmosphere. Carbon dioxide enters the plant through small openings in the surface of the leaf called **stomates** (stō-MAYTS). The hydrogen in glucose comes from water, which the plant absorbs through its roots. The leftover oxygen from the water is released into the atmosphere through the stomates.

The following equation summarizes what happens during photosynthesis:

stoma = mouth

$$\text{Raw Materials} \qquad\qquad \text{Products}$$

$$6CO_2 + 6H_2O \xrightarrow[\text{chlorophyll}]{\text{light}} C_6H_{12}O_6 + 6O_2$$

$$\text{carbon dioxide} + \text{water} \xrightarrow[\text{chlorophyll}]{\text{energy}} \text{glucose} + \text{oxygen}$$

During photosynthesis the bean plant stores light energy as chemical energy in the glucose molecule. In doing so, the bean plant provides itself with food and oxygen; however, it produces more glucose and oxygen than it actually needs. The bean plant stores the surplus glucose as starch or converts it into other food substances. The surplus oxygen goes into the atmosphere. All organisms that do not produce their own food and oxygen depend on this surplus.

It takes the world's plants and algae about two years to produce the amount of oxygen present in the earth's atmosphere.

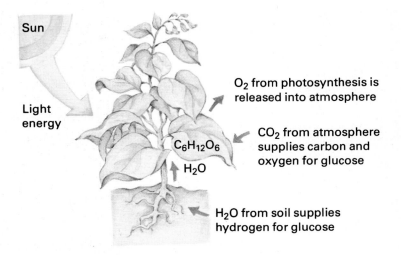

5-4 A photosynthesizing plant draws raw materials from the air and soil and manufactures food and oxygen. This process is powered by the sun.

The Leaf: Site of Photosynthesis

The leaf is the main site of photosynthesis in plants. Leaves are well adapted for this function. For example, many leaves are broad and thin and expose large surfaces of the plant to light.

Light passes easily through the clear protective layer of cells on the top of the leaf, known as the upper **epidermis** (EP-uh-DUR-mis). The upper epidermis is often covered with a waxy layer called the **cuticle** (KYOOT-ih-kul), which helps prevent the leaf from drying out.

Beneath the upper epidermis lies a dense layer of cells, known as the **palisade layer**, which contains chloroplasts. The arrangement of cells in the palisade layer exposes many cells to light. This layer carries out most of the leaf's photosynthesis.

Beneath the palisade layer lies the **spongy layer**. The cells in this layer are more loosely arranged than those in the palisade layer. **Veins** in the spongy layer bring water to the cells in the leaf and also carry food from the leaf to other parts of the plant.

The bottom layer of the leaf, the lower epidermis, is a protective layer containing most of the leaf's stomates. These stomates are less likely to become clogged by dirt particles than the few stomates on top of the leaf. Carbon dioxide enters the leaf through the stomates, and the oxygen produced during photosynthesis passes out through them. Specialized cells, called **guard cells**, open and close the stomates. Guard cells usually open the stomates when there is light and photosynthesis is taking place. In addition, the guard cells of some plants close the stomates during dry conditions, preventing water loss.

epi- = upon
derma = skin

A palisade is a fence made of stakes. The cells in the palisade layer have a similar appearance.

Most of the stomates of water plants are in the upper epidermis.

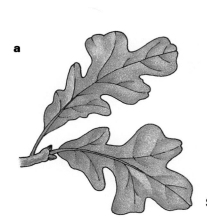
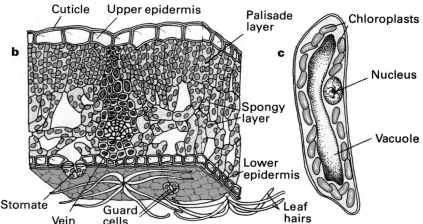

5–5 Leaf structure: **a.** upper leaf surface; **b.** cross section of leaf, showing leaf tissues and underside of leaf; **c.** single cell from palisade layer, which contains many chloroplasts.

Section 5.1 Food Production

The Structure of the Chloroplast

Most photosynthesis occurs inside the chloroplast. The outer membrane of a chloroplast resembles the cell membrane in structure. Both are made up of layers of lipid and protein molecules (see Chapter 3). Within the chloroplast are groups of flattened membrane sacs called **thylakoid** (THĪ-luh-koyd) **discs**, which are stacked in piles called **grana** (GRAN-uh). A fluid or gel-like substance called the **stroma** (STRŌ-muh) surrounds the grana.

thylax = pouch
-oid = like

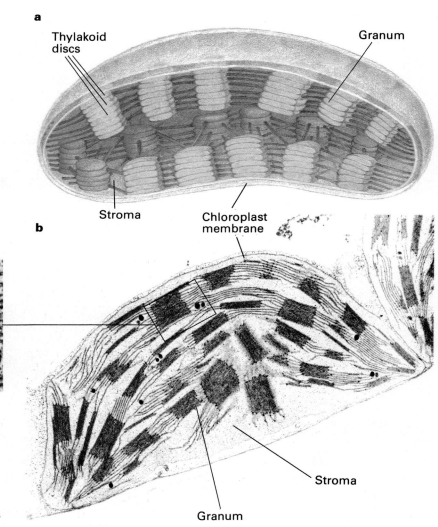

5–6 Structure of a chloroplast: **a.** artist's view; **b.** electron micrograph; **c.** close-up of a granum.

Plant Pigments

Leaves, flowers, and fruits contain many colored substances called **pigments**. The green pigment chlorophyll, which is found in all plants, is necessary for photosynthesis to occur. There are several kinds of chlorophyll, which vary slightly in their chemical composition. Plants also contain other kinds of pigments that are responsible for the red color of tomatoes, the orange color of carrots, and the red, yellow, and orange colors of autumn leaves. The three important pigment types that occur in plants are listed in Table 5–3.

The various colors of algae are due to the presence of different pigments.

Table 5–3 Common Plant Pigments

Pigment Type	Color
Chlorophylls	Green
Carotenoids	Yellow and orange
Anthocyanins	Red and purple

5–7 Carotenoids and anthocyanins are responsible for the pigmentation of black raspberries (**a**) and tomatoes (**b**).

Section 5.1 Food Production

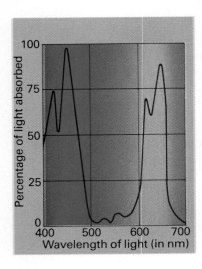

5-8 Combined absorption spectrum of chlorophyll *a* and chlorophyll *b*. Chlorophyll absorbs red and blue light better than green light. Absorption of light of different wavelengths enables more efficient use of available sunlight.

The color of a pigment depends on the color of light that it reflects. White light is composed of all colors of light; thus something that appears white reflects all light. Chlorophyll appears green because it strongly reflects green light. It absorbs all other colors of light, especially red and blue.

Chlorophyll and other pigments line both sides of the thylakoid membrane in chloroplasts. The chlorophyll plays an important role in photosynthesis by helping convert the light energy in red, blue, and other light into chemical energy. How this conversion occurs is discussed in the next section.

Building a Science Vocabulary

stomate	spongy layer	thylakoid disc
epidermis	vein	granum
cuticle	guard cell	stroma
palisade layer	pigment	

Questions for Review

1. What led Helmont to the conclusion that plant growth was due solely to water intake? Was this a reasonable conclusion? Explain.

2. Name the substance or substances described by each of the following.
    a. products of photosynthesis
    b. raw materials of photosynthesis

3. State one function of each of the following structures in photosynthesis.
    a. palisade layer
    b. lower epidermis
    c. cuticle
    d. spongy layer
    e. upper epidermis

5–9 The green hydra has chloroplasts and performs photosynthesis.

5.2 THE MECHANISM OF PHOTOSYNTHESIS

During this century researchers have learned that photosynthesis occurs in two stages: the **light reactions** and the **dark reactions**. Each stage includes a series of many chemical reactions that are catalyzed by many different enzymes. During the light reactions light energy is changed to chemical energy. During the dark reactions the chemical energy produced in the light reactions is used to make glucose. Photosynthesis is essentially the same in all photosynthetic organisms.

The Light and Dark Reactions: An Overview

As the name suggests, the light reactions occur in the light. In fact, they cannot occur without light. The light reactions take place in the grana of chloroplasts where chlorophyll and other photosynthetic compounds are found. During the light reactions:

ATP stands for adenosine triphosphate. NADPH stands for nicotinamide adenine dinucleotide phosphate.

1. Light energy is changed into chemical energy that is temporarily stored in two nucleotides: **ATP** and **NADPH**. These are unstable molecules that release energy when they break down.

Recall that a hydrogen atom is made up of one proton and one electron.

2. Water is split into protons, electrons, and oxygen. The protons and electrons eventually form hydrogen, which is used in the synthesis of glucose during the dark reactions. The oxygen is released into the atmosphere.

The dark reactions are so named because they can occur in darkness as well as in light. The dark reactions take place in the stroma of chloroplasts. During the dark reactions:

1. Glucose is synthesized from carbon dioxide and hydrogen. The energy to make glucose is supplied by ATP and NADPH produced in the light reactions.

5–10 Algae release oxygen bubbles while they photosynthesize. Oxygen is produced during the light reactions.

5-11 Major events of photosynthesis: light and dark reactions. The dark reactions depend on and follow the light reactions.

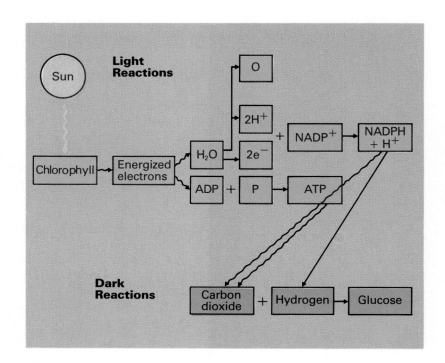

A Closer Look at the Light Reactions

In the light reactions, light energy is converted into chemical energy that is stored in molecules of ATP and NADPH. To understand how this occurs, it is necessary to know more about these molecules.

The Nature of ATP As you learned in Chapter 4, ATP consists of a sugar, a nitrogen base, and three phosphate groups. ATP is a fairly unstable compound, because the phosphate group on the end of the molecule can easily break off in a chemical reaction. For example, when ATP is put into water with a certain enzyme, the bond holding the last phosphate breaks, and a molecule of **ADP** (adenosine diphosphate) is formed, along with phosphate. This reaction releases energy.

The reaction of ATP with water is a hydrolysis reaction.

$$ATP + H_2O \rightleftharpoons ADP + P + energy$$

Cells undergo similar reactions in which ATP breaks down into ADP and phosphate. They use the energy that is given off in these reactions for their activities. Since energy is given off when ATP breaks down, energy must be added in order to make ATP from ADP and phosphate. In photosynthesis, the energy to make ATP comes from light.

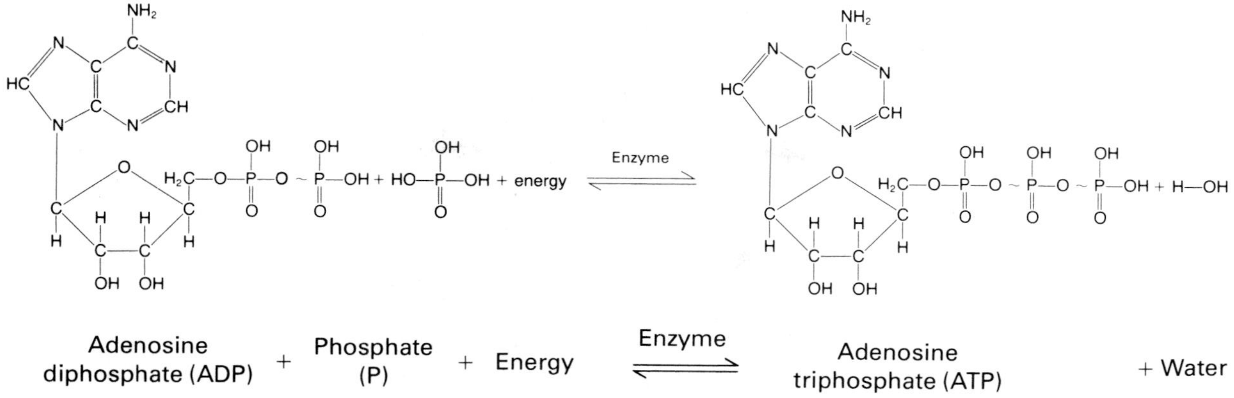

Adenosine diphosphate (ADP) + Phosphate (P) + Energy ⇌ Enzyme ⇌ Adenosine triphosphate (ATP) + Water

5-12 Formation of ATP, showing structural formulas. The adding of a phosphate group to ADP requires energy. The reaction is reversible, as is indicated by the double arrow.

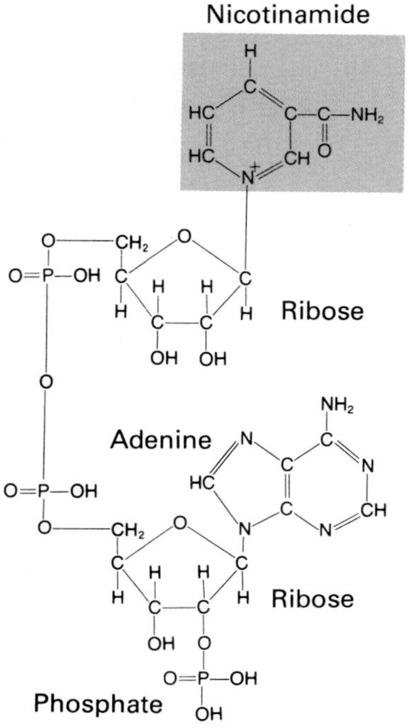

The Nature of NADPH The nucleotide NADPH belongs to a group of biological molecules known as **electron carriers**. Electron carriers transport electrons (e^-), protons (H^+), and energy within a cell.

During photosynthesis, NADPH is formed from $NADP^+$, electrons, and protons. This process requires energy.

$$NADP^+ + 2e^- + H^+ + energy \rightleftharpoons NADPH$$

NADPH changes back into $NADP^+$ by giving off its proton and electrons. When this occurs, energy is also released.

You can think of $NADP^+$ and NADPH as two parts of a system that shuttles electrons, protons, and energy from one place to another.

The energy to form both ATP and NADPH comes from light. Chlorophyll is responsible for trapping light energy and converting it into chemical energy.

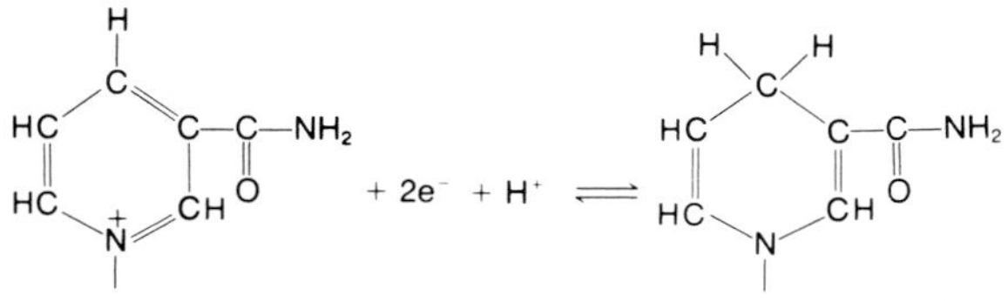

$$NADP^+ + 2e^- + H^+ \rightleftharpoons NADPH$$

5-13 Structure of $NADP^+$ (left), highlighting the functional part of the molecule. Formation of NADPH is shown above.

Section 5.2 The Mechanism of Photosynthesis

Light energizes electrons in chloroplyll.

Light Absorption by Chlorophyll Chlorophyll molecules line the thylakoid membrane in the chloroplast. When light shines on a leaf, some of the light is absorbed by the chlorophyll. The light energy is transferred to certain electrons in the chlorophyll. This causes the electrons to have much more energy than they usually have. These electrons are often referred to as "excited" or "energized" electrons.

Formation of ATP Excited electrons have enough energy to "jump off" certain chlorophyll molecules, leaving an "electron hole." Compounds in the thylakoid membrane of the chloroplast trap the energized electrons. These compounds and the electrons undergo a series of enzyme reactions in which the energy of the electrons is used to make ATP. In this way, light energy is converted into chemical energy.

Energy from some of the energized electrons is used to make ATP.

As ATP is made, the energized electrons lose most of the energy they received from light. The de-energized electrons are then captured by chlorophyll molecules other than the ones they originally jumped off.

Other energized electrons are used to make NADPH.

Formation of NADPH Light also strikes other chlorophyll molecules, causing the release of energized electrons. These electrons, along with protons in the thylakoid membrane, combine with $NADP^+$ forming NADPH. The extra energy in the energized electrons is then carried by the NADPH.

The NADPH goes through the thylakoid membrane and enters the stroma of the chloroplast. There it breaks down, releasing the energy, protons, and electrons. The electrons and protons form hydrogen which is used in the synthesis of glucose during the dark reactions.

Replacement of Electrons Energized electrons are used in the formation of ATP and NADPH during the light reactions. You may be wondering where the electrons come from that replace the missing ones in the original chlorophyll molecules. They come from water molecules in the chloroplast.

Electron holes in chlorophyll are filled with electrons from water.

The first set of chlorophyll molecules, which developed electron holes when the chlorophyll was struck by light, are unstable because they are missing electrons. These chlorophyll molecules fill the holes by pulling electrons from molecules of water. This action causes water molecules to break apart, releasing oxygen and protons. The oxygen is released into the atmosphere. The protons enter the thylakoid membrane of the chloroplast, replacing the ones taken in the formation of NADPH.

$$2H_2O \rightleftharpoons 4H^+ + 4e^- + O_2$$

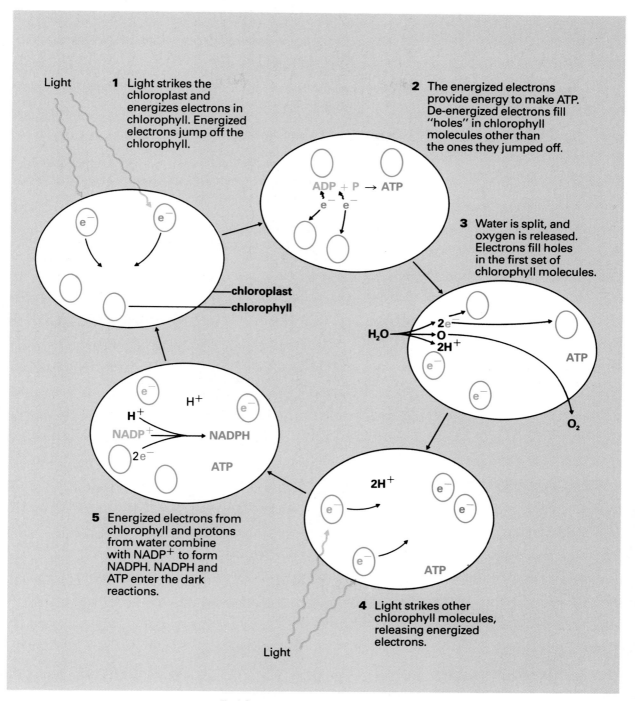

5–14 Events of the light reactions: red indicates high energy level; blue indicates low energy level.

Section 5.2 The Mechanism of Photosynthesis

A Closer Look at the Dark Reactions

The dark reactions involve many compounds and many complex enzyme reactions. The ATP and NADPH produced during the light reactions in the grana of the chloroplast move into the stroma. Carbon dioxide from the atmosphere enters the chloroplast. Hydrogen from the NADPH combines with the carbon dioxide when glucose is formed. The energy for this process comes from ATP and NADPH.

Unlike ATP and NADPH, which are unstable, glucose is a stable compound. It can be easily transported, stored, and broken down at a later time for energy. Glucose is also the subunit of many important polysaccharides, such as cellulose.

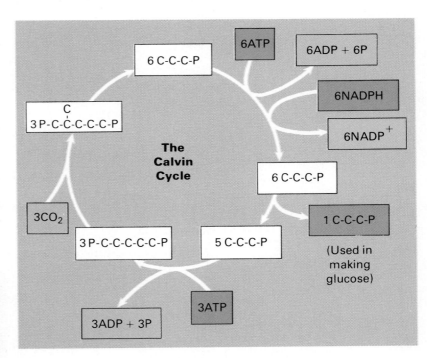

5-15 Dark reactions. The Calvin cycle is the major pathway by which carbon dioxide and hydrogen are combined to form glucose. Compounds in the cycle appear in short form, showing only their carbons and phosphates.

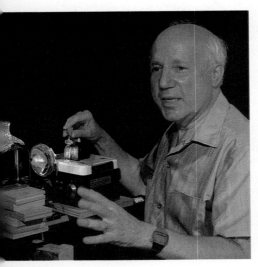

5-16 Melvin Calvin, for whom the Calvin cycle is named, shown working on the development of an artificial "chloroplast." The chloroplast is able to carry out the steps of the light reactions, splitting water using light energy and releasing hydrogen. Hydrogen may be usable someday as fuel.

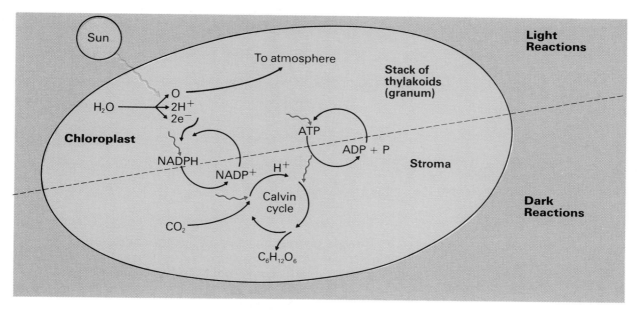

5-17 Summary of photosynthesis. The light reactions occur in the grana; the dark reactions take place in the stroma.

Other Nutrients Are Made from Glucose

Producers use the glucose they manufacture during photosynthesis for many purposes. Glucose is not only a source of energy; as a component of cellulose, it makes up cell walls.

Glucose is one of many substances known as **nutrients**. Nutrients are organic food molecules and minerals that living things need to maintain life. Producers synthesize all the organic nutrients they need from glucose and nitrogen and minerals from the soil.

Some nutrients are used immediately; others are stored. Many plants use stored nutrients during the winter when they do not photosynthesize. Plants that lose their leaves or die back in the winter survive on stored food in stems and roots.

Many plants, such as corn, rice, and potatoes, store glucose as starch. Green plants also convert glucose to amino acids which are then synthesized into proteins. Plants that store proteins include beans, such as the soybean, and many nut-producing trees. Other plants store oils, which can also be made from glucose. Corn, peanuts, soybeans, walnuts, and cashews are plant products that contain large amounts of oils. Plants also synthesize vitamins. Carrots, for example, are a good source of vitamin A.

Soybeans are a good source of protein.

Section 5.2 The Mechanism of Photosynthesis

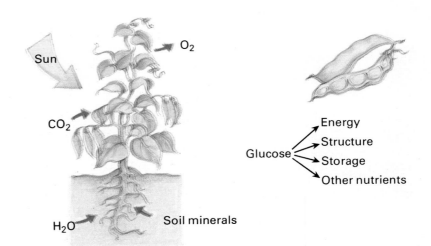

5–18 Plants use the glucose they produce by photosynthesis for many purposes.

Through photosynthesis, producers manufacture the food that all consumers eat. Every bit of food you eat contains chemical energy originally converted from light energy by producers.

Building a Science Vocabulary

light reactions   ADP   nutrient
dark reactions   NADPH/NADP$^+$
ATP   electron carrier

Questions for Review

1. State whether the following events occur in the light reactions or the dark reactions.
    a. Glucose is produced.
    b. Chlorophyll gathers energy.
    c. Oxygen is released into the atmosphere.
    d. ATP is converted into ADP, releasing energy.
2. What does NADPH carry from the light reactions to the dark reactions?
3. How are each of the following substances produced by plants?
    a. oxygen   b. amino acids
4. Explain how a plant uses light energy to produce ATP.

5.3 FOOD AND HUMAN NUTRITION

Organisms obtain energy and materials for growth and repair from food. This section looks at the substances in food that are important to all organisms and takes a closer look at our food needs. Eating the proper foods is vital to health. If you eat the wrong foods, or insufficient food, you can become tired, thin, overweight, or even sick. Understanding what a proper diet includes is a first step toward solving problems of human hunger and health.

Nutrients and Nutrition

The major classes of nutrients that the body needs to maintain life and health are proteins, carbohydrates, lipids, vitamins, and minerals. Different foods contain different combinations of these nutrients.

The process of obtaining and using food is called **nutrition**. Individuals who do not eat properly suffer from a condition known as **malnutrition**. Often a person suffering from malnutrition lacks energy and the raw materials for growth and repair of body parts. In other cases, a malnourished person is harmed by an oversupply of nutrients. Too few nutrients in the diet, digestive illnesses, and overeating are all causes of malnutrition.

All nutrients fall into two categories. Nutrients that the body can synthesize from other substances are called **nonessential** (NON-eh-SEN-shul) **nutrients**. Nutrients that must be obtained from food because the body cannot synthesize them are

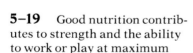

mal- = badly
nutrire = to nourish

5-19 Good nutrition contributes to strength and the ability to work or play at maximum ability.

Table 5-4 Essential and Nonessential Nutrients

Nutrients	Essential	Nonessential
Proteins	8 amino acids: Isoleucine Leucine Lysine Methionine Phenylalanine Threonine Tryptophan Valine	12 amino acids
Lipids	2 fatty acids: Linoleic acid Arachidonic acid	All other fats and oils
Carbohydrates	None	All carbohydrates
Vitamins	5 vitamins or vitamin groups: A B-complex C E K	All other vitamins
Minerals	All required minerals	None

called **essential nutrients**. There are nearly fifty essential nutrients. Eight of the twenty amino acids, two fatty acids, many vitamins, and all minerals are essential nutrients in your diet.

Proteins We obtain protein from food. When food proteins are broken down in the body, the amino acids they contain can be used to build new proteins. Cells use the new proteins for growth, replacement, and repair of cell parts. Sometimes people eat more protein than they need for these functions. In such cases, the excess amino acids may be converted to carbohydrate, which can be stored or used for energy, or they may be stored as fat.

Many animal proteins, including meat, fish, eggs, and dairy products, contain all eight essential amino acids. Gelatin is also a good source of animal protein; however, it lacks the essential amino acid tryptophan (TRIP-tuh-fan). If gelatin were the only source of protein in your diet, malnutrition would result.

Few foods by themselves provide the full complement of essential amino acids; one that does is whole milk.

Chapter 5 Food Production and Nutrition

5–20 Eggs are an excellent source of protein.

Table 5–5 Functions and Sources of Proteins

Major Functions	Some Food Sources	Examples
Growth and repair of tissues	Dairy products	Milk, cheese, cream, yogurt
Major component of enzymes	Meat	Beef, lamb, pork, veal
Source of energy	Poultry	Chicken, turkey, duck
	Fish	Flounder, cod, salmon, tuna
	Legumes	Beans, peas
	Grains	Wheat germ, oats, rice

Plant proteins rarely contain all eight essential amino acids. Most peas, beans, and lentils have all but two essential amino acids. Most grains, cereals, and nuts also lack two amino acids, but not the same ones lacking in peas, beans, and lentils. You can get all the essential amino acids from plants if you eat the right combination of them during a meal. For example, rice and beans eaten together provide all the essential amino acids.

Lipids Animal fats and plant oils are sources of lipids. The body uses the fatty acids in lipids in building membranes and other body structures. Fatty acids help membranes function normally and are an important source of energy. In fact, fats provide more energy per gram than any other nutrient. If you eat more lipids than the body needs, the excess is stored as body fat.

5–21 Peanuts contain a high proportion of lipids.

Table 5–6 Functions and Sources of Lipids

Major Functions	Some Food Sources	Examples
Source of energy	Animal fats	Butter, lard
Component of cell membranes and other structures	Vegetable fats and oils	Margarine, peanut oil, soybean oil, corn oil
Important in function of cell membrane		

Section 5.3 Food and Human Nutrition

Glucose is the only nutrient used for energy by the brain.

Carbohydrates The body breaks down disaccharides and polysaccharides into glucose and other simple sugars. The most important function of simple sugars, especially glucose, is to provide energy. Simple sugars supply energy faster than any other nutrient. If you eat more carbohydrates than you need for energy, the excess is either converted to fat and stored, or it is stored as glycogen in the liver and muscles.

Potatoes, rice, bread, and fruit are all good sources of carbohydrates; however, your cells can synthesize carbohydrates from lipids and from proteins. For this reason, carbohydrates are considered nonessential nutrients.

Table 5-7 Functions and Sources of Carbohydrates

Major Functions	Some Food Sources	Examples
Source of energy	Fruits	Bananas, apples, grapes
	Vegetables	Potatoes, peas, beans, corn, carrots, squash, beets
	Grains	Bread, spaghetti, rice, cereals
	Desserts	Cake, candy, ice cream, cookies

5-22 Grains, such as wheat, are high in carbohydrate content.

vita = life
amine = nitrogen-containing compound

Vitamin D is one of the few vitamins that our bodies can produce. This occurs when sunlight strikes the skin.

Vitamins Vitamins are organic substances that are needed in small amounts. They are part of the structure of many enzymes. When a person does not consume enough of a certain vitamin, a **vitamin deficiency** may result. This, in turn, causes an enzyme deficiency. For example, vitamin D forms part of the enzyme that controls how much calcium is deposited in bone. A severe deficiency of vitamin D in children results in a disease called **rickets**. Young children with rickets do not have enough calcium in their bones, so their bones become soft and deformed. Adults with vitamin D deficiency experience deterioration of bones.

Another severe vitamin deficiency disease is **beriberi**, which results from the lack of vitamin B_1. Beriberi affects the nerves, heart, and digestive system and weakens muscles. It was first identified among people whose diets consisted mainly of

Table 5-8 Vitamins Needed for Human Nutrition

Vitamin	Major Functions	Results of Deficiency	Some Food Sources
A	Needed to maintain vision and to maintain epithelial (covering) tissue	Night blindness, blindness, hardening of eye tissues	Green and yellow vegetables, milk, butter, cheese
B_1 Thiamin	Regulates reactions that remove CO_2	Beriberi (nerve disease with muscle weakness)	Pork, liver, whole grains, legumes
B_2 Riboflavin	Required to get energy from food	Eye sores, cracks at corner of mouth, digestive, skin, and nervous disorders	Dairy foods, meat, fish
B_3 Niacin	Needed for reactions in which energy is obtained from food	Pellagra (disease of skin, digestion, and nerves)	Liver, lean meat, grains, legumes
B_6 Pyridoxine	Needed for body's use of amino acids	Convulsions, irritability, kidney stones	Meats, vegetables whole grains
Folic acid (Folacin)	Needed for use of amino acids and nucleic acids	Anemia, digestive disorders	Legumes, green vegetables, whole wheat
B_{12}	Needed for use of amino acids and nucleic acids	Anemia, nerve disease	Muscle meats, eggs, dairy products
C Ascorbic acid	Needed to build cartilage, bone, and teeth	Scurvy (breakdown of skin, teeth, blood vessels)	Citrus fruits, tomatoes, green peppers
D	Needed for growth and deposit of calcium and other minerals in bones	Rickets (deformed bones in children), breakdown of bones in adults	Cod liver oil, eggs, dairy products, fortified milk
E	Prevents damage to cell membranes	Possibly anemia	Green, leafy vegetables
K	Needed for blood clotting	Severe bleeding if injured	Green, leafy vegetables

Section 5.3 Food and Human Nutrition

5-23 Even though the body produces vitamin D when sunlight strikes the skin, many people do not get enough sun. As a result, they do not produce sufficient amounts of vitamin D. Milk is usually fortified with vitamin D to provide a supplementary source.

polished rice. Polished rice has the husks removed and does not contain the vitamin B_1 found in whole-grain rice. It is ironic that many of us now eat polished rice to which synthetic vitamin B_1 has been added.

Vitamins are often added to foods to ensure that people do not develop a deficiency. For example, vitamin D is added to milk. Many grains and cereals are also fortified with vitamins to make them more nutritious and to replace ones lost during processing. Many vitamins are available as supplements in tablet form. Care should be taken in using them, however, because excessive amounts of some vitamins can have harmful effects.

Minerals Minerals are inorganic substances that are important for the formation and functioning of body parts. Sodium, which is found in table salt, helps maintain the proper blood pressure. Potassium, found in bananas, keeps nerves and muscles working properly. Calcium, found in milk and dairy products, is needed to build strong bones. Iron, which is obtainable from meat and whole-grain vegetables, is a vital part of red blood cells.

Although water is not a mineral, it is included here because it is a vital inorganic substance in the diet. Water acts as the solvent for all body fluids and functions in the transport of materials, temperature regulation, and chemical synthesis and hydrolysis.

Energy in Food

One way to measure the amount of energy in food is to burn the food and measure the amount of heat produced. This is done inside a **calorimeter** (kal-uh-RIM-eh-ter), which consists of a container surrounded by water. When food burns in the calorimeter, energy is released as heat, which causes the water temperature in the calorimeter to rise. Foods containing large amounts of energy cause a greater rise in the water temperature than foods containing little energy. The amount of heat produced, which equals the amount of energy in the food that was burned, can be calculated from the rise in temperature.

calor = heat
metron = a measure

Table 5–9 Minerals Needed for Human Nutrition

Minerals	Major Functions	Results of Deficiency	Some Food Sources
Calcium (Ca)	Needed for healthy bones, teeth, blood, nerves	Stunted growth, rickets, convulsions	Milk, cheese, dark green vegetables
Phosphorus (P)	Needed for forming bones and teeth, regulating acid-base balance	Weakness, loss of minerals (including calcium) from bones	Milk, cheese, meat, poultry, grains
Sulfur (S)	Makes up part of muscle tissue, cartilage, tendons	Failure to grow, weakness	Protein-containing foods
Potassium (K)	Regulates acid-base balance, body balance, nerve function	Muscle weakness, paralysis	Milk, meats, fruits
Chlorine (Cl)	Needed for formation of certain enzymes used in digestion, regulating acid-base balance	Muscle cramps, tiredness	Table salt
Sodium (Na)	Regulates acid-base balance, body water balance, blood pressure, nerve function	Muscle cramps, tiredness	Table salt
Magnesium (Mg)	Activates enzymes in protein synthesis	Failure to grow, disturbed behavior	Whole grains; green, leafy vegetables
Iron (Fe)	Needed to synthesize hemoglobin (in red blood cells) used in obtaining energy from food	Iron deficiency anemia	Eggs, lean meats; legumes; whole grains; green, leafy vegetables
Iodine (I)	Makes up part of thyroid gland products	Goiter (enlarged thyroid gland in neck)	Saltwater fish, vegetables, dairy products, iodized salt
Water (H_2O)	Transports materials in body, regulates temperature, takes part in dehydration synthesis and hydrolysis reactions	Thirst, dehydration	Solid foods, liquids, drinking water

5-24 A calorimeter is used to measure the amount of energy in a substance by burning the substance completely and measuring the rise in water temperature.

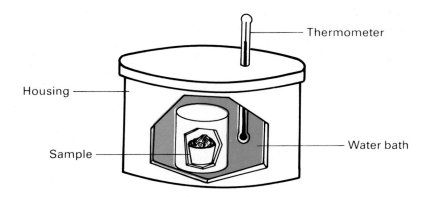

1 Cal = 4.2 kJ

Nutritionists use two units of measurement to describe the amount of energy in food. In the English system, nutritionists use a unit called the **Calorie** (Cal). A Calorie is the amount of heat needed to raise the temperature of 1 kg of water 1°C. In the metric system, a unit called the **kilojoule** (KIL-uh-jool), (kJ), is used. There are 4.2 kJ in 1 Cal.

Some foods contain much energy, others contain little. For example, one tablespoon of peanut butter contains 400 kJ. A boiled egg contains about 320 kJ. A hamburger on a bun has about 1500 kJ.

The energy in food is released quickly as heat when it is burned. The process of burning food in a calorimeter can be described by the following equation:

$$\text{food} + O_2 \longrightarrow H_2O + CO_2 + \underset{\text{(energy)}}{\text{heat} + \text{light}}$$

Of course, we do not burn the food that we eat. Instead, we and all other organisms break down food gradually, without burning it. During this gradual breakdown, energy is released in a form cells can use. The chemical energy in food is transformed into chemical energy in ATP.

$$\text{food} + O_2 \longrightarrow H_2O + CO_2 + \underset{\substack{\text{(chemical} \\ \text{energy)}}}{\text{ATP}}$$

Different Activities Require Different Amounts of Energy

The amount of energy we need each day is related to what we do. Swimming, running, playing the piano, eating, sitting, and sleeping all require different amounts of energy. The amount of

People lose weight when their activities use more energy than they take in. The difference is made up by fat stored in the body.

energy required for different activities varies from person to person, depending on their age, sex, mass, and other factors. Estimates can be made based on a person's mass. For example, for a 50 kg person, playing table tennis for one hour requires about 900 kJ; running for one-half hour requires nearly 750 kJ; eating dinner can use 60 kJ or more.

The amount of energy we have available depends on the food we eat and the food stored in our bodies as glycogen or fat. It is interesting to compare how much "activity" certain foods contain. To use up the 400 kJ in a tablespoon of peanut butter, a 45 kg person would have to walk fast for one-half hour. The same person would have to ride a bicycle at a moderate speed for three hours to use up the energy in a hamburger.

Our bodies require energy even when it does not seem as if we are doing anything. We need this energy to breathe, circulate blood, maintain body temperature, keep cells functioning, and perform other functions. The minimum amount of energy needed to keep these vital internal processes going is known as **basal metabolic energy**. Estimates of basal metabolic energy can be made based on mass, but, as with other activities, the amount of energy varies greatly with sex, age, and body build. In general, a person's basal energy needs are 4 kJ per kilogram per hour. Therefore, a 50 kg person would require a minimum of 200 kJ of energy to stay alive for one hour, assuming the person was not engaged in any activity.

5–25 To work off the energy in the hamburger shown above, this swimmer will have to swim at a moderate speed for about an hour.

Unbalanced Diets

Some people eat more carbohydrate, fat, or protein than their bodies need. Overeating can be a serious form of malnutrition. It may cause heart disease, high blood pressure, diabetes, and other disorders, and it usually results in weight gain.

Often, people take in excessive quantities of vitamins. Taking in too much of some vitamins can be harmful. For example, an overdose of vitamin A may be fatal, and too much vitamin D can cause headaches and kidney damage.

Some people put themselves on unbalanced diets to lose weight or because they enjoy only one or two foods. Unbalanced diets consisting of nothing but pizza and soft drinks or grapefruit and eggs, for example, leave out whole categories of essential nutrients and often lead to malnutrition. Often, the problem is a lack of information about good nutrition. A carefully planned diet can allow people to eat foods they enjoy or lose weight without becoming malnourished.

5–26 Diets that encourage eating only a few foods often lead to malnutrition.

An Adequate Diet

Different people have different food requirements. The amount and kinds of food a person needs depends on a number of factors, including age, sex, health, and exercise. A newborn baby needs twice as much food per kilogram as an adult. When you are sick, you may require less food than when you are well. If you are recovering from an illness, you may need extra protein or more of some other nutrient. A 70 kg man who lies in bed all day doing nothing but eating only requires about 7500 kJ. Active people need more.

There still is much to learn about human nutrition; however, we do know that there is a proper amount of each kind of food for each individual. Regardless of individual differences, it is important to include all five classes of nutrients and all essential nutrients in the diet.

Nutritionists believe that roughage, or fiber, found in whole grains is also an important part of the diet.

Food for the World Population

At the beginning of this chapter several important questions were raised. The first question, "Where does our food come from?", led to the discussion of photosynthesis. The questions about our own nutrition led to the study of the nutrients in food. The last questions about worldwide food needs are more difficult to answer.

The development of high-yield crops is known as the "Green Revolution."

The scientific approach to world food problems is one of the ways that these problems are being met. Food scientists work to increase the food supply by searching for methods to improve the yield and quality of crops and livestock. They also work in the development of new food crops, such as sunflowers. In 1960 sunflowers were not significant as a farm product in the United States, but in 1980, 4 million acres of sunflowers were planted. The variety of sunflower being grown contains a large amount of oil in its seeds. Since sunflower oil contains essential fatty acids, it is valuable in the diet. The oil is used in making margarine, salad oil, and cooking oil, and it is also used in the preparation of feed for animals.

A low-oil variety of sunflower is grown for its seeds. The seeds are used as snack food and bird feed.

In addition to scientific research, education plays a major role in improving public nutrition. Educational programs help people improve the quality of their diets by helping them understand the body's needs and ways that these needs can be met. Such programs also encourage people to conserve food.

5–27 Varieties of corn: **a.** modern hybrid corn is bred for yield, flavor, and disease-resistance; **b.** different varieties of maize are grown, some of which are used to breed hybrids.

Section 5.3 Food and Human Nutrition

It will take more than one or two groups of people to solve the world's food problems; scientists and educators cannot do it alone. Large and small agencies and the governments of many nations must work together to bring adequate food to large numbers of people. It will take planning, work, and cooperation to improve the poor nutritional conditions that exist in many places on earth.

Building a Science Vocabulary

nutrition
malnutrition
nonessential nutrient
essential nutrient
vitamin
vitamin deficiency
rickets

beriberi
mineral
calorimeter
Calorie (Cal)
kilojoule (kJ)
basal metabolic energy

Questions for Review

1. What condition results from a deficiency of each of the following?
   a. vitamin D
   b. iron
   c. vitamin B_1
   d. vitamin C

2. Assume that the four foods listed in Column B are available in equal amounts. Which food would you eat to best satisfy each of the needs in Column A?

   Column A
   a. most energy
   b. quickest energy
   c. protein
   d. vitamin E

   Column B
   glucose
   butter
   spinach
   chicken

3. Why can eating one food and nothing else lead to malnutrition?

4. How can research aimed to improve soil fertilizers help world nutrition?

Perspective: A New View
Is It Real Food?

You have learned that food is made up of proteins, carbohydrates, and fats, as well as vitamins and certain minerals. You know that protein is important in building muscle and other tissues, and that carbohydrates and fats are "fuel foods." The different vitamins and minerals are important in a wide variety of body activities.

On your way home from school you decide to stop at a fast-food restaurant and have a pizza. Since it has a crust, tomato sauce, cheese, and sausage, you assume you are getting some of the proteins, carbohydrates, and fats you need in a normal diet. That is not necessarily so. In some pizzas the cheese is not made from milk but from wheat and corn. The tomato paste is often made mostly from sugar, and the sausage from soybeans. Perhaps only the crust is made from real flour. However, in processing wheat into white flour, up to 80 percent of the vitamin and mineral content is lost.

With the large number of processed foods—the pizza, bacon bits, and fast-food "shakes"—figuring out what your diet really contains can be puzzling.

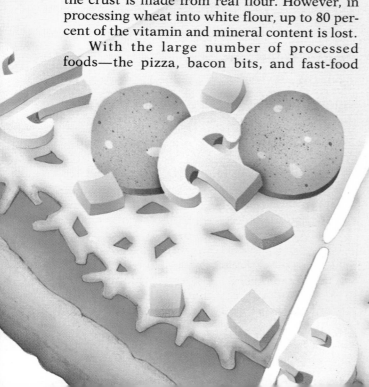

For many years people followed the "rule of four." Every day they ate foods from the four food groups: dairy products, meats, fruits and vegetables, and grains. These foods contain all the nutrients people need for good health. Processed foods, however, often contain foods from only some of the four food groups. In the above example the pizza is made from vegetable substances (soybean and corn sugar) and grains (corn and wheat). Vegetables and grains do contain some of the proteins, fats, carbohydrates, minerals, and vitamins that you need. But perhaps you should not trust the pizza to provide you with all the nutrients you need.

You might want to investigate the foods you buy to see what nutrients they contain. Check the labels on packages in the market. Those fast-food products that do not list their nutritional value may not be such a good buy. After all, you are what you eat.

Chapter 5

Summary and Review

Summary

1. The world's food and oxygen supplies depend on producers (plants and unicellular algae).

2. Photosynthesis is the process in which producers manufacture glucose and oxygen from carbon dioxide and water. This process requires light and chlorophyll, and can be expressed as:

$$6CO_2 + 6H_2O \xrightarrow[\text{chlorophyll}]{\text{light}} C_6H_{12}O_6 + 6O_2$$

3. The leaf is the major site of photosynthesis in plants.

4. Producers contain colored substances called pigments. These include green chlorophylls, yellow and orange carotenoids, and red anthocyanin. Chlorophylls reflect green light. They absorb and use red and blue light in photosynthesis.

5. Chloroplasts contain flattened sacs called thylakoid discs that are grouped into stacks called grana. Stroma surrounds the grana inside the chloroplast.

6. Photosynthesis occurs in two distinct stages: the light reactions and the dark reactions.

7. During the light reactions light energizes electrons in chlorophyll. These electrons are used in the formation of ATP and NADPH. Water is split, releasing oxygen into the atmosphere and supplying protons and electrons for the formation of NADPH.

8. Glucose is synthesized from carbon dioxide and hydrogen during the dark reactions. The ATP and NADPH produced in the light reactions supply the energy for the dark reactions; NADPH also supplies the hydrogen.

9. Producers manufacture organic nutrients from glucose and soil minerals. These nutrients provide food to consumers.

10. Organisms use nutrients for energy and as materials for the growth, replacement, and repair of cell parts. The main classes of nutrients are carbohydrates, proteins, lipids, vitamins, and minerals.

11. Nutrients are either essential or nonessential. Essential nutrients cannot be synthesized by the body; therefore they must be obtained from food.

12. Scientists determine how much energy a food contains by measuring the amount of heat the food releases when it burns. In the metric system, energy is measured in kilojoules (kJ).

13. A healthful diet includes all five classes of nutrients and supplies all essential nutrients. An improper diet results in malnutrition.

Review Questions

Application

1. What food is stored in potatoes? Explain briefly where this stored food comes from.

2. Sulfur bacteria carry out a special type of photosynthesis. In the presence of light they take in carbon dioxide (CO_2) and hydrogen sulfide (H_2S). From these raw materials, sulfur bacteria produce glucose ($C_6H_{12}O_6$).
   a. What substance is used by sulfur bacteria during photosynthesis that is not used by plants?
   b. What substance is used by plants during photosynthesis that is not used by sulfur bacteria?

c. What is the difference in composition between the two substances in questions *a* and *b*?

d. Plants give off oxygen during photosynthesis. Do you think this is also true of sulfur bacteria? Explain your answer.

3. Why is vitamin A an essential nutrient for humans but not for plants?

4. Scientists developed a low cost, highly nutritious food to help people in countries where nutrition was poor. Although the food tasted good to the scientists (it tasted like noodles), it was not familiar to the residents of the area that received it. As a result, the food was not eaten and nutrition was not improved. What could have been done, in addition to the scientific discovery of the food, to prevent failure of the nutrition program?

Interpretation

1. Biochemists have found that temperature changes affect the dark reactions more than the light reactions. Which of the two sets of reactions do you think uses more enzymes? Explain your reasoning.

2. ATP is formed from ADP and a phosphate in the presence of an enzyme by dehydration synthesis. What substance other than ATP is produced as a result of this process?

3. Outline one process that a scientist could use to separate the pigments in a solution of mixed leaf pigments.

Extra Research

1. Make a list of several food additives (they are listed on food packages). Find out what each one is used for and decide whether or not you think the additive is needed.

2. Keep a record of everything you eat for one week. Does your diet include all five classes of nutrients?

3. Keep a record of everything you eat for one day and every activity that you do. Refer to tables in nutrition books that give the energy content of different foods and the energy expenditure of different activities. Calculate your energy input and your energy output in kilojoules or Calories (do not forget your basal metabolic energy). How do your energy input and your energy output compare?

4. What effects might there be on industry if hydrogen could be mass-produced with synthetic chloroplasts?

Career Opportunities

Cooperative extension agents help farmers manage their land and solve their crop and livestock problems. Extension agents also advise home gardeners, work on community development, or sponsor youth activities. Write the Canadian Federation of Agriculture, 111 Sparks St., Ottawa, Ontario K1P 5B5, or the Canadian Society of Extension, 151 Slater St., Ottawa, Ontario K1P 5H4.

Dieticians plan meals to help people maintain or regain good health. They work in hospitals, schools, universities, or large businesses. Their job may also include teaching, managing large kitchens, or supervising food service workers. Write to the Canadian Dietetic Association, 385 Yonge St., Toronto, Ontario M5B 1S1.

Chapter 6

Obtaining Energy from Food

Focus *We depend on natural gas, oil, and coal for energy to power cars, heat homes, and run machinery. Cells, on the other hand, depend on food for energy. Cells do not burn food, as we burn natural gas, oil, and coal. Instead, they break down food gradually. This gradual breakdown releases energy, which cells use to make ATP. The ATP then provides energy for almost all life processes.*

6.1 THE PROCESS OF OBTAINING ENERGY FROM FOOD

On occasion, you may say something like, "I have no energy." Of course, you mean that you feel tired and need to rest. From a strictly biological point of view, however, it is impossible for a living organism to have no energy. Organisms must have a constant supply of energy just to stay alive.

All organisms obtain energy from food by a process called **cellular respiration**. Cells can get energy from several carbohydrates, proteins, and fats. Most often glucose is the food that supplies energy during cellular respiration. The first two sections of this chapter concentrate on how cells obtain energy from glucose.

6-1 On a truck assembly line, the machinery transforms electrical energy into motion and heat, while the worker transforms chemical energy in food into motion and heat.

181

Cellular Respiration: An Overview

In cellular respiration, most cells gradually break down glucose molecules into carbon dioxide. The chemical potential energy in glucose is released during this process. Much of the energy that is released goes off as heat; however, some of the energy is used to make ATP. The overall process of cellular respiration can be summarized by the following equation:

The formation of ATP is shown above the main reaction with a curved arrow. This indicates that ATP is formed at the same time glucose is broken down.

$$36ADP + 36P \longrightarrow 36ATP$$

$$\underset{\text{glucose}}{C_6H_{12}O_6} + \underset{\text{oxygen}}{6O_2} \xrightarrow[\text{many enzymes}]{\text{many steps}} \underset{\text{carbon dioxide}}{6CO_2} + \underset{\text{water}}{6H_2O}$$

ATP: Energy Go-between in Cells   The ATP produced during cellular respiration acts as an energy "go-between." It takes energy from glucose and transfers it to chemical processes in cells. For example, ATP supplies energy to your muscle cells when you move; it supplies energy to transport many substances into or out of cells; it also supplies energy to build body parts. Molecules of ATP provide energy for chemical reactions by breaking down into ADP and phosphate.

$$ATP \rightarrow ADP + P + \text{energy}$$

Cells reverse this process when they make ATP. Using energy, they add a phosphate to ADP. By synthesizing ATP in this way, cells develop a temporary supply of stored energy.

$$ADP + P + \text{energy} \rightarrow ATP$$

6-2 As they storm across the plain, these wildebeests rely on ATP produced by their muscle cells for energy.

Chapter 6 Obtaining Energy from Food

6–3 ATP acts as an energy go-between. Energy stores in food, such as glucose, is transferred to ATP, which provides energy to power cell activities.

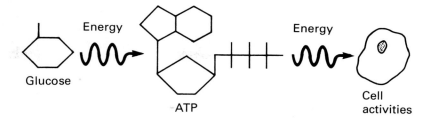

During photosynthesis, the energy needed to make ATP comes from light. During cellular respiration, the energy to make ATP comes from glucose.

Cellular Respiration Is Universal Cellular respiration occurs constantly in the cells of producers, as well as in the cells of consumers and decomposers. It may not be obvious why cellular respiration is important to producers in obtaining energy. Although plants and algae make some ATP during photosynthesis, they use all this ATP in the formation of glucose. To obtain energy for growth, maintenance, and reproduction they must break down some of the glucose they make.

Cellular Respiration Involves Biochemical Pathways

Some of the chemical processes that take place in cells are quite simple, requiring only one reaction and one enzyme. One such reaction is the formation of maltose from two glucose molecules in the presence of maltase (see Chapter 4). Processes such as the dark reactions of photosynthesis, however, require a complex series of many chemical reactions. A series of chemical reactions in a biological process is called a **biochemical pathway**.

In a biochemical pathway, reactions follow one another in a sequence. The product of the first reaction is the substrate of the next reaction. An example of a biochemical pathway follows. The compounds produced in the pathway are referred to by letters. Note that each reaction requires its own enzyme.

$$\underset{\text{initial substrates}}{A + B} \xrightarrow[\text{enzyme 1}]{\text{reaction 1}} C + D \xrightarrow[\text{enzyme 2}]{\text{reaction 2}} E + F \xrightarrow[\text{enzyme 3}]{\text{reaction 3}} \underset{\text{final products}}{G + H}$$

In this biochemical pathway, three reactions transform compounds A and B into compounds G and H. Compounds C, D, E, and F are intermediate compounds.

Section 6.1 The Process of Obtaining Energy from Food

6–4 This biochemical pathway, transforming succinic acid into oxaloacetic acid, is part of one of the pathways in cellular respiration.

glyko- = sugar
lysis = a loosening

an- = without
aēr = air
bios = life

A biochemical pathway can have as few as two or three steps or it can have more than 30 steps. Cellular respiration involves biochemical pathways that have many steps catalyzed by many enzymes.

The Main Stages of Cellular Respiration

Cellular respiration takes place in three main stages. Each of these stages is a complex biochemical pathway. **Glycolysis** (glī-KOL-uh-sis) is the first stage. Glycolysis takes place in the cytoplasm of the cell and does not require the presence of molecular oxygen (O_2). Because it does not use O_2, glycolysis is often referred to as the **anaerobic** (AN-ayr-Ō-bik) stage of cellular respiration. The second and third stages are the **citric acid cycle** and **electron transport**. Unlike glycolysis, these stages are **aerobic**; O_2 must be present for them to occur.

During glycolysis, the six-carbon glucose molecule is broken down into two molecules of a three-carbon substance. Energy is also given off in this process. Some of this energy is lost as heat, some energy is used to make two ATPs, and some energy is stored temporarily in molecules of the nucleotide **NADH**.

In the citric acid cycle, the three-carbon fragments are broken down into one-carbon molecules of carbon dioxide. In addition, one ATP is produced for each three-carbon molecule (two per glucose). More NADH is also made during the citric acid cycle.

During electron transport, the energy stored in the NADH made in glycolysis and the citric acid cycle is used to produce 32 more ATP molecules.

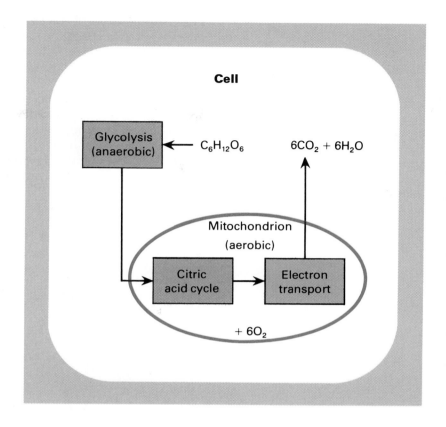

6–5 Three main stages of cellular respiration. Glycolysis occurs in the cytoplasm; the citric acid cycle and electron transport take place in mitochondria. A total of 36 ATPs are produced.

Mitochondria are called the "powerhouses" of the cell because they produce most of the cell's ATP.

crista = a comb, crest

The Mitochondrion: Site of the Aerobic Stages While glycolysis occurs in the cytoplasm, the citric acid cycle and electron transport take place in the mitochondria. A mitochondrion is a sausage-shaped organelle with a double membrane (see Chapter 3). The inner membrane of the mitochondrion is folded into shelflike structures called **cristae** (KRIS-tee), (see Fig. 6–6). Both the inner and outer membranes contain enzymes used in the production of ATP and H_2O.

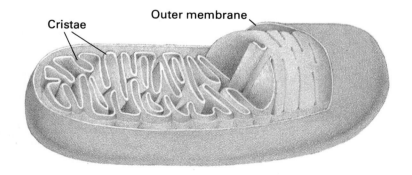

6–6 Structure of the mitochondrion, showing the outer membrane and the inner membrane, which is folded into cristae.

The Importance of Oxygen It is interesting to note how important oxygen is in obtaining energy from food. Even though some ATP is produced during glycolysis, which is anaerobic, most of the cell's ATP is made during the aerobic stages.

Most cells die if they are deprived of O_2. Some cells, however, such as yeast and vertebrate muscle cells, are capable of surviving anaerobic conditions. These cells cannot carry out the citric acid cycle and electron transport because of the lack of O_2. They can carry out glycolysis, though, since it does not require O_2. By doing this, they can generate enough ATP to function. This form of incomplete cellular respiration is called **fermentation**.

For many organisms, such as anaerobic bacteria, fermentation is the only method of obtaining energy from food. Yeast and vertebrate muscle cells, on the other hand, only ferment if there is no O_2 present. When there is plenty of O_2, they carry out all the stages of cellular respiration.

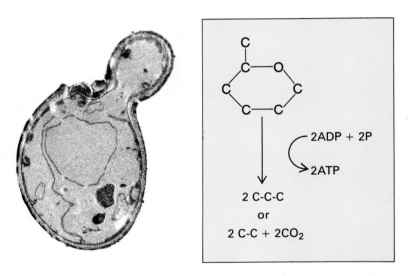

6–7 In fermentation, two ATPs are obtained by breaking down glucose into smaller compounds. In the absence of oxygen, yeast (shown with bud) ferment to stay alive.

The Balance Between Cellular Respiration and Photosynthesis

Cellular respiration and photosynthesis are similar in some ways. For example, both take place in specific organelles: much of cellular respiration occurs in mitochondria; photosynthesis occurs in chloroplasts. In other ways, the two processes differ. The energy to drive photosynthesis comes from light. The energy to make ATP in cellular respiration comes from glucose.

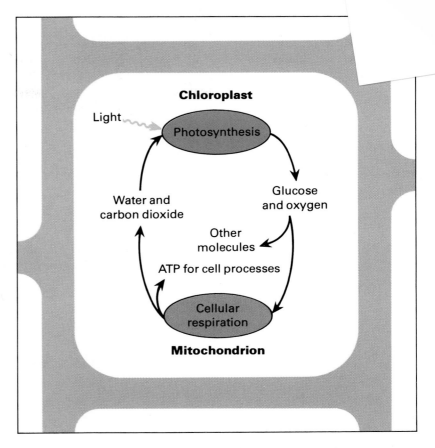

6–8 Photosynthesis and cellular respiration are complementary processes. The products of each process are the raw materials of the other process.

Photosynthesis and cellular respiration are not simply the reverse of each other since different metabolic pathways and different enzymes are involved.

The equations describing cellular respiration and photosynthesis are shown below for comparison. You can see that the products of photosynthesis are the raw materials for cellular respiration and that the products of cellular respiration are the raw materials for photosynthesis.

Photosynthesis:

$$6CO_2 + 6H_2O \xrightarrow[\text{enzymes}]{\text{chlorophyll} \atop \text{light}} C_6H_{12}O_6 + 6O_2$$

Cellular Respiration:

$$C_6H_{12}O_6 + 6O_2 \xrightarrow{\text{enzymes}} 6CO_2 + 6H_2O$$

On a global scale, each of these processes depends on the other. The carbon dioxide given off during cellular respiration is

Section 6.1 The Process of Obtaining Energy from Food

6–9 Cycle of energy production and use in nature. Producers harness energy and make food; all organisms break down food to make ATP.

used by producers in photosynthesis. The oxygen given off during photosynthesis is used by most organisms in cellular respiration. Thus, a cycle exists between the production and breakdown of food.

Building a Science Vocabulary

cellular respiration	citric acid cycle	crista
biochemical pathway	electron transport	fermentation
glycolysis	aerobic	
anaerobic	NADH	

Questions for Review

1. What is the relationship between ATP and food? between ATP and life processes?
2. What are the raw materials of cellular respiration? What are the products of cellular respiration?
3. What are the three stages of cellular respiration?
4. How do yeast and vertebrate muscle cells obtain energy in the absence of oxygen?
5. Explain how carbon dioxide and oxygen are in balance because of the processes of cellular respiration and photosynthesis.

6.2 THE MECHANISM OF CELLULAR RESPIRATION

Each of the three main stages in cellular respiration is a complex biochemical pathway. This section takes a closer look at these pathways and the process of fermentation.

Glycolysis

The breakdown of glucose begins with glycolysis. Glucose is a good compound for storing energy. The glucose molecule is fairly stable and does not break down easily by itself, even if enzymes are present. For this reason, energy is needed to start the process. At the beginning of glycolysis two ATP molecules are used to energize glucose so that it may be broken down.

During glycolysis, glucose molecules in the cytoplasm are split in half. Each half of the six-carbon glucose molecule eventually becomes a molecule of a three-carbon substance called **pyruvic** (pī-ROO-vik) **acid**.

As glucose is split and made into pyruvic acid, some energy is released. This energy is used to make ATP. A cell produces four ATPs for every molecule of glucose that goes through glycolysis. Two of these ATPs replace the ATP used to start glycolysis. Therefore, two net ATP molecules are produced.

6–10 In glycolysis, glucose is broken down into two molecules of pyruvic acid. Two net molecules of ATP are produced, along with four protons and four electrons.

The Formation of NADH Energized electrons are also released during glycolysis. Remember that during photosynthesis electrons are energized when light strikes chlorophyll. The energy from some of the energized electrons that "jump off" chlorophyll is used to make ATP. During cellular respiration, energized electrons from glucose are used to make ATP.

The energized electrons from glucose are picked up by two molecules of the electron carrier NAD^+, which is similar to $NADP^+$, the electron carrier in photosynthesis. Two energized electrons and a proton combine with each NAD^+, forming NADH.

$$NAD^+ + 2e^- + H^+ + energy \rightleftharpoons NADH$$

Summary of Glycolysis The following equation summarizes what happens to one molecule of glucose during glycolysis:

NAD stands for nicotinamide adenine dinucleotide.

In glycolysis, glucose is broken down into pyruvic acid. Two ATPs and two NADHs are made.

$$glucose \xrightarrow[\text{many steps}]{\text{enzymes}} 2 \text{ pyruvic acid}$$

$$2ADP \rightarrow 2ATP$$

$$2NAD^+ \rightarrow 2NADH$$

If O_2 is present in the cell, the pyruvic acid formed during glycolysis moves into the mitochondria and enters the citric acid cycle.

The Citric Acid Cycle

The citric acid cycle takes places in the fluid that fills the inside of mitochondria. Each molecule of pyruvic acid combines with other compounds present in the mitochondrion to produce a larger molecule called **citric acid**. Enzymes in the mitochondrion then break down the citric acid bit by bit. As enzymes tear down each molecule of citric acid, carbon dioxide is given off and one ATP is produced. Energized electrons are also released. Molecules of NAD^+ pick up these energized electrons and protons, forming NADH.

The energized electrons held by the NADH are used during electron transport to produce many ATPs. The carbon dioxide leaves the cell and enters the atmosphere as a gas.

citri- = citrus fruit
-ic = derived from

The citric acid cycle is often called the Krebs cycle after Hans Krebs, who won the Nobel Prize in 1953 for his research on this pathway.

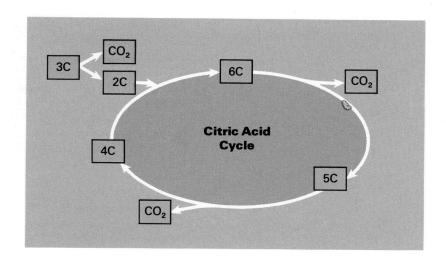

6–11 In the mitochondrion, molecules of pyruvic acid (3C) break down and combine with another molecule (4C) to form citric acid (6C). Citric acid molecules are then broken down, releasing CO_2. One ATP and five NADHs per pyruvic acid are also produced.

The citric acid cycle can be summarized as follows:

In the citric acid cycle, pyruvic acid is broken down into carbon dioxide. Two ATPs and ten NADHs are made.

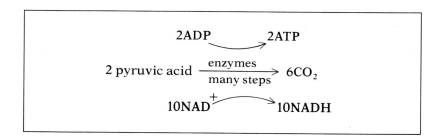

Electron Transport

So far, two ATPs per glucose have been produced during glycolysis, and two more ATPs per glucose have been produced during the citric acid cycle. During electron transport, 32 additional ATPs are produced. Electron transport takes place in the inner membrane of the mitochondrion. The energy to drive the formation of ATP from ADP and phosphate comes from the energized electrons held by the NADH produced in glycolysis and the citric acid cycle.

In electron transport, energy from twelve NADHs is used to make 32 ATPs.

$$32ADP \longrightarrow 32ATP$$
$$6O_2 + 12NADH \longrightarrow 12NAD^+ + 6H_2O$$

Section 6.2 The Mechanism of Cellular Respiration

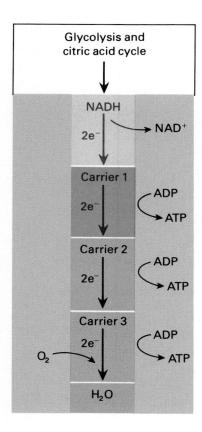

6–12 Electron transport (simplified). Energized electrons from NADH are passed from carrier to carrier in the chain. At each transfer, the electrons lose energy, which is used to produce ATP.

Cells cannot store oxygen for future needs.

As the name "electron transport" implies, electrons are passed from one compound to another. Imagine that these compounds are arranged in a chain. The energized electrons are passed from NADH to another carrier molecule and then from this carrier to the others in the electron transport chain. As the electrons are passed from one carrier molecule to the next, they lose energy. This energy is used to make ATP. By the time they reach the end of the chain, the electrons are de-energized. At this point, they combine with oxygen and hydrogen ions from NADH to form water.

The Role of Oxygen   Oxygen is the last molecule in the electron transport chain. In other words, it is the final acceptor of electrons. If there is no oxygen, electron transport gets backed up, because there is nowhere for the electrons to go. As a result, NAD^+ is not remade from NADH. This is a problem because the cell has a limited amount of NAD^+. Without NAD^+, both glycolysis and the citric acid cycle stop, and no ATP is made.

Something else happens when there is a lack of NAD^+ in the cell. The concentration of hydrogen ions increases because no carrier is available. As the quantity of H^+ in the cell increases, the pH of the cell decreases—the cell becomes more acidic (see Chapter 4). If the pH of the cell is too low, many enzymes stop working and the cell dies. This is why many cells cannot survive without a steady supply of oxygen.

Summary of Cellular Respiration

In cellular respiration, glucose is broken down into carbon dioxide and water. During this process, the energy stored in glucose is released and used by the cell to make ATP.

$$C_6H_{12}O_6 + 6O_2 + 36ADP + 36P \rightarrow 6CO_2 + 6H_2O + 36ATP$$

The first stage of cellular respiration, glycolysis, is anaerobic and takes place in the cytoplasm. In glycolysis, glucose is broken down into pyruvic acid; in addition, some NADH and ATP are made. Pyruvic acid moves into the mitochondria where the aerobic stages of cellular respiration occur. In the citric acid cycle, pyruvic acid is further broken down into CO_2. Some NADH and ATP are also made. In electron transport, energized electrons in NADH provide the energy to make much ATP. Oxygen plays a vital role in this process by acting as the final acceptor for electrons. Without oxygen, the citric acid cycle and electron transport cannot occur.

Many cells cannot survive very long without O_2. Some cells can, however, by undergoing fermentation.

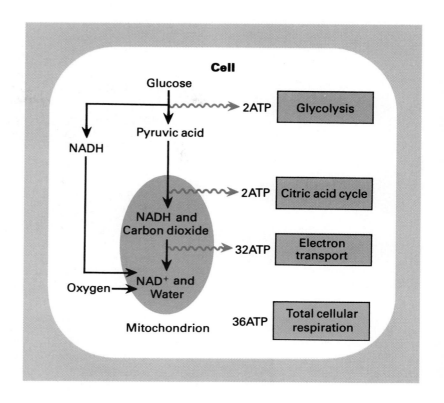

6–13 Summary of cellular respiration. The total output of all three stages is 36 ATPs per glucose.

Fermentation

When yeast, muscle cells, and certain bacteria do not obtain a constant supply of oxygen, they carry out the process of fermentation. Fermentation is an incomplete form of cellular respiration. First, glucose is broken down into pyruvic acid by glycolysis. Then the pyruvic acid from glycolysis is converted into another substance. In yeast, pyruvic acid is converted into **ethanol** (ethyl alcohol) and carbon dioxide. In muscle cells and in some bacteria, pyruvic acid is changed into **lactic acid**.

lac = milk
-ic = derived from

6–14 Products of fermentation: yeast convert pyruvic acid into ethanol; vertebrate muscle cells convert pyruvic acid into lactic acid.

Section 6.2 The Mechanism of Cellular Respiration

6–15 The pale bloom on grape skins is yeast. In winemaking, yeast in an airless container ferment the sugars in grape juice, producing ethanol and carbon dioxide.

In the absence of air yeast ferment the sugars in fruit juice and produce wine. If fresh air is then allowed to reach the wine, certain bacteria produce acetic acid from the ethanol made by the yeast. This conversion turns the wine to vinegar.

Fermentation in Yeast In making wine, yeast is added to fruit juice (usually grape juice) in large, closed vats. No air is allowed into the vats. The yeast carries out fermentation using the sugar in the fruit juice for energy. As fermentation proceeds, ethanol is made and carbon dioxide bubbles through the liquid. The product of fruit juice fermentation is wine.

Bakers also make use of yeast fermentation in preparing bread and cake dough. Yeast cells ferment the sugar in the dough. The carbon dioxide they produce makes the dough rise. As the bread or cake bakes, the yeast cells die and the alcohol they produced evaporates.

The formation of ethanol and carbon dioxide from pyruvic acid can be shown by the following equation:

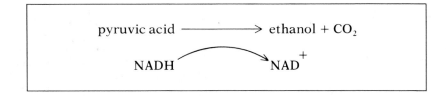

$$\text{pyruvic acid} \longrightarrow \text{ethanol} + CO_2$$
$$NADH \longrightarrow NAD^+$$

The NADH provides electrons, a proton, and energy for this process. The NAD^+ that is produced goes back to glycolysis.

The formation of NAD^+ in fermentation is very important. Recall that there is a limited amount of NAD^+ in the cell. If all the NAD^+ is used up glycolysis stops, no ATP is made, and the cell dies. By converting pyruvic acid to alcohol and remaking NAD^+ from NADH, it is possible for glycolysis to continue. When no O_2 is available, yeast cells survive on the two ATPs per glucose made during glycolysis.

Fermentation in Muscle Cells Sometimes you exercise so much that your muscle cells use up more O_2 than your breathing can supply. In the absence of O_2, muscle cells temporarily carry out fermentation. Instead of ethanol, however, they produce lactic acid. The muscle soreness that you may experience after strenuous exercise can result from a buildup of lactic acid.

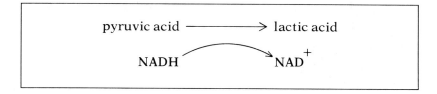

Pyruvic acid breaks down into lactic acid. Lactic acid accepts energized electrons and a proton from NADH. This results in the regeneration of NAD^+. As in yeast, this NAD^+ goes back to glycolysis and picks up more energized electrons. In this way, a muscle cell can continue functioning by making some ATP.

When you stop exercising, the lactic acid is converted back into pyruvic acid, which then goes through the aerobic stages of cellular respiration.

6-16 A track contestant suffers from cramps and fatigue caused by lactic acid accumulation in muscle tissue during strenuous exercise.

Fermentation and Cellular Respiration Compared

As you know, under anaerobic conditions muscle cells, yeast, and other cells are unable to perform the citric acid cycle and electron transport. They keep ATP production going by means of fermentation. During this process, two ATPs are produced from every glucose molecule.

Under aerobic conditions, organisms obtain many more than two ATPs per glucose molecule. During the citric acid cycle and electron transport, 34 ATPs are produced. A total of 36 new ATPs are made during all three stages of cellular respiration.

Complete cellular respiration yields 18 times as much ATP as fermentation yields. This difference is vital to organisms that require large amounts of energy.

Building a Science Vocabulary

pyruvic acid   citric acid   lactic acid
NAD^+ ethanol

Questions for Review

1. What happens to glucose during glycolysis?

2. During glycolysis more than two molecules of ATP is produced from each molecule of glucose. Why are two molecules of ATP considered the net yield of glycolysis?

3. What is the role of NAD^+ during glycolysis?

4. Only two ATP molecules are produced per glucose molecule during the citric acid cycle. What happens to the rest of the energy released from citric acid during the citric acid cycle?

5. Explain what happens to the energized electrons in NADH during electron transport.

6. What is the function of molecular oxygen, O_2, in cellular respiration?

7. What products are formed in fermentation? Compare the energy yields of fermentation and complete cellular respiration.

6.3 CELLULAR RESPIRATION AND METABOLISM

Cellular respiration is only one of the many important chemical processes that occur in organisms. Living cells constantly build up and break down many substances. The sum of all the chemical reactions of an organism is called its **metabolism**. Photosynthesis and cellular respiration are parts of metabolism. So are all the other chemical reactions that release or use energy in cells.

Energy from Nutrients

Obtaining energy is an important part of cell metabolism. The energy that cells use for movement, growth, and cell maintenance comes from the breakdown of nutrients. As you know, cellular respiration and fermentation are the processes cells use for this, and glucose is the most common substance broken down. Other nutrients are also used for energy. These nutrients are other carbohydrates, lipids, and proteins.

Cells break down large nutrient molecules into small ones that can enter the biochemical pathways of cellular respiration. Carbohydrates are hydrolyzed into simple sugars, proteins are hydrolyzed into amino acids, and lipids are hydrolyzed into glycerol and fatty acids. Cells convert some simple sugars into glucose, which enters glycolysis. They convert many other carbohydrates into six-carbon molecules similar to glucose that can also enter glycolysis. Cells convert amino acids, glycerol, and fatty acids into two-carbon and three-carbon compounds that enter the citric acid cycle.

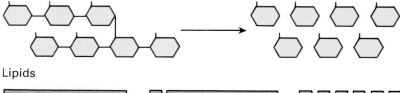

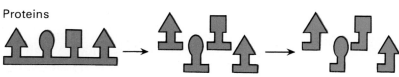

6–17 Chemical breakdown of nutrients into their usable components: carbohydrates to simple sugars; lipids to fatty acid fragments; and proteins to amino acid fragments.

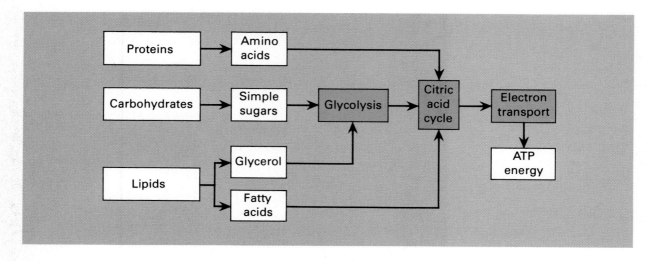

6–18 Different nutrients enter cellular respiration at different points.

Cells obtain roughly the same amount of energy per gram from amino acids and carbohydrates. Lipids, on the other hand, yield twice as much energy per gram as do carbohydrates or proteins.

Organisms usually store excess nutrients as starch or glycogen, and as fats or oils. When an organism cannot obtain or make food, these carbohydrate and lipid energy reserves are gradually used up. Under conditions of starvation, an organism will start breaking down its proteins for energy. This damages and eventually kills cells. It can cause the death of the organism if too many cells are harmed.

Cells do not usually depend on proteins for energy.

The Metabolism of Organisms

The energy that organisms obtain from nutrients is used for movement, growth, and cell maintenance. The rate at which organisms obtain and use energy is known as the **metabolic rate**.

Imagine that you could look at a living cell in a dry lima bean. If you could see the chemical reactions taking place, you might see just a few each hour. In a growing bean plant, however, each cell would be full of activity. Thousands of reactions would take place each minute.

The metabolic rate is low when chemical reactions take place slowly in an organism. It is high when chemical reactions take place rapidly. A high metabolic rate enables an organism to grow, move, and maintain itself quickly.

Table 6–1 Metabolic Rates of a 70-kg Person Performing Various Activities

Activity	Oxygen Consumed (litres per hour)	Kilojoules Used (per hour)
Sleeping	13	270
Sitting at rest	21	420
Typing rapidly	29	600
Walking slowly	41	840
Walking down stairs	73	1500
Swimming	104	2100
Running fast	118	2400
Walking up stairs	228	4600

Measuring Metabolic Rates Unfortunately, biologists cannot see the reactions that take place in cells. In order to measure the metabolic rate of an organism, they measure the rate of cellular respiration. This is just one of the many chemical processes in cells, but it is an important one. Since cellular respiration provides the energy for all cellular activities, it is directly related to the overall metabolic rate of the organism.

During aerobic cellular respiration, organisms consume oxygen and produce carbon dioxide. Biologists measure the amount of oxygen an organism consumes each minute (or the amount of carbon dioxide it produces). As you would expect, the more active an organism is, the more oxygen it consumes. Therefore, the metabolic rate increases with activity (see Table 6–1).

Diabetes is a disease that disturbs the metabolic rate.

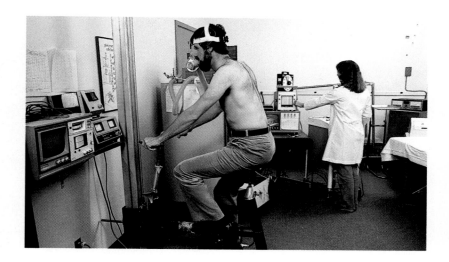

6–19 Metabolic rate can be determined by measuring the rate of oxygen consumption.

Section 6.3 Cellular Respiration and Metabolism

Table 6–2 Basal Metabolic Rates of Various Animals

Animal	Basal Metabolic Rate (mL O$_2$/kilogram body mass/hour)
Elephant	75
Horse	120
Human	210
Dog	300
Rabbit	480
Mouse	880

Growing organisms also have high metabolic rates. For example, young people have a metabolic rate that is at least twice that of adults. This explains why people in their teenage years usually need more food than adults.

In order to compare the metabolic rates of different organisms, biologists use the **basal metabolic rate**. This is the metabolic rate of a resting organism. From this, biologists have found that most organisms of a single species and age have similar metabolic rates. They have also found that the metabolic rates of various species are different (see Table 6–2).

Heat: A Metabolic By-product All chemical reactions involve the transfer of energy from one substance to another. In cellular respiration, for example, chemical energy in food is changed into chemical energy in ATP. Biologists, chemists, and

Smaller organisms have higher metabolic rates than larger animals. A mouse's metabolic rate is nearly eight times as high as a horse's and three times as high as a dog's.

6–20 When wood combines with oxygen in burning, chemical energy is converted into heat and light. Heat is also given off when chemical energy is transferred from food to ATP in cellular respiration.

6-21 Physical activity warms the body. As the level of activity increases, the rate of cellular respiration increases, and more heat is produced.

A person suffering heatstroke becomes dizzy, nauseous, and faint.

physicists have all found that some energy is lost to the surroundings when it is transferred. Usually the energy is lost as heat. For example, in a light bulb, electrical energy is changed to light energy. At the same time, however, some energy is lost as heat.

As you might expect, some energy is lost in the process of cellular respiration. The energy escapes as heat when glucose is broken down and ATP is formed. All metabolic activities that require ATP also produce heat. Here energy escapes as ATP transfers its energy to power these processes. You feel this when you become warm during strenuous exercise. Your active metabolism produces a large amount of heat. Sometimes this extra heat is harmful (for example, it can make you sick). At other times, the heat cells produce helps them function.

Body Temperature and Metabolic Rate

Most of the metabolic processes of cells take place with the help of enzymes, which speed up the rates of reactions. The rate of an enzyme reaction, however, also depends on the temperature of the organism. At 0°C (the freezing point of water) most reactions occur very slowly or not at all. On the other hand, at 30°C most reactions occur quite rapidly. (At temperatures over 40°C many enzymes are damaged so that most cells can no longer function.) Thus, body temperature is very important for metabolism.

All metabolic processes occur slowly when an organism's body is cold. Both the production and use of ATP is slow. A similar balance occurs when an organism's body is warm. For

6–22 Two very different climates: temperate forest and cold tundra. The difference between the two settings in type and abundance of vegetation is related to temperature.

During sleep the body temperature drops one or two degrees and the metabolic rate slows slightly.

example, at 35°C, most organisms use ATP quickly for their activities. Cellular respiration also occurs rapidly, providing the needed ATP. A good example of the effect of temperature on metabolism is the luxuriant growth of plants in moist, warm climates.

Warmblooded Animals Birds and mammals are warmblooded animals. They normally maintain a high body temperature using the heat from metabolism. Humans, for example, keep the core of their body at about 34°C, while in birds the temperature is usually a few degrees higher. The high body temperature of a warmblooded animal both requires and makes possible a high metabolic rate. The warmth of the body allows the chemical reactions of each cell to happen quickly. This, in turn, allows the organism to be alert and active in both warm and cold places. Birds and mammals can survive in cold areas where many other animals cannot.

The high metabolic rate of birds and mammals also has disadvantages. These animals must have a relatively large and constant supply of food. Particularly during cold parts of the year, they must eat and process extra food simply to keep warm.

A few animals use **hibernation** as a solution to the problems of being warmblooded in cold times of the year. During hibernation, the body temperature of warmblooded animals decreases. This causes all the body systems to function much more slowly. For example, the heartbeat of a groundhog may slow from 88 beats per minute to 15 beats per minute during hibernation.

6–23 In winter a coyote expends up to 30 percent of its energy budget to maintain a constant body temperature. Frequent meals provide energy to keep it warm and fuel its active life as a predator.

Because their metabolic rate is slower than when they are active, hibernating animals need less food. During hibernation they live off fat stored in their bodies. Later, when their bodies warm up, they begin actively eating again.

Hibernation is similar to sleep but much deeper. During hibernation the body temperature may drop 20 degrees or more, and the metabolic rate may drop to one-tenth the normal rate. The animal may seem to be sleeping, but it cannot be awakened. After a given amount of time, a hibernating animal will start to wake. Its body may shake to produce enough heat to warm itself. The warming process usually takes several hours before the animal is fully alert.

Only some mammals and a few birds can hibernate. When other warmblooded animals are in very cold surroundings, they must stay active and eat a lot of food. Their body coverings of feathers, fur, or blubber prevent heat loss and therefore help to keep the body warm. Sometimes, however, an animal cannot produce heat quickly enough and becomes cold. This leads to a condition known as **hypothermia**. Hypothermia results when the vital organs become too cool for the body to function properly.

hypo = under
thermē = heat

Hypothermia is dangerous to humans because it causes a person to lose the ability to think clearly.

Normally the skin, arms, or legs of a person can become 10 to 15 degrees cooler than the normal temperature without threatening body functions. This, however, is not true for the brain and internal organs. If the brain cools even 5°C, without medical help, a person will lose consciousness and eventually die. Accident victims are particularly susceptible to hypothermia. So are campers who become stranded in cold, wet weather.

Section 6.3 Cellular Respiration and Metabolism

6–24 The Eastern hog nose snake does not need to eat very often. Its energy requirements are low because it has a low metabolic rate and is inactive for long periods.

Coldblooded Animals Animals that do not keep a constant body temperature are known as coldblooded animals. Actually, their bodies are not always cold, but they stay close to the same temperature as their surroundings. Sitting in the sun, a coldblooded turtle may have the same body temperature as a warmblooded animal. In a cooler location, however, the turtle's body will be cooler.

The metabolic rate of coldblooded animals changes according to the temperature. This is because their body functions slow down when their cells are cold. "Singing" insects, such as crickets, are a good example of this. On a warm night they sing faster than on a cool night. They are less alert and move more slowly in cool surroundings. During very cold times of the year these animals either die or become inactive.

Most animals are coldblooded. This includes all invertebrates and three vertebrate groups: fish, amphibians, and reptiles. None of these animals keeps a constant body temperature, though a few of them can change their behavior to warm themselves. For example, bees keep their hive warm with the heat released by their constant activity, and some insects warm their bodies before flight by exercising their wings.

Building a Science Vocabulary

metabolism   basal metabolic rate   hypothermia
metabolic rate   hibernation

Questions for Review

1. How are nutrients other than glucose broken down for energy?
2. How do biologists compare the metabolic rates of different species?
3. How does hibernation differ from sleep?
4. Do coldblooded animals have cold blood? Explain.

Perspective: Technology
Petroleum from Plants

Hundreds of thousands of years ago, gigantic swamps and marshes covered vast areas of land on the earth. It is hard to imagine that some of those marshes and swamps thrived where today there is water, ice, or sand. We know that is the case because petroleum and natural gas are now found in such places.

We often refer to petroleum and natural gas as fossil fuels. Fossil fuels are compounds called hydrocarbons that are made up of carbon and hydrogen. They are the fossilized remains of the plants that inhabited those ancient swamps and marshes.

The supply of petroleum and natural gas is limited and will eventually be exhausted. Some authorities predict that all the oil and natural gas will be used up in 50 years or less. Not only is our supply of oil and natural gas limited, it is also nonrenewable. It took hundreds of thousands of years for plants to be converted into the various forms of fossil fuel.

On the other hand, maybe our supply of petroleum is not limited, but renewable. If plants did it once, why not again? The starting point of the petroleum-making process is photosynthesis. The important energy-containing products of photosynthesis are carbohydrates and hydrocarbons. The carbohydrates can be fermented (as in making wine or beer) to produce alcohol, which can be used as a fuel. The hydrocarbons may also be of some use. Considerable experimental work is being done on a number of plants. In particular, the hydrocarbons of the gopher plant (*Euphorbia lathyrus*) look very promising as a substitute for oil.

The potential of these "artificial petroleum" fuels is significant considering that the amount of solar energy falling on the surface of the earth in ten days is equivalent to the energy content of all the known fossil fuel reserves.

Obtaining petroleum from plants, without having to wait thousands of years, is a vitally important challenge for modern science. It could play an important role in solving the world energy crisis.

Chapter 6
Summary and Review

Summary

1. All organisms perform cellular respiration. During cellular respiration, glucose is gradually broken down into CO_2. The energy stored in glucose is released and used to make ATP.

2. The three main stages in cellular respiration are glycolysis, the citric acid cycle, and electron transport. Glycolysis takes place in the cytoplasm and is anaerobic; the citric acid cycle and electron transport take place in mitochondria and are aerobic.

3. During glycolysis, glucose is split in half, and two molecules of pyruvic acid, two ATPs, and two NADHs are made. During the citric acid cycle, the pyruvic acid is broken down into six molecules of CO_2, and two ATPs and eight NADHs are made. In electron transport, energized electrons from the 10 NADHs provide the energy to make 32 ATPs. Oxygen picks up the electrons and combines with protons, forming water.

4. The citric acid cycle and electron transport cannot occur without O_2.

5. Certain cells carry out fermentation when O_2 is lacking. During fermentation, muscle cells and certain bacteria convert pyruvic acid made in glycolysis into lactic acid. Yeast cells convert pyruvic acid into ethyl alcohol and carbon dioxide.

6. Complete cellular respiration produces 36 ATPs per glucose. Fermentation produces two ATPs per glucose.

7. The raw materials and products of cellular respiration are opposite to those of photosynthesis.

8. In addition to glucose, cells can obtain energy from other carbohydrates, lipids, and proteins. These nutrients are modified to enter either glycolysis or the citric acid cycle.

9. Metabolism includes all the chemical reactions that occur in cells. The metabolic rate is the rate at which organisms obtain and use energy. In active organisms the metabolic rate is high.

10. The basal metabolic rate is the metabolic rate of a resting organism. It is used to compare the metabolic rates of different organisms.

11. The metabolic rate is directly related to body temperature. Up to a certain limit cell processes occur more quickly at higher temperatures than at lower temperatures.

12. Warmblooded animals maintain a fairly constant body temperature. Coldblooded animals do not keep a constant body temperature. Their body temperature depends on the temperature of their surroundings.

Review Questions

Application

1. Why does cellular respiration not use up all the oxygen in the atmosphere?

2. For the pairs of substances listed below, state whether the substance on the left contains *more*, *less*, or the *same amount* of energy as the substance on the right.
   a. pyruvic acid—glucose
   b. citric acid—carbon dioxide
   c. ethyl alcohol—glucose
   d. ADP—ATP
   e. NADH—NAD^+

3. Why is cellular respiration a necessary process in plant cells?

4. How is being warmblooded an advantage over being coldblooded?
5. Why is the rate of oxygen consumption an accurate measure of the metabolic rate of an organism?
6. Cyanide is a poison that prevents electrons from combining with oxygen during electron transport. Explain why cyanide kills many living things.

Interpretation

1. Compare the processes of cellular respiration and photosynthesis. Include where they occur and their raw materials and products. What similarities and differences do you see between the two processes?
2. Anaerobic organisms, which can survive on fermentation, are very small and almost always unicellular. Aerobic organisms, however, can be as large and complex as whales and redwood trees. How might these differences between anaerobic and aerobic organisms be explained by their means of obtaining energy?
3. During photosynthesis, plant cells get the ATP that runs the dark reactions from the light reactions. Would it be an advantage to plants to use ATP produced during cellular respiration to run the dark reactions? Explain.

Extra Research

1. Make bread or cake following a recipe that uses yeast. Watch for signs of fermentation. How do you know fermentation is occurring? Why do you not taste alcohol in bread or cake made with yeast?
2. Look up the work of Louis Pasteur on fermentation. Write a short paragraph explaining how he saved the wine industry in France.
3. The electric eel produces electricity. Fireflies produce light. Use reference books to find out how electric eels and fireflies produce and use these forms of energy.
4. Look up the composition of the atmosphere of some of the other planets. Could life as we know it exist on these planets? Explain your decisions in light of what you know about cellular respiration and photosynthesis.

Career Opportunities

Laboratory technicians are members of the research teams in most science laboratories. They help design and run experiments, maintain equipment, and analyze data. For information, write to the Canadian Society of Laboratory Technologists, Box 830, Hamilton, Ontario L8N 3N8, or your local community college.

Biochemists study the complex chemicals in foods, drugs, pesticides, and cells. They design and supervise experiments to identify, synthesize, or study the effects of new compounds. For information, write to the Canadian Biochemical Society, University of Saskatchewan, Saskatoon, Saskatchewan S7N 0W0, or the Chemical Institute of Canada, 151 Slater St., #906, Ottawa, Ontario K1P 5H3.

Physicians must have a thorough knowledge of cells, tissues, and the human body in order to diagnose and treat people with injuries and diseases. Most physicians specialize in one of many different fields, such as family practice, surgery, or pediatrics. For information, write the Medical Council of Canada, 1867 Alta Vista Dr., Box 8234, Ottawa, Ontario K1G 3H7.

Unit II
Suggested Readings

Asimov, Isaac, *Photosynthesis*, Science and Discovery Series, New York: Basic Books, Inc., 1969. *Looking through a chemist's eyes, Asimov explains the vital process of photosynthesis and efforts of scientists to understand its role in the fundamental chemistry of life.*

Dethier, Vincent G., *To Know a Fly*, San Francisco: Holden-Day, Inc., 1962. *A biologist explains how to investigate the behavior and physiology of the common fly. Some experimental approaches of science are clearly illustrated.*

Dubos, René, *Louis Pasteur: Free Lance of Science*, New York: Charles Scribner's Sons, 1976. *Depicts an eminent scientist and the role he played in the development of microbiology.*

Ewald, Paul W., "The Hummingbird and the Calorie," *Natural History*, August-September, 1979. *Describes some of the main concerns of the hummingbird as it competes for food: limited supplies of nectar, dangers of chilling or overheating.*

Gore, Rick, "The Awesome Worlds Within a Cell," *National Geographic*, September, 1976. *Well illustrated article on many aspects of cells. Includes structure, chemistry, functions, and diseases.*

Jerome, Norge W., Judith G. McCleery, and Isabel D. Wolf, *Help Yourself: Choices in Foods and Nutrition*, New York: Butterick Publishing, 1981. *Includes interesting discussions of the total nutrient contributions of specific foods.*

Lechtman, Max D., Bonita Roohk, and Robert J. Egan, *The Games Cells Play: Basic Concepts of Cellular Metabolism*, Menlo Park, CA: The Benjamin/Cummings Publishing Co., Inc., 1979. *Clues and a game plan are used to translate abstract concepts of cellular metabolism into concrete analogies and images. Written in a clear, concise, lively style.*

Ledbetter, M. C., and Keith R. Porter, *Introduction to the Fine Structures of Plant Cells*, New York: Springer-Verlag, 1970. *Excellent collection of micrographs of plant cells taken under the electron microscope. Includes detailed descriptions.*

Miller, Julie Ann, "Microscopy's Bright Side," *Science News*, October 28, 1978. *Describes how fluorescent dyes are used in conjunction with the electron microscope to aid in viewing organelles and substances in cells.*

Porter, Keith R., and Mary A. Bonneville, *An Introduction to the Fine Structures of Cells and Tissues*, 4th ed., Philadelphia: Lea & Febiger, 1973. *Excellent collection of animal cell electron micrographs. Includes detailed scientific explorations.*

Rosenfeld, Albert, "The Great Protein Hunt," *Science '81*, January-February, 1981. *Describes the structure and functions of proteins in the body and explains how proteins are separated and identified in the laboratory.*

Thomas, Lewis, *The Lives of a Cell: Notes of a Biology Watcher*, New York: The Viking Press, Inc., 1974. *Twenty-nine essays by a pathologist who communicates his delight in scientific discovery as he discusses genetic and molecular biology.*

Weintraub, Pamela, "Splitting Water," *Discover*, February, 1981. *Describes Melvin Calvin's research on a synthetic chloroplast designed to produce hydrogen fuel.*

Wohlrabe, Raymond A., *Exploring the World of Leaves*, New York: Thomas Y. Crowell Co., 1976. *A comprehensive treatment of leaf biology, with several chapters on tissues of leaves, emphasizing photosynthesis and transpiration. Includes suggestions for experiments in plant physiology.*

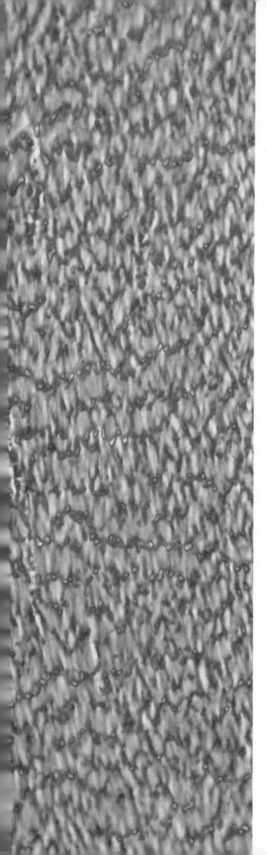

Unit III

Basic Body Processes

As you study biology, you learn about many different kinds of living things. Some of them are quite unusual in form and life history. Yet, every organism on earth is similar in fundamental ways. The bodies of amebas, buttercups, jellyfish, bats, and humans perform similar functions in the course of staying alive. They exchange gases with the environment. They obtain and use food. They transport materials within their bodies. They get rid of wastes. They exhibit movement. Different organisms perform these processes in different ways, but patterns common to all organisms can be identified.

How an organism performs the processes basic to life depends greatly on its size. Unicellular and small multicellular organisms are in close contact with their external environment. Their bodies are simple in comparison to larger organisms. For these reasons, the mechanisms they use in performing basic life processes are relatively uncomplicated. In large multicellular organisms, on the other hand, most of the organism's cells are not in direct contact with the external environment. In these organisms complex systems for carrying out life's processes are found.

Smooth muscle tissue plays a vital role in the body processes of animals. Each cell contains a single nucleus, appearing as small grains in this phase contrast micrograph.

Chapter 7

Gas Exchange

Focus *It is impossible for humans to live without breathing. Each time you breathe you move from one-half to two litres of air into and out of your body. The incoming air provides your cells with oxygen. The outgoing air carries away the carbon dioxide that cells produce. This exchange of gases is critical not only for humans, but for all aerobic organisms.*

7.1 AN OVERVIEW OF GAS EXCHANGE

Cells use oxygen and produce carbon dioxide during cellular respiration (see Chapter 6). In order for cellular respiration to occur, oxygen must reach cells and carbon dioxide must leave cells. This exchange of gases is called **respiration**.

The term *respiration* is used several ways in biology. Cellular respiration means the synthesis of ATP (from ADP and phosphate) by using the energy in food. Respiration also has two other meanings. It can mean taking in oxygen and giving off cabon dioxide through the cell membrane, or it can mean moving air into and out of the body. To avoid confusion, this book uses the terms *cellular respiration, respiration,* and *breathing* to distinguish these three processes. All organisms carry on some form of cellular respiration, most organisms (aerobic ones) carry on respiration, and many organisms breathe. These processes all occur simultaneously.

7–1 Whales are air-breathing mammals. When they surface for air they first "blow," clearing their air passage, and then inhale.

Table 7–1 Terms in Respiration

Term	Process
Cellular Respiration	Producing ATP from food
Respiration	Taking in oxygen and giving off carbon dioxide through the cell membrane
Breathing	Inhaling and exhaling air

General Plan of Respiration

re- = again
spirare = to breathe

Respiration occurs in many single-celled organisms as well as in plants and animals. It occurs in organisms that live in water and in those that live on land and are surrounded by air. Different kinds of organisms use different organs in respiration. Each of them, however, follows the same four basic steps. These are:

1. The organism takes in oxygen through a moist membrane.
2. The organism transports the oxygen to all parts of its body.
3. The organism moves carbon dioxide from all parts of its body to the same moist membrane through which oxygen enters.
4. The organism expels the carbon dioxide through the moist membrane.

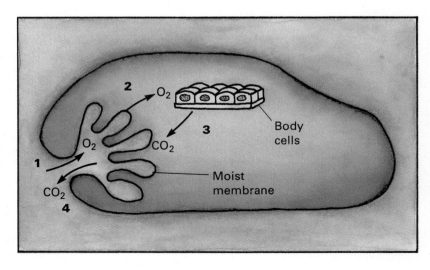

7–2 The four basic steps of respiration.

Chapter 7 Gas Exchange

Gas Exchange in Individual Cells

Oxygen and carbon dioxide move into and out of a cell because the number of gas molecules inside the cell is different from the number of gas molecules outside. If there is more oxygen outside the cell than inside, oxygen molecules enter the cell. If there is more oxygen inside the cell than outside, oxygen molecules leave the cell. The same is true for carbon dioxide.

In order for oxygen and carbon dioxide to cross the cell membrane, the membrane must be moist. The cell membranes of many unicellular organisms remain moist because they live in water. Oxygen dissolved in the water moves into the cell, and carbon dioxide produced in the cell moves into the water. In large organisms, body fluids keep cell membranes moist. Blood carries oxygen to the cells and removes the carbon dioxide that the cells produce (see Fig. 7–3).

Inside individual cells oxygen and carbon dioxide are dissolved in the watery cytoplasm. These gases move inside the cell in the same way they entered. Oxygen moves to areas where there is little oxygen, especially toward the middle of the cell. Carbon dioxide moves to areas where there is little carbon dioxide, especially toward the cell membrane.

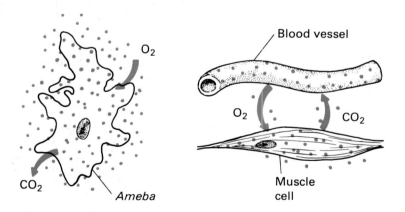

7–3 Unicellular organisms exchange O_2 and CO_2 directly with their environment. In multicellular organisms, gas exchange occurs between blood vessels and body cells.

Gas Exchange in Multicellular Organisms

Many multicellular animals are small and live in water. The cells of these organisms are in constant contact with the water, and oxygen and carbon dioxide can easily cross cell membranes.

Gas exchange is more of a problem, however, for large, multicellular organisms. Fish and other complex organisms that live in water are so large that their cells cannot be in constant

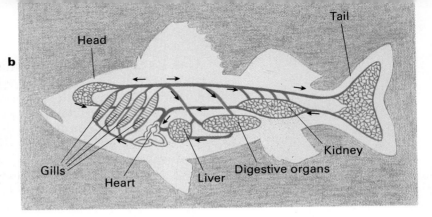

7-4 Respiration in fish. **a.** An open gill cover reveals the gills with their many filaments. **b.** The circulatory system is shown with the vessels of the major organ areas. Note the many small blood vessels in the gills.

contact with the water. In addition, land-dwelling organisms are surrounded by air, not water. The cells of large, multicellular organisms do not exchange gases directly with their external environment. Instead, they have specialized **respiratory systems**. A respiratory system is made up of tissues and organs (for example, gills, lungs, and blood) that carry out the four basic steps of respiration.

Consider the respiratory system of a fish (see Fig. 7–4). First, the fish takes in oxygen from its external environment (water) through its gills. Second, the fish's blood transports the oxygen to all the cells in its body. Third, the blood collects carbon dioxide from the cells and returns it to the gills. Last, the fish expels the carbon dioxide through its gills. In this way, the respiratory system of a fish enables all its cells to exchange oxygen and carbon dioxide with the external environment.

Building a Science Vocabulary

respiration   respiratory system

Questions for Review

1. What is the difference between respiration and cellular respiration?
2. How do oxygen and carbon dioxide move into and out of a cell?
3. What are the four parts of the general plan of respiration?
4. Where does the exchange of oxygen and carbon dioxide with ocean water occur in the shark?

7.2 RESPIRATION IN THE REPRESENTATIVE ORGANISMS

All five representative organisms are aerobic—their cells use oxygen during cellular respiration. In the paramecium and hydra, respiration is relatively simple. In the earthworm, grasshopper, and bean plant, which are more complex than the other two, respiration is more complicated. Notice that no matter how respiration is accomplished by these organisms, each of them follows the four steps in the general plan.

Respiration in the Bean Plant

Unlike the other representative organisms, the bean plant is a producer. Many of its cells use carbon dioxide during photosynthesis. In addition, all of its cells use oxygen during cellular respiration. These gases move in and out of air spaces inside the plant.

The stems and leaves of beans and other plants are covered with a waxy cuticle. This thin layer prevents the movement of liquids and gases into or out of plant parts except through special openings called **stomates** and **lenticels** (LEN-tih-selz). A stomate is an opening between two guard cells on a stem or leaf (see Chapter 5). Lenticels are openings between several cells on stems and bark.

Recall that cell membranes must be moist in order for oxygen and carbon dioxide to move into and out of cells. The membranes of the cells inside the plant always have a thin film of water covering them. Gas molecules entering a cell dissolve

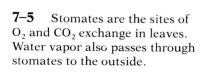

Unlike stomates, lenticels are always open.

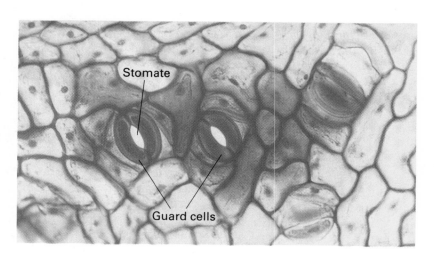

7–5 Stomates are the sites of O_2 and CO_2 exchange in leaves. Water vapor also passes through stomates to the outside.

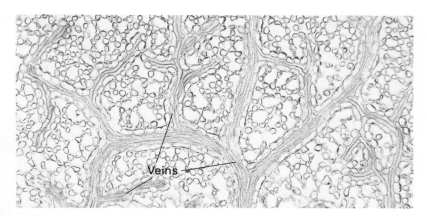

7–6 The leaves of many plants have netlike veins.

7–7 The leaf hairs of the vinegar weed trap moisture, protecting it against drying.

Air fills most of the spaces between cells in plants, whereas fluid fills the spaces between cells in animals.

into the water and move through the membrane into the cell. Similarly, gas molecules leaving a cell move through the cell membrane into the air.

The water that coats cells in the bean plant is carried from the roots through veins to all parts of the stem and leaves. Water evaporates from the plant through the stomates. The bean plant has several adaptations that limit the amount of water it loses. The most important adaptation is the waxy cuticle. Another important adaptation is the ability of the leaf to close the stomates, preventing direct water loss from the interior of the leaf. Tiny hairs on the outside of some leaves are a third adaptation. They help to cut down water loss by slowing the movement of moist air away from the surface of the leaf. These adaptations are of great importance in keeping membranes inside the plant moist.

Gas Exchange in the Leaf Air containing oxygen and carbon dioxide enters the leaf through the stomates, most of which are located on the underside of the leaf. Once inside the leaf, air moves through the loosely packed cells of the spongy layer. Air spaces between the cells allow the air to reach all the cells in the leaf.

During the day, the cells of the palisade layer actively photosynthesize. They take in carbon dioxide from the air spaces and produce oxygen. They use some of this oxygen in cellular respiration and release the rest into the air spaces. Cells in the leaf that do not play a major role in photosynthesis absorb oxygen, which they use in cellular respiration, and give off carbon dioxide. The air exits through the stomates containing less carbon dioxide and more oxygen than when it entered the leaf.

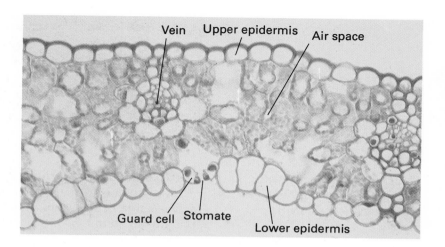

7-8 Cross section of wheat leaf. Air passes through the open stomate and between air spaces to reach all the cells in the leaf.

At night, when the leaf is not photosynthesizing, oxygen is taken up by all the cells and used in cellular respiration. The carbon dioxide produced moves into the air spaces and out of the leaf.

Opening and Closing of Stomates The amount of gas that can move into or out of a leaf depends on how wide open the stomates are. Opening and closing the stomates involves two things. One is the structure of the cell wall in the guard cells, which surround the stomate. The other is the amount of water in the guard cells. In Figure 7–9, you can see that the part of the cell wall facing the stomate is thicker than the part of the cell wall opposite the stomate. When the guard cells are filled with water, the water exerts pressure against the cell wall. This causes the thin part of the cell wall to "give," bending the cell wall outward. As the cell wall bends, it pulls the rest of the guard cell with it, opening the stomate.

When the guard cells are not full of water, there is little pressure on the outside walls, and the guard cells do not bend very much. As a result, the inner sides of the guard cells stay together, keeping the stomate closed.

The opening and closing of the stomates are linked to photosynthesis. The guard cells produce glucose from carbon dioxide and water during photosynthesis. Water moves into the guard cells, replacing the water used in photosynthesis. This water exerts pressure against the walls of the guard cells, causing them to open. When the plant is not photosynthesizing, water does not move into the guard cells and the stomates close.

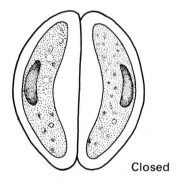
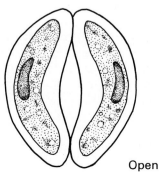

7-9 Increased water pressure against the thinner outer walls of guard cells causes them to pull apart, opening the stomate.

Section 7.2 Respiration in the Representative Organisms

7–10 The lenticels of this white poplar allow direct exchange of gases between living stem tissues and the outside.

Water plants, such as water lilies, have particularly large air spaces in their stems and leaves.

The absence of a cuticle also enables roots to exchange gases easily as well as perform their main function of absorbing water and nutrients from the soil.

Gas Exchange in Stems and Roots Bean stems have numerous lenticels, which allow the passage of air from the outside to the inside of the stems. Air moves through spaces between cells in the stems just as it does in leaves. The air spaces are interconnected, allowing air to reach all the living cells in the stems.

Gas exchange is even simpler in roots than in leaves and stems, because roots are not covered by a cuticle. Gases move directly across the moist membranes of the outer cells of roots. Oxygen moves through intercellular air spaces to cells, and carbon dioxide moves through the same air spaces to the outer cells and then passes to the outside.

The soil in which roots grow must be well aerated for them to obtain enough oxygen. This is also important for getting rid of carbon dioxide. Many plants die when they are flooded because the air spaces in the soil become filled with water, blocking gas exchange. One of the benefits of cultivating the soil is that it increases the circulation of air in the soil. This improves gas exchange with the roots.

7–11 Flooding killed these trees by filling the air spaces in the soil with water, blocking gas exchange in the roots.

Respiration in the Paramecium

Respiration is a simple matter for the paramecium, which lives in water. Since a paramecium is composed of only one cell, every part of its cell membrane is in direct contact with the water. Oxygen dissolved in the water moves directly across the cell membrane into the paramecium. Carbon dioxide within the paramecium moves directly into the water.

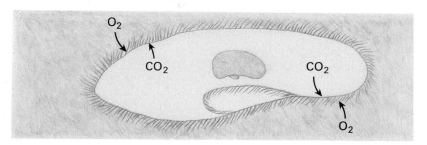

7–12 Respiration in the paramecium.

Respiration in the Hydra

The hydra is a multicellular coelenterate whose body is composed of two layers of cells. Cells on the outside of the body are in direct contact with the water in which the hydra lives. Cells lining the saclike digestive cavity are also in contact with water, which flows into and out of the hydra's mouth. As in the paramecium, respiration in the hydra is simple. Oxygen from the water moves directly into each cell, and carbon dioxide produced in each cell moves directly into the water.

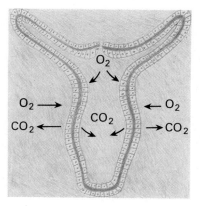

7–13 Respiration in the hydra.

Respiration in the Earthworm

Unlike most multicellular land animals, the earthworm does not have special organs for respiration. Instead, gas exchange occurs across the earthworm's moist skin. The skin is kept moist by water in the soil and by a thick fluid called **mucus** (MYOO-kus),

7–14 Mucus glistens on the earthworm's skin.

Section 7.2 Respiration in the Representative Organisms

Capillaries are the smallest vessels in the transport system.

which the earthworm secretes. Oxygen from the air spaces in the soil around the earthworm moves into the earthworm's skin, where it enters microscopic blood vessels called **capillaries** (KAP-uh-LAYR-eez). The capillaries and other blood vessels carry blood containing oxygen to all the earthworm's cells. Carbon dioxide produced by the cells enters the capillaries and eventually moves out of the body through the moist skin.

As long as the skin remains moist and there are air spaces in the soil around the earthworm, the life-supporting exchange of gases can occur. If an earthworm's skin dries out, oxygen cannot move across it and the worm dies.

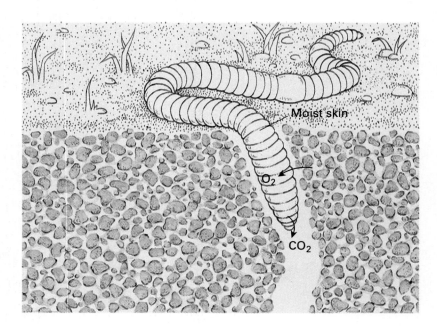

7–15 Respiration occurs across the earthworm's moist skin. An earthworm in its underground burrow is surrounded by small pockets of air and water.

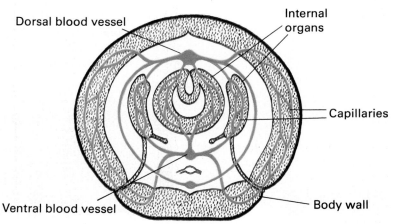

7–16 Cross section of an earthworm. Gases are exchanged through capillaries in the skin.

Chapter 7 Gas Exchange

7–17 Spiracles, visible along the side of the grasshopper, are the outer openings of air passages, known as tracheas, within the insect.

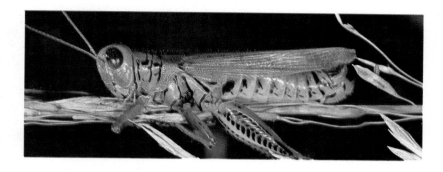

Respiration in the Grasshopper

The grasshopper is better adapted for life on land than the earthworm. The grasshopper has a hard body covering, the **exoskeleton**, that protects it and helps prevent it from drying out.

Air enters and leaves the grasshopper's body through ten pairs of holes along its side. These holes, called **spiracles** (SPEER-uh-kulz), are opened and closed by valves. The spiracles connect to a system of branching tubes, called **tracheas** (TRAY-kee-uhz), which carry air to and from all cells in the grasshopper. Tracheas are so small that there is a trachea near almost every cell in the body.

Oxygen in the tracheas dissolves in the thin film of water covering each of the grasshopper's cells and moves through the cell membranes into the cells. Carbon dioxide moves out of the cells into the tracheas and is expelled through the spiracles.

The air spaces in the tracheal system of some insects take up to 40 percent of the body volume.

Breathing in the Grasshopper The tracheas are connected to tiny balloonlike structures called **air sacs**. Air sacs help the grasshopper breathe by pumping air. The grasshopper inhales

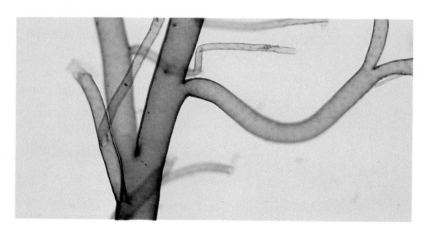

7–18 Insect tracheas (magnified 20 times) have spiral thickenings that keep the tube open. Gas exchange takes place at the thin, moist ends of the tubules.

Section 7.2 Respiration in the Representative Organisms

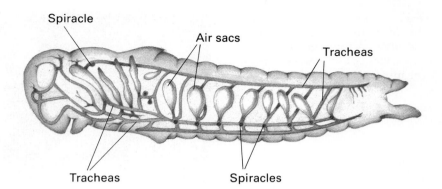

7–19 Respiratory system of a grasshopper. The tracheas also extend into the appendages (not shown).

by moving its body wall outward. The valves open the first four pairs of spiracles when the grasshopper inhales. Air rushes in and fills all the air sacs and tracheas. The grasshopper exhales by moving its body wall inward. When this happens, valves open the last six pairs of spiracles and air in the tracheas and air sacs is squeezed out. In this way, the grasshopper pumps air into and out of its body, delivering oxygen to its cells and getting rid of carbon dioxide.

Building a Science Vocabulary

stomate	capillary	trachea
lenticel	exoskeleton	air sac
mucus	spiracle	

Questions for Review

1. Why does a stomate open when its guard cells are filled with water?
2. What structures carry oxygen and carbon dioxide throughout the body of the grasshopper? the earthworm?
3. Describe breathing in the grasshopper. How is this different from respiration?
4. What is the source of oxygen for the hydra?
5. Why is a moist skin necessary for respiration in the earthworm?

7.3 BREATHING AND RESPIRATION IN HUMANS

The function of the human respiratory system is to supply oxygen to the cells of the body and to remove carbon dioxide. The major organs of the respiratory system are the lungs. In addition, the blood and blood vessels of the transport system play a vital role. In the lungs, blood vessels pick up oxygen and release carbon dioxide. The blood moves through the vessels continuously, transporting oxygen to all the cells of the body and returning the carbon dioxide produced by the cells to the lungs.

Breathing

The diaphragm separates the chest and the abdominal cavity.

You move air into and out of your lungs by breathing. This is accomplished by changing the size of your chest cavity. When you inhale, a dome-shaped muscle called the **diaphragm** (DĪ-uh-FRAM) moves down. At the same time, rib and neck muscles lift your ribs upward and outward. This movement of the diaphragm and other muscles enlarges your chest cavity, and air rushes into your lungs. When you exhale, the diaphragm and other muscles relax. This restores the diaphragm to its original position and pulls your ribs back. The size of the chest cavity becomes smaller, and air is forced out of your lungs.

7–20 Balloon models of breathing.

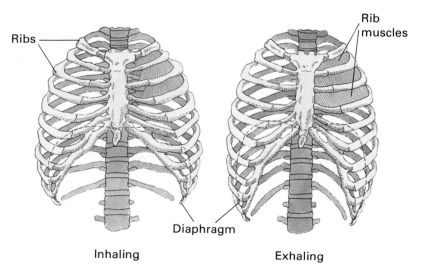

7–21 During breathing the volume of the chest cavity is changed by the action of the diaphragm and the ribs.

The Path of Air into the Lungs When you breathe, air first moves into the nose and **nasal passages**. These passages are lined with membranes that are coated with a moist, sticky mucus. Some of the membranes have moving cilia. As air passes over the membranes, it is warmed and moistened. Dust particles falling on the sticky surface of the membranes are swept by the cilia toward the **pharynx** (FAYR-anks), or throat, where they are either swallowed or coughed up. This removes most of the large foreign particles from the air.

The pharynx is also the tube through which food passes on its way to the stomach.

The air passes through the pharynx and enters the **trachea**, or windpipe, which is in the middle of the neck. The wall of the trachea has rings of the flexible support tissue, cartilage, that prevent it from collapsing. Cells lining the trachea also have cilia, which move small particles (such as the fine dust) away from the lungs.

You can feel the cartilage of your larynx in your throat.

The **larynx** (LAYR-anks), or voice box, is located in the trachea, a short distance from the throat. The larynx contains vocal cords that vibrate, making sound as air passes over them.

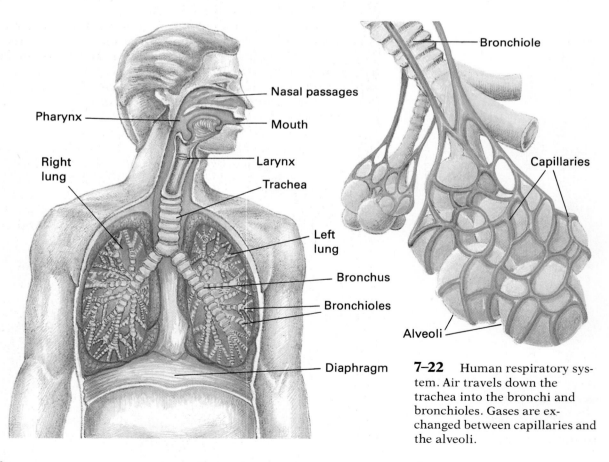

7–22 Human respiratory system. Air travels down the trachea into the bronchi and bronchioles. Gases are exchanged between capillaries and the alveoli.

Chapter 7 Gas Exchange

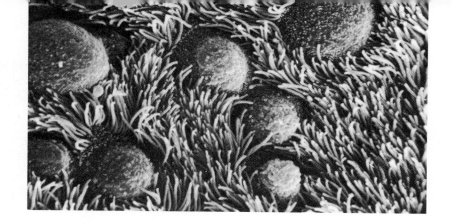

7-23 Cilia in air passages help filter out dust and other small particles, sweeping them toward the pharynx.

From TISSUES AND ORGANS: A TEXT ATLAS OF SCANNING ELECTRON MICROSCOPY by Richard G. Kessel and Randy H. Kardon. W. H. Freeman and Company. Copyright © 1979

After air has passed through the trachea, it enters two smaller tubes called **bronchi** (BRON-kee). Each bronchus enters a lung. Air is pulled into smaller and smaller branches of the bronchi and eventually enters microscopic branches called **bronchioles**. Cilia lining the bronchi and bronchioles also help remove small particles. Because of the branching of the air passages in the lungs, some people refer to this portion of the respiratory system as the **bronchial tree**.

The bronchioles connect to tiny ducts that lead into air sacs called **alveoli** (al-VEE-uh-lī). Alveoli are usually only one cell thick and are surrounded by capillaries. It is in the alveoli of the lungs that gas exchange occurs.

Prolonged exposure to substances such as silica, asbestos, cigarette smoke, and air pollution can damage cilia and other lung tissues.

alveus = a cavity

Gas Exchange in the Lungs

The membranes around the bronchioles, ducts, and alveoli provide a large area for gases to move across the lung surface into and out of the capillaries. It has been estimated that if you stretched out the lung membrane, it would cover 70 m², which is about equal to the size of the floor in an average classroom. This extensive surface area allows the large amount of gas exchange necessary to meet our substantial oxygen needs.

The millions of air pockets formed by the alveoli give the lungs a spongy texture.

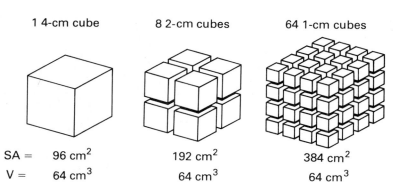

7-24 The many tiny alveoli give the lungs a much larger surface area than they would have if they were two large balloonlike pockets. This can be shown with cubes. Notice how the surface area of the cubes increases as the number of cubes increases, while the volume remains the same.

7–25 Gas exchange in an alveolus takes place between capillary walls and the moist membrane of the alveolus.

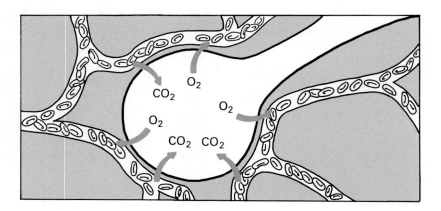

Table 7–2 Composition of Air Before and After Breathing

	Inhaled Air (percent)	Exhaled Air (percent)
O_2	20.71	14.60
CO_2	0.04	4.00
H_2O	1.25	5.90
N_2	78.00	75.50

When you inhale, oxygen in the alveoli moves across the moist alveolar membrane into the capillaries. This occurs because the concentration of oxygen is higher inside the alveoli than in the blood in the capillaries. At the same time as the oxygen enters the capillaries, carbon dioxide from the cells moves from the capillaries into the alveoli. This occurs because the concentration of carbon dioxide is higher in the capillaries than in the alveoli. When you exhale, the air leaves the alveoli, passes through the bronchial tree, and is expelled through the nose or mouth. This exhaled air contains more carbon dioxide and less oxygen than the air that first entered the lungs.

The Role of Blood One of the many functions of blood is to transport oxygen and carbon dioxide to and from all the cells in the body. Blood contains a number of different ions, proteins and other organic compounds, and specialized cells. In particular, the **red blood cells** play a major role in gas transport.

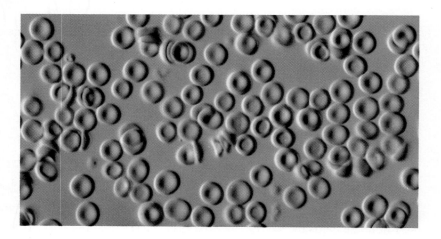

7–26 Red blood cells are specialized to transport oxygen in the blood.

7-27 The hemoglobin molecule consists of four polypeptide chains bonded together. Each chain has an iron atom associated with it. Oxygen binds with the hemoglobin in the lungs and unbinds in the tissues.

Red blood cells contain large amounts of the red pigment **hemoglobin** (HEE-muh-GLŌ-bin). The major function of hemoglobin is to carry oxygen in the blood. The hemoglobin molecule is a complex protein containing four iron atoms. In the lungs, a molecule of oxygen bonds with each of these iron atoms, forming **oxyhemoglobin**. Blood containing oxyhemoglobin is bright red in color. Because of its high oxygen content, it is said to be **oxygenated**.

The transport system carries oxygenated blood to all the cells in the body. As the blood reaches cells that are low in oxygen, oxygen unbonds from hemoglobin. It then moves out of the red blood cells, through the capillaries, and into the cells. Carbon dioxide from the cells moves into the capillaries, which transport it back to the lungs.

Regulation of the Respiratory System

The body is able to control the levels of oxygen and carbon dioxide in the blood. It responds to any change in physical activity by providing the amount of oxygen needed for the new activity. For example, when you begin running, the respiratory system increases the supply of oxygen to meet your increased oxygen needs. When you stop running, your oxygen needs decrease, and the respiratory system decreases the supply of oxygen. The body controls the levels of oxygen and carbon dioxide even when your physical activity remains the same. Whether you are sleeping, sitting, or riding a bicycle, your body maintains the proper levels of oxygen and carbon dioxide in the blood.

7-28 A diver could not stay underwater for more than a few minutes without a source of oxygen.

homos = same
stasis = a standing

Homeostasis The ability of the body to adjust and maintain the levels of oxygen and carbon dioxide is an example of **homeostasis** (HŌ-mee-uh-STAY-sis). Homeostasis is the maintenance of a balanced internal environment.

To better understand homeostasis, picture a car with a device that automatically controls its speed. You set the device to the speed you desire and the device maintains it. As you drive up a hill, the car will tend to slow down. The speed control device detects this and accelerates the car to maintain the set speed. After you pass the top of the hill, the car will tend to pick up speed as you drive down the hill. As before, the speed control device detects the increase in speed and slows the car down to the set speed.

An organism maintains its internal environment in a way similar to the operation of the speed control device. At each level of activity (sitting, walking, running), a certain amount of a particular substance is needed. Structures within the organism detect the amount of the substance in the body and adjust it to the proper level.

In the respiratory system, the human body adjusts the levels of oxygen and carbon dioxide by changing the amount of air inhaled in each breath and by changing the breathing rate. When you rest, for example, you breathe about 10 to 14 times a minute, taking in and giving off about 5 L to 7 L of air. When you exercise strenuously, however, your breathing rate increases to 40 or more times a minute and you take in and give off between 80 L and 120 L of air. This deep, rapid breathing keeps your cells supplied with oxygen and removes the carbon dioxide produced during exercise.

7–29 Regulating the oxygen supply to meet the body's needs is similar to regulating car speed by adjusting the flow of gas to the engine.

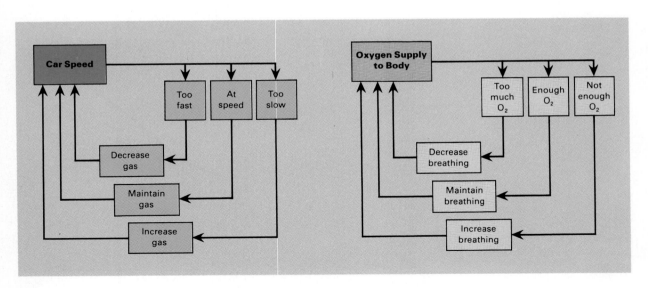

Control of the Breathing Rate The most important thing that controls the breathing rate is the amount of carbon dioxide in the blood. Two types of **respiratory centers** in the brain respond to the level of carbon dioxide in the blood. One type of center controls inhaling; the other controls exhaling. When the respiratory centers detect increases in the level of carbon dioxide, they signal the muscles involved in breathing to work more quickly. This increases the number of breaths per minute and lowers the carbon dioxide level. When the respiratory centers detect decreases in the level of carbon dioxide, they decrease the breathing rate.

The oxygen level in the blood also affects how fast you breathe, but it is not as important as the carbon dioxide level. A low level of oxygen in the blood stimulates centers in certain blood vessels. These centers signal the respiratory centers in the brain, which increase the breathing rate. When the proper level of oxygen is restored, breathing slows down.

As the body changes its level of activity, the respiratory system responds by changing the amount of oxygen it takes in. The respiratory system also responds to changes in the amount of oxygen in the air.

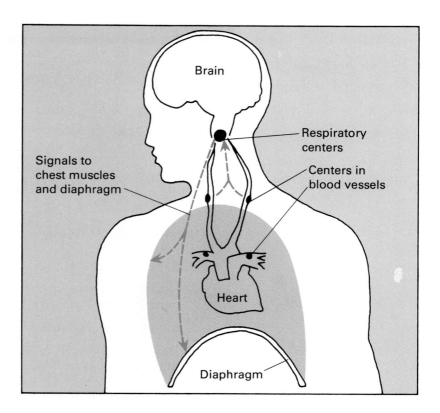

7–30 Breathing rate is mainly controlled by respiratory centers in the brain which monitor the CO_2 level in the blood. The O_2 level is monitored by centers in certain blood vessels and also affects the breathing rate.

7–31 Hikers at high altitudes often carry the air they breathe with them to ensure that they obtain enough oxygen.

The Effects of Environment If you have ever hiked in high mountains, you have probably noticed how much faster and deeper you breathe at higher elevations than at lower elevations. That is because there is less oxygen in the air at higher elevations than at lower elevations. Although you may not be hiking faster or going over rougher ground, you need to take in more air to obtain the same amount of oxygen as at a lower elevation.

At high altitudes, the air is thin and the air pressure is low. At very high altitudes, the air pressure is so low that it does not allow enough oxygen to cross the membranes of the lungs. At 7000 m, the brain does not get enough oxygen to function properly. At 7900 m, a person loses consciousness. At 19 500 m, the air pressure is less than the pressure of the moist air in the lungs. Under such conditions, the water inside the body begins to boil and death results. To avoid this, high altitude jet pilots and astronauts wear special equipment and work in pressurized cabins.

Building a Science Vocabulary

diaphragm	bronchus	hemoglobin
nasal passage	bronchiole	oxyhemoglobin
pharynx	bronchial tree	oxygenated
trachea	alveolus	homeostasis
larynx	red blood cell	respiratory center

Questions for Review

1. Explain the role of the diaphragm in breathing.
2. How does the respiratory system get rid of dust and other particles in the air we inhale?
3. Explain how respiration occurs in the alveoli of the lungs.
4. How does the body respond to an increase in the carbon dioxide level of the blood?

Perspective: Careers
So You'd Like to Work Outdoors?

Not all biologists work in laboratories, dealing with microscopes, preparing experiments, and conducting research. Some biologists have careers that allow them to work outdoors in all kinds of weather.

Biologists who work outdoors perform such diverse activities as wildlife management, teaching, environmental planning, and basic research. To work in these areas a person usually needs four years of college. Some positions as technicians, however, require only two years of college.

Wildlife managers strive to protect and preserve wildlife. Their work may involve breeding hardier stocks of pheasants, rabbits, or other animals, or determining when the size of a deer herd needs to be reduced by making the hunting season longer. Wildlife managers also plan the use of land, monitor the effects of pollution, and develop and manage refuges for animals fleeing natural disasters or the encroachment of people.

Teaching outdoor biology takes many forms. Many public school systems have outdoor programs in which students spend a week or more studying at a special camp. Most summer camps have a nature program. Parks and recreation agencies have interpretive programs that feature slide shows, nature walks, displays, and talks.

The federal governments of Canada and the United States are the two largest single employers of outdoor biologists in North America. They hire many people to work as wildlife managers, teachers, and park rangers. Park rangers in the national park systems give illustrated lectures to campers and are also involved in many other activities. They study the effects of people on parks, protect endangered plants and animals, and plan new and alternative uses of parks. Sometimes park rangers even have to tranquilize bears and transport them away from campsites.

Outdoor biology can be very rewarding. On the other hand, if you don't like getting your hair wet or your shoes soggy, you might be better off as an indoor biologist.

Chapter 7
Summary and Review

Summary

1. In respiration, aerobic organisms: (1) take in oxygen through a moist membrane, (2) transport the oxygen to all parts of their bodies, (3) move carbon dioxide to the moist membrane, and (4) expel carbon dioxide through the membrane.

2. Oxygen moves into a cell because there is more oxygen outside the cell than inside it. Carbon dioxide moves out of a cell because there is more carbon dioxide inside the cell than outside it. These gases also move within the cell by the same principle.

3. Unicellular organisms exchange gases directly with the water in which they live. Most large multicellular organisms have respiratory systems that carry out respiration.

4. The bean plant exchanges gases with the atmosphere through stomates in the leaves, lenticels in the stems, and the epidermis of the roots. Gases move through air spaces to all the cells of the plant.

5. Respiration in the paramecium involves the movement of gases between it and the water in which it lives.

6. Respiration in the hydra occurs when its cells exchange gases with the water around it and in its digestive cavity.

7. Respiration in the earthworm involves the exchange of gases between its moist skin and air spaces in the soil. Capillaries transport oxygen to the cells and carbon dioxide to the skin.

8. Gases move into and out of the grasshopper through spiracles and are transported in tracheas to and from the cells. The grasshopper breathes by moving its body wall.

9. In humans, breathing involves changing the size of the chest cavity by the action of the diaphragm and other muscles. Air moves into and out of the lungs through the nose or mouth, trachea, larynx, and bronchial tree.

10. Gases are exchanged in the alveoli of the lungs. Oxygen enters capillaries around the alveoli and combines with hemoglobin in red blood cells. The blood carries oxygen to body cells and picks up carbon dioxide, which it returns to the lungs.

11. Mechanisms in the body adjust the intake of oxygen to meet the body's needs by changing the breathing rate and the amount of air inhaled in each breath.

12. Respiratory centers in the brain maintain the proper level of carbon dioxide in the blood by adjusting the breathing rate.

13. The air pressure is low at high elevations, making it difficult to obtain enough oxygen.

Review Questions

Application

1. Explain why cellular respiration depends on respiration.

2. Would an earthworm live if it were put into an aerated fish tank with water? Why or why not?

3. How is the grasshopper adapted to prevent water loss by evaporation? Why is this adaptation important to respiration?

4. How is the grasshopper's respiratory system better adapted for life on land than the earthworm's?

5. How does respiration occur in a cell in your toe?

6. A person breathes faster at high altitudes than at low altitudes. What signal do you think stimulates this increase in the breathing rate? Explain your answer.

7. Spirogyra is an aquatic alga that is composed of a filament of cells. How does spirogyra perform respiration? (Remember the four steps of the general plan.)

8. What is the most important difference between the way oxygen reaches the cells of a grasshopper and the way it reaches the cells of a bear? (A bear has the same type of respiratory system as a human.)

Interpretation

1. What is the source of the carbon dioxide that is given off during respiration?

2. In what process does an organism use the oxygen it takes in during respiration?

3. Oxygen and carbon dioxide move across the cell membrane during respiration. How does the structure of the cell membrane allow this to occur?

Extra Research

1. Make a model of the respiratory system of one or more of the representative organisms. Use arrows to show the movement of oxygen and carbon dioxide.

2. Make a working model of the human respiratory system like the one pictured. Use a heavy plastic jar with the bottom removed, a Y-shaped tube, a one-holed stopper, a piece of rubber sheeting, two balloons, and a piece of cord. The rubber sheeting acts as the human diaphragm. Show how pulling the rubber sheeting increases the size of the cavity and causes the balloons to fill with air.

3. Design and carry out an experiment that tests the effects of exercise on the number of breaths a person takes each minute. Record your data.

4. Test the idea that earthworms can survive under water. Place earthworms in a fish tank that has an aeration system.

Career Opportunities

Inhalation therapists care for people whose breathing is temporarily limited due to illness or injury. They also instruct patients in the use of special equipment, such as respirators and breathing machines. For information, write the Canadian Society of Respiratory Technologists, 504 Main St., Winnipeg, Manitoba R3B 1B8.

Paramedics often work as ambulance drivers and attendants, providing emergency medical care to accident victims. They must be able to think and act quickly and carefully in life-and-death situations. To become a paramedic, a person must first complete an emergency medical technician training course. For more information, write the Canadian Hospital Association, 410 Laurier Ave. W., Ottawa, Ontario K1R 7T6.

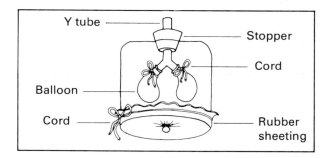

Chapter 8

Food Processing

Focus "In a zoo at midnight a good-sized live pig was pushed inside the door of the bachelor apartment of the python. The pig was nervous. After the conventional preliminaries that began with squeezing the breath of life out of the pig, the python dislocated his jaw, which is part of the ritual for any of his better meals, and in went the pig, slowly. Whereupon at his upper end the python bulged. Next morning he bulged less and farther down. The pig was melting. Life was making life out of life as the pig advanced through the great snake's digestive tube."

Gustav Eckstein

8.1 GETTING AND USING FOOD: A GENERAL PLAN

If you listed in order of importance the things that are necessary for your life, food would probably appear high on your list. As you know, all living things require a constant supply of food for energy, growth, and repair. Producers make their own food by photosynthesis, but all other organisms (consumers) must obtain food from their surroundings. After obtaining food, consumers change it into a form that their cells can use and distribute it to all parts of their bodies.

8–1 Food is necessary for all living things. Here a cheetah drags its meal into the shade to eat.

8–2 The abdomen of a female mosquito swells as she sucks blood from a person's arm.

Consumers have different ways of using food, but there are usually five steps in the process. They are:

1. Obtaining food
2. Breaking down food into nutrients
3. Absorbing and delivering nutrients to the body
4. Getting rid of food wastes
5. Using and storing nutrients

This general plan can be applied to most consumer organisms. Consider the mosquito as an example. After landing on a person's arm, a female mosquito 1) bites the person, 2) digests the person's blood in its digestive system, and 3) absorbs the digested nutrients into its bloodstream, which carries the nutrients to its cells. The mosquito then 4) eliminates the food wastes from its digestive system and 5) either uses the nutrients the cells receive from the blood or stores them.

Food Acquisition

in- = into
gerere = to carry

After locating food in its environment, a consumer takes food into its body by a process called **ingestion**. During ingestion, food is taken into a **food cavity** in the consumer's body. Your food cavity is a tubelike **digestive tract**. An amoeba's food cavity is a pouchlike **food vacuole** that forms around food after the amoeba's pseudopods surround it.

Most food cavities are either pouches or tubes.

8–3 An amoeba is about to engulf a paramecium (bottom) by extending pseudopods in a circle around the sides of the protist. As it engulfs the paramecium, the amoeba forms a food vacuole around it. Another paramecium is visible at top right.

Organisms that filter water for food are called filter feeders. They include whales, which eat tiny crustaceans called krill.

Consumers obtain food in several ways. Often, they do not actively capture food. Many decomposers, such as fungi, actually live and grow directly on their food. A bracket fungus, for example, lives on dead branches. It produces enzymes that break down, or decay, the branches, and then it absorbs the nutrients. Some animals, such as the tapeworm, obtain food from the organisms in which they live. Tapeworms live in the digestive systems of dogs and other mammals. They have no mouth or digestive system themselves; food molecules from the host's digestive tract enter the tapeworm directly through its body wall.

Other animals depend on a food supply that comes to them. Clams, scallops, and oysters filter the surrounding water to obtain microscopic bits of food. They do this by moving water across their fine, netlike gills. The food bits are caught in mucus and are moved into the mollusk's mouth. Many aquatic insects also obtain food by filtering water.

Many animals actively search for food by moving from place to place. Such animals have special adaptations that help them find and capture food. For example, jellyfish swim around and use their tentacles to seize small invertebrates. Stingers in the tentacles shoot a paralyzing poison into their prey. The disabled animals are ingested as the tentacles direct them toward the jellyfish's mouth.

a

c

b

8-4 Food-getting: **a.** A lichen absorbs water and minerals from the tree on which it lives. **b.** Filter-feeding mosquito larvae sweep tiny aquatic plants and animals into their mouths. **c.** This jellyfish paralyzes its prey with stinging tentacles.

Section 8.1 Getting and Using Food: A General Plan

8–5 A European bee eater feasts on a dragonfly, another favorite food.

Producing a poison helps jellyfish and other animals capture food. Yet these poison-producing animals also serve as food for others. For example, the bee eater is a bird that feeds on poisonous bees. After it captures a bee, it returns to its branch in a tree. Before eating the bee, the bee eater rubs it against the branch to squeeze out the poison. Good eyesight, a sharp beak, and poison-removing behavior adapt the bee eater for getting its food.

Keen senses help many animals locate food. Otters cannot detect a motionless fish, but once it stirs, the fish has little chance of escape. Sharks are aided by their sense of smell when tracking smaller fish, which are their prey.

Sharks use their efficient teeth and jaws to tear and cut their prey into pieces that can be swallowed.

Food Breakdown into Nutrients

The ultimate destination of the food that an organism ingests is the organism's cells. Pieces of ingested food, however, are usually too big to cross membranes and enter cells. Therefore, most organisms must break down their food. This process is called **digestion.** Big chunks of food are broken down into small pieces by **mechanical digestion**, and large food molecules are broken down into smaller ones by **chemical digestion**.

di- = apart

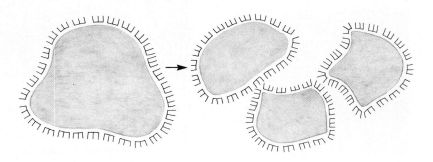

8–6 The mechanical digestion of food into small pieces exposes more surface area for enzymes to attack the food molecules.

Chapter 8 Food Processing

The reason for including gravel in bird cages is to provide the necessary particles for the birds to use in their gizzards.

Snakes (and other organisms that swallow whole prey) depend solely on chemical digestion.

8–7 Chemical digestion of starch. Amylase hydrolyzes bonds between every other pair of glucose units, forming maltose. Maltose is broken down into glucose by maltase.

Mechanical digestion involves grinding, mixing, and mashing food. In many animals this is done with the teeth, tongue, and stomach muscles. Chickens and many other birds carry out mechanical digestion in their **gizzards**. This muscular portion of the digestive tract often contains bits of sand which help grind the food.

In chemical digestion, enzymes break apart large food molecules by hydrolysis. During hydrolysis water is "added" at the site where a bond is broken (see Chapter 4). Different enzymes are used in the breakdown of different food molecules. Carbohydrates are broken down into simple sugars by a variety of enzymes. Two enzymes used in carbohydrate digestion are **amylase** (AM-ih-lays) and **maltase**. Amylase breaks starch into maltose molecules. Then maltase breaks maltose into glucose. Proteins are hydrolyzed into amino acids by a class of enzymes called **proteases**. Similarly, lipids are broken down into fatty acids and glycerol by **lipases**, and nucleic acids are broken down into nitrogen bases and simple sugars by **nucleases**.

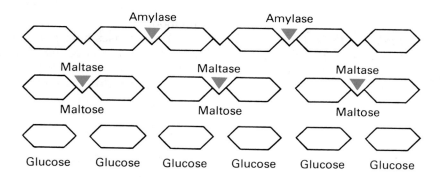

Table 8–1 Chemical Digestion of Food

Food	Enzyme	End Product of Digestion
Carbohydrates*		
Starch	Amylase	Maltose
Maltose	Maltase	Glucose
Proteins	Proteases	Amino acids
Lipids		
(Fats and oils)	Lipases	Fatty acids and glycerol
Nucleic acids	Nucleases	Nitrogen bases and simple sugars

*Carbohydrates not listed are also digested by specific enzymes.

8–8 In intracellular digestion (left), enzymes break down food particles in food vacuoles in the cell. In extracellular digestion (right), digestion takes place in a digestive cavity or tract.

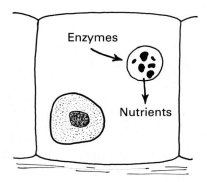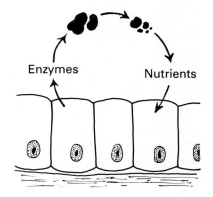

intra- = within
cella = cell

extra- = outside

ab- = from
sorbere = to drink in

Chemical digestion that takes place within a cell is called **intracellular digestion**. In the amoeba, for example, enzymes produced by the cell enter the food vacuole and digest the food contained within it. Intracellular digestion is used, at least in part, to break down ingested food in protozoa, sponges, coelenterates, and flatworms. It also accounts for the chemical breakdown of stored food present in the cells of most organisms, including plants.

In most animals, chemical digestion takes place in a digestive tract that is lined by cells. This type of chemical digestion is called **extracellular digestion**. Enzymes produced by cells and special organs in the digestive system are secreted into the digestive tract, where they digest the food.

Nutrient Absorption and Delivery to Cells

After food is digested, it must be distributed to all parts of the organism's body. The first step in this process is the movement of nutrients out of the food cavity and into the cytoplasm of cells. This process is called **absorption**.

In the amoeba and other one-celled consumers, nutrient absorption occurs when the end products of digestion move across the food vacuole membrane into the cytoplasm. The food vacuole becomes smaller and smaller as absorption occurs. Food vacuoles move throughout the amoeba, delivering nutrients to all parts of the cell.

In multicellular organisms with a digestive tract, nutrient absorption begins when food molecules move into cells in the wall of the digestive tract. The nutrients are then transported, usually by a bloodstream, to the cells of the body. Nutrient absorption continues as food molecules leave the blood and enter body cells.

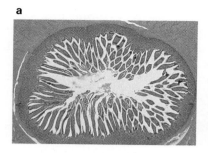

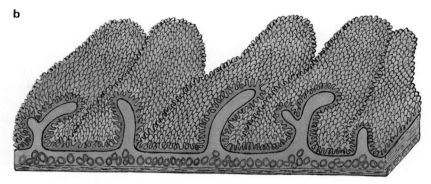

8–9 Intestinal folding: **a.** cross section of small intestine; **b.** illustration of villi on folded surface.

Nutrient absorption in the digestive tract occurs quickly if the digestive tract has a large surface area. A large surface area allows digested food to be absorbed into many cells at once. Most animals have a digestive tract that is curved and folded, presenting a large surface for nutrient absorption. Mammals also have **villi** (VIL-lī), which are microscopic, fingerlike projections on some of the cells lining the digestive tract. The villi increase the surface area for absorption hundreds of times.

Removal of Food Wastes

Organisms rarely digest all parts of their food. Usually, a portion of the ingested food cannot be made usable. Most vertebrates, for example, lack the enzymes to digest cellulose. Cellulose remains in the digestive system of vertebrates when they eat plants. Organisms rid themselves of undigested material, such as cellulose, by the process of **egestion** (ee-JES-chun).

e- = out
gerere = to carry

Egestion is accomplished simply in the amoeba. The food vacuole gets smaller and smaller as digestion and absorption proceed. Finally, only the undigested material remains. The food vacuole moves to the cell surface, and the waste is passed to the outside. In animals with a digestive tract, contractions of the muscles in the wall of the tract force the undigested waste material out of the digestive tract through the **anus** (AY-nus).

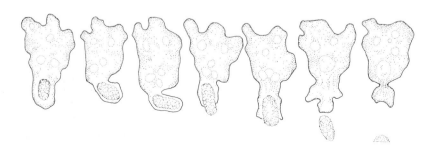

8–10 Egestion in an ameba. Undigested material is pushed to one end of the cell. The cell membrane draws away from the pellet until it is released.

Section 8.1 Getting and Using Food: A General Plan

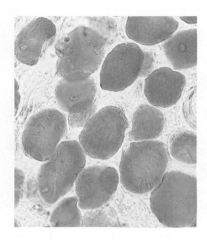

8–11 Excess fats are stored in large droplets in cells.

Nutrient Use and Storage

Nutrients that have been absorbed into the cytoplasm of cells can be either used or stored. Simple sugars and fatty acids in the cytoplasm can undergo cellular respiration, a process that yields ATP. Some ATP is used as amino acids are transported to the ribosomes and are built up into proteins and enzymes. More ATP is used as fatty acids and glycerol are combined to form parts of the cell's membranes.

Excess nutrients are changed to a form in which they can be stored for later use. In vertebrates, for example, glucose is stored as glycogen in the liver and in muscle, while fat is stored in fat tissue. When a cell needs food, intracellular digestion breaks down stored food molecules, making them available for use.

Building a Science Vocabulary

ingestion	gizzard	intracellular digestion
food cavity	amylase	extracellular digestion
digestive tract	maltase	absorption
food vacuole	protease	villus
digestion	lipase	egestion
mechanical digestion	nuclease	anus
chemical digestion		

Questions for Review

1. List the five steps consumers use to process food.
2. Name an organism that does not actively search for food and one that does. Describe ingestion in each.
3. State one similarity and one difference between chemical digestion and mechanical digestion.
4. What happens to the amino acids that a cell absorbs?
5. State whether each of the following is an example of extracellular digestion or intracellular digestion.
   a. Enzymes in a shark's digestive tract break down oil molecules in a fish the shark has eaten.
   b. A protozoan changes cellulose into glucose.
   c. A mosquito digests blood in its digestive tract.

8.2 FOOD PROCESSING IN THE REPRESENTATIVE ORGANISMS

Food Processing in the Paramecium

Digestion in the paramecium is intracellular.

Although it is only a single cell, the paramecium has highly specialized organelles that help it obtain and digest food. The beating of small hairlike cilia that cover its surface sets up a food-gathering current. The cilia sweep bacteria, protozoa, and other food particles into the **oral groove**, which is a channel located on the side of the paramecium. The cilia lining the oral groove push food along to the mouth where a food vacuole forms.

Once food is trapped in the food vacuole, ingestion is complete, and the process of intracellular digestion begins. Digestive enzymes from a lysosome inside the paramecium are released into the vacuole. The enzymes break down the large food particles into small nutrient molecules, which move across the membrane of the vacuole into the cell. The food vacuole moves around inside the parmecium, distributing the end products of digestion to all parts of the cell.

Eventually, the food vacuole reaches a tiny, specialized opening in the cell membrane called the **anal pore** (see Fig. 8–12). The membrane of the food vacuole bursts, egesting the undigested particles of food from the cell through the anal pore.

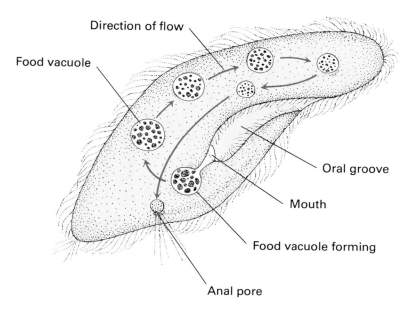

8–12 Paramecium digestion. A food vacuole forms around ingested food. Digestion takes place in the vacuole. Nutrient molecules pass through the vacuole membrane and are distributed throughout the cell as the vacuole moves. Wastes are egested through the anal pore.

Food Processing in the Hydra

Digestion is both intracellular and extracellular in the hydra.

While feeding, hydras usually remain fixed to rocks or plants in the stream or pond where they live. Sometimes, however, they do eat while floating freely. Hydras and other coelenterates use tentacles to capture their food, which includes small crustaceans and worms. The tentacles are equipped with specialized stingers, called **nematocysts** (neh-MA-tō-sists), which shoot poisonous threads into their prey. The hydra's tentacles then draw the immobilized animal to the mouth. From there the food enters the pouchlike digestive cavity, which has only one opening. In all coelenterates, food enters the digestive cavity and waste products are egested through the mouth.

Large pieces of food are broken down by extracellular digestion. Cells lining the digestive cavity secrete digestive enzymes into it. When the hydra moves its body, the food and enzymes in the cavity mix together. This mixing is aided by the movement of the flagella of the cells lining the cavity. The enzymes break down the food into smaller particles; however, the food is not completely broken down in the digestive cavity. The smaller particles of food are engulfed by some of the cells lining the cavity and digestion continues in food vacuoles inside these cells.

Digested nutrients pass from the vacuoles into the cells lining the digestive cavity. Some nutrients then move from these cells to those on the outside of the body. Meanwhile, undigested materials in the food cavity are egested through the mouth. The hydra sometimes loses some food in the outgoing flow.

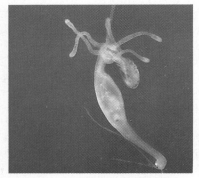

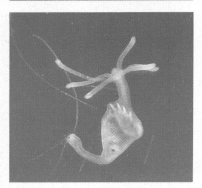

8–13 When a hydra feeds, it first paralyzes its prey with its nematocysts and then draws the prey into its mouth.

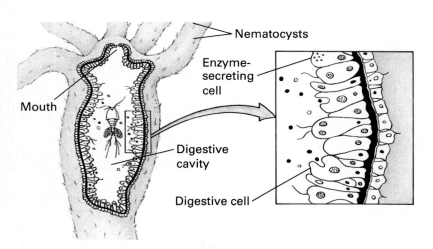

8–14 Digestion begins in the hydra's digestive cavity and is completed in digestive cells lining the cavity.

Chapter 8 Food Processing

Food Processing in the Earthworm

The earthworm's digestive cavity is a tube with two openings. At one end of the tube is the mouth. At the other end is the anus. This type of "tube-within-a-tube" digestive system is found in many kinds of animals. The outer tube is the animal's body. The inner tube is the digestive tube, also called the digestive tract.

Advantages of a Digestive Tract The earthworm's digestive tract has several advantages over the pouchlike digestive cavity of the hydra. Different regions of a digestive tube can be specialized to carry out different functions. This specialization results in greater efficiency, because the digestive tract can process food in an assembly-line fashion.

The earthworm moves food through its body using the muscles in the wall of its digestive tract. The muscles produce wavelike contractions called **peristalsis** (PAYR-ih-STOL-sis). As peristalsis moves food through the digestive tract, mechanical and chemical digestion take place. The end products of digestion are absorbed at various points along the length of the digestive tract. Any undigested wastes that reach the anus are egested.

Specialization in Earthworm Digestion Figure 8–16 shows the specialized parts of the earthworm's digestive tract. Each part has a specific function in the digestive process. The pharynx is just inside the mouth. The muscles of the pharynx create a sucking action that pulls food into the earthworm's digestive tract. Earthworms ingest materials such as decayed bits of plants and animals, along with soil and grains of sand. Peristalsis pushes this ingested material along the digestive tube. From the pharynx, the food moves through the narrow **esophagus** (ih-SAHF-uh-gus) to the thin-walled **crop**. The crop is a food storage organ found in many animals. It allows animals to take in food when it is available and to use it gradually.

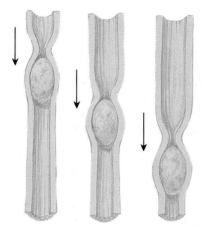

8–15 Peristalsis. Food is pushed down a digestive tract by waves of muscular contraction.

peri- = around
stallein = to place

Digestion in the earthworm is extracellular and occurs in a digestive tract.

oisein = to carry
phagein = to eat

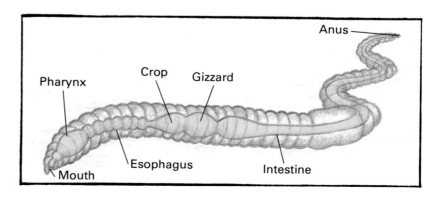

8–16 An earthworm's digestive tract is essentially one tube that stretches the entire length of its body. Specialized segments of the tract perform different functions.

Section 8.2 Food Processing in the Representative Organisms

Food leaves the crop and enters the gizzard, which has muscular walls and a protective lining. As the muscles of the gizzard contract, soil particles help to grind the food. This is a form of mechanical digestion. Food then leaves the gizzard and enters the **intestine**, where chemical digestion takes place. Some cells in the wall of the intestine secrete enzymes onto the food. Other cells lining the intestine absorb the end products of digestion and pass them to the bloodstream. Blood transports the nutrients to cells throughout the earthworm's body.

If you examine the inside of the earthworm's intestine, you will notice that the intestinal wall has a deep fold along the top. This fold dips into the digestive cavity and provides extra surface area, which allows for greater absorption.

Bits of sand, rock, and soil are egested through the anus, along with any undigested food. The material the earthworm egests is called a **casting**. Castings fertilize the soil and are therefore beneficial to plant growth.

The burrowing of the earthworm aerates the soil.

Food Processing in the Grasshopper

The digestive tract of the grasshopper follows the tube-within-a-tube design. Grasshoppers eat leafy plants. A grasshopper grasps food in its forelegs and puts it into its mouth. The mouth has special lips for holding food and powerful jaws for chewing. Taste organs in the mouth help a grasshopper choose its food. **Saliva** enters the mouth and moistens the food. It also contains enzymes that begin chemical digestion.

Chewed food from the mouth passes into the crop where it can be stored. Small quantities of food are released from the crop to a muscular gizzard, which is lined with a set of sharp teeth. A valve at the end of the gizzard closes, trapping food for a thorough grinding. The valve opens when grinding is completed, and the food enters the **stomach**. Six pairs of pouches are attached to the stomach. These pouches secrete digestive enzymes and aid in absorption. Food is digested in the stomach by the enzymes and absorbed by cells lining the stomach and the pouches.

The waste products of digestion enter the intestine. In the **rectum**, at the end of the intestine, water is absorbed from the undigested food and returned to the bloodstream. The remaining wastes are egested through the anus. The nutrients that were absorbed by the stomach pouches enter the blood and are transported to all the grasshopper's cells.

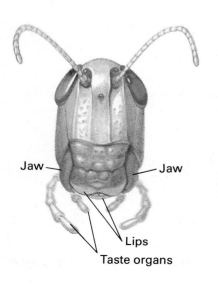

8–17 Head of a grasshopper, showing the mouthparts.

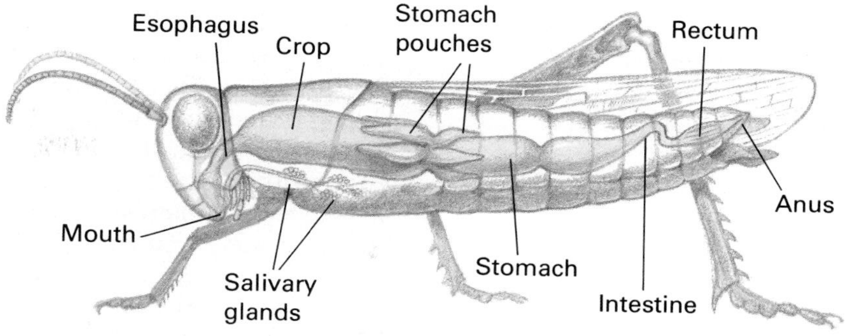

8–18 Digestive system of the grasshopper. Between the mouth and the anus are specialized digestive organs.

Building a Science Vocabulary

oral groove
anal pore
nematocyst
peristalsis

esophagus
crop
intestine
casting

saliva
stomach
rectum

Questions for Review

1. Select the organism from Column B that is best described by each of the statements in Column A.

    Column A
    a. It has well-developed mouth parts.
    b. Its digestive cavity is a tube-within-a-tube.
    c. It egests food through a mouth.
    d. It moves food by peristalsis.
    e. Intracellular digestion is its only method of food breakdown.

    Column B
    hydra
    paramecium
    grasshopper

2. State the function of each of the following.
    a. crop in the earthworm
    b. oral groove in the paramecium
    c. folds in the earthworm's intestine
    d. gizzard in the grasshopper

3. What are two advantages an earthworm's digestive tract has over a hydra's digestive tract?

Section 8.2 Food Processing in the Representative Organisms

8.3 FOOD PROCESSING IN HUMANS

Mammals and most other vertebrates have a digestive system similar to that of humans.

Like the earthworm and the grasshopper, humans have a tube-like digestive tract (see Fig. 8–20). The digestive tract has six major parts: the mouth, esophagus, stomach, small intestine, large intestine, and anus. In addition, the digestive system includes other structures called **digestive glands** (see Table 8–2). The digestive glands produce secretions that contain enzymes that help digest foods.

Digestion in the Mouth

Mechanical and chemical digestion in humans begin as soon as food enters the mouth. The teeth and tongue break up food and mix it with saliva from the **salivary glands**. Saliva is a digestive fluid containing water, enzymes, and mucus. The water in saliva moistens and dissolves particles of food, thereby aiding chemical digestion and the ability to taste. The enzyme amylase in saliva begins the digestion of starch. Even while food is still in the mouth, some of the starch you eat is broken down into maltose. The mucus in saliva helps make chewed food smooth and easy to swallow.

The breathing reflex stops automatically when you start to swallow.

A swallowed ball of moistened, chewed food passes through the pharynx at the back of the mouth, where a small flap of tissue, known as the **epiglottis** (EP-uh-GLOT-is), sticks out. The epiglottis directs food to the esophagus and keeps it out of the air passage of the trachea. Peristalsis moves food through the esophagus to the stomach.

Table 8–2 Digestive Glands in the Human

Gland	Location in Body	Site of Release of Secretion
Salivary glands	Below and in front of ear, below jaw, and in mouth	Mouth
Gastric glands	Wall of stomach	Stomach
Liver	Right side of abdominal cavity	Small intestine
Pancreas	Near small intestine	Small intestine
Intestinal glands	Wall of small intestine	Small intestine

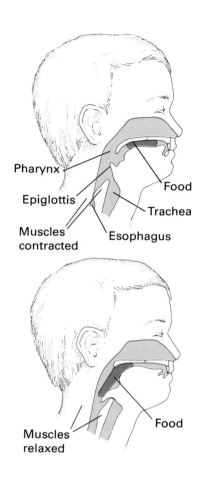

8-19 Swallowing. A ball of chewed food, moistened by saliva, is directed into the esophagus by the epiglottis. Muscles at the upper esophagus relax, allowing food to enter.

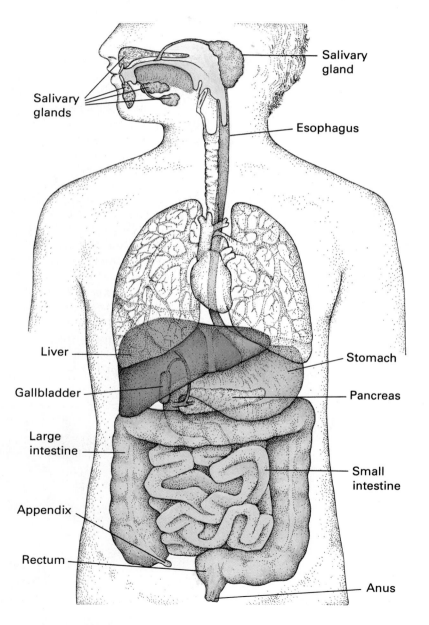

8-20 Human digestive system, including major glands.

Section 8.3 Food Processing in Humans

Digestion in the Stomach

It takes about seven seconds for swallowed food to reach the stomach. Waves of peristalsis, produced by muscles in the stomach wall, churn and break up the food. **Gastric glands** in the stomach wall secrete **gastric juice**, which mixes with the food and promotes chemical digestion.

Three of the important substances in gastric juice are proteases, hydrochloric acid, and mucus. Proteases are secreted into the stomach in an inactive form. They become activated in an acidic environment. The hydrochloric acid in the gastric juice lowers the pH of the stomach contents and activates the proteases. This enables them to begin protein digestion. The mucus in gastric juice coats the food and makes it smooth.

The hydrochloric acid in the stomach is strong enough to injure cells, but the membranes of the cells lining the stomach are adapted to withstand it. Occasionally, however, the natural barrier breaks down and a sore or inflammation called an **ulcer** develops. Stomach ulcers can be quite painful and slow to heal, because the acidic gastric juice irritates them. Stomach ulcers may be treated with surgery, special diets, or medicines that neutralize stomach acid.

In addition to beginning protein digestion, an important function of the stomach is to store food. This allows people to live on three meals a day or fewer instead of eating continually. Sometimes a person's stomach must be removed due to disease. Then that person must eat many small meals each day.

Digestion in the stomach can take up to four hours. During this time, mucus, acid, and proteases mix with the food until it is soft and amost liquid. Gradually, the partially digested food moves into the small intestine through a valvelike muscle called the **pylorus** (pī-LOR-us).

Protein digestion begins in the stomach.

Aspirin penetrates stomach cell membranes, frequently causing bleeding. When aspirin and alcohol are taken together, the aspirin enters stomach cells more quickly. In combination, aspirin and alcohol can cause serious damage to the stomach lining.

pylōros = gatekeeper

8–21 Stomach ulcer. Cells in the stomach lining are continually irritated by the stomach's acidic secretions. Mucus balls have formed around the ulcer.

Digestion in the Small Intestine

Food moves into and along the small intestine by peristalsis. During its journey, food is mixed with the secretions of digestive glands. The digestion that began in the mouth and stomach is completed in the small intestine. There, the nutrients released by digestion are also absorbed into the bloodstream.

The small intestine is called "small" because of its narrow diameter—not because of its length. In fact, in the average adult, the small intestine is about 300 cm long, whereas the large intestine is about one third that length. The coils in the small intestine allow it to fit in a relatively small space. The inner surface of the small intestine also has many folds and villi. The coils, folds, and villi give the small intestine a large surface area, which aids digestion and absorption.

Three types of digestive secretions enter the small intestine: **bile**, **pancreatic** (PAN-kree-AT-ic) **juice**, and **intestinal juice**. Bile and pancreatic juice enter near the pylorus; intestinal juice is secreted along the entire length of the small intestine.

Bile is a secretion of the **liver**, the body's largest gland. As bile is produced, it is stored in a sac called the **gallbladder**. Bile contains **bile salts**, water, and other substances dissolved in a basic solution. The bile salts break up fats and oils into small globules that can dissolve in water. Bile salts are not enzymes, but their action prepares lipids for chemical breakdown by the lipases in pancreatic juice. Bile salts also help prepare lipids for absorption. The basic solution that contains the bile salts is also important. It changes the pH of the food mixture from acidic to slightly basic.

The secretions of the **pancreas** contain several different enzymes that work best in a basic solution. Proteases continue the digestion of proteins that began in the stomach. A lipase completes the digestion of lipids to fatty acids and glycerol. Amylase changes any starch not digested in the mouth to maltose, and nucleases begin the digestion of nucleic acids.

The surface area of the small intestine is approximately 300 m², about the size of one and one-half tennis courts.

The digestive tract in animals that eat meat is about four times their body length. In animals that eat plants, it is about 20 times their body length.

Most of the digestive enzymes in the small intestine work best in basic conditions (pH value above 7).

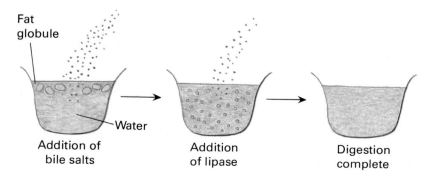

8–22 Digestion of fats and oils. Bile salts, secreted by the liver, help break up fats and oils so that they can dissolve in water.

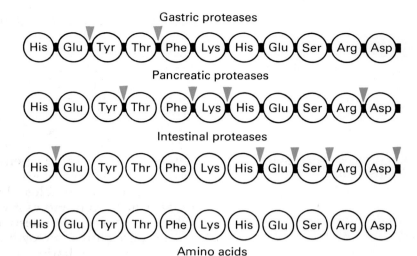

8–23 Digestion of protein. Proteases secreted in the stomach begin the breakdown of proteins into amino acids. Pancreatic and intestinal proteases complete the process in the small intestine.

Chemical digestion is completed in the small intestine.

The **intestinal glands** are located in the wall of the small intestine throughout its length. Intestinal juice also contains several enzymes that complete the digestion of proteins, carbohydrates, and nucleic acids. This yields a souplike mixture of nutrients including simple sugars, amino acids, glycerol, fatty acids, vitamins, and water (see Table 8–3).

Table 8–3 Chemical Digestion in the Human Digestive Tract

Location	Digestive Secretion	Type of Enzyme	Action
Mouth	Saliva	Amylase	Begins digestion of starch to maltose
Stomach	Gastric juice	Protease	Begins digestion of proteins
Small intestine	Bile (bile salts)	(no enzymes)	Breaks up lipid globules
	Pancreatic juice	Amylase	Completes digestion of starch to maltose
		Protease	Continues digestion of protein
		Lipase	Completes digestion of lipids to fatty acids and glycerol
		Nuclease	Begins digestion of nucleic acids
	Intestinal juice	Protease	Completes digestion of proteins to amino acids
		Carbohydrate-enzymes	Completes digestion of carbohydrates to simple sugars
		Nuclease	Completes digestion of nucleic acids to sugars, phosphates, and nitrogen bases

Absorption in the Small Intestine

The small intestine loses and replaces about 17 billion cells each day.

The small intestine is the main organ of absorption. Nutrients released by digestion are absorbed through cells lining the small intestine (see Table 8–4). At the same time, water and other useful materials from the digestive juices are absorbed for re-use in the body. Blood vessels in the wall of the intestine collect these substances for transport to body cells.

The coils, folds, and villi of the small intestine create a large surface area for the absorption of nutrients. In addition, the cells lining the intestine have surface folds called **microvilli**. The microvilli and other folds allow the intestine to absorb 600 times as many nutrients as it would if it were a straight tube.

A microvillus is about 1 μm long.

Nutrients from the intestine are absorbed into the transport system through the microvilli. Each villus contains two blood vessels, many capillaries, and a single **lymph** (LIMF) **vessel**. (Lymph vessels are part of a system of tubes that return fluids from the tissues to the blood.) The fatty acids from digested fats or oils enter the lymph vessels in the villi. Other nutrients (simple sugars, amino acids, most vitamins, and minerals) are absorbed into the capillaries. Nutrients are transported throughout the body in the capillaries and other blood vessels.

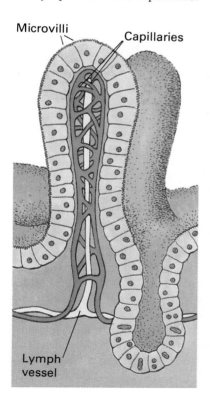

8–24 Section through a villus. Nutrients are absorbed into the lymph vessel and capillaries.

Table 8–4   Absorption in the Digestive Tract

Substance Absorbed	Source	Where Absorption Occurs
Simple sugars	Carbohydrates Nucleic acids	Small intestine
Amino acids	Proteins	Small intestine
Fatty acids and glycerol	Lipids	Small intestine
Nitrogen bases	Nucleic acids	Small intestine
Bile salts	Bile	Small intestine
Vitamins	Ingested food and colon bacteria	Large intestine
Minerals and ions (calcium, iron, Na^+, H^+, Cl^-, SO_4^{2+})	Ingested food and digestive secretions	Small intestine and large intestine
Water	Ingested food and digestive secretions	Mostly in large intestine

Section 8.3 Food Processing in Humans

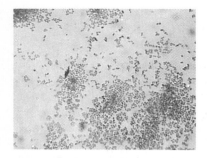

8–25 *Escherichia coli* is the most numerous colon bacterium.

Absorption in the Large Intestine

Food enters the large intestine about 19 hours after it is eaten. The two major functions of the large intestine, also called the **colon** (KŌ-lun), are absorption and egestion. The inner wall of the colon is adapted for absorption with folds, however, it does not have villi as the small intestine does. Water, ions, and some vitamins are absorbed in the colon. These materials then enter the blood. The colon's absorption of water is particularly important, because water is used by cells during life processes. If we egested much water with our undigested food, the body would dry out and we would need to drink water more often.

The large intestine contains an enormous number of bacteria, which live and grow on nutrients we are unable to digest. Usually these bacteria are harmless. In fact, certain colon bacteria produce important vitamins, such as **folic** (FŌ-lik) **acid** and vitamin K. Some vitamins made by the colon bacteria are absorbed along with the water and ions.

a 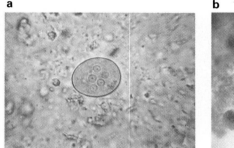 b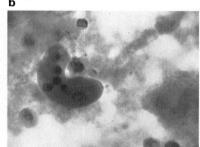

8–26 Intestinal bacteria: **a.** *Entamoeba coli*, shown encased in a cyst, is a harmless inhabitant of the large intestine. **b.** *Entamoeba histolytica* causes amebic dysentery. Note the ingested red blood cells.

Egestion of Wastes

About 24 hours after food is ingested, the first waste products, or **feces** (FEE-seez), may be egested from the anus by peristalsis. The feces are made up of undigested food (mostly cellulose plant fibers), some inorganic material, waste products from the bile, water, and bacteria. About 30 percent of the feces are bacteria. Most of the waste products from a given meal are removed during the next 48 hours.

Cellulose, often referred to as dietary fiber, helps activate egestion from the colon.

Of course the time in digestion varies with factors such as meal size, type of food, and state of health. Biologists have timed the movement of food through the digestive system. In one test, they asked people to swallow small colored beads with a meal. They found that 70 percent of the beads were recovered in the undigested wastes in 72 hours. In took more than a week, however, to recover all the beads.

Regulation of the Digestive System

The body provides a constant supply of food to all of its cells by controlling the digestion of food. For example, the control of muscles that cause peristalsis ensures that food passes through at the proper rate. Controls on the digestive glands ensure that the proper kinds and amounts of enzymes are secreted at the appropriate times.

Methods of Control The body has two ways of controlling digestion. Nerves provide one kind of control. Substances called **hormones** provide the other. Hormones are chemicals carried in the blood that affect specific target organs in the body. A target organ may respond to a hormone by increasing or decreasing its activity. A digestive gland, for example, can be "turned on" or "turned off" by a particular hormone in the blood.

Control of Saliva Production Nerves control the secretion of saliva by the salivary glands. As soon as you put food in your mouth, a signal travels on a nerve to your brain. The brain's immediate response is to send nerve signals to the salivary glands, stimulating them to produce saliva. Your brain responds the same way when you see or smell food. In fact, your brain even sends a nerve message to the salivary glands when you think about food.

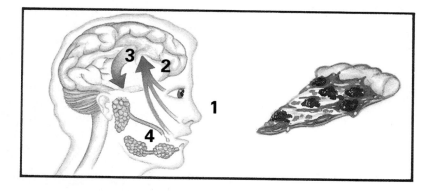

8–27 Regulation of the salivary glands. Sense organs see, smell, taste, or feel food (**1**) and send nerve signals to the brain (**2**). The brain signals the salivary glands (**3**) to release saliva (**4**).

Control of Stomach Activity The stomach is controlled by both nerves and hormones. The brain sends signals to your stomach when you taste, see, smell, or think about food. The signals stimulate the muscles in the stomach wall to begin peristalsis, and they stimulate the gastric glands to produce gastric juice. This helps prepare the stomach for its job of food breakdown. Food reaching the stomach stretches the stomach wall, thereby setting off a nerve signal to the brain. The brain

8-28 Regulation of stomach activity involves both nerves and hormones.

1. Thought or sensation of food begins peristalsis and gastric juice secretion.
2. Food stimulates peristalsis and gastrin production, which stimulates gastric juice secretion.
3. Food enters the small intestine. Lack of food in stomach and intestinal hormone decrease peristalsis and gastric juice secretion in stomach.

then sends more nerve signals to the stomach wall to continue peristalsis and the secretion of gastric juice. Food in the stomach also brings about the production of a hormone called **gastrin**. Stomach cells secrete gastrin into nearby blood vessels, which carry the gastrin to the gastric glands (the target). The gastric glands then produce more gastric juice.

As food moves out of the stomach and enters the small intestine, the signal triggered by the stretched stomach wall stops. In addition, as food enters the small intestine, the intestinal wall secretes a hormone that slows the production of gastric juice. Both of these factors serve to slow down mechanical and chemical digestion in the stomach.

Control of Pancreatic Secretions When the acidic contents of the stomach empty into the small intestine, the acid stimulates certain cells in the small intestine to produce a hormone called **secretin** (sih-KREET-in). Secretin enters the blood and is carried throughout the body. When secretin reaches the pancreas, it stimulates the production of pancreatic juice. The pancreatic juice enters the small intestine and helps digest the entering food. When the pH of the food undergoing digestion in the small intestine becomes basic, secretin production stops. Without secretin, pancreatic secretions also stop.

Hormones also control the activity of the liver, gallbladder, and intestinal glands. In all, eight hormones that control digestion have been identified.

Nutrient Uptake and Storage by Cells

Blood transports the end products of digestion from the digestive system to every cell in the body. The nutrients enter cells by absorption from the blood. Absorption, like other body functions, is a controlled process. The absorption of glucose by cells, for example, is controlled by the hormone **insulin**. Insulin is produced by specialized cells in the pancreas and is released into the blood. Insulin's target is the cell membrane. When insulin is present, glucose crosses the cell membrane. Inside the cell, glucose and the other end products of digestion are used for cell needs.

Nutrients that cells do not use for energy or synthesis are stored. The ability to store nutrients is an adaptation that helps an organism survive if food is scarce. Fat and glycogen are the two forms of stored nutrients in the body.

Glycogen is formed from glucose units and is stored primarily in the liver. When the blood contains a high level of glucose, liver cells convert some of it into glycogen. Another hormone produced by the pancreas, **glucagon**, is released when the glucose level of the blood is low. It raises the glucose level by stimulating the breakdown of glycogen in the liver to glucose. The opposite effects of insulin and glucagon help to maintain a constant level of glucose in the blood.

Fat is stored in fat tissue under the skin, around body organs, and between muscle fibers. Each cell in fat tissue contains a drop of fat. The cells in fat tissue take in excess nutrients from the blood and convert them to fat. Stored fat can be digested and used when cells need energy. The storage and use of fat is also under hormonal control.

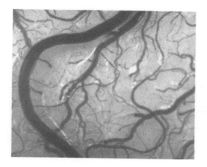

8–29 Body cells obtain nutrients by absorption from capillaries.

Most excess carbohydrates are converted to glycogen and stored. Other excess carbohydrates, fats, and proteins are converted to fat and stored.

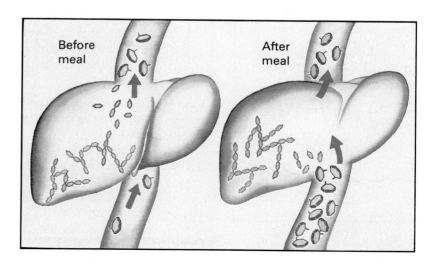

8–30 Before a meal the liver breaks down glycogen and releases glucose to maintain the glucose level in the blood. After a meal excess glucose in the blood is stored as glycogen in the liver.

Building a Science Vocabulary

digestive gland	intestinal juice	folic acid
salivary gland	liver	feces
epiglottis	gallbladder	hormone
gastric gland	bile salt	gastrin
gastric juice	pancreas	secretin
ulcer	intestinal gland	insulin
pylorus	microvillus	glucagon
bile	lymph vessel	
pancreatic juice	colon	

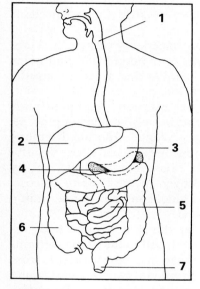

8-31

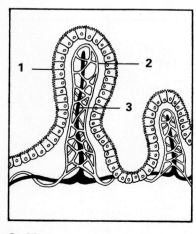

8-32

Questions for Review

1. Write the number and the name of the structure in Figure 8-31 that best performs each of the following functions.
   a. begins protein digestion
   b. absorbs water
   c. absorbs digested nutrients
   d. secretes pancreatic juice
   e. has an acidic pH
   f. secretes bile

2. The following questions refer to the pancreas.
   a. What hormone stimulates the pancreas to begin secreting pancreatic juice?
   b. Where is this hormone produced?
   c. How does the hormone get to the pancreas?
   d. What causes the pancreas to stop producing pancreatic juice?

3. The following questions refer to Figure 8-32.
   a. What are the name and function of the structure shown?
   b. In what organ is the structure found?
   c. What is the number of the structure amino acids and glucose enter?
   d. What is the number of the structure in which microvilli can be found?

Chapter 8 Food Processing

Perspective: A New View
The Evil Spirit in Milk

For many people, cow's milk is a perfect food. It contains carbohydrates, proteins, fats, and most of the vitamins and minerals that people need. Yet many people suffer headaches and intestinal distress after drinking milk. For them, cow's milk is far from being a perfect food.

Milk and milk products contain lactose (milk sugar), a disaccharide made up of two simple sugars, glucose and galactose. Normally, the enzyme lactase in the intestine breaks down lactose into glucose and galactose, which are then absorbed. Some babies and most adults, however, do not have enough lactase.

People who lack lactase are unable to digest lactose. Instead, the lactose passes into the colon along with water and other undigested material. The presence of sugar in the colon prevents water from being absorbed into the blood. Diarrhea, or watery egestion, results and can lead to dehydration of the body. Certain bacteria in the colon use lactose as a food, producing large amounts of painful gas as a by-product of fermentation.

In certain Asian and African countries, about 90 percent of the people over the age of five lack lactase. The Kanuri people in Nigeria say the milk contains an "evil spirit." Actually, these people simply do not have enough lactase.

People who cannot digest lactose because they do not have enough lactase are said to be lactose intolerant. Children and adults who are lactose intolerant usually can eat some milk products such as hard cheese and yogurt. During the production of these milk products, bacteria ferment most of the lactose. Babies who are lactose intolerant can be fed a formula that does not contain lactose, yet includes all the other nutrients they need for healthy growth.

Now it is possible for most lactose intolerant people to drink milk without becoming ill. Recently, lactase has been made available as a nonprescription product in many pharmacies. A few drops of the enzyme preparation are added to a carton of milk. After 24 hours in the refrigerator, most of the lactose has been "digested." Milk treated in this way can be used in any way ordinary milk is used. The only difference is that lactase-digested milk tastes sweeter than ordinary milk.

Scientists believe the ability to produce lactase is inherited like eye color and hair color. If you have problems drinking milk or eating milk products you might be lactose intolerant. Thanks to the availability of lactase, milk may still be a perfect food for you.

Chapter 8
Summary and Review

Summary

1. Consumers must obtain food from other living things. Food processing includes: ingestion, digestion, absorption, egestion, and nutrient use and storage.

2. Digestion is the process of breaking down food into particles small enough to cross cell membranes. Mechanical digestion involves mashing and grinding food. Chemical digestion occurs with the aid of enzymes.

3. In most organisms, a cavity holds food while digestion occurs. Intracellular digestion takes place in a vacuole within the cell. Extracellular digestion usually occurs in a digestive cavity or a digestive tract.

4. Digestive systems may have one opening into a digestive cavity, as in the hydra, or two openings in a digestive tract, as in the earthworm, grasshopper, and human.

5. Peristalsis moves food through the digestive tract, while digestion and absorption occur. Curves and folds in the digestive tract increase the surface area for absorption. Undigested food is egested through the anus.

6. The main parts of the human digestive tract are the mouth, esophagus, stomach, small intestine, large intestine, and anus. In addition, digestive glands produce secretions that function in chemical digestion.

7. The digestive glands that secrete mucus and enzymes into the digestive tract are the salivary, gastric, and intestinal glands, and the liver and pancreas.

8. Most digestion and absorption of food occurs in the small intestine. Nutrients are absorbed into the capillaries and lymph vessels in the villi.

9. Minerals, vitamins, and water are absorbed by the large intestine. The absorption of water helps prevent dehydration.

10. Nerves and hormones control the digestive system. They coordinate peristalsis and the secretions of glands.

11. The transport system delivers nutrients to cells, where they are used or stored.

Review Questions

Application

1. How is each of the following important in food processing?
   a. peristalsis   b. hormones   c. villi

2. How do certain intestinal parasites survive without their own digestive system?

3. A person can live without a stomach because (choose one)
   a. the stomach has no function.
   b. most digestion goes on in the small intestine.
   c. other organs take over all of the stomach's functions.
   d. food is stored in the gallbladder.

4. Explain how the body maintains a constant amount of sugar in the blood.

5. Suppose the nerves to the salivary glands were cut. What do you think would happen to saliva production if food were placed in the mouth? Explain.

Interpretation

1. Explain the following observations, which were made on an animal whose digestive tract is similar to that of humans.
   a. Gastric juice is produced when food enters the animal's stomach.

b. The animal's gastric juice is produced when there is no food in the animal's stomach, and food is shown to the animal.
c. The nerve to the animal's stomach is removed. Food enters the stomach and gastric juice is produced.

2. The information used to construct the graph shown below was obtained by studying digestive enzymes.

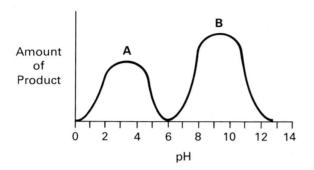

a. In what part of the human digestive tract would you expect to find enzyme A? enzyme B?
b. Enzymes A and B work on the same nutrient. What nutrient might it be?
c. What digestive secretion contains enzyme A? Explain why that digestive secretion works solely in one part of the digestive tract.

Extra Research

1. Build a model of the digestive system for one or more of the organisms described in this chapter.

2. Design a food for astronauts that would require a minimum of egestion. What substances would you put in the food? What would you leave out?

3. Find out how the digestive system of a newborn infant is different from the digestive system of an adult.

4. Use reference books and magazines to find information about digestive diseases. Write an article that could be used by a newspaper giving information about one of them. Include:
   a. a description of the disease
   b. the number of people the disease affects
   c. cause (if known)
   d. treatment and prevention

5. Write a science column article that could be used in your school newspaper. A reader asks: What is dietary fiber and what does it do? Research the answer to this question and prepare a response.

Career Opportunities

Dental hygienists work under the supervision of dentists to promote good oral health. For this they must have two or more years of training. Their duties may range from cleaning teeth and taking X-rays to instructing patients in the care of teeth and preparing them for dental work. Write the Canadian Dental Hygienists' Association, 2136 Fillmore Cres., Ottawa, Ontario K1J 6A4, or the Canadian Dental Nurses & Assistants Association, #204, 542 7th St. S., Lethbridge, Alberta T1J 2H1.

Nurses provide health care to sick and injured persons. Most work on one of the three daily shifts in hospitals and nursing homes. Others work in physicians' offices, private homes, schools, or in industry. The actual work that nurses do depends on their training. Many nurses develop specialties such as pediatrics, intensive care, or psychiatrics. For information, write the Canadian Nurses Association, 50 The Driveway, Ottawa, Ontario K2P 1E2, or the Canadian Hospital Association, 410 Laurier Ave. W., Ottawa, Ontario K1R 7T6.

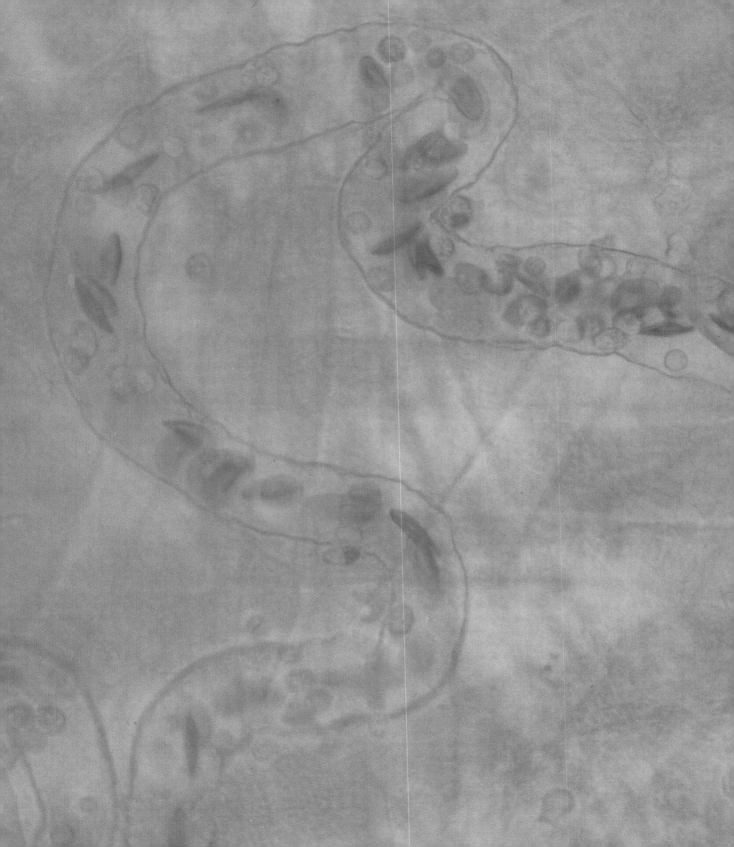

Chapter 9

Transport

> **Focus** *A cell is similar to a factory in many ways. Like a factory, a cell uses raw materials, manufactures products, and accumulates wastes. Transport mechanisms exist that deliver food, oxygen, and other necessities to cells and pick up cellular products and wastes. As in a factory, when transport ceases, the work of the cell comes to a grinding halt.*

9.1 CHARACTERISTICS OF TRANSPORT

The beating heart is a sign of life. Suppose you found an unconscious person at an accident. To determine if the person were alive, you would try to detect the person's pulse. If you felt a pulse, you would know that the person's heart was still beating. This would be a good sign, because a beating heart indicates that the transport system is functioning.

All living things depend on some kind of transport mechanism. Transport provides cells with needed substances and removes cellular wastes. Both unicellular and multicellular organisms die when transport to and from cells ceases.

One type of transport involves the movement of materials from a cell's environment into the cell. This occurs through the cell membrane, which acts as a kind of gatekeeper by permitting only certain substances to enter or leave the cell. To understand how materials pass through the cell membrane, it is necessary to understand how molecules move.

9–1 Blood cells and other materials course through a capillary.

The Movement of Molecules

Imagine that you are sitting in your room at home waiting for dinner. Gradually you become aware of the smell of food. You may wonder how the smell of food reaches you from the kitchen. Odor-bearing molecules from the food spread out from their source. They move from the kitchen, where they are concentrated, to other parts of the house, where they are less concentrated. Eventually the food molecules become evenly distributed throughout the house. This is an example of **diffusion**, which can be defined as the movement of a substance from a place of high concentration to a place of low concentration. Once the concentration of the substance is the same everywhere, diffusion stops. Molecules of the substance do, however, move randomly back and forth.

The movement of molecules of a substance by diffusion depends on a difference in the concentration of the substance in adjacent places.

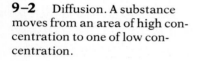

9–2 Diffusion. A substance moves from an area of high concentration to one of low concentration.

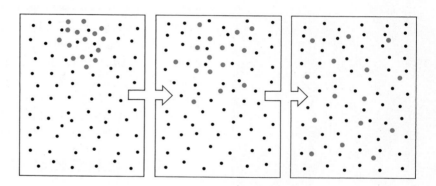

Diffusion of Molecules Through Air The diffusion of food molecules through the house is an example of the diffusion of molecules through air. Diffusion of molecules through air is important in the biological world. Some substances diffusing through air act as chemical signals. For example, odor-bearing molecules from flowers diffuse through the air and are sensed by bees, which follow the scent to its source. During their mating period, male dogs find mates by following chemicals from females that diffuse through the air. Other molecules that diffuse through air are necessary for basic life processes. Oxygen produced by plants during photosynthesis diffuses through the air and is picked up by organisms and used during cellular respiration. The carbon dioxide that organisms give off diffuses through the air and is used by plants during photosynthesis.

9–3 A dog's world is primarily one of smells, which carry information about food, possible mates, and territorial boundaries.

9–4 Diffusion of a dye through water. When the dye has completely dissolved the water will be the same color throughout the beaker.

Molecules move about 10 000 times faster in air than in water.

Diffusion of Molecules Through Water You can observe diffusion of molecules through water by placing a tea bag into a cup of water. As the tea dissolves in the water, it begins to diffuse outward into the water in the cup. At the same time, water molecules diffuse into the tea. Eventually the concentration of tea is the same throughout the cup.

Diffusion through water is also important in living things. Nutrients, waste products, and other compounds are dissolved in water solutions in the cytoplasm of cells. These molecules move into, out of, and within cells by diffusion through water. The glucose produced in the chloroplast, for example, dissolves in the watery cytoplasm and diffuses throughout the cell and to other cells.

Diffusion Across a Permeable Membrane To help you understand the movement of molecules during diffusion, picture the following demonstration. You fill a beaker with a solution of sugar and water. Into the beaker you place a bag filled with pure water. The bag is made from a membrane that is permeable to sugar and water, that is, the membrane allows both sugar and water to pass through it. After waiting 24 hours, you examine the contents of the bag and the beaker. What do you think you will find?

After 24 hours, the bag and beaker will both contain identical sugar-water solutions. During the 24 hours, the sugar molecules moved from a place of high concentration (outside the bag) to a place of low concentration (inside the bag). Similarly, the water molecules moved from a place of high concentration (inside the bag) to a place of low concentration (outside the bag). Eventually, the concentrations of sugar and water became the same in both the bag and the beaker. At this point, the concentrations do not change, even though sugar molecules and water molecules continue to move randomly back and forth.

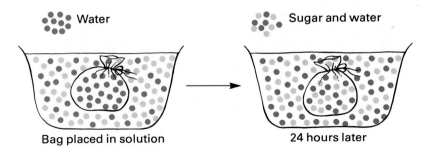

9–5 Diffusion across a permeable membrane. Water molecules move out of the bag and sugar molecules move into the bag until the concentration is the same on both sides of the membrane.

Section 9.1 Characteristics of Transport

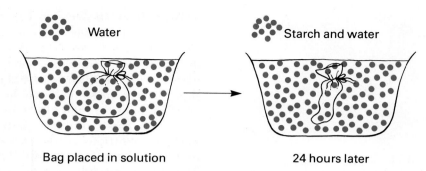

9-6 Osmosis. The membrane is permeable to water but not starch. The water molecules move from an area of high concentration inside the bag to an area of low concentration outside the bag.

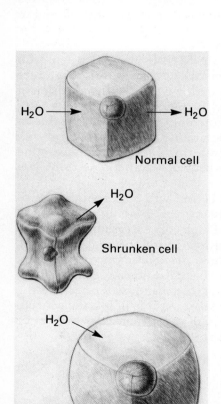

9-7 Normally, the amount of water that moves into and out of a cell is equal. If the concentration of salts is higher outside than inside, the cell shrinks. If the cell is placed in pure water, it swells.

Osmosis: Diffusion of Water Across a Semipermeable Membrane A special case of diffusion involves the movement of water alone. Consider another demonstration. You fill a beaker with a solution of starch and water. As before, you place a bag filled with pure water into the beaker. The bag is made from the same kind of membrane as was used in the previous demonstration. Although the membrane is permeable to water and sugar, it is not permeable to starch. The membrane is referred to as a semipermeable membrane, because it allows some substances to pass through it, but not others. After placing the bag into the beaker you again wait 24 hours. What do you think you will find this time?

The first thing that you will observe after 24 hours is that the bag has collapsed. The reason for its collapse is the outward diffusion of water through the semipermeable membrane. In this case, the starch molecules could not diffuse through the membrane. Only water molecules were able to diffuse. Since water molecules were more concentrated inside the bag than outside it, they diffused outward across the membrane until the bag was empty.

The diffusion of water across a semipermeable membrane is called **osmosis** (os-MŌ-sis). Osmosis accounts for the movement of soil water into root cells. It is also the process by which the water that you drink passes into cells of the colon and then into the blood. In fact, the movement of water into and out of cells generally occurs by osmosis.

The concentrations of substances inside and outside the cell are critical to the life of a cell, since they affect the flow of water into and out of the cell. If too much water enters a cell, the cell may burst. For example, a red blood cell, which contains many ions and proteins, will burst if it is placed in pure water. On the other hand, a plant cell will shrivel if it is placed in a strong salt solution. If the plant cell continues to be deprived of water, it will die.

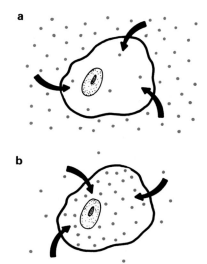

9-8 Transport in cells: (**a**) passive transport; (**b**) active transport.

Transport into and out of Cells

Living cells are covered by a semipermeable cell membrane. Some substances move freely through the cell membrane; others move through it slowly, or not at all. The movement of materials through the cell membrane involves two types of transport: passive transport and active transport.

Passive Transport   In **passive transport**, the difference in the concentration of a substance inside and outside a cell causes the substance to move across the cell membrane. The direction of flow is from an area of high concentration to an area of low concentration. The cell does not supply energy to make the substance move. Diffusion and osmosis are examples of passive transport.

Many substances are moved passively into and out of cells by special molecules in the cell membrane called **carrier molecules**. Carrier molecules bind to substances and move them from a place of greater concentration to a place of lesser concentration. For example, specific carrier molecules in the cell membranes of animals combine with glucose and transport it from the blood, where the glucose concentration is high, into cells, where the glucose concentration is low.

Active Transport   Some substances move across the cell membrane from an area of low concentration to an area of high concentration. Such substances are actually pushed or pulled across the cell membrane. Movements of this kind require the use of energy from ATP. The transport of materials through a cell membrane using ATP is called **active transport**.

Active transport is usually accomplished by carrier molecules; however, these are different carrier molecules than those that perform passive transport. Sodium ions and potassium ions are examples of substances that undergo active transport. They are carried across the cell membrane by carrier molecules in a direction that is opposite to diffusion.

9-9 Models of transport. **a.** Passive transport of some substances uses a carrier molecule, which may simply rotate in the membrane. **b.** Active transport requires ATP and a carrier, which changes shape as it channels the substance across the membrane.

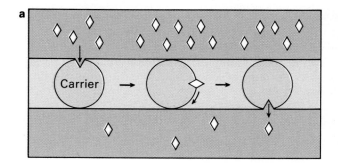
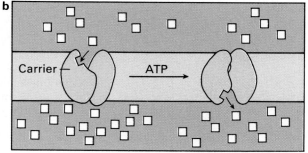

Section 9.1 Characteristics of Transport

endon- = within
-cyte = a cell
-osis = action

exō- = without

Endocytosis and Exocytosis Large molecules and particles are brought into the cell by the process of **endocytosis** (EN-dō-sī-TŌ-sis). In an ameba, for example, the cell membrane forms a pocket around a solid piece of food. Later, the pocket pinches off and the food is trapped in a vacuole in the cell. Once inside the vacuole, the food is digested. The digested substances cross the vacuole membrane by passive transport and active transport.

Large molecules may be exported from cells by **exocytosis**. During exocytosis, a vacuole fuses with the cell membrane and expels its contents from the cell. Cellular products are released from cells in this way. For example, digestive enzymes made in the pancreas are secreted into the pancreatic duct by exocytosis.

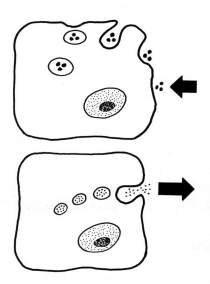

9–10 General plan of endocytosis (above) and exocytosis (below).

9–11 Endocytosis. A unicellular didinium swallows a paramecium nearly its own size. A membrane completely surrounds the paramecium once it is inside, forming a food vacuole. One didinium can eat 12 paramecia in a day.

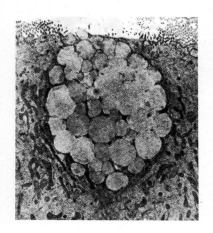

9–12 Exocytosis. A mucus-producing cell in the lining of the intestine releases mucus from a vacuole into the digestive cavity.

Characteristics of Transport Systems

Water is the major component of the liquid of transport systems.

All cells obtain nutrients and rid themselves of wastes. In unicellular organisms, this exchange of materials occurs directly between the cell and its watery environment by diffusion, osmosis, and active transport. In more complex, multicellular organisms, however, specialized transport systems carry materials between the cell and the external environment. All transport systems have three features in common.

1. They move materials through the organism in some kind of liquid.
2. They move the liquid and materials in pathways. These pathways are usually specialized tubes.
3. They have some mechanism for moving the liquid and materials within the organism.

These three features apply to the transport systems of vascular plants as well as to the transport systems of complex animals.

Building a Science Vocabulary

diffusion   carrier molecule   exocytosis
osmosis   active transport
passive transport endocytosis

Questions for Review

1. What two important functions are accomplished by transport in all living things?
2. Explain the difference between the terms in each of the following pairs.
   a. diffusion and osmosis
   b. active transport and passive transport
   c. endocytosis and exocytosis
3. What three features do all transport systems have in common?
4. Explain the function of carrier molecules in the cell membrane.

9.2 TRANSPORT IN THE REPRESENTATIVE ORGANISMS

Transport in the Bean Plant

Roots absorb water by osmosis and minerals by active transport.

The bean plant, like all other organisms, requires water. In addition, it needs minerals from the soil. In the bean plant, water and minerals enter the roots and move upward, against gravity, through the stem to the top of the plant.

Most seed plants and ferns have a transport system similar to that of the bean plant.

Water and Mineral Absorption by Roots Roots are well-adapted for absorbing water and minerals. Many root epidermal cells have fine projections that extend outward into the soil. These projections, called **root hairs**, increase the surface area of the root epidermal cells. This increase in surface area greatly increases the amount of water and minerals that roots can absorb (see Fig. 9–13).

Water molecules move by osmosis into the epidermal cells of the root when the concentration of water is higher in the soil than in the cells of the root. Minerals, including nitrogen, potassium, calcium, phosphorus, and others, pass from the soil into root cells by active transport. This occurs even when the minerals are more concentrated in root cells than in the surrounding soil.

Water and Mineral Transport After water enters the cells of the root epidermis, it moves toward the middle of the root by osmosis. Minerals absorbed by the root move inward by active transport. Water and minerals eventually enter narrow tubes called **xylem** (ZĪ-lum), which consist of dead, tubelike cells ar-

 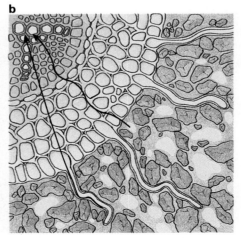

9–13 Root hairs. **a.** Radish seedling with many root hairs. **b.** Drawing of root cross section shows root hairs penetrating into soil. Movement of water and minerals occurs through cells or through spaces between cells.

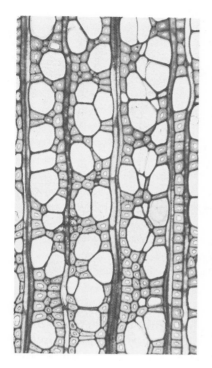

9–14 Xylem tissue of sweet gum tree, longitudinal section. The long, narrow xylem cells have perforated end walls, which allow passage of water from cell to cell.

ranged end to end. Xylem runs through the plant from the roots to the leaves. The walls of xylem tubes are made of cellulose and therefore are very strong. In addition to transporting water and minerals upward, xylem also provides support for the plant.

Water and minerals entering the roots of a large plant must often move great distances to reach cells in the stems and leaves. In some California redwoods, for example, water moves 100 m upward against the force of gravity to the leaves. The mechanism of water transport in plants is not fully understood. Many scientists believe, however, that three interacting forces move water through a vascular plant.

One of the forces is pressure caused by the flow of water by osmosis into the root. This pressure is called **root pressure**. Root pressure occurs partly because the cells toward the center of the root have a lower water concentration than those on the outside. In addition, various root cells actively transport minerals toward the center of the root. The result is that minerals become more concentrated at the center of the root, and water becomes less concentrated. Because of the difference in water concentration, water moves into the root. Root pressure results, therefore, from both active transport of minerals and passive transport of water. The pressure of incoming water "pushes" the water in the xylem up the root.

A second force that moves water upward through the xylem is the tendency of liquids to cling to the sides of narrow vessels. This tendency of liquids is called **capillarity**. Capillarity occurs as water enters the narrow xylem vessels. You can observe capillarity if you place a narrow piece of glass tubing into a small beaker that contains colored water. The water in the tube rises slightly above the water level in the container. The narrower the tube, the higher the liquid rises by capillarity (see Fig. 9–15).

In the sugar maple and other plants, sap containing sugars moves up the xylem in the spring from food storage areas in the roots. People tap the sugar maple sap to make maple sugar and maple syrup.

9–15 Capillarity. The chart shows the relationship between tube diameter and the height to which the colored water rises (measured from the base of each tube).

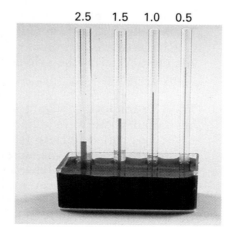

Tube diameter (mm)	Water height (mm)
2.5	25
1.5	38
1.0	52
0.5	66

9–16 Three forces apparently interact in moving water upwards through a plant: root pressure, capillarity, and cohesion-tension. The evaporation of water through stomates (transpiration) pulls water up the xylem.

trans- = through
spirare = to breathe

In sumac and mock orange, sugary sap is conducted upward in the phloem in the spring from storage areas in the roots.

Root pressure and capillarity alone could account for the rise of water to the top of a bean plant. But, they do not explain the rise of water to the top of tall plants. The third force is believed to play the greatest role in water transport in tall plants. This force is explained by the **cohesion-tension theory.**

According to the cohesion-tension theory, water is *pulled* upward through the xylem tubes, which are long, unbroken columns from the roots to the leaves. Water in the leaves evaporates into the air through the stomates in a process called **transpiration** (TRAN-spuh-RAY-shun). As water molecules evaporate during transpiration, water molecules from below move up to replace them. This occurs because water molecules are attracted to each other (this mutual attraction is called *cohesion*). As a result of cohesion, water molecules that evaporate from the leaves exert a pull, or *tension*, on the molecules below them. These molecules in turn pull molecules below them. The result is that the entire column of water is pulled upward as an unbroken unit. Water in the xylem is then transported to cells in the plant by osmosis.

Food Transport Sugars and other nutrients produced by the bean plant dissolve in water and are transported through narrow tubes called **phloem** (FLŌ-um). Like xylem, phloem runs through the roots, stems, and leaves of plants. Phloem is made up of tubular cells that have cell walls strengthened by cellulose. Unlike xylem cells, however, phloem cells are alive.

Nutrients are transported rapidly upward or downward through the phloem. Sometimes food substances move as

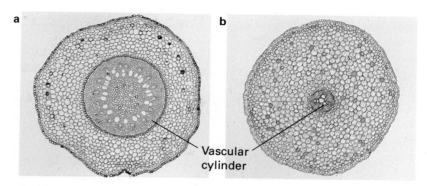

9–18 Cross section of a monocot root (**a**) and a dicot root (**b**).

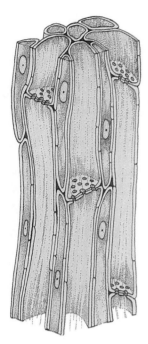

9–17 Phloem tissue, longitudinal section. Phloem is made up of tubular cells that have perforated end plates and narrow cells called companion cells.

quickly as 100 cm/h through the phloem. This is too rapid to occur by diffusion alone. Active transport also plays an important role in the movement of food through phloem. Food is transported laterally out of the phloem to other cells of the plant by diffusion.

The root and stem are the main organs of transport in the bean and other seed plants. In the stem, the xylem and phloem exist together in structures called **vascular bundles.** These structures provide support for the plant in addition to transporting materials. In the root, the xylem and phloem are surrounded by one or more layers of cells in a structure called the **vascular cylinder** (see Fig. 9–18).

Transport in the Paramecium

Since the paramecium lives in water, substances enter by diffusion across the cell membrane. Water enters by osmosis, and dissolved oxygen enters by diffusion. Waste products, such as carbon dioxide and ammonia, move out of the paramecium by diffusion.

Inside the paramecium, substances are transported to all parts of the cell by diffusion and by the "streaming" movement of cytoplasm within the cell. This streaming movement is known as **cytoplasmic streaming.** Materials are transported around the cell much faster by cytoplasmic streaming than by diffusion. Both molecules and organelles are transported in this way. The movement of food vacuoles around the cell is one example of the transport of organelles. As food vacuoles circulate within the paramecium, digested food moves out of them by diffusion and active transport. In this way, nutrients are distributed throughout the cytoplasm.

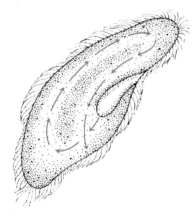

9–19 Cytoplasmic streaming in a paramecium. Food vacuoles are transported from mouth to anus along a moving path of cytoplasm.

The hydra's cells exchange materials with the environment mainly by diffusion.

Transport in the Hydra

Transport in the hydra is very much like transport in the paramecium. Each of the hydra's two layers of cells is in contact with the watery environment. Water enters the cells by osmosis. Dissolved oxygen diffuses into the cells, and waste products (carbon dioxide and ammonia) diffuse out. As food is broken down in the digestive cavity by extracellular digestion, small particles of food enter cells lining the digestive cavity by diffusion and endocytosis. During intracellular digestion, nutrient molecules are transmitted across vacuole membranes into the cytoplasm by diffusion and active transport. In addition, food molecules pass from the inner cell layer to the outer cell layer by diffusion.

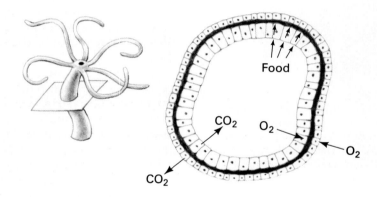

9–20 Cross section of a hydra, showing movement of materials between the hydra and its internal and external environments.

Transport in the Earthworm

The cells of the earthworm are not surrounded by water from the outside environment. This is a problem for land animals and large water animals because materials can diffuse across a cell membrane only if the membrane is moist. In such organisms, cells are bathed with the fluids of the transport system. In complex animals, such as the earthworm, the transport system circulates blood near each cell through blood vessels. Capillaries, which are microscopic branches of these vessels, are in contact with all the cells of the organism. Materials are exchanged with cells through the thin walls of capillaries.

The earthworm's hearts are muscular portions of the main blood vessel.

Five pairs of muscular hearts pump the earthworm's blood in one direction through the blood vessels. The blood vessels and hearts form a closed circuit through which blood keeps circulating. This type of transport system is called a **closed transport system**.

9–21 The earthworm has a closed transport system. Its five tubular hearts pump blood in one direction through the blood vessels and capillaries.

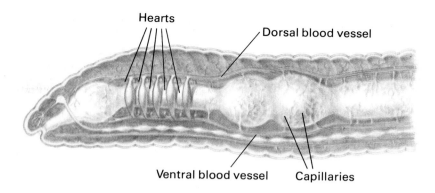

Blood circulating inside the earthworm picks up oxygen from the skin and nutrients from the intestine. These materials, along with water, are transported to all the cells of the body. The transport system carries carbon dioxide from cells to the capillaries of the skin where it diffuses out. Other waste products are transported to special excretory organs for removal from the body.

Transport in the Grasshopper

Blood enters tissue spaces in the body of the grasshopper and bathes the cells directly with fluid.

In the grasshopper, as in the earthworm, blood transports materials to and from cells. Unlike the blood of an earthworm, however, the blood of a grasshopper does not circulate through a closed circuit of blood vessels. Instead, blood circulates through blood vessels that open into spaces between the body organs. This type of transport system is called an **open transport system**.

The grasshopper has a large blood vessel that runs along the top of its body. This blood vessel enlarges at the tail, or *posterior*, end forming a tubelike heart. Blood moves through the grasshopper by the pumping of the heart and by the action of its

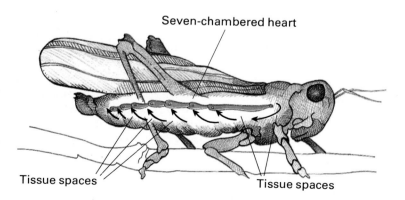

9–22 The grasshopper has an open transport system. Its seven-chambered heart pumps blood into tissue spaces.

Section 9.2 Transport in the Representative Organisms

muscles when it jumps and flies. From the heart, blood moves forward into spaces at the head, or *anterior*, end of the grasshopper. From the anterior end, the blood moves back through the body spaces toward the posterior end. Eventually, blood reenters the blood vessel at the posterior end of the grasshopper's body.

Nutrients and water from the blood diffuse into cells, and cell wastes diffuse into the blood. The wastes are carried from the cells to the excretory system, where they are removed from the body. Carbon dioxide and oxygen are not exchanged through the blood. Instead, they pass directly between the cells and the outside of the grasshopper through the tracheas (see Chapter 7).

Building a Science Vocabulary

root hair
xylem
root pressure
capillarity
cohesion-tension theory
transpiration

phloem
vascular bundle
vascular cylinder
cytoplasmic streaming
closed transport system
open transport system

Questions for Review

1. Explain what is meant by the statement that plants must lose water in order to get water.

2. State one similarity and one difference between the terms in each of the following pairs.
   a. xylem and phloem
   b. open transport system and closed transport system
   c. diffusion and cytoplasmic streaming

3. What function is accomplished by the transport system of the earthworm that is not accomplished by the transport system of the grasshopper?

4. State whether each of the following occurs by active transport, diffusion, or osmosis.
   a. Minerals enter the root cells of the bean plant.
   b. Water enters the skin cells of an earthworm.
   c. Water molecules in leaf cells evaporate into the air.

Perspective: Careers
Butterflies and Beetles

Many biologists showed their first interest in biology by collecting butterflies, beetles, shells, seeds, or some other organism. James Watson started that way. He was one of the scientists who received the Nobel Prize for discovering the structure of DNA.

But for their life work—their careers—some biologists stopped collecting butterflies and beetles and, like Watson, shifted to other kinds of biology—to genetics, anatomy, physiology, or ecology. Others continue to collect some kind of "butterfly," which they use in their research.

People who continue to collect organisms do so to understand the organisms' relationships to other species, their physiology, and their ecology. These biologists tend to specialize in insects (entomologists), birds (avian biologists), amphibians and reptiles (herpetologists), mammals (mammologists), algae (algologists), and so on. They usually conduct their research in universities and in museums, where there are many large collections to study. Most of these specialists have doctoral degrees, although some have master's degrees, and a few have only bachelor's degrees.

Not all biologists are involved in research; many prefer photographing, painting, or drawing biological subjects. This book has many examples of such work. Writing is another outlet for biologists. Nearly every newspaper carries articles on science, and there are several science magazines that regularly publish articles on biology.

Making life science and nature films is yet another artistic activity that revolves around living things. To make a good film, the film producers, script writers, exhibit designers, and sculptors all need to understand biology.

The opportunities to use biology in a career are as varied as there are people with different interests.

9.3 TRANSPORT IN HUMANS

Each year more than half a million people in North America die from heart attacks.

Heart disease and other ailments of the transport system are major causes of death to people living in North America. In the United States alone, three out of five deaths are caused by these diseases. These statistics show that it is important to understand the structures and functions of the transport system. If we know how the system functions, we will have a better idea of how to keep it healthy.

Like the transport systems of other animals, the human transport system has vessels that carry blood to all parts of the body and a heart that pumps the blood through the vessels.

The Composition and Function of Blood

Blood has many functions. It picks up digested food from the small intestine, water and minerals from the colon, and oxygen from the lungs and transports them to all the cells in the body. It transports the carbon dioxide that cells produce to the lungs and other cellular waste products to organs that remove them from the body. Blood also carries hormones and other substances and helps to regulate body temperature.

The body of an average adult contains about 5 L of blood, which accounts for 6 to 8 percent of the person's mass. About 45 percent of the blood volume consists of cells and cell fragments. The rest is composed of a liquid called **plasma**.

Plasma Plasma is straw-yellow in color. It contains many kinds of substances, including water, sugars, wastes, inorganic salts (such as sodium chloride), carbon dioxide, enzymes, hormones, and proteins. Large proteins in plasma, called **plasma proteins**, have a number of different functions. Some of them help form blood clots, which stop bleeding; others fight infections. Some help maintain a constant pH level in the blood; others are carrier molecules. For example, *albumen*, a plasma protein, is a carrier for fatty acids, amino acids, and enzymes.

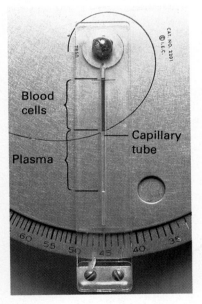

9–23 Blood is centrifuged to separate plasma and cells. The percentage of cells (hematocrit), determined using the device shown, is a measure of the blood's oxygen-carrying ability. Adult males average 40-52; females 37-47.

Red Blood Cells The red color of blood is due to the presence of the protein hemoglobin in the **red blood cells**, or **erythrocytes** (ih-RITH-ruh-sīts). The major function of these cells is to carry oxygen to all body cells (see Chapter 7). As red blood cells circulate through the capillaries of the lungs, hemoglobin combines with oxygen, forming oxyhemoglobin.

$$\text{Hemoglobin} + O_2 \rightleftharpoons \text{Oxyhemoglobin}$$

9-24 Red blood cells (magnified 1500 times).

Certain white blood cells help provide long-lasting protection against disease. These can live for many years.

White blood cells engulf bacteria, dead tissue, and foreign material by endocytosis.

Red blood cells containing oxyhemoglobin release oxygen when they reach the body cells. The deoxygenated hemoglobin then returns to the lungs, where it picks up more oxygen.

Red blood cells are shaped like doughnuts without holes. There are about 30 trillion of them in the body, which is more than any other kind of cell. Red blood cells live an average of 120 days and then are destroyed. Some are destroyed in the spleen (a small organ in the abdomen); others are destroyed in the liver. Most of the remains of old red blood cells become part of the bile produced by the liver. Much of the iron in the hemoglobin, however, is reused in the synthesis of more hemoglobin.

Hemoglobin and new red blood cells are produced by cells in the **bone marrow**. Bone marrow is composed of cells that are located in the cavities of the long bones in the body. When red blood cells are first formed, each one has a nucleus. Before they are released into the transport system, however, they lose their nuclei.

White Blood Cells There are several types of cells that function in body defense. These are known collectively as **white blood cells**, or **leukocytes** (LOO-kuh-sīts). The blood contains approximately one white blood cell for every 500 red blood cells. The white blood cells are produced in the bone marrow and in tissue called lymph tissue. Most white blood cells live only a few hours but some live several years.

When bacteria or certain foreign substances enter the body, the bone marrow is stimulated to produce and release large numbers of specific types of white blood cells. These seek out, ingest, and destroy the foreign bodies. Some white blood cells even squeeze out through the walls of the blood vessels to get to the site of the foreign substance. You have probably noticed that watery pus sometimes forms at the site of an infection. This pus contains dead white blood cells and bacteria.

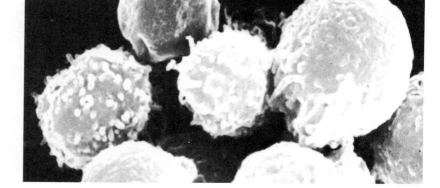

9-25 Scanning electron micrograph of white blood cells. Certain kinds migrate out of blood vessels into tissues, moving like amoebas.

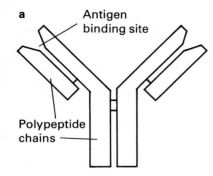

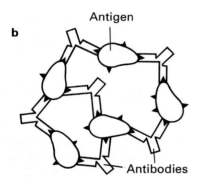

9-26 Most antibody molecules (**a**) are made up of polypeptide chains. They "fight" antigens by binding to them (**b**).

Many foreign substances are called **antigens** (AN-tih-jenz). Foreign proteins, viruses, foreign tissues, and some bacteria are examples of antigens. To combat antigens, certain white blood cells produce chemical substances called **antibodies**. A specific antibody is produced for each type of antigen. For example, if measles viruses enter the body, white blood cells produce measles antibodies to fight them.

Platelets In addition to red blood cells and white blood cells, the plasma carries cell fragments called **platelets** (PLAYT-lets). Platelets are much smaller than red blood cells and there are fewer of them (1 platelet for every 17 red blood cells). Platelets are also formed in the bone marrow and live an average of ten days.

Platelets function in the process of clotting blood. Platelets at the site of a punctured or broken blood vessel break down into smaller pieces and release certain chemicals. These chemicals react with plasma proteins, forming **fibrin** (FĪ-brin), which is made up of tiny protein threads. Fibrin threads form a mesh at the site of the damage in the blood vessel wall. The threads pull the edges of the opening together and trap blood cells, forming a clot. The clot stops the flow of blood out of the vessel. Eventually, cells in the wall of the blood vessel grow over the clot and the vessel heals.

The ability to form a blood clot is a life-saving adaptation that prevents us from losing great quantities of blood as a result of an injury.

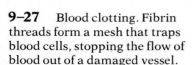

9-27 Blood clotting. Fibrin threads form a mesh that traps blood cells, stopping the flow of blood out of a damaged vessel.

From TISSUES AND ORGANS: A TEXT ATLAS OF SCANNING ELECTRON MICROSCOPY by Richard G. Kessel and Randy H. Kardon. W. H. Freeman and Company. Copyright © 1979

Table 9–1 Blood Cell Types

Blood Cell Type	Where Formed	Number Per Microlitre (µL)	Approximate Average Life	Function
Red blood cells (erythrocytes)	Bone marrow	47 million (males) 42 million (females)	120 days	Carry oxygen
White blood cells (leukocytes)	Bone marrow Lymph tissue	4000-11 000	Some, hours; others, years	Destroy dead cells and bacteria and other foreign matter
Platelets	Bone marrow	300 000	10 days	Aid in clotting blood

Blood Vessels

Three types of blood vessels make up the closed human transport system. **Arteries** carry blood away from the heart. **Veins** transport blood back to the heart. **Capillaries** connect arteries with veins. Differences in the three types of blood vessels relate to the function each performs.

Arteries The walls of arteries are thicker and stronger than the walls of veins and capillaries. Artery walls contain large amounts of muscle and elastic tissue. The elastic tissue enables an artery to stretch when the heart pumps blood into it and to recoil when the heart relaxes. The blood exerts pressure against the blood vessel walls as it moves through the vessel. This pressure is called **blood pressure**. The blood exerts a large amount of pressure against artery walls when the heart contracts. The thickness, strength, and elasticity of artery walls enable them to withstand this pressure.

The action of muscle and elastic tissue allows an artery to change its diameter. When an artery widens, more blood can flow through it, and the blood pressure decreases. When an artery contracts, less blood can flow through it and the blood pressure increases. By changing diameter, an artery helps regulate blood flow and blood pressure.

The largest artery in the body, the **aorta** (ay-OR-tuh), leads from the heart and branches into smaller and smaller vessels. Eventually, tiny arteries, called **arterioles**, feed blood into capillaries.

9–28 Structure of the major kinds of blood vessels. Note that arteries are thicker than veins.

9–29 Blood enters a capillary bed under pressure. The pressure pushes blood plasma through capillary cell walls into tissue spaces. Water and wastes from the tissue fluid move into capillaries at the vein end, where blood pressure is lower.

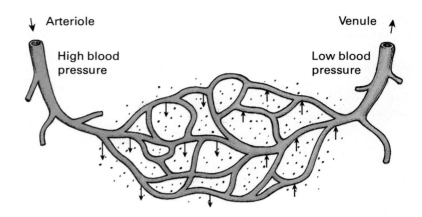

Capillaries connect arteries and veins. Materials are exchanged through capillary walls.

Capillaries Each cell in the body is close to a capillary. Capillaries are so small that red blood cells move through them in single file. (The diameter of a capillary can be as small as 8 μm.) Unlike other blood vessels, the walls of capillaries are composed of only one cell layer.

The exchange of materials with cells takes place through the thin capillary walls. In the artery end of the capillary, blood pressure pushes some of the plasma through the capillary wall and into the spaces around the cells. In the tissue spaces, this plasma is called **tissue fluid**. Tissue fluid contains food, oxygen, and other materials that cells need. It does not contain plasma proteins, however, because they are too large to pass through the capillary wall. Cells absorb materials from the tissue fluid by passive and active transport, and their waste products diffuse into it. In the vein end of the capillary, the blood pressure is lower than in the artery end. Without blood pressure to oppose its movement, water containing carbon dioxide and other wastes from the tissue fluid moves into the capillary by osmosis.

In the artery end of a capillary, blood is rich in oxygen and has a bright-red color. This bright-red color changes to a dark-red color when blood gives up its oxygen to cells and the oxygen-poor blood leaves the capillary.

Veins Blood leaving capillaries flows into tiny vessels called **venules**. These in turn unite to form the larger vessels known as veins. The veins come together into two large veins, the **vena cavae** (VEE-nuh KAY-vee), which return blood to the heart.

Veins have larger diameters and thinner walls than arteries. They also have less muscle and elastic tissue in their walls than arteries. The blood pressure in veins is lower than it is in arteries and capillaries. Because the blood pressure is low, blood has a tendency to flow backward. There are, however, cuplike valves in the walls of veins in the legs and arms, which stop blood from

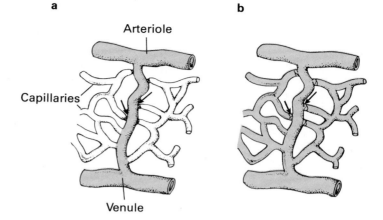

9–36 The supply of blood to the capillaries in an organ depends on how active the organ is. When an organ is not active, tiny valves shut off the blood supply to most capillaries (**a**). Major capillaries are left open. When an organ is active, the valves open and all the capillaries carry blood to the cells (**b**).

The total volume of the blood in the body could be contained in the capillaries alone if they were all open at once. Usually, only about 25 percent of the capillaries have blood in them at any one time.

gan. Blood pressure is affected by heart action and the diameter of blood vessels. These are controlled by nerves that have their center in the brain. If the blood pressure in certain vessels increases, the brain sends nerve signals to the heart and to the blood vessels, causing the heart rate to slow and the blood vessels to widen. The result is that the blood pressure decreases. If the blood pressure becomes too low, the brain sends nerve signals that cause the heart rate to increase and the blood vessels to narrow, thereby increasing the blood pressure. In a healthy person, this regulatory mechanism helps keep the blood pressure at safe levels.

The capillaries in a tissue or organ are not always filled with blood. Only certain capillaries in a particular organ are always open. In a resting muscle, for example, these capillaries supply enough blood to the cells of the muscle to keep it healthy in its resting state. Once the muscle becomes active and its need for oxygen and nutrients increases, the arterioles carrying blood to the tissue widen. This increases the blood supply to the muscle. At the same time the smaller capillaries open and distribute the increased blood supply to the muscle cells. When the muscle returns to its resting state, the arterioles contract and the smaller capillaries close.

The ability of blood vessels to change diameter also helps regulate body temperature. As blood circulates, it carries heat throughout the body. When body temperature increases above normal levels, the small arteries in the skin widen. As a result, more blood flows into them. The skin becomes reddened, or flushed, and the blood loses heat through the surface of the skin. The small arteries in the skin contract when body temperature drops. Then less blood flows to the skin, the skin pales, and body heat is conserved.

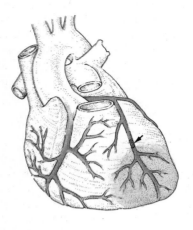

9-37 Blood vessels of the heart. A heart attack results from blockage of a heart vessel (see arrow), depriving a portion of the heart of oxygen. Lasting damage to or destruction of the affected area frequently occurs.

Diseases of the Heart and Blood Vessels

More than half the people who die from diseases of the transport system are the victims of heart attacks. During a heart attack, a portion of the heart muscle is damaged due to lack of oxygen. Heart attacks are brought about by a clogged blood vessel in the heart. A heart vessel may be clogged by a clot formed in the vessel or by one that was carried in the blood from a vessel elsewhere in the body. Scientists are trying to develop an artificial heart that could replace a badly damaged or diseased heart. The advantage of an artificial heart over a donor heart is that the plastic substance it is made of is not rejected by body tissues.

Arteriosclerosis, or hardening of the arteries, is a common disease in old people that increases the chances of a heart attack. This disease is characterized by deposits of cholesterol and other substances on the inner walls of arteries. These deposits harden, and the arteries lose their elasticity and become narrower. Often, the flow of blood slows down in the narrowed vessel, and the blood supply to an organ is reduced. In severe cases, pressure on the rigid wall can cause it to break. Sometimes pieces of the blood clots that form at the site of the break or portions of the hardened wall itself break off. A clot or a piece of artery wall traveling in the bloodstream can block a blood vessel and cut off the supply of blood to tissues. If the block occurs within a blood vessel in the heart, a heart attack can occur. A blocked blood vessel in the brain results in a "stroke." Many people reduce the amounts of saturated fats and cholesterol in their diet in an attempt to prevent arteriosclerosis.

9-38 In arteriosclerosis, fatty deposits build up over time in artery walls, causing the walls to thicken. The walls become less flexible, and the arteries become narrower.

hyper- = over
tension = pressure

Another example of a disease involving the heart and blood vessels is **hypertension**, or high blood pressure. In hypertension, the pressure of the blood against blood vessel walls is high, and regulatory mechanisms cannot bring down the blood pressure to normal levels. If arteriosclerosis exists, hypertension can cause a clot or piece of artery wall to enter the bloodstream. In many cases, hypertension can be controlled with medical treatment.

The Lymph System

Tissue fluid from the blood plasma surrounds every cell in the body. It moves between the cells and keeps cell membranes moist. Tissue fluid moves freely into and out of capillaries through capillary walls. Some tissue fluid returns to the transport system by osmosis at the vein end of capillaries. Most tissue fluid returns to the transport system through a special network of vessels, which are part of the **lymph system**.

9–39 Lymph system: **a.** microscopic view of relationship between lymph vessels and capillaries. **b.** lymph vessels reach every part of the body, with nodes concentrated in the head, neck and torso.

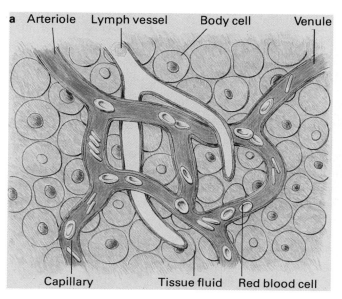
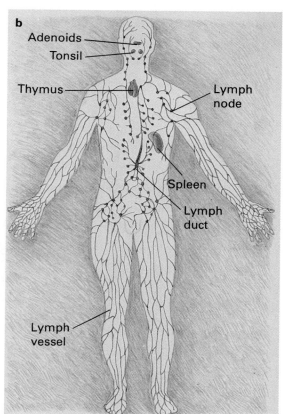

Section 9.3 Transport in Humans

Microscopic branches of the lymph system are located in the tissue spaces throughout the body. Tissue fluid seeps into the walls of these lymph vessels. Tissue fluid taken into the vessels of the lymph system is called **lymph**. The contraction of muscles when you move squeezes the lymph into larger and larger lymph vessels. In addition, contraction of the lymph vessels themselves helps transport lymph. Lymph travels in the direction of the heart. Valves in the walls of lymph vessels prevent back flow. Eventually, the lymph flows into two very large lymph vessels that empty into veins in the chest near the heart. Lymph is once again called plasma when it returns to the transport system.

The Lymph System and Fatty Acid Absorption In addition to circulating tissue fluid, the lymph system absorbs fatty acids from the small intestine. The small intestine has tiny projections called villi. Each villus contains a microscopic lymph vessel called a **lacteal**, which absorbs fatty acids into the lymph (see Fig. 8–24). The fatty acids are carried by the lymph system to the blood.

The Lymph System and Body Defense The lymph system contains clumps of tissue called **lymph nodes** (see Fig. 9–39). The nodes filter bacteria, dead cells, and foreign substances from the lymph. White blood cells are also produced in lymph nodes. If you have an infection, the lymph nodes that drain the affected area may become enlarged. This occurs because cells in the lymph node become very active in combatting the disease. In the body of someone with cancer, lymph nodes often collect cancer cells that are moving in the body. For this reason, during cancer surgery, the surgeon often removes the lymph nodes near the cancer, along with the cancer itself.

Transport Helps Maintain Homeostasis

The human brain takes in an average of 49 mL of oxygen every minute.

Suppose the supply of freshly oxygenated blood to someone's brain were cut off. In five seconds that person would be unconscious. After five minutes, some brain cells would be damaged forever. This example illustrates how cells depend on the transport system. The transport system helps maintain homeostasis in several ways. It prevents wastes produced by cells from accumulating and poisoning the cells; it replaces the food and oxygen that cells use as they function; it regulates body temperature; and it keeps the pH of the tissue fluid constant, thereby allowing enzymes to function properly. All of these functions create a balanced environment, which enables cells to function normally.

Building a Science Vocabulary

plasma	vein	heart valve
plasma protein	capillary	pulmonary circulation
red blood cell	blood pressure	systemic circulation
erythrocyte	aorta	sinoatrial (S-A) node
bone marrow	arteriole	pacemaker
white blood cell	tissue fluid	adrenalin
leukocyte	venule	arteriosclerosis
antigen	vena cava	hypertension
antibody	varicose vein	lymph system
platelet	atrium	lymph
fibrin	ventricle	lacteal
artery	septum	lymph node

Questions for Review

1. State the part of the body described by each of the following functions.
   a. carries blood to the heart
   b. produces red blood cells
   c. prevents back flow of blood in the heart

2. Explain the difference in function between the left ventricle and the right ventricle of the heart. How is each one adapted for its particular function?

3. A heart removed from the body continues to beat. Which structure makes this possible?

4. Why is hypertension dangerous to a person suffering from arteriosclerosis?

5. Describe two ways in which tissue fluid returns to the transport system.

Chapter 9
Summary and Review

Summary

1. Transport provides cells with needed substances and removes cellular wastes.

2. Transport of small molecules across cell membranes occurs by passive transport (including diffusion and osmosis) and active transport. Cells engulf large particles by endocytosis and export them by exocytosis. Transport within cells occurs mainly by diffusion.

3. Transport systems in complex organisms include a liquid for carrying materials, pathways through which the liquid moves, and a mechanism for moving the liquid.

4. In plants, such as the bean, water moves up the xylem by a combination of three forces: root pressure, capillarity, and cohesion-tension. Food moves from the leaves down the phloem by diffusion and active transport. Food and water move laterally by diffusion.

5. In the paramecium and the hydra, transport occurs mainly by diffusion and osmosis.

6. In the earthworm, five pairs of hearts pump blood through a closed system of blood vessels. Materials are exchanged between cells and capillaries.

7. In the grasshopper, a tubular heart pumps blood through vessels into open tissue spaces. Food is delivered to cells by diffusion and wastes are picked up by the blood. Muscle action squeezes blood from the tissue spaces back into vessels and moves it to the heart.

8. In humans, blood consists of plasma and cells or cell parts. The plasma transports food, wastes, salts, hormones, large proteins, and enzymes. Erythrocytes carry oxygen; leukocytes help protect the body and destroy dead body tissue; platelets aid in clotting.

9. Arteries carry blood away from the heart; veins carry blood toward the heart; capillaries connect arteries and veins. The exchange of materials between the blood and body cells occurs in the capillaries.

10. The heart consists of four chambers. Two atria receive blood and two ventricles pump blood. The left ventricle pumps blood to the body; the right ventricle pumps blood to the lungs. The heart rate and strength of the heartbeat is set by the S-A node, which is controlled by nerves and hormones.

11. Blood pressure is affected by the action of the heartbeat and by changes in the diameter of blood vessels.

12. The lymph system returns tissue fluid to the transport system and aids in nutrient absorption and body defense.

Review Questions
Application

1. Trace a drop of blood from the right atrium to the big toe and back to the right atrium. Name every structure through which the drop of blood passes.

2. How do control mechanisms in the body lower a person's blood pressure when it rises above normal levels?

3. Diffusion and osmosis across cell membranes occur only when the membranes are moist. How are cell membranes kept moist in the leaves of trees that are located 90 m above the ground? in your skin cells?

4. How do the human transport and lymph systems combat disease-causing bacteria?

Interpretation

1. When an animal dies, why do all the body cells not die the instant the heart stops beating?

2. In an experiment, all the stomates of a geranium plant were clogged with petroleum jelly. Do you think the plant wilted from lack of water? Explain.

3. For years scientists searched for capillaries in the grasshopper. Why do you think they were unable to locate any?

4. Urea (a cellular waste product) diffuses out of liver cells into the tissue spaces. Would you expect that urea enters a capillary at its artery end or its vein end? Explain.

5. What would happen if a bag permeable to water and not to sugar were filled with a sugar-water solution and placed in a beaker of pure water?

Extra Research

1. Stand the cut end of a celery stalk in water that has been colored with a few drops of ink or food coloring. After one day, cut the stalk in half crosswise and observe the route of the colored water through the xylem. If a microscope is available, squash one of the celery strings and observe the xylem vessels under the microscope.

2. In the library, look up the effects of smoke, alcohol, and rich food on the heart and blood vessels. Write a short essay on your findings.

3. Buy a beef or sheep heart from the butcher shop or supermarket. Study its structure. Find valves, septum, chambers, and blood vessels. Draw and label a picture of your observations.

4. Hold one arm up in the air and let the other arm hang down at your side for two minutes. Compare the color of each hand and how much your veins stand out. What do you think causes the difference?

5. Using reference books, trace the path of a drop of blood from the right ventricle to each of the following organs and back to the right ventricle.
   a. heart (coronary circulation)
   b. kidneys (renal circulation)
   c. small intestine (hepatic-portal circulation)

6. If you have a goldfish and a microscope, you can observe blood cells circulating in capillaries. Cover the goldfish's gills with wet absorbent cotton and observe the thin part of the tail under the microscope. You can make similar observations on a tadpole's tail and on the web of a frog's foot.

Career Opportunities

Physical education instructors must have a good understanding of the human body and how it functions in order to instruct people of all ages in physical activities. Depending on their interests, physical education instructors work in schools, colleges, universities, public recreation programs, or private health clubs. Most instructors have attended four years of university. Consult university calendars, or write Canadian Association for Health, Physical Education & Recreation, 333 River Rd., Vanier, Ontario K1L 8B9.

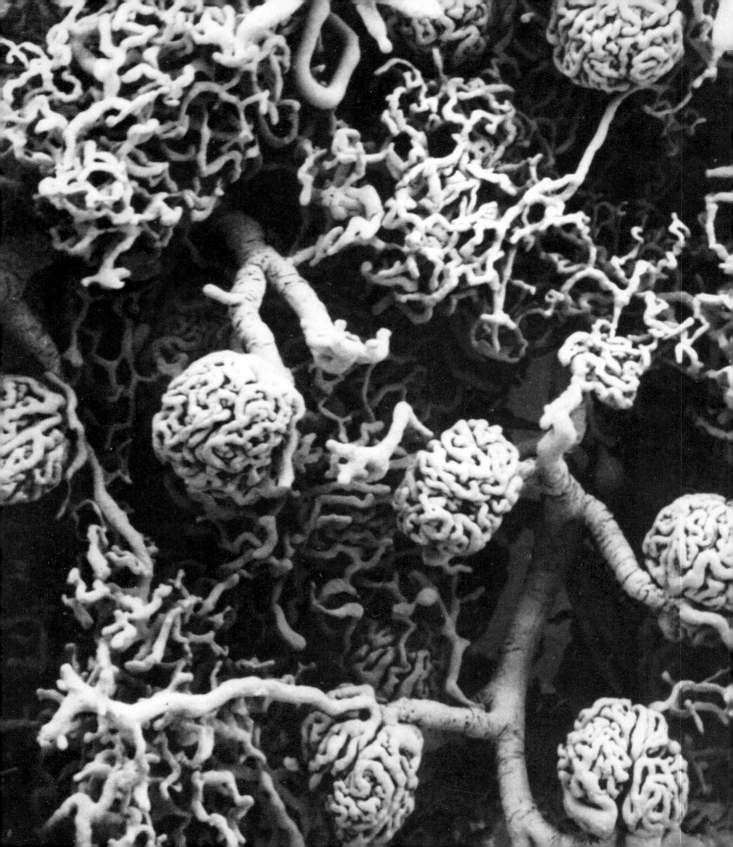

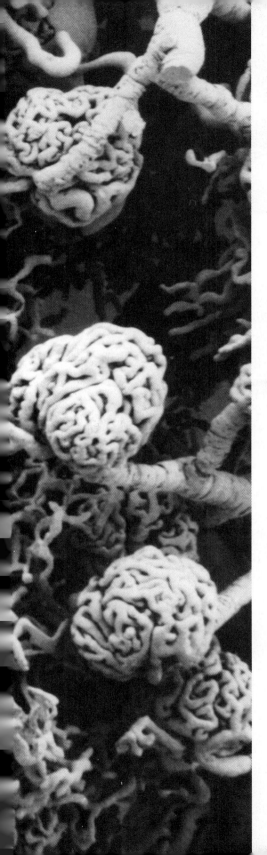

Chapter 10

Excretion

Focus *The outside environment of an organism undergoes change. Yet, the environment within the body remains relatively constant. Mechanisms of homeostasis guard this internal stability. Some of these mechanisms involve excretion. Excretion protects cells from being poisoned by their own wastes. It also maintains the delicate balance of salts and water that exists between cells and the fluids around them.*

10.1 AN OVERVIEW OF EXCRETION

A wrecked ship lies adrift in the middle of the ocean. The crew's prospects do not look good if they are not rescued. Once the supply of fresh water runs out, the sailors cannot survive for long. They cannot drink sea water because it contains salt. If they drink it, salt will accumulate in the tissue fluid around their cells. A large volume of fresh water will be needed to flush out the salt. If they can not get fresh water, the sailors' tissues will dehydrate, since water will move by osmosis out of their cells and into the salty tissue fluid. This will have a dramatic effect on their brain cells, which are easily damaged by dehydration. The sailors will become confused and disoriented. If dehydration continues, the sailors will die. Their immediate cause of death will probably be brain damage.

10–1 Tiny capillary networks and their associated arterioles and venules play a major role in excretion in the kidney.

FROM TISSUES AND ORGANS: A TEXT ATLAS OF SCANNING ELECTRON MICROSCOPY by Richard G. Kessel and Randy H. Kardon. W. H. Freeman and Company. Copyright © 1979

10–2 Though surrounded by salt water, stranded sailors will die from dehydration without fresh water to drink.

You might think that the crew can survive by eating raw fish. Fish are not salty, and their tissues contain water. Fish, however, contain large amounts of protein. Poisonous waste products are made when the body breaks down the amino acids in protein. The body gets rid of these wastes by flushing them out with fresh water. If the sailors eat fish without drinking water, wastes will accumulate in the tissue fluid and in the blood. The sailors will lose energy, vomit, and have convulsions. Without water, they will eventually lose consciousness and die.

Dehydration and the accumulation of poisonous wastes are drastic conditions that disrupt homeostasis. Under normal conditions, however, an organism's cells are cleansed of wastes, and the salt and water balance is maintained. These functions are accomplished by **excretion**.

Excretion is one of the principle mechanisms by which homeostasis is maintained.

The Waste Products of Cells

Cells cannot function properly if cell wastes accumulate. Two of the most important kinds of waste products made by cells are nitrogenous (nitrogen-containing) wastes and carbon dioxide.

Nitrogenous Wastes Most nitrogenous wastes are produced during the breakdown of amino acids for energy. Unlike carbohydrates and lipids, amino acids contain nitrogen. The nitrogen is part of the amino group ($-NH_2$). Before a cell uses an amino acid for energy, the amino group is removed. The remaining part of the molecule is then converted into pyruvic acid or

10–3 When amino acids are used for energy, ammonia, a nitrogenous waste, is produced.

$$H_2O + \underset{\text{Water}}{} \underset{\text{Amino acid}}{\begin{array}{c}HRO\\||\\H-N-C-C\\||\\HH\end{array}} \longrightarrow \underset{\text{Ammonia}}{\begin{array}{c}HH\\|\\N\\|\\H\end{array}} + \underset{\begin{array}{c}\text{Amino acid}\\\text{fragment}\end{array}}{\begin{array}{c}RO\\|\\O=C-C\\OH\end{array}} + 2H$$

Chapter 10 Excretion

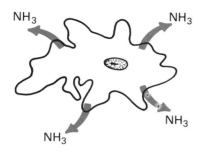

10-4 Ammonia diffuses across the cell wall of an amoeba directly into the watery surroundings.

some other molecule that enters the citric acid cycle. The amino group removed from the amino acid is converted to **ammonia** (NH_3). Ammonia is extremely poisonous. It dissolves rapidly in water and forms a base that can destroy enzymes.

In unicellular organisms that live in water, such as the ameba, removal of ammonia is relatively simple. Ammonia produced in the cells diffuses rapidly out into the surrounding body of water. In large multicellular organisms, however, disposal of ammonia is not as simple. Most of the cells of a large organism are not in direct contact with the outside environment. As a result, ammonia produced in one part of the body has to be carried elsewhere in the organisms for removal to the outside.

Ammonia is too poisonous to remain within an organism for long. In humans, a thousandth of a milligram of ammonia per litre of blood will kill a person. Large multicellular organisms convert ammonia to less poisonous substances. Many organisms convert ammonia to **urea** (yoo-REE-uh). One urea molecule is constructed from two ammonia molecules and one carbon dioxide molecule. Urea dissolves in water and is therefore easily transported by the blood. Urea is far less poisonous than ammonia. Yet, like ammonia, it takes quite a bit of water to flush urea out of the body.

Many animals that survive in places where there is little fresh water excrete **uric acid** as a nitrogenous waste. Uric acid is a complex molecule built in part from four ammonia molecules and three carbon dioxide molecules. Uric acid is even less poisonous than urea and is excreted as a powder or paste. It does not dissolve easily in water and requires almost no water to be excreted from the body. Reptiles, insects, and birds all produce uric acid.

10-5 Much of the fresh water the impala drinks will be used to flush urea out of its body. A coat of guano covers an island where sea birds have excreted uric acid in a thick, white paste.

Section 10.1 An Overview of Excretion

Table 10–1 Nitrogenous Waste Products

Waste Product	Advantages	Disadvantages	Some Organisms that Produce It
Ammonia	Dissolves easily in water (easily transported)	Highly poisonous, must be removed immediately to outside watery environment, requires large amounts of water for removal	Paramecia, hydras, tadpoles
Urea	Dissolves easily in water (easily transported) Less poisonous than ammonia	Requires water for removal, cannot accumulate for prolonged period of time	Humans, whales, adult frogs
Uric acid	Much less poisonous than ammonia or urea Does not require water for removal	Does not dissolve easily in water (harder to transport)	Insects, reptiles, birds

The ability to excrete uric acid is an adaptation to a dry environment.

As a waste product, uric acid has several advantages over ammonia and urea. Consider a young reptile or bird developing within an egg on dry land. The water supply is limited, and ammonia or urea would poison the animal in the closed environment of the egg. Uric acid, on the other hand, is converted into dry crystals and stored in a small sac until the animal hatches out of the egg. Uric acid has another advantage for birds and flying insects. By reducing the amount of water that the body must hold, uric acid helps keep the animal light in weight, thereby helping it to fly.

Carbon Dioxide Cells produce carbon dioxide during cellular respiration. Carbon dioxide combines with water and forms carbonic acid. As carbonic acid accumulates in cells, they become more acidic. If cells become too acidic, enzymes cannot function properly. In unicellular organisms, such as the ameba, carbon dioxide diffuses directly from the cell into the water outside. In more complex organisms, carbon dioxide is removed by the respiratory system.

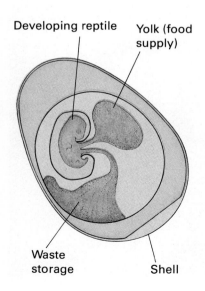

10–6 A developing reptile stores nitrogenous wastes as uric acid crystals in a sac in its egg.

Salt and Water Balance in Organisms

The salt and water levels in an organism are closely related. The amount of salt in the tissues and cells regulates the movement of water by osmosis. It is important for cell chemistry that the balance between salt and water be carefully maintained.

Life in Salt Water Humans are not adapted to survive the intake of salt water. We cannot rid our bodies of the excess salt in the water without becoming severely dehydrated. Drinking salt water overloads human mechanisms that maintain the salt and water balance in the body.

Some organisms are adapted to take in salt water. Consider the problem of surviving on salt water. If there is a higher concentration of fresh water in the body of an organism than there is in the surrounding ocean, water tends to leave the body by osmosis and dehydration results. In some animals, this problem is solved by removing the excess salt from water that is taken into the body. Bony saltwater fish, such as mackeral or bluefish, drink ocean water and do not dehydrate because they have special gland cells on their gills that gather salt and excrete it. Large sea turtles excrete salty tears, thereby ridding themselves of the excess salt they ingest. Gulls, penguins, and other sea birds have a salt gland in the head that drips a concentrated solution of salt off the tip of the beak. Grasslike plants that grow in salt marshes excrete excess salt from salt glands in their leaves. These salt-removing mechanisms maintain the proper level of salt and water in the organisms' cells.

10-7 Saltwater fish tend to lose too much water and take in too much salt. They compensate by drinking large amounts of water and by excreting salts across their gills.

10-8 Gulls get rid of excess salt by producing salt "tears," which drip off the tip of the beak.

Section 10.1 An Overview of Excretion

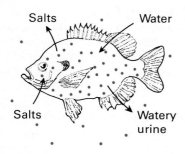

10-9 Freshwater fish tend to take in too much water and lose too much salt. They compensate by drinking little water, by absorbing salts across their gills, and by excreting watery urine.

Kangaroo rats die of thirst on a diet of soybeans, which are high in protein.

10-10 Water balance. To avoid dehydration the amount of water lost by an organism must equal the amount it takes in. A horse loses water when it breathes, sweats, and gets rid of wastes. It takes in water when it eats and drinks and produces water by cellular respiration.

Sharks and other nonbony fish maintain the salt and water balance in a different way. Urea, the amino acid breakdown product, is held in the animal's blood. This makes the concentration of water inside the body similar to the concentration of water in the ocean. As a result, water is not lost to the environment by osmosis.

Life in Fresh Water Organisms living in fresh water have the opposite problem to that of saltwater organisms. Since there is a higher concentration of fresh water in the environment than inside the organism, water tends to enter the organism by osmosis. In addition, salts tend to diffuse out. Freshwater fish, such as brook trout and perch, have kidneys that excrete large amounts of water and retain important salts. Some freshwater fish also have gland cells on their gills that accumulate salts from the water.

Life on Land The greatest threat to survival on land is dehydration. Land organisms tend to lose water by evaporation and by excretion of wastes. In most land animals, water needs are met by ingesting water and food and by cellular respiration (see Chapter 6). Some land animals survive with little or no drinking of water. For example, the kangaroo rat can survive on a diet of seeds that are high in fats and low in protein. The low level of nitrogenous waste and high level of water produced when these seeds are used for energy make possible survival without drinking water.

On a normal day, the average human takes in about 2.2 L of water in food and drink and obtains about 0.2 L from cellular respiration. A person loses some water during exhalation, perspiration, and egestion. The major water loss from the body, about 1.4 L per day, is due to the excretion of urea in the urine.

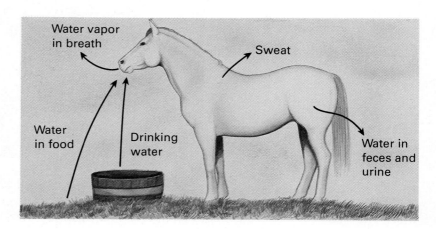

Chapter 10 Excretion

Table 10-2 Daily Loss of Water in Humans (in mL)

	Normal Weather	Hot Weather	Prolonged Heavy Exercise
Lungs	350	250	650
Urine	1400	1200	500
Sweat	450	1750	5350
Feces	200	200	200
Total	2400	3400	6700

Even though our intake of water varies, the salt and water balance in the body is maintained partly by varying the amount of water we excrete. If we consume large amounts of fluids, we excrete the excess water. If we consume small amounts of fluids, we excrete less water.

Excretion in Complex Organisms

In unicellular and small multicellular organisms that live in water, excretion occurs simply by diffusion. In complex multicellular organisms, wastes are removed from the body by the excretory system. Organisms with excretory systems carry out four steps in excretion. They are:

1. Wastes are removed from cells.
2. Wastes are transported to the excretory system.
3. Wastes are separated from the transport system by excretory organs.
4. Wastes are eliminated from the body.

Consider the excretory system of a frog (see Fig. 10-11). First, wastes move from cells into the frog's blood by diffusion. Then, they are carried in the blood to the frog's kidney. The kidney takes urea and other wastes out of the blood and makes urine. Urine is stored in the bladder until it is excreted through the **cloaca** (KLŌ-ay-kuh). If the frog spends most of its time in water, its blood may contain excess water. In this case, the kidney excretes a large amount of watery urine. If the frog spends most of its time on land, its blood contains less water,

10–11 A frog's kidneys respond to low water intake by producing concentrated urine. They respond to high water intake by producing dilute urine.

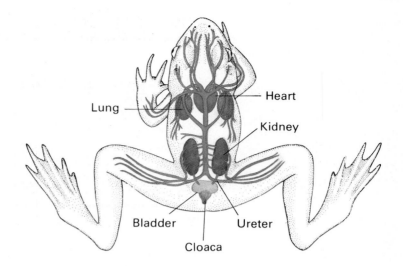

and its kidneys produce more concentrated urine. In this way, the excretory system of the frog removes cell wastes and maintains the balance of salt and water in the frog's body.

Building a Science Vocabulary

excretion   urea   cloaca
ammonia   uric acid

Questions for Review

1. What would be the effect of the following situations on the organisms described?
   a. A saltwater fish is placed in fresh water.
   b. A person on a diet eats only meat and does not drink anything at all.

2. Why would an animal that produces uric acid as its main excretory product be more likely to survive in the desert than an animal that produces urea?

3. What two functions are accomplished by excretion?

4. What process in cells results in the production of each of the following?
   a. water   b. ammonia   c. carbon dioxide

10.2 EXCRETION IN THE REPRESENTATIVE ORGANISMS

Excretion in the Bean Plant

Although beans and other plants do excrete carbon dioxide and oxygen, they do not normally produce and excrete nitrogenous wastes.

The bean and other plants use most of the substances that they produce during metabolism. Beans use the carbon dioxide from cellular respiration for photosynthesis. They also use some of the oxygen they produce during photosynthesis for cellular respiration; the rest of the oxygen is excreted. When photosynthesis is not occurring, bean plants excrete carbon dioxide.

Bean plants rarely use amino acids for energy. Normally, they rely on the carbohydrates produced in photosynthesis. As a result, nitrogenous wastes, such as ammonia, are not produced by the bean plant. Research has been performed to see whether plants are capable of producing nitrogenous wastes from amino acids, and, if produced, whether they are excreted. In one experiment, leaves were removed from a plant and observed. Photosynthesis slowed down in the cutoff leaves and eventually stopped. The starving leaves then used amino acids for energy. Ammonia was formed as a result; however, the leaves did not excrete the ammonia. Instead, they used it to synthesize more amino acids.

In the bean and other plants some waste products, such as salts or acids, are dissolved within vacuoles in the cells. These remain until the above-ground part of the plant dies at the end of the year.

Water balance is maintained in the bean plant when the water that enters the roots by osmosis equals the amount lost by the leaves during transpiration. If more water evaporates from the top of the plant than enters from the soil, wilting occurs.

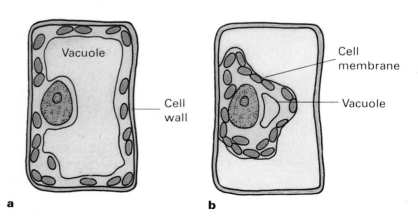

10–12 Under conditions of water balance, a plant cell's central vacuole is full of water, and the cell membrane presses tightly against the cell wall (**a**). Under conditions of insufficient water, the vacuole shrinks and the cell membrane pulls away from the cell wall (**b**).

Excretion in the Paramecium

The water concentration outside the paramecium is greater than the water concentration inside. As a result, water constantly enters the paramecium by osmosis. If the paramecium could not excrete excess water, it would burst. The paramecium has two **contractile vacuoles** that fill up with excess water (see Fig. 10–13). The contractile vacuoles alternately pump water out of the cell through pores in the cell membrane. The pumping action of the contractile vacuoles requires energy from ATP.

Carbon dioxide and salts diffuse directly out of paramecia into the surrounding water. Ammonia also diffuses directly into the surrounding water; however, some of it is excreted by the paramecium's contractile vacuoles.

10–13 A paramecium pumps out excess water and some wastes with its contractile vacuoles (only one is shown). The vacuole and the canals that lead to it contract once they are full, forcing water out of the cell.

Excretion in the Hydra

Each cell of the hydra is in direct contact with the watery environment. Ammonia, salts, and carbon dioxide leave the cells of the hydra by diffusion. Since hydras live in fresh water, excess water tends to enter their cells by osmosis. It is not known exactly how the hydra excretes this excess water.

Excretion in the Earthworm

Carbon dioxide produced in the earthworm is transported by the blood to the skin, where it diffuses out. Urea and other wastes that the earthworm produces diffuse into the body fluid around the cells. Water and excess salts also enter the body fluid. These wastes enter excretory organs called **nephridia** (neh-FRID-ee-uh), which are located in pairs on most segments of the earthworm's body. The beating of cilia in the nephridia creates a current that draws in the body fluid. The nephridia filter the body fluid, separating wastes and returning useful materials (water, sugar, salts) to the body fluid and the blood. The nephridia help regulate the earthworm's water balance by controlling the amount of water returned to the body. Waste products leave the earthworm's body through ducts connecting the nephridia with pores in the skin.

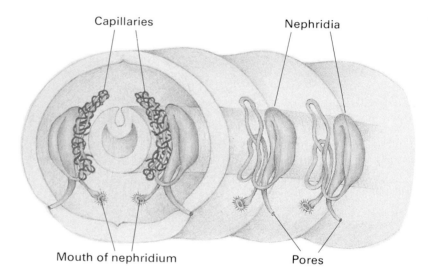

10-14 Excretory system of an earthworm. Segmental pairs of nephridia filter wastes from the body fluid. Wastes pass to the outside through pores in the skin.

Excretion in the Grasshopper

In the grasshopper, carbon dioxide leaves the cells and enters the tissue fluid located in the spaces between cells. It is then absorbed into the tracheas and expelled through spiracles. Uric acid and salts also enter the tissue fluid. There they mix with blood from the transport system. The excretory organs, called **Malpighian** (mal-PIG-ee-un) **tubules**, open into the tissue spaces. The uric acid and waste salts enter the Malpighian tubules by active transport. Water enters them by osmosis. Most of the water is then reabsorbed into the grasshopper from the tubules.

Section 10.2 Excretion in the Representative Organisms

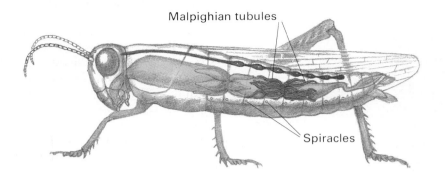

10–15 Excretory system of a grasshopper. The main excretory organs are the Malpighian tubules. They remove wastes from the blood and empty them into the digestive tract.

In the tubules, uric acid is changed to a powdery, crystalline form. The uric acid crystals and waste salts then pass through the tubules into the digestive tube. These wastes pass out of the body with the feces.

Grasshoppers live in a dry environment and take in little water. Their bodies maintain water balance by conserving the water they do take in. Producing uric acid as a nitrogenous waste is an efficient means of conserving water, since almost no water is used in its formation or excretion.

Building a Science Vocabulary

contractile vacuole   nephridium   Malpighian tubule

Questions for Review

1. State one difference between the items in each of the following pairs.
   a. nephridia and Malpighian tubules
   b. the way a paramecium gets rid of excess water and the way it gets rid of ammonia
   c. excretion of carbon dioxide in the bean plant and excretion of carbon dioxide in the hydra
2. How does the grasshopper's excretory system adapt it to a dry environment?
3. How do nephridia help maintain water balance in the earthworm?

10.3 EXCRETION IN HUMANS

The human body and the bodies of most mammals are about 60 percent water. Two thirds of the water is in cells; the rest is in fluid around the cells. You can think of the fluid around cells as a kind of internal "sea." The chemical composition of this "sea" is critical for the life of the cells. If it becomes too salty, the cells can become dehydrated. If it becomes too watery, the cells can bloat. If too much acid or base accumulates, the cells' enzymes can stop working. If urea and other wastes collect, the cells can die of poisoning. Homeostatic mechanisms within the body keep the chemical composition of the internal "sea" constant when conditions change.

Excretion aids homeostasis by removing wastes and regulating the salt and water balance. The blood picks up carbon dioxide, excess salts, urea, and any excess water that may be present and carries them to excretory organs. The excretory organs then channel the wastes out of the body.

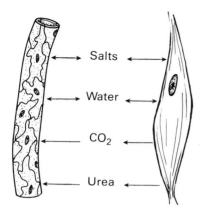

10–16 Tissue fluid provides a medium of transport for waste products from cells to capillaries. Water and salts are also kept in balance through the tissue fluid.

The Excretory System

Several organs take part in human excretion. The lungs, skin, liver, and transport system play important roles; however, the main excretory organs are the **kidneys**.

Excretion in the Lungs Carbon dioxide and water are produced by all body cells during cellular respiration. The blood carries these to the lungs. Carbon dioxide diffuses into the alveoli and is removed from the body when we exhale. Some water also leaves the lungs as water vapor.

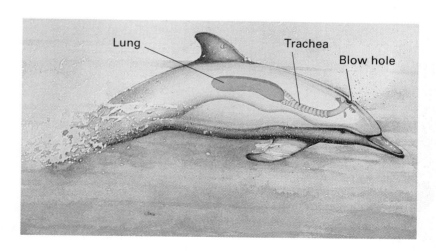

10–17 Excretion in the lungs. Aquatic mammals such as the dolphin must surface regularly to blow out carbon dioxide. Some water is also expelled.

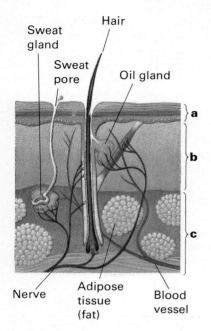

10–18 Structure of the skin, showing its three layers: **a.** epidermis; **b.** dermis; and **c.** subcutaneous (beneath the skin) layer. Salts and other wastes are excreted through the sweat pores.

Excretion by the Skin Most salts are carried by the transport system to the kidneys; however, some are excreted by the skin. When we perspire, sweat glands in the skin excrete salts, water, and some urea. This does not merely aid excretion. It also helps to regulate body temperature, since evaporation of the water in perspiration helps cool the body.

The Role of the Liver in Excretion Although we obtain most of our energy from carbohydrates and fats, we sometimes use amino acids for energy. To extract energy from amino acids, our bodies first convert them to pyruvic acid or other substances used in the citric acid cycle. This conversion occurs mainly in the cells of the liver. This process also produces ammonia, which the liver converts to urea. The products resulting from the conversion of amino acids in the liver are picked up by the blood. The blood delivers the energy-yielding molecules to cells, where they are used during cellular respiration to produce ATP. The blood delivers urea to the kidneys, where it is excreted.

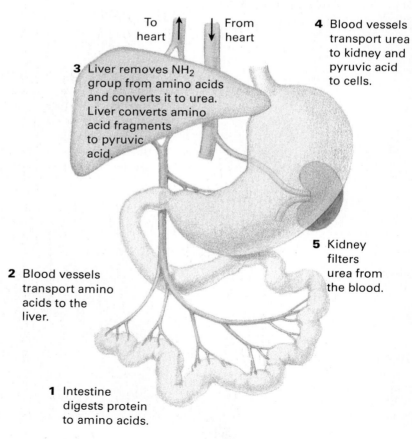

10–19 The liver converts amino acids to urea and pyruvic acid. The urea is excreted; the pyruvic acid is broken down for energy.

10–20 Liver tissue. **a.** Spokes of cells radiate from central veins. **b.** Close-up view, showing cells, blood vessels, and bile ducts.

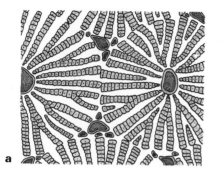
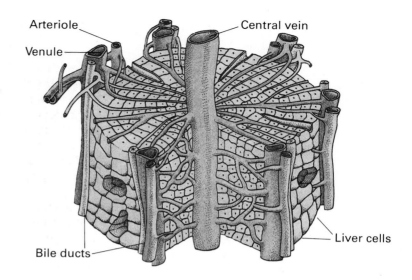

The liver also breaks down red blood cells. About 120 to 130 days after they are formed, erythrocytes are destroyed by special cells in the liver. These cells engulf red blood cells and break down their hemoglobin. Iron released from the hemoglobin is carried by the blood to various parts of the body. In the bone marrow the iron is used in the manufacture of new hemoglobin. Another part of the broken down hemoglobin is a pigment that is a waste product. This pigment enters the bile, passes into the small intestine, and is eventually removed from the body with the feces.

In addition to breaking down amino acids and red blood cells, the liver breaks down harmful and foreign substances. It breaks down many ingested poisons and drugs after they leave the digestive system and before they reach the rest of the body. It also breaks down excess hormones and enzymes produced by the body. The poisons that the liver breaks down often have a chemical structure similar to some naturally occurring materials. Since the liver has the enzymes to break down the natural materials, it is also able to destroy certain drugs. This function can be life-saving, because it makes destructive substances harmless and prevents hormones, medications, and other chemicals from accumulating to harmful levels. Cortisone and sulfur-containing medicines are examples of drugs that the liver breaks down.

The liver's ability to make substances harmless has limits. Materials can accumulate in the liver and destroy its tissue. For example, alcohol taken in excess over a period of time can deteriorate the liver. Pollutants, such as PCBs (polychlorinated biphenyls) and pesticides, such as DDT, can also harm the liver.

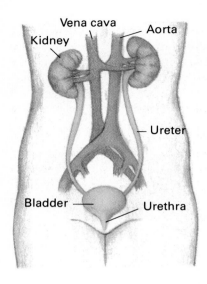

10–21 Urinary system. Urine passes through the ureters to the bladder, where it is stored until it passes to the outside through the urethra.

The Kidneys: Major Organs of Excretion The two bean-shaped kidneys are located on either side of the spinal column in the lower back (see Fig. 10–21). Each kidney is about 10 cm long. The kidneys produce **urine**, which consists of urea, water, and other wastes removed from the blood. Urine leaves each kidney through a tube called a **ureter** (yoo-REE-ter). The ureters empty urine into the **urinary bladder**, which stores it until it can be excreted from the body. Urine from the bladder leaves the body through a duct called the **urethra** (yoo-REE-thruh). Together, the kidneys, ureters, bladder, and urethra make up the **urinary system**.

Kidney Structure

The working unit of the kidney is a microscopic structure called a **nephron**. A nephron is a system of tubes that acts as a filter. Each kidney contains nearly a million nephrons. A nephron consists of a cupped end, called **Bowman's capsule**, and a long coiled **tubule**. Each tubule enters a **collecting tubule**. Each collecting tubule joins with other tubules, eventually forming the ureter, which leads out of the kidney.

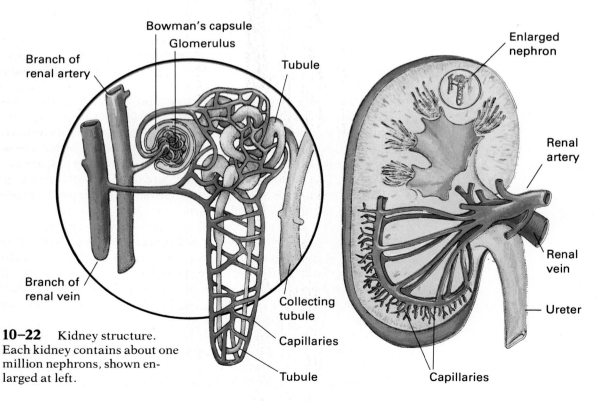

10–22 Kidney structure. Each kidney contains about one million nephrons, shown enlarged at left.

Chapter 10 Excretion

renalis = kidney

glomus = a ball

Blood enters each kidney through a large artery called the **renal (REE-nul) artery**. In the kidney, the renal artery branches into smaller and smaller arterioles. Each tiny arteriole enters a Bowman's capsule, where it branches into a microscopic network of capillaries. This capillary network resembles a small clump. The capillary clump inside the capsule is called a **glomerulus** (glah-MER-yoo-lus).

The capillaries of the glomerulus unite, forming a tiny artery that leaves Bowman's capsule. After leaving the capsule, the artery branches again into another capillary network, which surrounds the tubule. The capillaries reunite forming a small venule. Each venule unites with others; eventually, small veins from all the nephrons join, forming the **renal vein**.

Kidney Function

Blood entering the kidney in the renal artery contains a large amount of wastes. Blood leaving the kidney in the renal vein contains almost no wastes. The wastes are removed in the kidney by the nephrons. The removal of wastes by the nephrons involves three processes: filtration of the blood, active secretion of certain wastes, and reabsorption of useful materials.

Filtration When blood reaches the glomerulus, blood pressure forces plasma through the capillary walls into Bowman's capsule. The plasma entering Bowman's capsule contains waste materials, as well as materials that the body needs, including glucose, water, and certain salts. Cells and large protein molecules in the blood remain in the capillary, however, because they are too large to be pushed through the capillary walls. The movement of plasma from the glomerulus into Bowman's capsule is called **filtration**. The plasma in Bowman's capsule is called the **filtrate**. The filtrate contains essentially the same substances as plasma, but its contents change as it moves through the tubule.

Uremia (urine in the blood) is a kidney disease in which urea and other wastes are not filtered out of the blood. As a result, body cells are poisoned.

Secretion Some ions and molecules in the blood are selectively moved from the capillaries into the tubule by active transport. This process is called **secretion**. Penicillin is an example of a substance that is removed from the blood in this way. In addition, secretion helps maintain the pH of the blood by adding H^+ ions to the urine.

Reabsorption As the filtrate passes through the tubule, some of its contents pass back into the capillaries around the tubule. The passage of materials from the tubule back into the transport

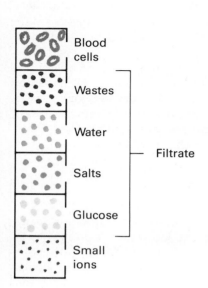
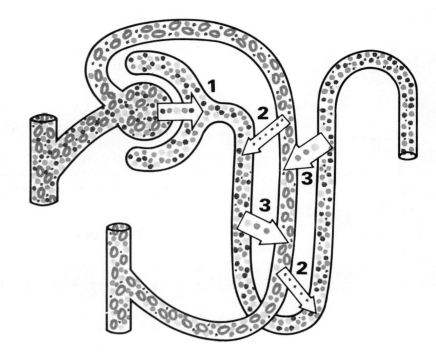

10–23 Nephron function: **1.** Filtration. Materials move from the blood plasma into Bowman's capsule. **2.** Secretion. Molecules are actively transported from capillaries into the tubule. **3.** Reabsorption. Water and salts move by active and passive transport from the filtrate into capillaries, leaving water and waste products in the tubule.

system is called **reabsorption**. Reabsorption occurs as a result of both passive and active transport. Under normal conditions all glucose is reabsorbed. The body also reabsorbs water and many salts, according to its needs. Both secretion and reabsorption occur at the same time in the tubule.

By the time the filtrate reaches the end of the tubule, it contains only water and waste products. At this point, the filtrate is called urine. Urine moves through the collecting tubules into the ureters and passes drop by drop down the ureter into the urinary bladder.

Control of Excretion

anti- = against
diuresis = increased excretion of urine

Thirst is a control mechanism for water intake.

The excretory system is largely under chemical control. For example, the regulation of the amount of water in the urine is controlled by a hormone called **antidiuretic** (AN-tih-dī-yoo-RET-ik) **hormone (ADH)**. This hormone is released by the pituitary gland, which is located at the base of the brain. Antidiuretic hormone affects the permeability of the membranes of tubule cells to water. If a person does not have sufficient water, the lowered level of water in the blood signals the pituitary to

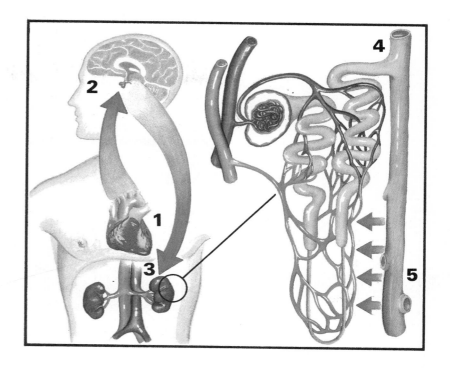

10–24 Water conservation by the kidney. Low water level in the blood (**1**) signals the pituitary to release more ADH (**2**). The blood carries ADH to the kidneys (**3**). The ADH increases the permeability of tubule cell membranes to water (**4**), causing more water to be reabsorbed from the tubule into the blood (**5**). The opposite occurs if the water level in the blood is high.

release more ADH. The ADH travels in the blood throughout the body. When it reaches the kidney tubule cells it stimulates the cell membranes to become permeable to water. As a result more water is reabsorbed into the blood and the urine becomes more concentrated. If, on the other hand, a person has too much water, less ADH is released. Then the cell membranes of tubule cells become impermeable to water. As a result, they do not allow water to return to the blood and the urine is very dilute.

Chemical control also plays a role in the reabsorption of glucose, amino acids, ions, and other substances. The active transport of these materials from the tubules back into the transport system depends on the presence of carrier molecules and ATP.

Kidney Disease

The proper functioning of a person's excretory system is vital. A person would die within two weeks without working kidneys. Kidney function can be reduced or halted by a number of causes, including shock, excessive blood loss, infections, birth defects, certain poisons, and hardening of kidney arteries. In the past, a

A person can live normally with only one working kidney.

person with severely damaged kidneys could not survive. Today's medical technology, however, can help people whose kidneys do not work properly. A machine known as an **artificial kidney** is sometimes used to clear a patient's blood of poisonous substances. In other cases, a patient's diseased kidney can be replaced surgically with a kidney from a donor. Such an operation is known as a **kidney transplant**. These techniques save thousands of lives each year.

Building a Science Vocabulary

kidney	Bowman's capsule	filtrate
urine	tubule	secretion
ureter	collecting tubule	reabsorption
urinary bladder	renal artery	antidiuretic hormone (ADH)
urethra	glomerulus	
urinary system	renal vein	artificial kidney
nephron	filtration	kidney transplant

Questions for Review

1. Match the organ in Column A with its function in Column B.

   Column A
   a. skin
   b. liver
   c. lung
   d. kidney
   e. ureter
   f. urethra
   g. urinary bladder

   Column B
   stores urine
   carries urine to urinary bladder
   produces urine
   produces urea
   helps regulate body temperature
   excretes carbon dioxide
   removes urine from the body

2. Write the number and name of the part of the nephron described by each of the following (refer to Fig. 10–25).
   a. It receives plasma leaving the blood.
   b. It carries urine to the ureter.
   c. It reabsorbs materials from the kidney.

3. Name two substances that are reabsorbed in the kidney.

4. List the events that occur in the body when water intake is too high.

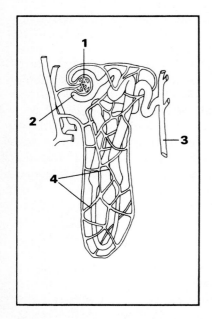

10–25

Perspective: Technology
A Growing Parts Catalog

As recently as fifty years ago, people who had suffered dismemberment or loss of eyesight from war injuries or from accidents had few choices about how to remedy their condition. Most often, the artificial parts available to them were cumbersome and conspicuous. Today, the combined efforts of medical specialists and creative inventors have provided an ever enlarging catalog of substitute body parts. Altogether, over 200 such parts are now on the market.

In the 1960s doctors performed the first heart transplants. In addition to using hearts from people who had just died, they also tried using hearts from higher primates. Although heart transplants did not prove to be too successful, the idea opened up the possibility of transplanting other organs. Kidneys and corneas have been transplanted, as have hips and knees. Unless the body's immune system is adjusted, such transplants are usually rejected. But scientists are finding ways to cut down the rejection of these foreign tissues.

In addition to transplanting actual body parts, a number of artificial substitute parts are now used. Plastic heart valves, plastic knee and hip joints, and steel plates in the skull are a few substitutes commonly used.

In cases where transplant surgery or the use of substitute parts doesn't work, a different kind of substitute can be used. If the kidneys are diseased, a kidney dialysis machine might be used. This machine is literally hooked up to the patient. The patient's blood passes through the machine, which filters the wastes from it.

Portable machines are also being developed to replace other organs. One serves as a "pancreas" for people with diabetes. These portable organs periodically and automatically pump needed substances, such as insulin for the diabetic, directly into the body. This eliminates the need for repeated injections.

Today, a person injured in a serious accident could survive with a transplanted heart and cornea, a plastic hip and steel skull plate, and a portable kidney and pancreas. Through the efforts of medical science, the person could continue to lead a useful life.

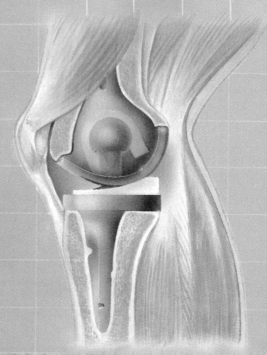

Chapter 10
Summary and Review

Summary

1. In the process of excretion, an organism's cellular waste products are removed from the body, and salt and water balance is regulated. These functions aid homeostasis.

2. Cellular waste products include nitrogenous wastes, salts, carbon dioxide, and sometimes excess water.

3. The nitrogenous waste that an organism excretes (ammonia, urea, or uric acid) is related to the organism's complexity and how much water is available in its environment.

4. Excretory mechanisms help regulate the salt and water balance between an organism and its environment. These mechanisms operate in organisms living in salt water, fresh water, and on land.

5. Complex organisms carry out four steps in excretion: wastes are removed from cells, wastes are transported to the excretory system, excretory organs separate wastes from the transport system, and wastes are eliminated from the body.

6. The bean plant uses most of its own products. Water vapor and excess carbon dioxide and oxygen leave through stomates.

7. Diffusion is the main mechanism for waste removal in the paramecium and the hydra. Contractile vacuoles remove excess water in the paramecium.

8. Nephridia excrete urea and help maintain water balance in the earthworm.

9. Malpighian tubules excrete uric acid in the grasshopper. Excretion of uric acid helps conserve water.

10. In humans, the lungs, skin, liver, urinary system, and transport system all aid in excretion.

11. The kidney's unit of function is the nephron. Filtration, secretion, and reabsorption go on in the nephron.

12. Antidiuretic hormone helps regulate the amount of water that the kidney releases into the urine.

Review Questions

Application

1. State one similarity and one difference between the items in each of the following pairs.
    a. contractile vacuole and food vacuole
    b. secretion and reabsorption
    c. perspiration and transpiration
    d. plasma in renal artery and plasma in renal vein

2. If you ate six candy bars would there be sugar in your urine? Explain.

3. If you drank three litres of water would you expect more water in your urine? Why or why not?

4. Does the human body excrete iron obtained from the breakdown of red blood cells? Explain.

5. Why does a kangaroo rat die on a diet of only soybeans?

Interpretation

1. A researcher injected a harmless blue dye into the bloodstream of a rabbit. The dye was not found in the Bowman's capsules of the kidneys but it did appear in the rabbit's urine. How did the dye get into the urine?

2. How does the liver "recognize" and break down some of the newly invented medicines administered to people? Explain the breakdown of medicines in terms of the lock-and-key theory.

3. Why is it that a saltwater fish such as a flounder does not taste salty when you eat it?

4. Would you expect a bird to have sweat glands? Why or why not?

Extra Research

1. Buy a beef or lamb kidney in the grocery store. Study its external and internal appearance. Use reference books to help identify the parts and learn the functions. Record your results in drawings and brief descriptions.

2. Make a model of the excretory system of the earthworm or grasshopper. Use clay for organs and wire, cord, or thread for ducts.

3. Investigate what happens to human nitrogenous wastes when they reach the ecosystem. Write a short report.

4. Make a list of discarded materials in your ecosystem. What beneficial uses are made or could be made of these discarded materials?

5. Try to find out how eating asparagus or beets changes the urine of some people.

Career Opportunities

Medical secretaries and **medical records clerks** work in hospitals, nursing homes, and other health agencies. They use typing and filing skills in collecting and maintaining records for physicians, administrators, and insurance companies. Specialized training in medical terminology is preferred, but not required, for many positions. For information, contact the Canadian Health Record Association, 187 King St. E., Oshawa, Ontario L1H 1C2, or the Ontario Medical Secretaries' Association, 240 St. George St., Toronto, Ontario M5R 2P4.

Biomedical engineers devise equipment and procedures for medical use. Their inventions include artificial limbs, kidney machines, and heart pacemakers. Biomedical engineers usually have special biomedical training in addition to a background in mechanical, electrical, or chemical engineering. Contact the Canadian Medical & Biological Engineering Society, #183, M-50, National Research Council of Canada, Ottawa, Ontario K1A 0R8.

Chapter 11

Movement and Locomotion

Focus *People associate movement with life. They ask, "Is it moving?" to determine if a caterpillar, fly, or fish is alive or dead. Most living things, except for plants and some fungi, move from place to place. Even if an organism does not move around, it usually does move parts of its body. In most organisms, movement has vital functions that aid survival.*

11.1 MOVEMENT IN LIVING THINGS

According to the *Guinness Book of World Records*, the fastest animal in the world is the spine-tailed swift. This bird's speed of flight has been clocked at 150 km/h. Its rapid motion is an example of **locomotion**, the movement of an organism from one place to another.

Locomotion is not the only kind of movement than an organism performs. A plant's leaves, branches, and roots move during growth and when they fill with water. Portions of your body move during chewing, breathing, and peristalsis. You move your arms when you write, wash, and stretch.

Functions of Movement

The movements of the green anole lizard illustrate the variety of functions movement and locomotion have in organisms. Anole lizards live on insects, which they catch with their long, sticky

11–1 Running is the result of many coordinated body movements.

11-2 Like the anole lizard, the three-horned chameleon catches flies and other prey with its sticky tongue.

11-3 The male anole lizard attracts a mate by puffing out his red dewlap while bobbing his head.

tongues. The lizard rapidly shoots out its tongue and traps prey that is either in flight or that is resting on a branch or leaf. Movement also helps these lizards find shelter. In autumn, they cluster in groups behind the loose bark on a tree or scamper under rocks or fallen logs. In addition to finding food and shelter, movement helps the anole lizard find a mate. As the male anole lizard moves toward a female during breeding season, it extends the red dewlap at its throat and bobs its head. These movements attract the female and initiate the mating process.

Movement has similar functions in many different animal species. Finding mates, communicating, escaping predators, and finding food and shelter are common functions of movement. In addition, movements of cells and organs within the body aid in carrying out life processes.

Cellular Movement and Locomotion

Since Leeuwenhoek first observed sperm cells under the simple microscope in the seventeenth century, scientists have known that cells move. Certain bacteria migrate within the body of their host, and many types of protozoa and unicellular algae swim freely in their watery environment. Individual cells that are part of multicellular organisms also exhibit locomotion. White blood cells migrate from the bloodstream to a wound, where they engulf incoming bacteria. Cells within developing organisms move around as organization of the body proceeds. Motion also occurs within individual cells: muscle cells contract; gland cells export membrane-covered products; cell membranes engulf molecules; and cytoplasm streams.

Most cellular movement and locomotion depends on special protein threads in the cell. Movement of both muscle and nonmuscle cells occurs when these threads contract. In general, three types of movement are found in cells.

Protozoa with cilia are called ciliates.

Protozoa with flagella are called flagellates.

Movement by Cilia and Flagella Cilia are microscopic, hairlike structures that extend from the surface of cells (see Chapter 3). The beating of cilia creates a current in the surrounding environment. This aids in capturing food and also results in locomotion in certain protozoa and many flatworms. The beating of the cilia on an individual cell is coordinated into a continuous wave of motion, which enables ciliates to glide through the water. In planaria and other flatworms, the cilia on different cells of the organism work together in groups, enabling it to move along in a trail of mucus that it secretes.

Flagella and cilia have the same structure, but flagella are longer. Locomotion in flagellated protozoa, such as *Euglena*, and in sperm cells is accomplished by the beating action of flagella. Flagellated cells in the internal cavities of sponges and coelenterates create a current that circulates food and water to cells in the organism.

The internal structure of cilia and flagella consists of nine pairs of microtubules in a ring with two microtubules in the center. The microtubules are made of proteins and create movement by contracting. This activity requires ATP.

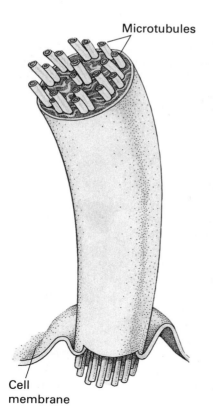

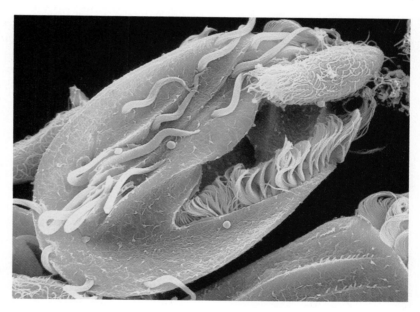

11–4 The one-celled ciliate *Euplotes* propels itself through water using cilia. A fringe of smaller cilia aids in food-getting.

11–5 Internal structure of a cilium or flagellum. Movement results when the members of a pair of microtubules slide past one another.

11–6 The pseudopods are the only parts of an amoeba that touch surfaces. They appear to pull the amoeba along in a creeping movement.

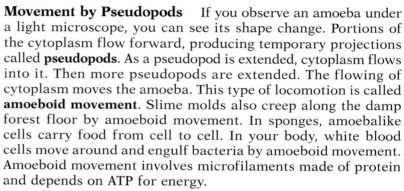

pseudēs = false
podion = foot

Movement by Pseudopods If you observe an amoeba under a light microscope, you can see its shape change. Portions of the cytoplasm flow forward, producing temporary projections called **pseudopods**. As a pseudopod is extended, cytoplasm flows into it. Then more pseudopods are extended. The flowing of cytoplasm moves the amoeba. This type of locomotion is called **amoeboid movement**. Slime molds also creep along the damp forest floor by amoeboid movement. In sponges, amoebalike cells carry food from cell to cell. In your body, white blood cells move around and engulf bacteria by amoeboid movement. Amoeboid movement involves microfilaments made of protein and depends on ATP for energy.

Protozoa that move by means of pseudopods are called sarcodines.

Cilia, flagella, and pseudopods provide movement for single cells and for small, simple organisms. Movement of larger, multicellular organisms, however, requires the coordinated and controlled action of complex systems.

Movement in Plants

Plants do not move from place to place, yet they do move. Plant movements occur as a result of growth and changes in water content. When plants move during growth, their parts do not just get bigger; they change position. It is hard to detect movement due to growth, because growth is a slow process.

A plant's movements due to changes in water content are called **turgor movements**. These movements are more noticeable than growth movements. An example of turgor movement can be found in the insect-eating plant, Venus' flytrap. When an insect walks on the leaf of a Venus' flytrap and brushes against the leaf hairs, the water content of the leaf changes. Then the leaf halves close, and the insect is trapped.

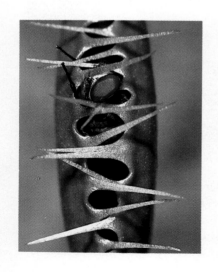

11–7 When insects brush against certain hairs on the leaves of a Venus' flytrap, the toothed leaves close quickly.

Chapter 11 Movement and Locomotion

Movement and Locomotion in Animals

Both cilia and muscles function in the movement and locomotion of Planaria *and other flatworms.*

Coelenterates and simple flatworms move by contracting protein fibers in their cells. In jellyfish, for example, fibers contract when the animal swims and when it moves its tentacles. Complex flatworms, such as planaria, and other complex animals have specialized muscle tissue, which contracts when the animal or its parts move.

Muscles Muscles consist of long, narrow cells containing protein fibers. A muscle has only two possible movements: it can contract or relax. Muscles cannot expand. Muscle contraction is an active process; that is, it requires ATP.

Muscles usually work in opposing pairs. For example, two opposing sets of muscles enable you to move your arm up and down. You raise (bend) your forearm by contracting one set of muscles and relaxing the other. You lower (straighten) your forearm by relaxing the first set of muscles and contracting the second.

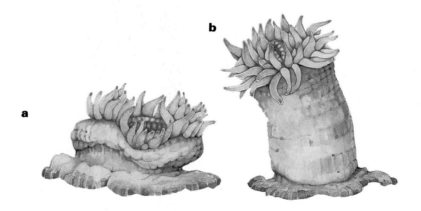

11–8 Opposing muscles. A sea anemone has longitudinal muscles running up and down its body wall and circular muscles surrounding its inner digestive cavity. When contracted, longitudinal muscles shorten the body (**a**), circular muscles lengthen it (**b**).

Skeletons Aid Movement Many animals have a skeleton, which is a hard framework that gives the animal's body shape and support. It also provides protection for body organs and aids in locomotion and moving body parts. In animals that have skeletons, including arthropods and vertebrates, muscles pull against bones and other skeletal tissues. This results in strong, efficient movement. By working together, skeletons and muscles enable arthropods and vertebrates to fly, run, jump, and swim quickly. In animals that lack skeletons, such as slugs, muscles pull against the fluid-filled body. This results in weak swimming or crawling movements. This type of muscle action is not as strong or effective as muscle action in animals that possess skeletons.

Mollusks have a system of muscles and skeleton that functions in movement. For example, muscles connect to the outside shells of the clam, and muscles connect to the internal shell and cartilage of the squid.

Section 11.1 Movement in Living Things

11–9 A molting damselfly has just emerged from its exoskeleton, which is still clinging to a reed.

Exoskeletons A skeleton that is found outside the body is referred to as an **exoskeleton**. The arthropod skeleton is an exoskeleton. It is a hard, nonliving body covering. All the living parts of the organism are inside, protected by the skeleton. The arthropod exoskeleton is mostly made of a complex carbohydrate called **chitin** (KĪT-in). In addition to chitin, the exoskeleton also contains protein and minerals. In land arthropods, wax covers the outside of the exoskeleton, making it waterproof. The wax prevents water from entering the body from the outside and also prevents evaporation of water from the inside.

To understand how an exoskeleton works, imagine a lobster's exoskeleton. It sticks closely to the skin. There are no bones in the lobster's body, and the muscles connect directly to the exoskeleton. The contraction of the lobster's muscles moves the parts of the exoskeleton.

Exoskeletons present two problems. One arises as the animal grows, because the hard exoskeleton does not grow with the organism. In arthropods this problem is overcome by a process of periodic **molting**. During certain times of the year, an arthropod sheds its exoskeleton and grows a new one. Young crayfish, for example, molt several times a year. Before the old hard exoskeleton is shed, a new soft one develops underneath. Then the old one breaks open and the animal works its way out, leaving its former covering behind. For the few days that it takes the new exoskeleton to harden, the crayfish is quite defenseless and usually hides.

The second problem of an exoskeleton is that it limits the size of the animal. An exoskeleton on a large organism would be extremely heavy and bulky. Exoskeletons are efficient body structures only in small organisms.

In starfish and other echinoderms, muscles attach to an internal skeleton composed of chalky plates.

Endoskeletons A skeleton that is found inside the body is called an **endoskeleton**. The vertebrate skeleton is an endoskeleton. It is a structural framework inside the body composed of bone or cartilage or both. A substance containing calcium and phosphorus between the cells of bone tissue makes it hard and strong. A substance containing protein and carbohydrates between the cells of cartilage tissue makes it flexible and rubbery. The skeletons of sharks and rays are composed wholly of cartilage. In most of the vertebrates, however, the skeleton is composed of bone with cartilage in the joints and between the vertebrae.

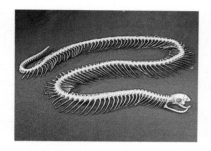

11–10 Endoskeleton of a snake. Notice the hinged jaw and numerous ribs attached to the spinal column.

11–11 Muscles of a vertebrate attach to internal bones; those of an arthropod attach to hinged parts of the exoskeleton.

The endoskeleton is moved by muscles. In vertebrates, most muscles are attached to two different parts of the skeleton. When a muscle contracts, it pulls the bone attached to one end toward the bone attached to the other end. For example, there is a muscle attached to the lower jaw on one end and the upper jaw on the other. Contraction of this muscle closes the jaws by pulling the lower jaw toward the upper jaw. Unlike exoskeletons, endoskeletons grow with the body. They are also light and flexible enough to support large animals.

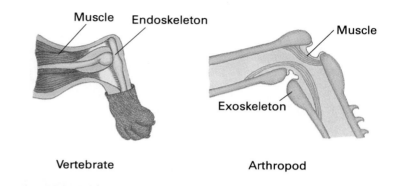

Building a Science Vocabulary

locomotion   turgor movement   molting
pseudopod   exoskeleton   endoskeleton
ameboid movement   chitin

Questions for Review

1. State whether each of the following is an example of locomotion, movement of a body organ, or movement within a cell.
   a. peristalsis
   b. flying
   c. cytoplasmic streaming
   d. turning of leaf toward light

2. What structures move cilia and flagella? How do they work?

3. Compare the following features of exoskeletons and endoskeletons.
   a. location in body
   b. composition
   c. size of organisms with them
   d. advantages and disadvantages

11.2 MOVEMENT AND LOCOMOTION IN THE REPRESENTATIVE ORGANISMS

Movement in the Bean Plant

If you watch a bean plant hoping to see it move, you probably will not see very much. If you take periodic photographs over a 24-hour period, however, you will detect changes in the position of plant organs. Leaves and shoots turn toward light; roots turn downward and toward water. Climbing varieties of bean plants cling to and climb up poles or other stakes. Young leaves and flower petals unfold during opening. Changes in position of plant organs are usually the result of unequal growth. For example, when petals open, the upper surface of the petal grows faster than the lower one. This causes the petal to bend outward horizontally. Such growth movements in plants are under chemical control.

In addition to growth movements, movement of plant organs can be due to changes in water content. Bean plants seem to "sleep" at night because their leaves fold up. During the day, the leaves lay flat. The differences in leaf position (often called sleep movements) are due to changes in the water content of special cells at the base of the leaves.

11–12 Certain varieties of bean plants twine their stems and tendrils around poles and other objects as they grow.

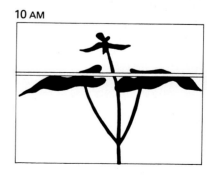

11–13 Sleep movements in a bean plant. Leaves fold up or open in response to the water content of certain cells at the base of the leaves.

Movement and Locomotion in the Paramecium

The streamlined body of the paramecium enables it to move quickly through the water. The paramecium has a hard outer covering that gives it a permanent slipperlike shape. As many as 17 000 cilia propel the paramecium through the water. In addition, cilia in the oral groove create a current that draws water and particles into the paramecium.

The propelling action of cilia resembles a swimming stroke in which the arm reaches forward and then pushes back against the water. The paramecium's cilia do not all beat at the same time. A wave begins at one end of the protist and works toward the other. The cilia beat at an angle, causing the paramecium to turn on its long axis as it swims. In swimming backwards, the paramecium's cilia reverse their stroke.

You can observe the range of a paramecium's movement by placing a foreign substance, such as dilute sodium hydroxide, in the water with a paramecium. The organism will swim toward the material, sample it by drawing it into the oral groove, back up, turn, and swim away. The coordinated action of cilia makes these movements possible.

11–14 In a paramecium, cilia beat one after another in sequence, creating a wave of movement that propels the animal.

11–15 A paramecium swims forward until it contacts an object. It then swims backwards, rotates, and swims forward again.

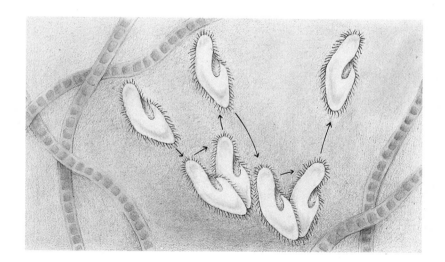

Movement and Locomotion in the Hydra

Hydras are usually attached to rocks or water plants by a sticky secretion of the base. While attached, a hydra may shorten and extend its body, twist about, and wave its tentacles. All these actions aid in the capture of prey. If no food appears, or if the environment becomes too warm or lacking in oxygen, the hydra

11–16 Somersaulting is a hydra's fastest means of locomotion.

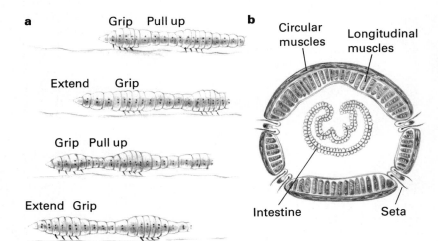

11–17 Contractile fibers in the inner (left) and outer (right) cell layers of the hydra are responsible for movement.

may move to another location. In locomotion, a hydra glides along slowly on its base by ameboid movement of cells in the base. It can also swim and somersault through the water.

All of the hydra's movements are due to the action of contractile fibers. These long, thin, threadlike structures are located in cells in the body wall and tentacles. In the hydra's outer layer, contractile fibers run lengthwise. When these fibers contract, the body shortens and the tentacles bend. Contractile fibers in the hydra's inner layer are arranged in a circular fashion. When the circular fibers contract, the body narrows and the tentacles straighten out. Contractions of fibers around the hydra's mouth and at the bases of the tentacles close the mouth. The hydra's movements and locomotion are controlled by specialized nerve cells, which stimulate the contractile fibers.

Locomotion in the Earthworm

The earthworm has a well-developed system of muscles. Under the skin, there is a thin layer of circular muscles and a thicker layer of longitudinal muscles. The earthworm becomes longer and thinner when the circular muscles contract, and shorter and fatter when the longitudinal muscles contract.

11–18 Earthworm locomotion: **a.** An earthworm moves forward by extending its front end and grabbing the soil with its setae. It then pulls up the segments behind the front end. **b.** Cross section of a segment, showing the muscles and setae.

The earthworm's body is divided into segments. Each segment of the earthworm has its own longitudinal and circular muscles. Having a separate muscle system in each segment allows the worm to elongate one part of its body while it shortens another. Four pairs of bristles or **setae** (SEET-ee) protrude from the sides and bottom of the body wall in each segment. Muscles attach to the setae and move them.

The earthworm is well adapted for burrowing into the earth. It has a streamlined shape and its setae anchor it to its burrow. It also lacks sense organs on the head, which would get in the way as it passes through the soil.

When the earthworm crawls, it extends its front end and grabs the soil with its front setae. Then the rear setae release their hold and the back part of the worm is drawn up. The muscle contractions pass in wavelike fashion from segment to segment along the worm. These contractions are controlled by nerves in each segment.

Locomotion in the Grasshopper

If you have ever tried to catch grasshoppers, you know that their locomotion is well developed. They jump, hop, fly, and even change direction in mid-flight. These actions are the result of the coordinated and controlled contraction of many muscles.

The grasshopper's brain and nerves control muscle contractions in movement and locomotion.

The grasshopper's body is divided into three parts: *head, thorax,* and *abdomen*. Muscles inside the grasshopper attach to the exoskeleton covering each body part. Head muscles move the grasshopper's jaws, feelers, antennas, and food handlers. Thorax muscles move the three pairs of jointed legs and two pairs of wings. Muscles in the abdomen contract during breathing. They also function in reproduction, such as when the female grasshopper digs a hole in which she lays eggs.

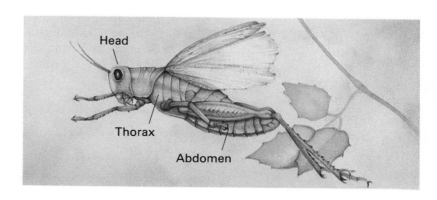

11–19 As in all insects, a grasshopper's body has three main sections: the head, thorax, and abdomen. Muscles in the thorax move the legs and wings.

Section 11.2 Movement and Locomotion in Representative Organisms

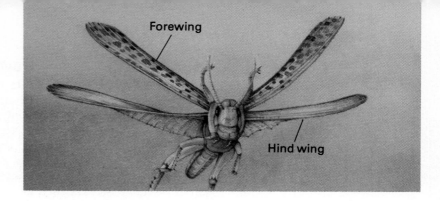

11–20 Both pairs of a grasshopper's wings are active in flight. When not in flight the hind wings fold up under the forewings.

Each of the grasshopper's legs ends in two claws with a small soft pad between them. The pad helps the grasshopper cling to surfaces. The first two pairs of legs are walking legs. The rear legs are used for leaping. Large powerful muscles located in the upper part of the rear legs contract when the grasshopper jumps or hops.

The grasshopper has two pairs of wings. The first pair of wings is narrow and hard. The hind wings, however, are broad and flexible. They open wide during flight. The thorax muscles that attach to the wings twist them as well as move them up and down. These actions keep the insect in flight and allow it to adjust to varying air conditions. They also enable grasshoppers to reverse direction while flying.

Building a Science Vocabulary

seta

Questions for Review

1. What causes each of the following movements?
    a. sleep movements in the bean plant
    b. petal opening in the rose
    c. bending of cilia in the paramecium
    d. extending of setae in the earthworm
    e. bending of a leg in the grasshopper

2. Explain how the hydra lengthens and shortens its body.

3. How is the earthworm adapted for burrowing?

4. How does being able to twist its wings aid the grasshopper in flight?

11.3 MOVEMENT AND LOCOMOTION IN HUMANS

Most of us are so accustomed to walking, running, throwing, and carrying that we tend to take these movements for granted. Yet the ability to throw a baseball with accuracy, walk around with packages in both arms, or run a race requires a great deal of coordination. We are adapted for locomotion on two legs. This type of locomotion allows our hands and arms to be free to carry objects or use tools. Two problems of locomotion are that balance and support of body weight are more difficult on two legs than on four. One foot at a time must support our entire body when we walk. In addition, our knees must be strong enough to support our weight when we stand, run, and walk; yet, they must be flexible enough to bend and turn.

Muscles contract and move parts of the skeleton. These movements are aided by the sense organs and controlled by the brain and nerves. Complex interactions occur that help us keep our balance, support our weight, rotate our bodies, and move gracefully from place to place. At the same time, internal body movements also occur in the heart, blood vessels, diaphragm, digestive tract, and white blood cells.

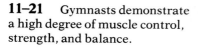

11–21 Gymnasts demonstrate a high degree of muscle control, strength, and balance.

The Human Skeleton

The human skeleton, like the skeleton of all vertebrates, is located inside the body. In addition to its important role in movement, the skeleton provides shape and support to the body and protects delicate internal organs. For example, the brain is protected by the skull, and the heart and lungs are protected by the ribs.

The adult human skeleton is made up of 206 bones.

Structure of the Human Skeleton The human skeleton is made up of two parts, as is the skeleton of other vertebrates. One part includes bones that run from the head to the base of the spine. These include the skull, backbone, and ribs. Bones in the other part of the skeleton attach to the head-through-spine portion. These bones make up the appendages (arms and legs). The collarbones, shoulder bones, and hipbones are considered parts of the appendages.

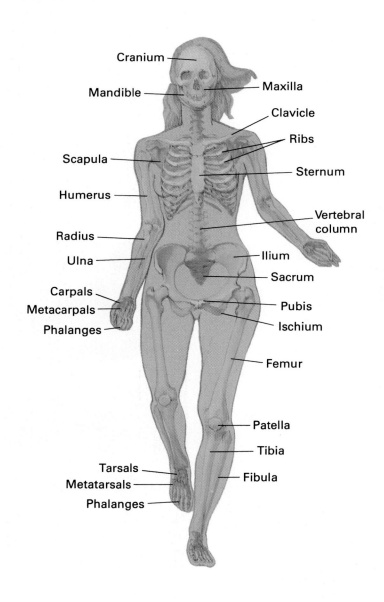

11–22 The human skeleton is strong yet versatile.

Bone marrow contains blood and fat. Blood cells in various stages of development are located in the bone marrow.

The shape of a bone in the body is related to the function it performs.

The bones in the human body have various shapes, which relate to their function. Long bones in the legs and arms are strong and able to support weight and withstand strain during movement and locomotion. Bone marrow in the long bones produces new blood cells. Short bones in the wrists, fingers, hands, and feet also support weight and, more importantly, enable us to make small, delicate movements. Flat bones in the shoulders, ribs, hips, and skull protect soft parts of the body. In addition, there are irregularly shaped bones in the spine and inner ear. The vertebrae in the spine protect the spinal cord. Inner ear bones help conduct sound through the ear. Finally, small rounded bones, such as the kneecap, are often found in **joints**, which are places where two bones come together.

Skeletal Tissues

Unlike the exoskeleton of arthropods, the human endoskeleton is composed of living tissue. It consists of cartilage, bone, and tissues that connect them.

Cartilage The material that supports the outer ear and gives it shape is **cartilage**. Cartilage is also found at the tip of the nose, in rings supporting the trachea and bronchi, in the larynx, and at the end of ribs. In addition, it lines and cushions surfaces of joints. Anyone who has ever suffered the pain of a "slipped disk" in the spine, or an injured knee, appreciates the function of cartilage.

Cartilage cells are large and rounded. They produce and secrete polysaccharides and protein fibers. These materials make up the firm, flexible **matrix** (MAY-triks) that surrounds the cells. Before birth, much of a person's skeleton is cartilage. It is gradually replaced by bone as the person grows.

The large flexible part of a chicken breastbone is made of cartilage.

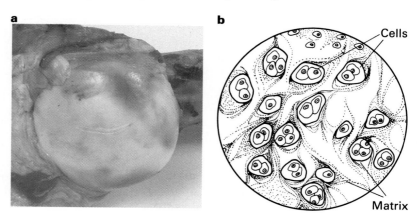

11–23 Cartilage lines the surfaces of joints, such as the pelvic end of this cow femur (**a**). Cartilage consists of cartilage cells embedded in a matrix made up of polysaccharides and proteins secreted by the cells (**b**).

Section 11.3 Movement and Locomotion in Humans

Bone makes up less than 20 percent of the weight of the human body.

Skeletons of birds are extremely light, because their bones have many spaces. In some birds, the skeleton weighs less than the feathers.

Bone Two types of bone tissue are found in the human body: **spongy bone** and **compact bone**. Spongy bone is found at the ends of the long bones. It has many spaces and is very light. Compact bone is found along the length of the bone. Canals in bone tissue contain blood vessels that transport materials to and from the bone cells. Grooves between the bone cells contain tissue fluid, which bathes bone cells, making the transport of materials possible.

Bone tissue is composed of bone cells and a hard matrix that is produced by specialized cells. The matrix contains ropelike fibers made of the protein **collagen** (KAHL-uh-JEN) and mineral crystals made of calcium and phosphorus. The mineral crystals cling to the collagen fibers. During a person's lifetime, the mineral content of bone is constantly replaced. In one year, 100 percent of the calcium in an infant's bones is replaced. In an adult, 18 percent is replaced each year. During replacement, bone cells break down calcium in the matrix and release it into the blood. At the same time, calcium from food in the intestine is delivered to the bone cells, which deposit it in the matrix. Vitamins play an important role in bone formation. Vitamin C is necessary for the production of collagen, and vitamins A and D are necessary for matrix formation.

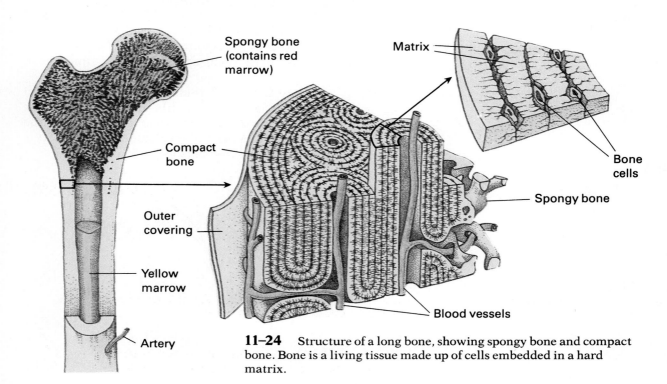

11–24 Structure of a long bone, showing spongy bone and compact bone. Bone is a living tissue made up of cells embedded in a hard matrix.

Ligaments attach bone to bone; tendons attach muscle to bone.

Skeletal Connections The bones that meet at some joints do not move. For example, in adults there is no movement between the bones that join in the skull. Other joints are movable, including the shoulder, elbow, hip, and knee. The bones in movable joints are attached to each other by **ligaments**. Ligaments are mainly composed of collagen, which provides strength, and **elastin**, a protein that permits ligaments to be stretched. Cartilage acts as a cushion between the bones in joints. Joints must be strong yet flexible. This dual purpose makes a joint such as the knee prone to injury. You can easily injure your knees if you twist your knee joints beyond their capacity to turn.

Muscles are attached to the bones by **tendons**, which are mostly made up of collagen. They are not very stretchable, because they do not contain elastin. Some tendons are quite long, such as the tendon that attaches the calf muscle to the heel bone.

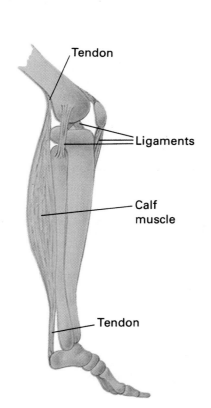

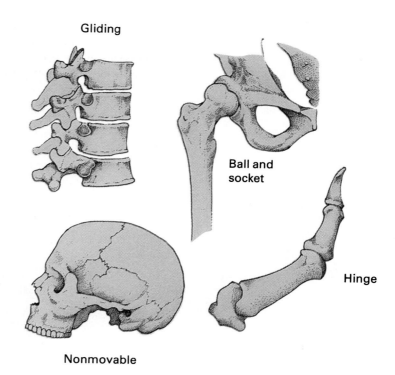

11–25 Types of joints: gliding (vertebrae); ball and socket (hip); nonmovable (skull); and hinge (finger).

11–26 Some of the ligaments and tendons of the leg and foot. Ligaments attach bone to bone; tendons attach muscle to bone.

Section 11.3 Movement and Locomotion in Humans

Human Muscle

Bones move as a result of muscle contraction. The body has three types of muscle: skeletal muscle, smooth muscle, and cardiac muscle.

Skeletal muscle is striated and voluntary.

Skeletal Muscle The muscles involved in locomotion are **skeletal muscles**. Skeletal muscle has stripelike markings, or striations. The striations are quite striking when the tissue is examined under the compound microscope. Skeletal muscle is composed of long **muscle fibers**. Each muscle fiber is a cell that contains several nuclei.

The nervous system controls the contraction of skeletal muscles. Many skeletal muscle contractions are automatic and we are unaware of them. For example, we do not have to think about moving the diaphragm and other muscles in order to breathe. We can, however, consciously control the action of skeletal muscles. For example, we can stop breathing for a while if we want to. Similarly, we decide when to walk, sit, or throw a ball. For this reason, skeletal muscles are also called **voluntary muscles**.

Skeletal muscles usually work in pairs. Remember how the arm moves. One set of muscles bends the arm, and another set of muscles straightens it. In general, muscles that bend a body part are called **flexors**. The biceps is a flexor in the arm. Muscles that straighten a body part are **extensors**. The triceps is an extensor in the arm. The combined action of flexors and extensors produce most of the movements of the skeleton.

11–27 Skeletal muscle has striations and is found primarily in the arms and legs.

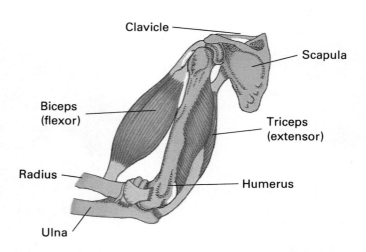

11–28 The biceps and triceps are a muscle pair. The biceps pulls arm bones toward each other; the triceps pulls them apart.

Chapter 11 Movement and Locomotion

Smooth muscle is not striated and is involuntary.

Smooth Muscle Most of the muscle in our internal organs is **smooth muscle**, which is composed of spindle-shaped cells that interlace with one another. Each smooth muscle cell has a single nucleus. Smooth muscle cells do not have striations.

Smooth muscles make up the wall of the digestive tract. Their contractions cause the wavelike movements of peristalsis. Smooth muscles are also found in the urinary bladder, gallbladder, arteries, and veins.

Smooth muscles are controlled by the nervous system and hormones. Under most circumstances, we cannot consciously control the action of smooth muscles. Therefore, they are often called **involuntary muscles**. Involuntary muscles are found in all vital organs where movement or contraction occurs. They even control the size of the opening through which light enters the eye.

Cardiac muscle is striated and involuntary.

Cardiac Muscle The heart's coordinated contractions are performed by cardiac muscle. Like smooth muscle, cardiac muscle is involuntary. Cardiac muscle resembles skeletal muscle in that it is composed of striated fibers that contain many nuclei. Cardiac and skeletal muscle differ, however, in the arrangement of fibers. The fibers in skeletal muscle are parallel to one another. In cardiac muscle, they form a kind of network. In addition, cardiac fibers are pressed close together where they meet at the ends. In places, the cell membranes of adjacent fibers are fused, forming junctions. Impulses pass from fiber to fiber across the junctions. The way the fibers are arranged in cardiac muscle helps coordinate the individual contractions of the heartbeat.

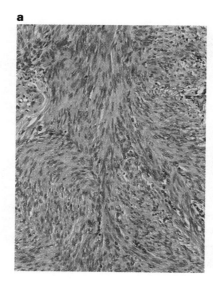 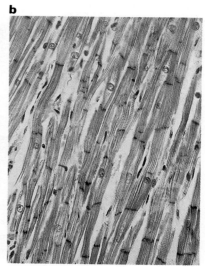

11–29 Smooth muscle from the uterus (**a**) and cardiac muscle (**b**). Smooth muscle is also found in the blood vessels, digestive tract, urinary bladder, and gallbladder. Cardiac muscle is found only in the heart.

Structure and Contraction of Skeletal Muscle

When you observe a small piece of skeletal muscle under the compound microscope, you can see that it is composed of long fibers. Each fiber is considered to be one cell. Further magnification shows that each muscle fiber is composed of even narrower parts called **fibrils** (FĪ-brulz).

Electron microscope examination of a piece of a single fibril shows that it is composed of **filaments**. The filaments in a fibril are made of protein and are arranged in an orderly manner. The thicker filaments contain the protein **myosin** (MĪ-uh-sin). The thinner filaments contain the protein **actin**. Tiny cross bridges of myosin connect the actin and myosin filaments.

The way skeletal muscle contracts is closely related to its structure. In addition, certain chemical substances are necessary for contraction to occur.

The Sliding Filament Theory of Muscle Contraction The **sliding filament theory** has been developed to explain how skeletal muscles contract. According to this theory, the actin and myosin filaments in the fibril do not get shorter when a muscle contracts. Instead, the filaments slide past each other. When this happens, the cross bridges are rearranged. When skeletal muscle relaxes, the filaments slide apart and return to their original position.

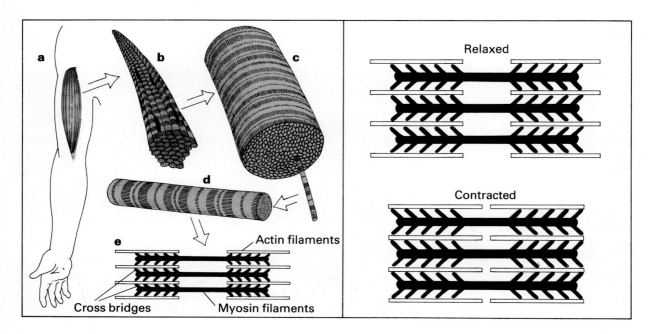

11–30 Skeletal muscle structure (left): **a.** biceps muscle; **b.** muscle fibers; **c.** single muscle fiber; **d.** fibril; **e.** fibril structure, showing actin and myosin filaments. Muscle contraction (right) appears to occur by the sliding of actin and myosin filaments past each other.

Substances Required for Contraction Two substances are required for the contraction of skeletal muscles. One is ATP, which supplies energy. The other is calcium, which is needed to start muscle contraction. Small membrane pockets around the muscle fibrils store calcium. Calcium is released when a nerve impulse reaches a muscle. The calcium flows into the muscle cells and causes the myosin bridges to become active. Then, the filaments slide together using energy from ATP. Some of the energy from ATP is given off as heat during muscle contraction.

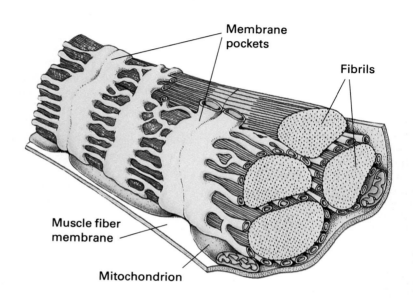

11–31 Muscle fibrils are surrounded by a system of membrane pockets, which release calcium when a nerve impulse reaches the fiber. Mitochondria, which supply ATP for contraction, lie between the fibrils and the muscle fiber membrane.

Structure and Contraction of Smooth and Cardiac Muscle

Most skeletal muscles are long and narrow. When skeletal muscle cells contract, they shorten the muscle. When cardiac muscle cells contract, however, they constrict the heart chambers.

The cells of smooth muscle contain contracting fibrils. These fibrils are made up of thick myosin filaments and thin actin filaments. The fibrils are not arranged in parallel fibers as in skeletal muscle. Instead, they interweave to form a sheet or mass of fibers. During contraction, the thick and thin filaments slide past each other. Contraction is slower and lasts longer in smooth muscle than in skeletal muscle. It also uses ATP as a source of energy.

The cells (fibers) of cardiac muscle are made up of actin and myosin fibrils. The fibrils slide past each other when the muscle contracts. Membrane pockets around the fibrils supply calcium needed for contraction. Large numbers of mitochondria close to the fibrils supply them with ATP.

Section 11.3 Movement and Locomotion in Humans

Muscle Action and Exercise

The action of a muscle fiber consists of a brief contraction followed by relaxation. This action is called a **twitch**. A fast twitch in skeletal muscle can last as little as 7.5 milliseconds (ms). A slow twitch in smooth muscle can last 100 ms. Individual fibers twitch when a muscle contracts. One twitch of one muscle fiber does not exert much force. Yet when many fibers in a muscle twitch, the force is considerable. A muscle contraction is made up of the individual twitches of many fibers. To sustain a contraction, such as when you flex your biceps, some fibers contract while others relax.

Muscles require food, oxygen, waste removal, and the other services of the transport system. Muscles generally have a good blood supply. Muscles with an ample supply of blood are red in color. You can see red muscle and white muscle when you eat chicken. The white muscle on the breast does not get sufficient blood for sustained exercise. The dark leg muscles get more blood, and are able to work longer and harder.

During sustained exercise the blood cannot always deliver enough oxygen to muscle cells. In this case, muscles perform lactic acid fermentation to obtain enough ATP to contract (see Chapter 6). The lactic acid accumulates in muscle cells, lowering the pH and reducing the muscles' ability to contract. This condition is known as **muscle fatigue**. When the exercise stops, less ATP is needed, fermentation stops, and the lactic acid is broken down.

Trained athletes can sustain exercise for a longer period and build up less lactic acid than untrained people. Through training their muscles are able to take in more oxygen than usual.

11–32 The effects of muscle stimulation by an electric impulse. When the stimuli are far apart simple twitches result (**a**). As the stimuli become more frequent, there is little time for relaxation of the muscle (**b**). When the stimuli are very close together, the muscle is constantly contracted (**c**). If such frequent stimulation continues, the muscle eventually becomes fatigued as lactic acid accumulates (**d**).

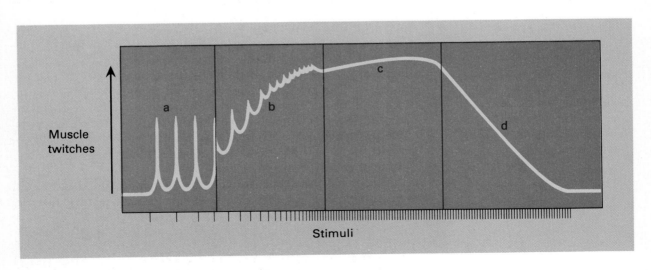

11–33 Muscle development in a weight lifter. The biceps is performing most of the work in this exercise.

Athletes also build up the size of their muscles. A muscle becomes larger when it is used strenuously and continually. You can see this in the muscular development of weightlifters and in the enlarged leg muscles of skaters, dancers, and runners. The enlarged muscle works more efficiently since it stores more glycogen and other nutrients and its fibers increase in diameter.

Muscles can become smaller and weaker when they are not used. This condition is referred to as **muscle atrophy**. One month in a cast, for example, can cause leg or arm muscles to decrease considerably in size. Lack of use may reduce a muscle's size, but the muscle can still function normally. The permanent loss of nerve stimulation, however, causes a muscle to deteriorate. Scientists believe that without nerve stimulation, cell membranes of skeletal muscle fibers may be unable to absorb nutrients.

Control of Muscle Movement

Skeletal muscle in the body responds to stimulation by nerves. Nerve fibers in the brain and spinal column carry signals that cause muscles to contract. For example, if you decide to pick up a pen, nerve signals travel along nerves to the proper muscles in your arm and hand. The signals make these muscles act in a coordinated manner, and you grasp and raise the pen.

Smooth muscle is controlled by nerves and hormones. For example, nerves regulate widening and narrowing of the arteries by controlling the smooth muscle in artery walls. Nerves also control smooth muscle in the intestine and help regulate peristalsis. The smooth muscles in the arteries and in the intestine are also affected by the hormone adrenalin. Adrenalin in the bloodstream causes arteries to narrow and peristalsis to slow down.

Cardiac muscle is affected by nerves and hormones; however, the heartbeat's strength and rate are set by the pacemaker tissue in the heart. Unlike skeletal muscle and smooth muscle,

11–34 A long, narrow nerve fiber is in close contact with a skeletal muscle fiber. Skeletal muscles contract when stimulated by nerves.

cardiac muscle that has been cut into pieces continues to beat. This occurs because pacemaker cells beat on their own and spread the contractions to nearby cells. Cardiac muscle is unique in its ability to create and synchronize its own contractions.

Building a Science Vocabulary

joint	tendon	fibril
cartilage	skeletal muscle	filament
matrix	muscle fiber	myosin
spongy bone	voluntary muscle	actin
compact bone	flexor	sliding filament theory
collagen	extensor	twitch
ligament	smooth muscle	muscle fatigue
elastin	involuntary muscle	muscle atrophy

Questions for Review

1. Which of the two main parts of the skeleton includes the skull? How does the shape of the skull help in its function?

2. State one difference and one similarity between the terms in each of the following pairs.
   a. cartilage and bone
   b. ligament and tendon
   c. voluntary muscle and involuntary muscle

3. Which of the three types of muscle would you find in
   a. fingers?
   b. the small intestine?
   c. the heart?
   d. arteries?

4. Describe the structure of a skeletal muscle fibril as seen under the electron microscope.

5. How do muscle cells get oxygen? What causes muscle fatigue?

6. If heart muscle is cut into pieces it continues to beat, but skeletal muscle does not. Explain the difference in the control of these two types of tissue.

Perspective: A New View
Biology as Art

Living things are depicted by artists throughout this book. There are line drawings, renderings, and photographs. Visual art brings us messages that words sometimes cannot convey as easily.

From our earliest beginnings, people have represented animal and plant life in art form. Drawings of buffalolike animals adorn the caves where Cro-Magnon lived more than 25 000 years ago. Drawings of animals and plants are the focus of the mosaic floors, paintings, and pottery of ancient Greece, Rome, and China. Making drawings of flowers and fruits is one of the earliest lessons in art classes today.

Biological art is usually representational. That is, it is designed to show details realistically, rather than in a fanciful way. Some of the finest early examples of such art were made by the herbalists of sixteenth-century Europe. Their drawings of plants are meticulous in detail and are accompanied by written descriptions of the plants and their medicinal uses. Unlike the herbalists, some early artists took considerable liberty in their work, resulting in bizarre creatures or stylized benign animals. Today, we expect great accuracy in portraying animals and plants. Among the most outstanding animal artists was John James Audubon, whose paintings of North American birds and mammals are now highly prized.

Biological organisms do not all sit still, and the artist is challenged to capture motion on canvas. In its crudest form, such motion is shown by lines and arrows. In its most sophisticated form it is shown by super-positioning of images. Motion can be caught by the photographic artist on movie or TV film, as well as by time lapse photography.

Whether on canvas or film, biological art is a creative activity, an exploration of form, color, and design; it is an expression of original thought. Science is also a creative activity since it, too, involves exploration of form and design and the development of original thought. Art and science are unique human experiences, worthy of both study and appreciation.

Chapter 11
Summary and Review

Summary

1. Movement in living things includes movement within cells, movement of body parts, and locomotion.

2. Movement helps animals find food, mates, and shelter. Movement also helps animals escape predators and other dangers; it aids animals in defense and in communication. Movement is an important part of many body functions.

3. Cilia, flagella, and pseudopods are three structures used in the movement of unicellular organisms and cells in multicellular organisms. All depend on the contraction of protein filaments.

4. Plants exhibit growth movements and turgor movements. Simple animals use contractile fibers in body cells for movement; more complex animals use muscles. Animals with muscles and a skeleton are best adapted for movement.

5. In the bean, organs change position during growth. This occurs when leaves turn toward light and roots turn toward the ground. In addition, changes in water content result in sleep movements of leaves.

6. Paramecia move using cilia, hydras move by contracting fibers in cells, and earthworms move using muscles and setae.

7. Movement in the grasshopper depends on muscles attached to the chitinous exoskeleton.

8. Humans move using muscles connected to an endoskeleton containing cartilage and bone.

9. Cartilage and bone tissue consist of cells embedded in a matrix. Bone matrix is hardened by mineral deposits; cartilage is flexible.

10. The three types of muscle in the human are skeletal, smooth, and cardiac. They differ in structure and function; however, contraction occurs by filaments sliding past each other. The source of energy for contraction is ATP.

11. Nerves and hormones control the function of muscle tissue. Cardiac tissue is also controlled by pacemaker cells.

Question Summary

Application

1. Why is a grasshopper unable to shorten and lengthen its body as an earthworm can? Why is an earthworm unable to jump and hop as a grasshopper can?

2. Why might it be difficult to pull an earthworm out of its burrow?

3. Movement in living things often depends on contractile fibers. Name a movable structure that contains contractile fibers in the following organisms.
   a. paramecium
   b. hydra
   c. snake
   d. grasshopper

4. What is the function of calcium in bone tissue? in muscle tissue?

5. The green anole lizard runs away quickly as predators approach. What kind of muscle is used during its locomotion? How is that muscle controlled? Briefly describe how a muscle bends the anole lizard's leg.

Interpretation

1. Would you expect to find wax in the exoskeleton of the crayfish? Why or why not?

2. Why are knee injuries common in skiers?

3. A student felt that the sliding filament theory of muscle contraction was incorrect. Instead, the student proposed that muscle fibrils fold up like an accordion when muscles contract. What kind of evidence would be needed to support the student's idea? Does this evidence exist? What evidence exists to support the sliding filament theory?

Extra Research

1. Study the skeletons and body movements of birds, monkeys, apes, and chimps. Write a paragraph explaining the differences in the ways these organisms walk on two legs.

2. Make a model of the knee joint. Use balsa wood for the bones, elastic bands for the ligaments and tendons, and string for the muscles. Show how the knee provides strength for standing and flexibility for turning. Also show why the knee is prone to accidents.

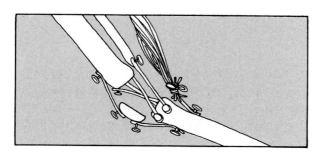

3. In the library try to find information about famous athletes who injured their knees (Joe Namath, Steve Podborski, Bobby Orr, and others). Determine the cause of their injuries and what was done to help them. Write up your findings in the form of a feature article for a magazine.

4. Snakes lack arms, legs, and wings; yet they are able to move efficiently from place to place. Using reference books in the library, find out how snakes move around.

Career Opportunities

Radiological technologists must have two or more years of training. They use radiation to help physicians diagnose and treat persons with injuries and diseases, such as tumors, ulcers, broken bones, and blood clots. Radiological technologists position instruments and patients for X-rays or radiation therapy and set the proper dosages. Most work in hospitals. Contact the Canadian Association of Medical Radiation Technologists, 280 Metcalfe St., #410, Ottawa, Ontario K2P 1R7.

Physical therapists and their assistants help rehabilitate persons with injuries and diseases of the nerves, bones, or muscles. They administer and interpret tests of physical abilities. Then, with a physician's prescription, they help patients with exercises and other treatments. A bachelor's degree, specialized training, and certification are required. For information, write the Canadian Physiotherapy Association, 44 Eglinton Ave. W., #201, Toronto, Ontario M4R 1A1.

Unit III
Suggested Readings

Buchsbaum, Ralph, *Animals Without Backbones*, Chicago: University of Chicago Press, 1975. *Beautifully illustrated and well-written classic text on invertebrates.*

Dunne, Robert, and Madeleine Livaudais, *Skeletons: An Inside Look at Animals*, New York: Walker and Co., 1972. *Structure and function are emphasized in these studies of 13 chordate skeletons, from snake to human.*

Eckstein, Gustav, *The Body Has a Head*, New York: Bantam Books, Inc., 1980. *A literary style conveys the poetry and wonder of the human body.*

Englebardt, Stanley L., *How to Get in Shape for Sports*, New York: Lothrop Lee and Shepard Co., Inc., 1976. *Discusses human nutrition and conditioning.*

Lenihan, John, *Human Engineering: The Body Re-Examined*, New York: George Braziller, Inc., 1975. *A study of the body through the eyes of an engineer. Answers such questions as, Why does a sprinter hold his breath? and, How does bone compare to fiberglass?*

Nilsson, Lennart, with Jan Lindberg, *Behold Man: A Photographic Journey of Discovery Inside the Body*, Boston: Little, Brown & Co., 1974. *Spectacular photography illuminates various internal structures and processes of the human body.*

Patton, John S., and Martin C. Carey, "Watching Fat Digestion," *Science*, April 13, 1979. *Describes what fat digestion in the small intestine looks like under the light microscope.*

Prince, J. H., *How Animals Move*, New York: Elsevier/Nelson Books, 1981. *Animal motion and how it takes place is presented in the context of variations among species.*

Riedman, Sarah R., *Heart*, Golden Guide Series, New York: Western Publishing Co., Inc., 1974. *Discusses anatomy and functioning of the heart, as well as heart disease, its causes, and treatment.*

Rutland, Jonathan, *Human Body*, Warwick Press/F. Franklin Watts, Inc., 1977. *A short book of essential facts well organized and amply illustrated. Covers cell structure, tissues, organs, breathing, speech, digestion, food, and vitamins, among other topics.*

Silverstein, Alvin, and Virginia B. Silverstein. *The Excretory System: How Living Creatures Get Rid of Wastes*, Englewood Cliffs: Prentice-Hall, Inc., 1972. *Compares process of excretion in humans, water and land animals, and plants. Outlines some simple experiments.*

Snider, Arthur J., "Kidney Patient's Dilemma—Dialysis or Transplant?," *Science Digest*, March, 1974. *Describes the pros and cons of two types of treatment for kidney diseases.*

Sonstegard, David A., Larry S. Matthew, and Herbert Kaufer, "The Surgical Replacement of the Human Knee Joint," *Scientific American*, January, 1978. *Fascinating, well-illustrated article explains the structure and function of natural and artificial knees.*

Walker, Dan B., "Plants in the Hostile Atmosphere," *Natural History*, June-July, 1981. *Describes a tree as a "vegetable skyscraper" and uses engineering concepts to explain plant and tree systems of support, gas exchange, conduction, and water retention.*

Zihlman, Adrienne, and Douglas Cramer, "Human Locomotion," *Natural History*, January, 1976. *Excellent article comparing human locomotion with locomotion in other animals.*

Unit IV

Body Regulation and the Senses

Imagine the many activities that are taking place at the same time in the body of a rabbit. It is breathing and transporting oxygen to all its cells, while it expels carbon dioxide produced by its cells. It is processing food, separating nutrients from wastes and delivering the nutrients to all parts of the body. It is filtering its blood, removing wastes and maintaining the proper balance of ions and other substances. It is moving, both voluntarily and involuntarily. It is constantly producing new cells and disposing of dead ones. How is it possible for all of these activities to occur simultaneously, as complicated as they are? The smooth functioning of the rabbit's body depends on its control systems.

At the same time as the rabbit's control systems coordinate the activities going on inside its body, it is obtaining information about both its internal and external environments. The rabbit can see and smell food. It hears the rustling of leaves and breaking of twigs that signal the approach of an enemy. It knows where its body parts are. All of these functions are made possible by the rabbit's sense organs. They connect the rabbit with its environment.

Functioning together, an organism's control systems and senses enable it to perform many complicated activities at once. They are vital to the survival of complex living things.

The instantaneous coordination of body systems enables a runner to sustain a high level of activity.

Chapter 12

Chemical Control

Focus *You are walking down a lonely street late at night. All at once, you hear a sound behind you. You turn and see that someone seems to be following you. Suddenly your heart begins pounding, and you start to inhale deep breaths. Although you are not aware of it, your blood pressure is increasing, your eyes are letting in more light, and glucose is pouring into your bloodstream. You begin to run faster than you ever thought possible. This series of events is known as a "fight or flight" reaction. In an emergency, your body undergoes changes that make you stronger and more alert. These changes are due to chemical signals sent out by specialized organs in the body. Emergency signals are one example of how chemical substances regulate the body.*

12.1 THE ROLE OF HORMONES IN ORGANISMS

Hormones are chemicals that regulate body functions in complex plants and animals. They are produced by special cells and glands and usually circulate through the body in the transport system. Each hormone regulates the activity of a specific **target structure**, which could be a muscle, another gland, or even individual cells. Hormones provide control at a distance, since they are made in one part of the body and control another.

12–1 Performing difficult athletic feats depends on hormones.

349

Hormone Production, Transport, and Function

The delivery of a hormone occurs in three steps. First, specialized cells produce the hormone. Second, the hormone is carried in the transport system throughout the body. Last, when the hormone reaches its target structure, it regulates the structure's activity.

| Specialized cells produce the hormone. | → Transport system carries the hormone throughout the body. → | Target is regulated by the hormone. |

Even though a hormone is transported to all parts of the body, its target structure is particularly sensitive to it.

Hormones are often referred to as "chemical signals" because they usually cause the target to either start or stop its activity. Hormones are extremely effective signals; only tiny amounts of a hormone are needed to turn a target on or off.

Hormones differ in their molecular structure. Some are amino acids, others are proteins. Some are **steroids** (STER-oydz), a kind of lipid. Many have characteristic ring structures. When a hormone reaches its target, it turns the target on or off by triggering a chemical reaction. The chemical reaction that a hormone initiates depends on the molecular structure of the hormone.

Hormones are very powerful. Only a tiny amount of a hormone is needed to stimulate a target structure.

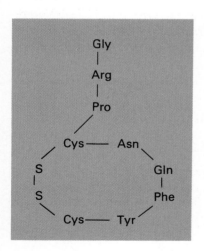

12–2 Structure of antidiuretic hormone (ADH). Antidiuretic hormone is a protein hormone made up of nine amino acids.

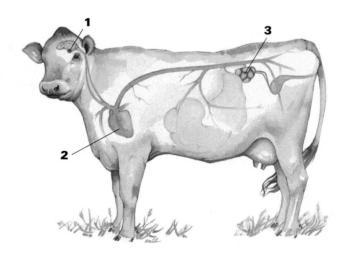

12–3 Hormone action occurs in three steps. In the case of ADH secretion in a cow: **1.** The pituitary gland releases ADH. **2.** The transport system carries ADH throughout the body. **3.** The kidney tubules are regulated by ADH.

Plant Hormones

Hormones in complex multicellular plants are produced by specialized cells in the growing regions of the plant. When the hormones are released, they travel through the phloem or pass from cell to cell by passive or active transport. When a plant hormone reaches its target, it is "recognized" by receptor cells in either the cell membrane or the cytoplasm of target cells. A receptor recognizes a hormone by its shape in much the same way that an enzyme recognizes its substrate. The hormone and receptor combine, and the regulating function of the hormone begins.

Plant hormones affect growth and development of plant organs and bring about flowering and fruit ripening. Some plant hormones also function in defending the plant from predators and the nearby growth of other plants.

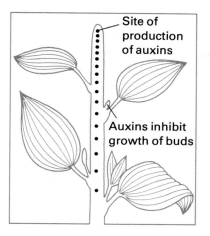

12-4 The high concentration of auxins inhibits the growth of buds near the growing tip of a plant.

Growth Hormones There are four types of hormones that regulate growth in plants. **Auxins** (OX-inz) are produced by cells in the tips of stems and travel throughout the plant. When auxins reach the growing cells of the stem, they cause the cell walls to soften. This allows more water to enter the cells, which then increase in size (grow) by elongation. If the auxin concentration is the same in all parts of the stem, the stem grows straight up. If the auxin concentration varies, however, the stem grows in a curve.

Auxins are responsible for the bending of leaves and stems toward light. This occurs because light inhibits the production of auxins in growing tips. Therefore, auxins become more concentrated on the dark side of a stem than on the light side. As a result, stem cells elongate more on the dark side than on the light side, causing the stem to bend toward the light.

Auxins were the first plant hormones discovered.

12-5 When sunlight is directly above a plant, the production of auxins is equal in all parts of the stem, and the stem grows straight. When sunlight strikes a plant from the side, the production of auxins is greater on the dark side of the plant than on the light side, and the stem grows toward the sun.

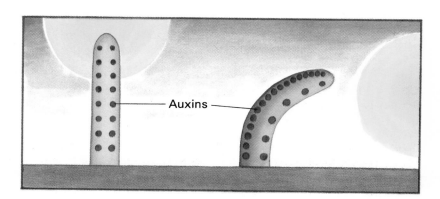

12-6 Effects of gibberellins on the growth of mung beans: normal growth (left); after treatment with gibberellins (right).

Gibberellins were first discovered in Japan. They were the product of a fungus that infected rice plants, causing them to grow tall but killing them before they reached maturity.

Large concentrations of auxins can inhibit the growth of certain plant parts, especially roots. A root lying horizontally on the ground grows downward. This occurs because so much auxin collects in the lower part of the root that growth almost stops. Growth continues in the upper part of the root, however, causing the root tip to grow downward.

Gibberellins (JIB-uh-REL-inz), the second type of growth hormone, are produced in growing tips and stems, leaves, and plant embryos. Gibberellins stimulate the growth of stems by cell elongation. They also help seed growth by stimulating the production of enzymes that break down stored food.

Cytokinins (sī-tuh-KĪ-ninz), the third type of growth hormone, stimulate the cells in growing areas of plants to reproduce, thereby increasing cell number. In addition, cytokinins and auxins work together in transforming unspecialized plant tissue into roots, stems, and leaves.

Abscisic (ab-SIS-ik) **acid**, the fourth type of growth hormone, affects growth in a negative way. It inhibits growth at the bases of leaves and fruit stalks. The presence of abscisic acid in these areas causes leaves to drop and fruit to fall off.

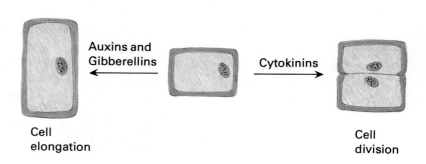

12-7 Effects of plant growth hormones. Auxins and gibberellins cause cells to elongate; cytokinins stimulate cell division.

Flowering and Fruit Ripening Hormones Biologists believe that the flowering of plants is influenced by hormones; however, these hormones have not yet been isolated. In addition, a specific chemical substance is known to cause fruit to ripen. This substance is **ethylene**, C_2H_4. Ethylene has a ripening effect on melons, oranges, lemons, and bananas; however, it does not appear to have this effect on many other fruits, including peaches and apricots. Ethylene is used commercially by many fruit producers. For example, honeydew melons are picked green, when they are easy to transport to market. The hard, green fruit is then placed into a chamber and treated with ethylene gas. When you buy the melons at the market, the ripening process has begun. This process is also used by banana producers.

If you place green tomatoes in a bag with apples and seal the bag, ethylene made by the apples will ripen the tomatoes.

Defense Hormones Some plants produce hormones that help defend them from the organisms around them. Yucca and agave plants produce steroids that cause an animal's red blood cells to burst. This destroys many of the plant's predators.

The balsam fir, an evergreen tree, produces a hormone that inhibits insect maturation. Immature insects that eat the needles of this tree do not become adults and therefore never reproduce. The hormone helps protect the tree from future predation. It is interesting to note that this hormone is identical to one of the insect's own hormones.

The brittlebrush, a desert plant, produces a growth inhibitor in its leaves that affects other species of plants. The substance reaches the soil when it rains and when the leaves fall. This substance stops the growth of nearby plants and reduces competition for needed resources.

12–8 Agave plants produce steroid defense hormones that help protect them from being eaten by animals.

12–9 Effects of synthetic plant growth hormones on a birds nest fern: normal growth (left); after treatment with growth hormones (right).

Synthetic Plant Hormones Scientists have synthesized auxins and other hormones in the laboratory. Some of these hormones are used commercially. One synthetic auxin, 2, 4-D, is used as a weed killer. It causes broad-leaved plants to grow so quickly that they die. It is often used to kill weeds in lawns because it does not harm the narrow-leaved grass. Synthetic hormones are also used to stimulate root growth. If you want to grow a plant such as a camellia or a lemon from a woody cutting, you can dip the end of the stem in a preparation of certain hormones and vitamins. This treatment will help stimulate roots to grow.

Building a Science Vocabulary

hormone	auxin	abscisic acid
target structure	gibberellin	ethylene
steroid	cytokinin	

Questions for Review

1. Outline how hormones carry out "control at a distance."
2. Auxins and cytokinins are both growth hormones. How do they differ in the way they affect growing cells in plants?
3. Give one example of the commercial use of plant hormones.
4. A student grew bean seeds at home in the kitchen. Regardless of how the student turned the pot, the plants always turned toward the lighted window. Explain why this occurred.

12.2 HORMONE FUNCTION IN ANIMALS

Little is known about hormonal control in most invertebrates.

Endocrine glands have been identified in insects, crustaceans, some mollusks, and all vertebrates.

Hormones in complex multicellular animals are produced by glands and are released directly into the blood. Hormone-producing glands have no tubes or ducts that carry the secretion to its target. Since they lack ducts, these glands are called ductless glands, or **endocrine glands**. The hormones produced by endocrine glands regulate many life processes, including cellular respiration, transport, growth, development, reproduction, behavior, and the secretion of other hormones.

Types of Animal Hormones

Proteins are produced in cells by the ribosomes along the rough endoplasmic reticulum.

Protein Hormones Some endocrine glands produce protein hormones. After a protein hormone is made in a gland cell, the Golgi apparatus surrounds it with a membranous sac. When it is needed, the hormone leaves the cell by exocytosis and enters a capillary. It is then transported to all parts of the body in the blood plasma.

Protein hormones are too large to pass through the cell membranes of their targets. As a protein hormone reaches its target, it is recognized by receptor molecules in the cell membranes of target cells. Upon recognition, the protein hormone and receptor combine. In the presence of ATP and an enzyme, a small molecule called **cyclic AMP** is formed. Cyclic AMP acts as a "second signal" that enters the cell and performs the regulating function of the hormone.

Cyclic AMP is closely related to ADP and ATP. It plays a major role in regulating many metabolic processes.

One example of a protein hormone is the digestive hormone secretin, which was discussed in Chapter 8. Secretin is produced by specialized cells in the wall of the small intestine. When secretin reaches the cells of the pancreas, receptor molecules combine with it. Cyclic AMP is formed and enters the cells, where it stimulates the production of digestive secretions.

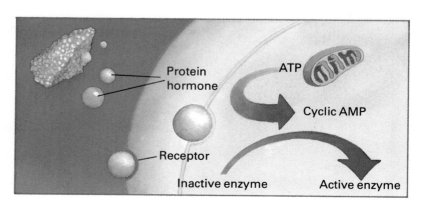

12–10 Protein hormone action. A protein hormone first combines with receptors in the target cell membrane, stimulating the production of cyclic AMP in the cell. Cyclic AMP acts as a second signal. In many cases, cyclic AMP functions by activating inactive enzymes.

Section 12.2 Hormone Function in Animals

12–11 Steroid hormone action. Steroid hormones pass through cell membranes and combine with receptors inside target cells. In many cases the steroid-receptor complex functions by stimulating the production of enzymes or hormones.

Steroids are produced in the smooth endoplasmic reticulum and in mitochondria.

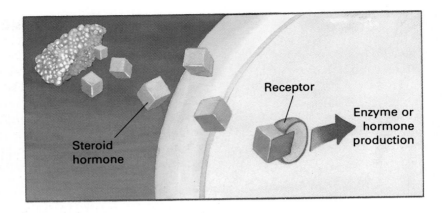

Steroid Hormones Steroid hormones are produced by different endocrine glands than those that produce protein hormones. Steroid molecules are small enough to pass easily through cell membranes. They diffuse out of the cells that produce them and into the blood. The blood carries them to all parts of the body, where they pass easily into most cells. Like protein hormones, steroid hormones combine with receptors; however, steroid receptors are located inside target cells. Once combined with their receptors, steroid hormones stimulate the production of enzymes or other hormones by target cells.

Consider a specific case of steroid control. When the weather becomes warm in spring, a small gland at the base of the brain of the male green anole lizard produces **gonadotropin** (gō-NAD-uh-TRŌ-pin), a steroid hormone. Gonadotropin is carried in the blood to all cells of the body. It combines with receptor molecules inside the cells of the sex organs. This stimulates the sex organs to produce sperm, preparing the lizard for breeding.

Control of Animal Hormone Levels

Hormones are very powerful substances. Very small quantities can produce tremendous changes in body metabolism. Too much of a hormone may even be harmful. Hormone levels are controlled in two ways. Excess hormones are destroyed or inactivated in certain organs, such as the liver. In addition, the production of hormones in most endocrine glands is regulated by **chemical feedback mechanisms**. In one type of feedback mechanism, a chemical signal turns the gland on or off, according to the body's need for the hormone.

Consider calcium regulation in mammals. Calcium is needed for proper nerve functioning, blood clotting, and muscle contraction. It is also needed to maintain bones and teeth. The level of calcium in the blood is regulated by **parathyroid** (PAYR-uh-THĪ-royd) **hormone (PTH)**, which is produced by the **parathyroid glands**. The chemical feedback mechanism that regulates the production of PTH is the level of calcium in the blood. As calcium is used in forming bone and in other activities, the level of calcium in the blood drops. The lowered level of calcium is a signal to the parathyroid glands to produce more PTH. When PTH is in the blood, calcium is removed from the bone matrix and sent to the blood. It is also absorbed from the small intestine and reabsorbed in the kidneys, further increasing the amount of calcium in the blood. The raised level of calcium in the blood is a signal to the parathyroid glands to cut back production of PTH.

The feedback mechanisms that control the production of many animal hormones are similar to the mechanism that controls PTH production by the parathyroid glands. In this type of feedback mechanism, increasing levels of a substance such as calcium or of the hormone itself cause a decrease in hormone production. Decreased levels of the substance or hormone cause an increase in hormone production. Hormone feedback mechanisms of this type play an important role in homeostasis.

12–12 Chemical feedback. When the parathyroid glands sense a low calcium level in the blood they increase production of PTH (**1**). The PTH stimulates the removal of calcium from bone (**2**). It also stimulates the reabsorption of calcium by the kidneys and the absorption of calcium by the small intestine (**3**). As a result, the calcium level in the blood increases (**4**). The increased level of calcium in the blood leads to a decrease in PTH production.

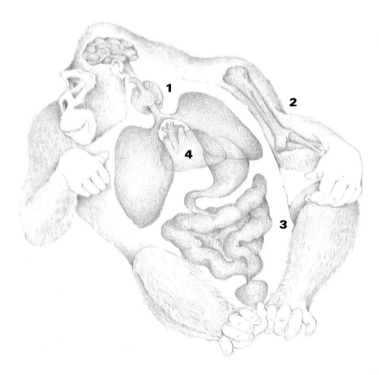

Control of Insect Development

As a grasshopper or other insect grows, it periodically sheds its exoskeleton (molts) and develops a new one (see Chapter 11). The grasshopper molts five times as it grows. Tiny wing pads appear after the second molt. These grow during each successive molt and eventually give rise to wings. After the fifth molt, the grasshopper is an adult with fully developed wings. Molting is under the control of hormones. Cells in the insect's brain secrete **brain hormone**, whose target is the **thoracic gland**, an endocrine gland in the grasshopper's thorax. When stimulated by brain hormone, the thoracic gland secretes **molting hormone**, which begins molting by stimulating the insect's epidermis.

As the insect grows, a tiny cluster of cells behind the insect's brain produces **juvenile hormone**. As long as juvenile hormone is present, molting hormone stimulates the production of another juvenile exoskeleton. When the insect reaches its adult size, the production of juvenile hormone stops. Without juvenile hormone present, molting hormone stimulates the production of an adult exoskeleton.

Molting also occurs in lobsters and other arthropods.

Increasing body size is a feedback signal for the production of juvenile hormone.

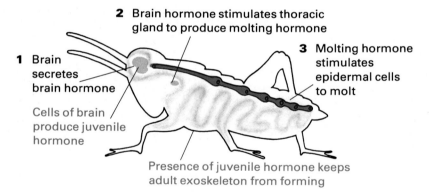

12–13 Hormonal control of molting in an immature grasshopper allows growth of the body by periodically shedding the exoskeleton.

Metamorphosis In most insect species, the immature or juvenile insects look different from the adults. As an insect matures from egg to juvenile to adult, its form changes. Changes in form are part of a process called **metamorphosis**.

In some insects, such as grasshoppers and cockroaches, metamorphosis occurs in three stages: **egg**, **nymph** (NIMF), and **adult**. The nymph is the juvenile form. A newly hatched grasshopper resembles an adult except that it has a relatively large head and no wings. It also lacks reproductive organs. By the time a grasshopper becomes an adult, its wings and reproductive organs are fully developed, and its head is in proper proportion to its body. Metamorphosis that occurs in three stages is called **incomplete metamorphosis**.

meta- = change in
morphē = form

12–14 Incomplete metamorphosis in the dragonfly.

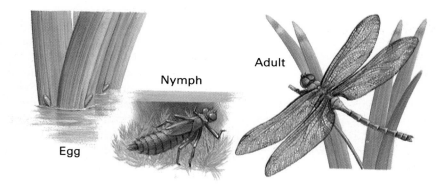

About 90 percent of all insect species undergo complete metamorphosis.

Other insects, including flies, beetles, butterflies, and moths, go through four strikingly different stages of development: **egg, larva, pupa** (PYOO-puh), and **adult**. This type of metamorphosis is called **complete metamorphosis**. The eggs of these insects hatch into wormlike, segmented larva. Larvae have different names in different species. For example, the larvae of butterflies and moths are known as caterpillars, fly larvae are called maggots, and mosquito larvae are called wrigglers. Larvae eat great quantities of food, and they grow rapidly. The larvae molt as they grow; the number of times they molt depends on their species.

When the larval stage is complete, the insect enters the pupal stage. During the pupal stage, an insect does not move from place to place or eat. It attaches itself to a twig or leaf and obtains all its nourishment from stored food. During this stage, the larva's tissues are rearranged, and it develops wings, legs, and other adult parts. For many insects, the pupal stage takes place inside a protective covering. The housefly pupa, for exam-

The life cycle of the gypsy moth consists of egg, caterpillar, pupa, and adult. The caterpillar devours leaves of trees.

12–15 Complete metamorphosis in the housefly.

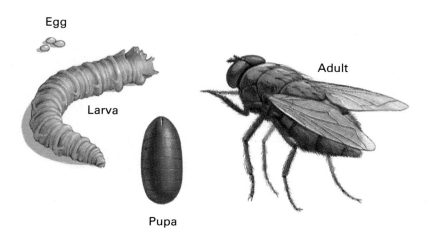

Section 12.2 Hormone Function in Animals

ple, is covered by a hard brown shell formed from the larval skin. Many moth larvae shed the larval skin and spin a cocoon made of silk threads. Most butterfly pupae, however, do not have protective coverings.

Metamorphosis is largely controlled by molting hormone and juvenile hormone. As the nymph or larva grows and molts, the presence of juvenile hormone causes the insect to produce another juvenile exoskeleton. When nymph or larval growth is complete, the production of juvenile hormone decreases. Nymphs then undergo their final molt and larvae enter the pupal stage.

Juvenile hormone, synthetically produced, is used in biological pest control to prevent insects from maturing and reproducing.

Building a Science Vocabulary

endocrine gland	juvenile hormone
cyclic AMP	metamorphosis
gonadotropin	egg
chemical feedback mechanism	nymph
parathyroid hormone (PTH)	adult
parathyroid gland	incomplete metamorphosis
brain hormone	larva
thoracic gland	pupa
molting hormone	complete metamorphosis

Questions for Review

1. Animal protein hormones are too large to enter cells. How do they exert their control?
2. Animal steroid hormones enter all cells of the body. How do they regulate cells in target structures?
3. How does a chemical feedback system work?
4. Explain what stops the grasshopper from molting after the fifth molt.
5. What is the difference between the metamorphosis of a grasshopper and that of a fly?
6. What stimulates the development of adult parts in the grasshopper?

12–16 Charles and Lavinia Stratton on their wedding day in 1863.

12.3 HORMONAL CONTROL IN HUMANS

Hormones play a vital role in regulating complex body functions. They are very powerful, considering that they circulate through the body in very small quantities. If a hormone is not produced in the right amount, drastic results can occur.

The case of Charles and Lavinia Stratton, two dwarfs who were famous stage performers of the nineteenth century, illustrates how powerful hormones are. The Strattons were dwarfs, because they lacked minute amounts of growth hormone. Charles Stratton stopped growing normally when he was six months old. As an adult, his full height was 1m, and he weighed 32 kg. Lavinia, his wife, was only 60 cm tall as an adult. Both the Strattons were talented and intelligent people, and their bodies, though small, were properly proportioned.

In this section, the major endocrine glands and their hormones are discussed. Endocrine hormones function in regulating not only growth, but development, reproduction, behavior, and other processes as well.

The Endocrine System

As in other complex multicellular animals, most hormones in humans are secreted directly into the blood by endocrine glands. The hormones are carried to all parts of the body; however, they particularly affect their target structures.

Hormonal control of metabolic functions is complex. The body makes a large number of hormones, and several hormones act together to control many body processes. Nearly all of the hormones in the body are regulated by some sort of feedback mechanism. In this way, the proper level of each hormone is maintained.

The Thyroid Gland

The **thyroid** (THĪ-royd) **gland** is an H-shaped structure located in the neck beneath the larynx. The thyroid has two lobes—one on each side of the trachea. The lobes are connected by a bridge of tissue that crosses in front of the trachea.

The thyroid gland produces a hormone called **thyroxin**. Thyroxin regulates many metabolic processes, including cellular respiration, growth, and development. Thyroxin stimulates

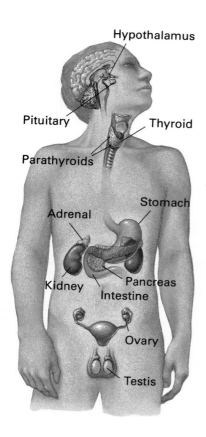

12–17 Major endocrine glands of the human body.

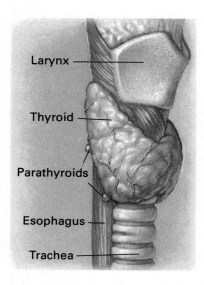

12–18 Thyroid gland (three-quarters view). The small parathyroid glands are located to the sides of and behind the thyroid gland.

Hypothyroidism is treated by administration of thyroxin.

guttur = throat

the intake of oxygen by cells. It also helps regulate the use and storage of lipids, and it increases the absorption of carbohydrates from the small intestine.

Careful regulation of thyroxin production is vital. If too little thyroxin is produced, many body processes slow down. Undersecretion of thyroxin is called **hypothyroidism**. People with this condition become sluggish and often gain weight. If too much thyroxin is produced, many body processes speed up. Oversecretion of thyroxin is called **hyperthyroidism**. The excess thyroxin causes nervousness, weight loss, increased heartbeat, and excess heat production. People with hyperthyroidism usually perspire more than other people.

Proper levels of thyroxin are especially important for normal development during infancy and childhood. A serious lack of thyroxin causes physical and mental retardation. These conditions can often be corrected if treated early. After a certain point, however, the retardation becomes permanent.

An important part of the thyroxin molecule is the essential mineral iodine. Insufficient iodine in the diet is one cause of a condition called **goiter** (GOYT-er). A goiter is an enlargement of the thyroid gland. Goiter caused by an iodine deficiency can be prevented and treated by adding iodine to the diet. In many locations, people obtain iodine from their drinking water. In locations where there is not sufficient iodine in drinking water, people often use table salt that has iodine added.

The Pituitary Gland

The pea-size **pituitary** (pih-TOO-ih-TAYR-ee) **gland** is located at the base of the brain. The pituitary gland is composed of two lobes: the posterior lobe and the anterior lobe (see Fig. 12–19). The posterior lobe stores and releases two hormones that are produced in a nearby part of the brain. *Oxytocin* starts the birth process by stimulating contraction of the muscles of the uterus at the end of pregnancy. *Antidiuretic hormone* (*ADH*) controls the reabsorption of water in the kidneys. (The function of ADH is discussed in more detail in Chapter 10.) The anterior lobe produces hormones that regulate growth, reproduction, and other endocrine glands.

The hormone that regulates growth, **growth hormone**, stimulates the growth of certain parts of the skeleton. When the

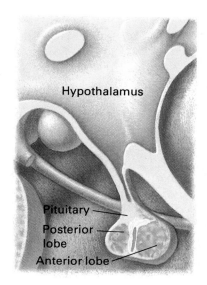

12–19 The pituitary gland has two lobes. Together they release eight different hormones.

pituitary produces too little growth hormone from birth, dwarfism results, as in the case of Charles and Lavinia Stratton. Sometimes the pituitary produces too much growth hormone from birth and giantism results. People with this condition grow very tall, although their bodies are properly proportioned. Occasionally, a person begins producing too much growth hormone as an adult. This causes **acromegaly** (AK-rō-MEG-uh-lee). The nose, hands, feet, and jaw of a person with acromegaly grow out of proportion to the rest of the body.

12–20 The progress of acromegaly through a lifetime. This condition, in which bones grow abnormally in width, is the result of excess growth hormone secretion in adulthood.

The Hypothalamus

The pituitary gland is attached by a stalk to a part of the brain called the **hypothalamus** (HĪ-pō-THAL-uh-mus), which is involved in regulating many body processes. As part of its function, the hypothalamus acts as an endocrine gland. The hypothalamus produces oxytocin and ADH, which pass along nerve tissue to the posterior lobe of the pituitary, where they are stored. Other hormones produced by the hypothalamus pass through a blood vessel to the anterior lobe of the pituitary. When these hormones reach the anterior lobe, they stimulate the release of the anterior pituitary hormones.

The hypothalamus plays an important role in controlling the levels of many hormones in the body. For example, the amount of thyroxin produced by the thyroid gland is controlled by a feedback system that involves the hypothalamus, the pituitary, and the thyroid gland. The hypothalamus produces *thyrotropic releasing factor* (*TRF*). Thyrotropic releasing factor stimulates the pituitary to produce *thyroid stimulating hormone* (*TSH*). Thyroid stimulating hormone travels in the bloodstream to the thyroid gland, where it stimulates the production of thyroxin.

The level of thyroxin in the blood affects the hypothalamus and the pituitary. If there is too much thyroxin in the blood, the hypothalamus reduces its production of TRF. As a result, the pituitary reduces its production of TSH. Lowered levels of TSH result in decreased production of thyroxin by the thyroid gland. Similarly, if there is too little thyroxin in the blood, the hypothalamus increases its production of TRF. This leads to increased production of thyroxin by the thyroid.

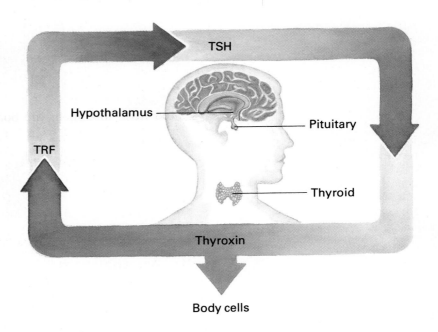

12–21 Feedback control of thyroxin production. If there is too little thyroxin in the blood, the hypothalamus produces more TRF, causing the pituitary to produce more TSH and the thyroid to produce more thyroxin. The reverse occurs if there is too much thyroxin in the blood.

The Parathyroid Glands

The four, tiny **parathyroid glands** are located just behind the thyroid gland. As mentioned in Section 12.2, parathyroid hormone regulates the amount of calcium in the blood. Control of the parathyroid glands is extremely important. If there is not enough PTH in the body, the calcium level in the blood drops. This condition can result in **tetany** (TET-un-ee). A person with tetany develops severe muscle spasms. Injections of either PTH or calcium stop tetany.

The Pancreas

The pancreas secretes enzymes that help the body digest food. It also functions as an endocrine gland that produces two important hormones: **insulin** and **glucagon**. Together, these hormones help regulate the amount of glucose in the blood.

The target for insulin is the cell membrane. In the presence of insulin, glucose leaves the blood and enters cells. When glucose enters liver and muscle cells, it is converted to glycogen and stored until it is needed. Glucose also enters fat cells, where it is converted to fat and stored.

The effect of glucagon is opposite that of insulin. Glucagon causes glycogen in liver cells to break down into glucose. The glucose leaves the liver cells and enters the blood, increasing the blood glucose level.

The combined action of insulin and glucagon help keep the level of glucose in the blood fairly constant.

Insulin and Glucagon Diseases As already noted, insulin tends to reduce the amount of sugar in the blood. Too much insulin in the body causes **hypoglycemia** (HI-pō-glī-SEE-mee-uh), a condition in which there is not enough sugar in the blood. Hypoglycemia can cause violent muscle spasms and unconsciousness.

hypo- = under
glykys = sweet
-aimia = condition of the blood

Too little insulin in the body can cause **diabetes mellitus** (DĪ-uh-BEET-is muh-LĪT-is). A person with diabetes mellitus has too much glucose in the blood. In the absence of insulin, glucose does not enter cells. Sometimes there is so much glucose in the blood of a diabetic that the kidney cannot reabsorb all of it. People with diabetes mellitus frequently lose weight because they excrete glucose in the urine. In addition, they become dehydrated easily, since much water is required to excrete the excess glucose in the urine. Severe diabetes can cause death if it

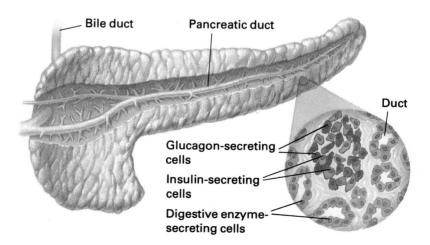

12-22 The pancreas. Cells lining ducts secrete digestive enzymes. Other cells secrete the hormones glucagon and insulin.

For medical use, insulin, thyroxin, and other hormones are made artificially or obtained from animals. Some hormones are now being produced by bacteria in the laboratory.

ad- = toward
renalis = kidney

Cortisone is a derivative of cortisol and is used as a medicine to treat allergies, arthritis, and other conditions.

Adrenalin and noradrenalin give people great strength during emergencies, enabling them to run faster or lift heavier objects than they ordinarily could.

is not treated. Most people with diabetes mellitus are helped by taking insulin on a regular basis.

Abnormalities also exist in the amount of glucagon in the body. Too little glucagon causes hypoglycemia; too much glucagon makes diabetes worse.

The Adrenal Glands

On top of each kidney is an **adrenal** (uh-DREE-nul) **gland**. Each adrenal gland has two parts. The outer part of the gland is called the **adrenal cortex**. The inner part is called the **adrenal medulla** (mih-DUL-uh).

The adrenal cortex secretes several steroid hormones. The secretion of these hormones is regulated by a hormone from the pituitary gland. This is part of a feedback system similar to the one for thyroxin. The principal hormone of the adrenal cortex is *cortisol*. It and other hormones of the adrenal cortex help regulate the body's use of carbohydrates and proteins. They also help maintain the body's salt and water balance.

The nervous system stimulates the adrenal medulla to secrete **adrenalin** (uh-DREN-ul-in) and **noradrenalin**. These hormones are sometimes called "fight or flight" hormones, because they function in emergency situations. Adrenalin and noradrenalin cause rapid changes in the body when a person is frightened or angry. Adrenalin, in particular, causes glycogen in the liver and muscles to be rapidly changed to glucose. This conversion makes glucose immediately available to cells for energy. Adrenalin and noradrenalin also increase the heart rate and blood pressure. They constrict blood vessels in many internal organs, such as the stomach, and increase the diameter of blood vessels to skeletal muscles and the liver. In this way blood is made available to the parts of the body that need it most. This enables the body to act quickly and forcefully in an emergency.

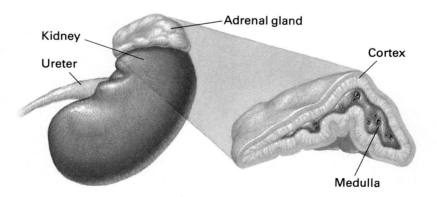

12–23 The adrenal gland, shown in position and in cross section, is actually two glands: the adrenal cortex, which secretes cortisol, and the adrenal medulla, which secretes adrenalin and noradrenalin.

Reproductive Hormones

Reproductive hormones play an extremely important role in the human body. They are produced by the reproductive glands, which in males are the **testes** (TES-teez) and in females are the **ovaries** (Ō-vuh-reez). The reproductive hormones influence the development of the reproductive cells—sperm and eggs—and also influence the development of **secondary sex characteristics**. These characteristics appear at the time of maturation from childhood to adulthood. In males, secondary sex characteristics include deepening of the voice, development of the beard and muscles, and growth of body hair. In females, they include enlargement of the breasts, widening of the hips, and the onset of menstruation (men-STRAY-shun). The growth of pubic hair and underarm hair is a secondary sex characteristic in both males and females.

In males, the testes produce the hormone **testosterone** (tes-TOS-ter-ōn). In females, the ovaries produce **estrogen** (ES-truh-JEN) and **progesterone** (prō-JES-tuh-RŌN). The production of testosterone in males is controlled by one of the anterior pituitary hormones. This hormone maintains the level of testosterone by chemical feedback. The reproductive system of females, unlike that of males, undergoes regular, cyclical changes. The cycle is called the **menstrual** (MEN-strul) **cycle** and lasts about 28 days.

Several events occur during a menstrual cycle. First, an egg matures in the ovary. Then, the egg is released from the ovary during **ovulation** (ah-vyoo-LAY-shun). The egg travels down the **oviduct** (Ō-vih-dukt) to the **uterus** (YOOT-er-us). This usually takes about three days. During this time the wall of the uterus

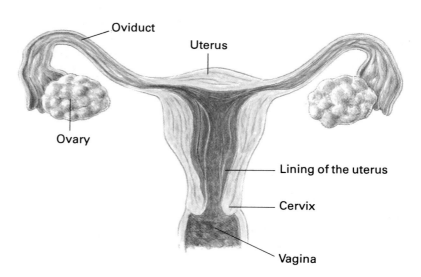

12–24 The human female reproductive organs.

The menstrual cycle usually begins in females between the ages of 10 and 14.

menstruus = monthly

begins to change. It becomes thick and spongy and filled with blood. These changes prepare the uterus to receive an embryo. If the egg is fertilized, the tiny embryo attaches to the uterus wall and pregnancy begins. If the egg is not fertilized, the lining of the uterus wall breaks down and part of the thickened lining and extra blood pass out of the body. This process is called **menstruation** and usually lasts four to five days.

The Menstrual Cycle: A Study in Hormonal Control

The pituitary gland produces hormones that function in reproduction. Two of these are *follicle stimulating hormone* (*FSH*) and *luteinizing hormone* (*LH*). The release of these hormones is controlled by hormones from the hypothalamus. Both FSH and LH play important roles in the menstrual cycle.

The menstrual cycle has three main stages: the *follicle stage, ovulation,* and the *corpus luteum* (KOR-pus LOO-tee-um) *stage*. The corpus luteum stage is followed by either menstruation or pregnancy, depending on whether fertilization has occurred.

Follicle Stage (10–14 days)   Hormones from the hypothalamus stimulate the pituitary to increase its production of FSH. The target of FSH is the ovary. The increased quantity of FSH in the blood stimulates an egg sac, or **follicle**, in the ovary to grow. The immature egg inside the follicle develops and enlarges, and the follicle increases in size.

The swollen follicle inside the ovary secretes estrogen, which has two main targets. First, estrogen stimulates the uterus wall to thicken. Then, when there is enough estrogen in the bloodstream, it causes the pituitary to cut back on the production of FSH. This prevents the growth of more follicles. Increased estrogen in the blood also stimulates the pituitary to increase its production of LH.

Ovulation   In the presence of increased estrogen and LH levels, the egg breaks out of the follicle and begins to travel down the oviduct.

Corpus Luteum Stage (10–14 days)   After ovulation occurs, LH stimulates the empty follicle to become a temporary endocrine gland, the **corpus luteum**. The corpus luteum secretes progesterone, which has two targets. One target is the uterus wall, which progesterone causes to thicken. The other target for progesterone is the pituitary. Progesterone causes the pituitary to decrease its production of FSH.

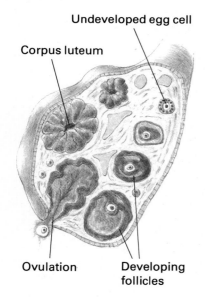

12–25   Development of follicles in the ovary, from maturation of eggs to formation of the corpus luteum.

Menstruation (4–5 days)   If the egg is not fertilized as it travels down the oviduct, it passes out of the body. The corpus luteum disintegrates and stops producing progesterone. Without progesterone, the spongy lining of the uterus breaks down, and blood and part of the lining pass out of the body. As the level of progesterone decreases, so does the level of LH. This causes the pituitary to increase production of FSH and the cycle begins again.

Pregnancy (9 months)   If the egg is fertilized as it travels down the oviduct, it begins developing. When it reaches the uterus, the tiny embryo attaches to the wall of the uterus and pregnancy begins. For about three months the corpus luteum does not disintegrate. It continues to produce progesterone, maintaining the thick wall of the uterus. After that, the **placenta** (pluh-SEN-tuh), which attaches the embryo to the uterus wall, produces progesterone and estrogen. These maintain the uterus wall until the baby is born. At birth, the placenta is expelled and the menstrual cycle starts over.

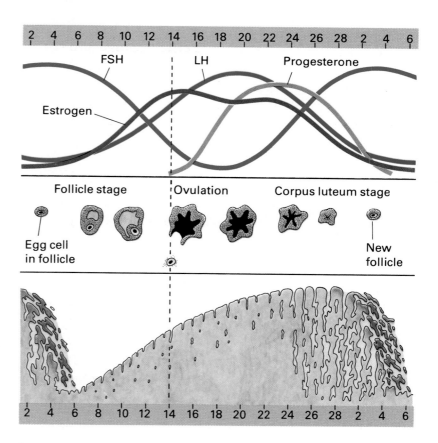

12–26   Stages in the menstrual cycle. Top: changes in pituitary secretion of LH and FSH and in ovarian secretion of estrogen and protesterone. Middle: cycle of egg development. Bottom: changes in the lining of the uterus. The numbers at the top and bottom refer to the days of the cycle.

Section 12.3   Hormonal Control in Humans

Coordination of Body Processes by Hormones

Hormones work together in regulating and coordinating many body processes. The menstrual cycle is just one example of this. Consider also the chemical coordination of muscle contraction. At least five different hormones are needed for a single muscle twitch. Muscle action requires calcium and ATP. Parathyroid hormone insures a supply of calcium to the muscle by regulating the level of calcium in the blood. In order to produce ATP, a muscle depends on the delivery of glucose and oxygen. The combined action of insulin, glucagon, and adrenalin insure delivery of glucose to the muscle by maintaining the glucose level of the blood. In addition, thyroxin regulates the rate at which cells use oxygen.

Hormones also coordinate the action of different body systems. This is evident in the work of the adrenal medulla in mobilizing the body for emergencies. During this process, adrenalin coordinates responses in the respiratory system, the digestive system, the nervous system, the transport system, the muscle system, and the sense organs. The coordination of many body processes and systems by hormones is an important key to the total working of the human body.

The major endocrine glands are listed in Table 12–1, along with the names and functions of their hormones.

12–27 Hormones play a vital role in body coordination.

Table 12–1 Human Endocrine Glands and Their Secretions

Endocrine Gland	Hormone	Major Functions
Hypothalamus	Hormones that regulate the pituitary	Regulate pituitary secretions of TSH, FSH, LH, GH, PRL, and ACTH
	Oxytocin	Stimulates contraction of uterine muscles during childbirth
	Antidiuretic hormone (ADH)	Stimulates water reabsorption in kidneys
Pituitary		
Posterior lobe		Stores and releases oxytocin and antidiuretic hormone
Anterior lobe	Adrenocorticotropic hormone (ACTH)	Stimulates adrenal cortex

Chapter 12 Chemical Control

Table 12–1 Continued

Endocrine Gland	Hormone	Major Functions
Anterior lobe, continued	Thyroid stimulating hormone (TSH)	Stimulates production of thyroxin by thyroid
	Growth hormone (GH)	Stimulates growth of skeleton
	Follicle stimulating hormone (FSH)	Stimulates growth of follicle in ovary in females and stimulates sperm production in males
	Luteinizing hormone (LH)	Stimulates ovulation and formation of corpus luteum in females and regulates testosterone production in males
	Prolactin (PRL)	Stimulates milk production in mammary glands
Thyroid	Thyroxin	Regulates growth and development; controls use of food and oxygen by cells
Pancreas	Insulin	Stimulates glucose uptake by cells
	Glucagon	Stimulates glucose release by liver cells
Adrenal cortex	Steroid hormones	Regulate use of carbohydrates and proteins; maintain salt and water balance
Adrenal medulla	Adrenalin	Mobilize body for emergency; increase blood sugar; increase rate of blood transport
	Noradrenalin	
Parathyroids	Parathyroid hormone (PTH)	Increases blood calcium
Stomach wall	Gastrin	Stimulates secretion of gastric juice
Intestinal glands	Secretin	Stimulates pancreas digestive secretions
Testes	Testosterone	Stimulates development of secondary sex characteristics in male; necessary for production of sperm
Ovaries	Estrogen	Stimulates development of secondary sex characteristics in female; helps regulate ovaries and uterus during menstrual cycle and pregnancy
	Progesterone	Helps regulate uterus during menstrual cycle and pregnancy

Building a Science Vocabulary

thyroid gland	adrenal medulla
thyroxin	adrenalin
hypothyroidism	noradrenalin
hyperthyroidism	testis
goiter	ovary
pituitary gland	secondary sex characteristic
growth hormone	testosterone
acromegaly	estrogen
hypothalamus	progesterone
parathyroid gland	menstrual cycle
tetany	ovulation
insulin	oviduct
glucagon	uterus
hypoglycemia	menstruation
diabetes mellitus	follicle
adrenal gland	corpus luteum
adrenal cortex	placenta

Questions for Review

1. Explain how a deficiency of insulin affects each of the following.
    a. cell membrane of body cell
    b. liver cell
    c. kidney tubule cell

2. Describe the feedback mechanism that lowers the output of thyroxin by the thyroid gland.

3. Why does hyperthyroidism cause loss of weight?

4. Describe a feedback mechanism in the menstrual cycle that involves FSH and estrogen.

5. Explain the relationship between the terms in each of the following pairs.
    a. hypothalamus and pituitary gland
    b. insulin and glucagon

Perspective: Careers
The ABCs of Teaching Biology

Teaching is a very old profession. It is also one of the basic human instincts—teaching the newborn and the young the dos and don'ts of surviving and getting along in the world.

Most people can recall some favorite teacher. Yours may be a kindergarten teacher, or a third-grade teacher, or a teacher you have right now. Something about this person has influenced your life. Perhaps your favorite teacher opened up a view of the world that had not been clear to you before.

Maybe teaching, perhaps biology teaching, is the right career for you. Consider the ABCs of a good teacher. *A* stands for attitude. A good teacher wants to see students learn and grow. *B* stands for biological knowledge. A good teacher loves the subject and is curious to learn more about it. *C* stands for communication. A good teacher has the skills to get the ideas across. Do

you have some of these traits, even if they are not yet fully developed?

Some elementary school teachers specialize in teaching science, so biology is a large part of their work. Others teach math, reading, and other subjects, along with science. High school biology teachers usually teach only that subject, but some also teach chemistry or physics, and some even coach a sport as well. To be a teacher, you must have at least four years of university.

For post-secondary teaching, a higher degree is usually required, meaning several more years of study. In community colleges the emphasis is on teaching only. In universities, teachers are usually expected to be involved in research as well as teaching. University teachers who work with students seeking advanced degrees must be involved in research if they are to do a good job of educating their students.

So remember the ABCs for teaching biology. They make a strong foundation upon which a career in teaching can be built.

Chapter 12
Summary and Review

Summary

1. Hormones are chemical regulators found in complex plants and animals. They control processes such as growth, development, and day-to-day functions.

2. Hormones are produced by specialized cells or glands and carried by the transport system to all parts of the body. They are "recognized" by receptor molecules in or on target cells. Upon recognition, they cause target cells to start or stop producing enzymes or other substances.

3. There are four major types of plant hormones. Auxins cause cell elongation, bending of stems and roots toward light, and growth of roots downward. Gibberellins cause cell elongation and seed growth. Cytokinins stimulate cell division. Abscisic acid stimulates leaf and fruit fall. In some plants, ethylene stimulates fruit ripening. Hormones also function in defense in some plant species.

4. In animals, protein hormones stimulate the production of cyclic AMP, which enters target cells as a "second signal." Steroids enter all cells, but they combine with receptor molecules only in target cells.

5. Feedback mechanisms regulate hormone production in animals. In a feedback mechanism, the level of a hormone or other substance in the blood stimulates a gland to start or stop producing its hormone.

6. Hormones control the metamorphosis of immature insects into adults. Metamorphosis may be incomplete, as in grasshoppers, or complete, as in butterflies. As they grow in size, immature insects shed their exoskeletons in a process called molting.

7. The human endocrine system helps regulate and coordinate body functions, including growth, development, metabolic rate, and reproduction. The production of hormones by endocrine glands is controlled by feedback mechanisms. One example of feedback is the regulation of thyroxin production, which involves the hypothalamus, the pituitary gland, and the thyroid gland.

8. Imbalances in hormone levels can result in severe conditions such as goiter, tetany, diabetes mellitus, and acromegaly.

9. Reproductive hormones are responsible for the development of secondary sex characteristics and the production of mature eggs and sperm. In females, the reproductive cycle is known as the menstrual cycle, which lasts about 28 days. The menstrual cycle is under hormonal control.

Review Questions
Application

1. How does the pancreas act both as a ducted gland and as a ductless gland?

2. Parathyroid hormone (PTH) is a protein molecule. Answer the following questions about PTH.
   a. Where is PTH formed?
   b. How does PTH get to its targets?
   c. What are three of the targets of PTH?
   d. How do target cells "recognize" PTH?
   e. How does PTH exert control inside a target cell?
   f. What action is stimulated by PTH in each of its targets?

3. When immature insects eat balsam fir leaves, they inject a chemical that prevents them from becoming adults. What is this chemical and how does it prevent immature insects from becoming adults?

4. A scientist wanted to cause a plant root to grow upward instead of downward. How could hormones be used to obtain this result?

The graph below represents the changing level of glucose in a student's blood from 12:00 noon to 8:00 P.M. on a given day. At 1:30 P.M. the student ran a race. At 5:00 P.M. the student ate a meal. Refer to the graph in answering questions 5–8.

5. Name one hormone whose output was increased during the race.
6. What was the student's probable source of glucose during and after the race?
7. Why did the glucose level rise and then fall after the meal?
8. What hormones were involved in the decrease in the glucose level after the meal?

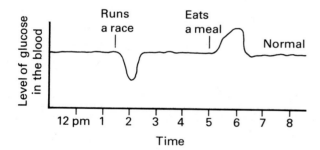

Interpretation

1. Why does someone with diabetes mellitus become very hungry? thirsty?
2. The tip of a stem is shown in the figure below. Where are auxins most concentrated—point *a*, *b*, or *c*?

Extra Research

1. Look up the history of diabetes mellitus. Write a short report describing advances in the treatment of this disease during the twentieth century.
2. Prepare a brief dictionary of hormonal diseases and conditions. For each disease or condition, include the name of the gland that is not functioning properly, and state whether the disease results from undersecretion or oversecretion of a hormone. Also give symptoms and treatment.
3. Find out how hormones control metamorphosis in amphibians, such as the frog. Describe your findings in words and pictures.
4. Look up commercial uses of hormones. Assemble the information in a brochure. Concentrate your research either on hormones used in agriculture and livestock production or on hormones used in medicine. Include information about how each hormone is obtained, its molecular structure, and its commercial use.

Career Opportunities

Pharmacists dispense medicines and advise doctors and patients on their use. Some run their own businesses, while others work in hospitals or for drug manufacturers. For more information, write the Canadian Pharmaceutical Association, 1815 Alta Vista Dr., Ottawa, Ontario K1G 3Y6.

Psychiatrists are medical doctors with additional training in psychiatry. They specialize in the diagnosis and treatment of people with mental disorders. Psychiatrists must have a thorough knowledge of the structure and function of the brain, including the emotional effects of hormones. The treatment they prescribe often includes medication, counseling, or group therapy. Write the Canadian Psychiatric Association, 225 Lisgar St., #103, Ottawa, Ontario K2P 0C6.

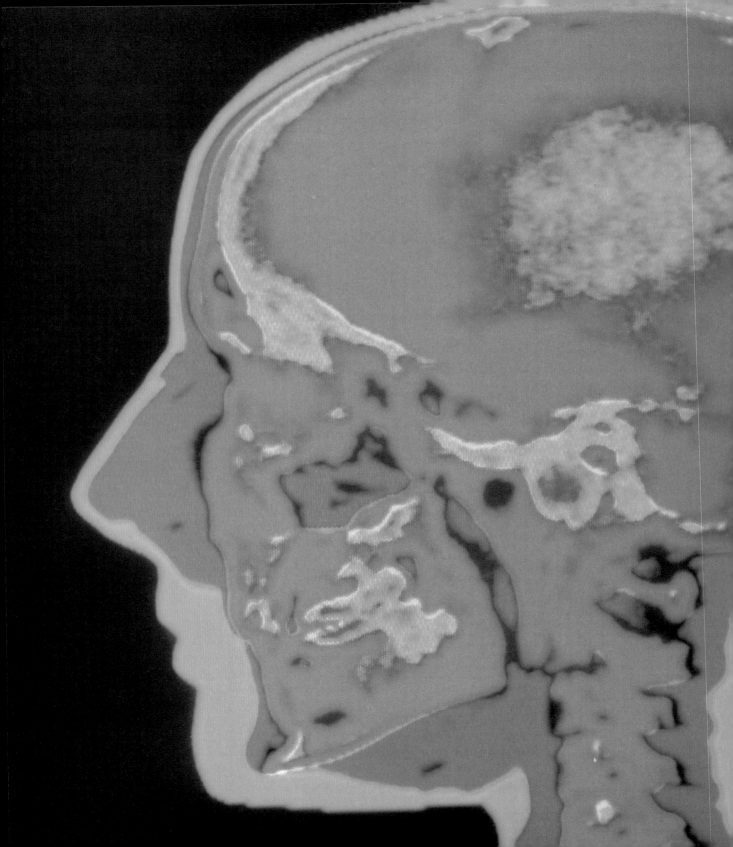

Chapter 13

Nervous Control

Focus *The most complex organ in the human body is the brain. It is capable of inventing, discovering, and creating. It processes information from all parts of the body and controls movement, thought, and speech. The brain is part of the nervous system, which in humans consists of more than 100 billion nerve cells. The nervous system in every complex animal controls and coordinates functions and behavior. This system links all areas of the body and enables the parts to work together as a single organism.*

13.1 CHARACTERISTICS OF NERVOUS SYSTEMS

One of the basic life functions of all living things is the ability to respond to a stimulus (see Chapter 1). A change in air temperature could be a stimulus; so could a change in the carbon dioxide level in the blood. In many organisms stimuli are detected by **receptors**. A receptor is a kind of sensor that picks up information about an organism's internal or external environment. A receptor is specialized to pick up a certain kind of stimulus. For example, the eye is a receptor that is sensitive to light, but not to odor.

13–1 Colored density scan. The densest parts of the head (bone) are shown in red. The least dense parts of the head (soft tissues and thin parts of the skull) are shown in green.

13–2 An octopus' eyes transmit visual information to the brain by nerve impulses.

Information from a receptor is transmitted to an **effector**, which is the part of an organism that responds. For example, glands are effectors; when stimulated by nerves, they secrete enzymes or hormones.

In animals, information is transmitted from receptors to effectors by nerve "signals," or **impulses**. Nerve impulses travel very quickly between receptors and effectors. In many animals, a **central nervous system** coordinates impulses from receptors with impulses to effectors. In vertebrates, the central nervous system consists of the brain and spinal cord.

Nerves work together with hormones in controlling an animal's life processes. In general, nerves provide rapid communication for actions such as escaping, feeding, and mating. Hormones, which act more slowly, generally control gradual processes, such as growth, maturing, and day-to-day cell functions.

Basic Units of the Nervous System

Nervous systems are largely made up of specialized cells called **neurons** (NER-onz). Each neuron has a **cell body**, which contains the nucleus and organelles. Projecting outward from the cell body are tubelike extensions of the cell membrane, which carry nerve impulses. Impulses are picked up and transmitted to the cell body by extensions called **dendrites**. A neuron can have many dendrites, which are usually short and may be highly branched. Impulses travel away from the cell body along extensions called **axons**. A neuron has only one axon. Axons vary in length depending on the type of neuron; however, they are usually much longer than dendrites.

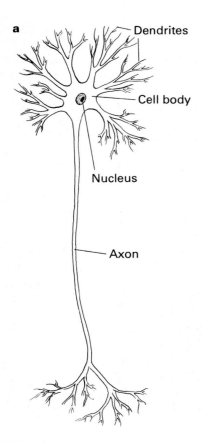

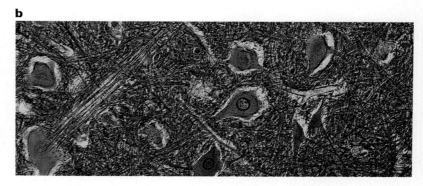

13–3 Neurons: **a.** generalized neuron, showing the main parts of all neurons; **b.** cross section of human spinal cord, showing neuron cell bodies and their axons surrounded by smaller glial cells.

Chapter 13 Nervous Control

Glial cells are the "housekeepers" of the nervous system.

Neurons are surrounded by special cells called **glial** (GLEE-ul) **cells**. Glial cells have cell bodies and branching extensions. They outnumber neurons ten to one. Although glial cells appear to help nourish and protect neurons, their function is not fully understood.

Types of Neurons Neurons are put into three groups based on the direction that their impulses travel. **Sensory neurons** transmit impulses from receptors. **Motor neurons** transmit impulses to effectors. **Interneurons** transmit impulses from sensory neurons to motor neurons. They also transmit impulses to other interneurons.

Most interneurons are located in the central nervous system.

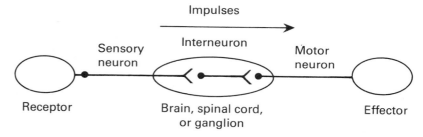

13–4 Most nerve impulses travel from a receptor to one or more interneurons to an effector.

Since cell bodies are thicker than axons and dendrites, ganglia resemble swellings.

Bundles of Neurons The cell bodies of neurons often occur in clumps or clusters called **ganglia**. Ganglia coordinate incoming and outgoing impulses.

Nervous system tissue in many animals is made up of billions of neurons. The axons of these neurons are gathered into bundles or cables called **nerves**. Nerves can contain hundreds or even thousands of axons. **Sensory nerves** carry impulses from receptors toward the central nervous system or toward ganglia. **Motor nerves** carry impulses from the central nervous system or ganglia toward effectors. **Mixed nerves** contain both sensory and motor axons.

13–5 A cluster of human ganglia (left) and cross section of a bundle of nerves, each containing numerous axons (right).

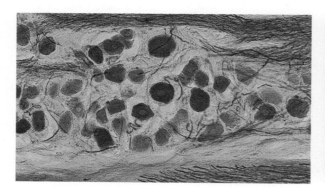
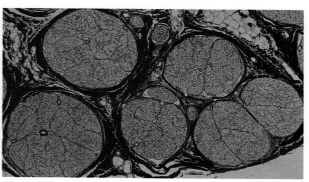

Section 13.1 Characteristics of Nervous Systems

The Action of Neurons

A nerve impulse is carried along the cell membrane of a neuron. When a neuron is sufficiently stimulated, it conducts an impulse. If the stimulus is not strong enough, the neuron does not conduct an impulse. For a single neuron, there is no weak or strong impulse; it either transmits an impulse or it does not. This characteristic of the nerve impulse is called the **all-or-nothing response**. Once an impulse starts down a neuron, it continues to the end without fading out.

A neuron in its resting state still actively carries on basic life processes.

The Resting State of the Neuron A neuron that is not transmitting an impulse is said to be in its **resting state**. At rest, the concentrations of certain ions inside the neuron and in the fluid surrounding the neuron are very different. The concentration of sodium ions (Na^+) is very low inside a neuron compared with the outside of the neuron. The concentration of potassium ions (K^+), chlorine ions (Cl^-), and negatively charged protein molecules is higher on the inside of the neuron than on the outside. The effect of the unequal distribution of ions on the two sides of the membrane is that the inside of the resting membrane is negatively charged relative to the outside. Such a membrane is said to be **polarized**.

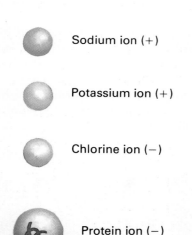

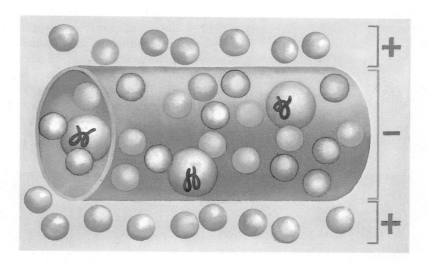

13–6 In its resting state the inside of an axon is negatively charged with respect to the outside of the axon. The distribution of positive and negative ions differs on each side of the membrane. The membrane is said to be polarized because of the charge differences on either side of the membrane.

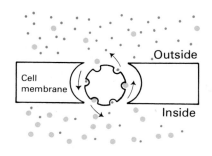

13–7 The sodium pump maintains and restores the resting state distribution of ions in a neuron by exchanging sodium ions (small dots) inside the cell for potassium ions (large dots) outside.

You might think that positive ions outside the neuron membrane would be attracted to negative ions inside and that they would enter and neutralize the charge. You might also think that the ions on either side would diffuse across the membrane and equalize concentration. The membrane prevents this from happening in two ways. First, the membrane is selectively permeable. Second, a mechanism that pumps sodium ions and potassium ions (usually called the **sodium pump**) exists in the cell membrane. The sodium pump counteracts any diffusion that takes place by actively transporting sodium ions out of the cell and potassium ions into the cell. In this way, the sodium pump helps maintain the polarized state of the resting cell membrane.

The Nerve Impulse When a strong enough stimulus reaches the cell body or dendrites of a neuron, the permeability of the cell membrane changes. At the point of stimulation, sodium ions begin crossing the membrane. The movement of sodium ions decreases the number of positive charges outside the membrane and increases the number of positive charges inside. Eventually, the membrane becomes positive on the inside and negative on the outside.

The change in permeability and movement of sodium ions at the point of stimulation changes the permeability of adjacent points of the membrane. This causes ions to move across the membrane all along the axon or dendrite. This movement of ions is called the nerve impulse.

The change in the membrane's permeability lasts about one half of one millisecond. Then the membrane again becomes impermeable to sodium ions. At the same time, the membrane becomes permeable to potassium ions, which move out of the cell. The immediate effect of potassium ions moving out is to make the outside positive and the inside negative again. As with the initial change of permeability, these events occur along the length of the axon or dendrite, following the initial movement of sodium ions.

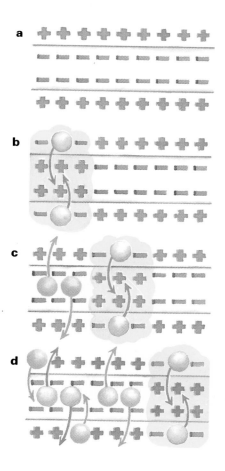

13–8 Nerve impulse: **a.** resting state; **b.** a stimulus causes the membrane to become permeable to sodium ions (green); **c.** as the impulse proceeds along the membrane, areas behind the impulse become permeable to potassium ions (orange); **d.** after the movement of potassium ions, the sodium pump restores the resting state.

Section 13.1 Characteristics of Nervous Systems

The nerve impulse is not a form of electricity. An electric current passes through a conducting wire at the speed of light, which is more than 300 000 km/s. The fastest impulse passing over an axon travels about 120 m/s.

The outward flow of potassium ions lasts only a few milliseconds. Finally, the sodium pump actively pumps the sodium ions out of the cell. As this occurs, the potassium ions rush back in and the normal resting state is restored. It takes about 40 ms for the neuron membrane to recover its polarized resting state.

The transmission of an impulse by the movement of ions in a neuron can be compared to putting a match to a trail of gunpowder. First, the beginning of the trail is lit. When the flame catches, it sets fire to the gunpowder directly in front of it. This keeps happening as the flame moves steadily down the trail of gunpowder to the end. Similarly, as an impulse travels along an axon, it excites the portion of the axon directly ahead. As a result, the impulse keeps moving down the neuron. Unlike the burned-up gunpowder trail, after the impulse has passed, each part of the neuron membrane changes back to its resting state and the process can begin again.

Transmission of the Impulse from One Neuron to Another

Most neurons do not touch each other directly. A microscopic space exists between the axon of one neuron and the dendrites of another neuron. This space is called a **synapse** (SI-naps). When an impulse travels from neuron to neuron, it moves along the axon of one neuron, crosses a synapse, and then continues along the dendrite of the next neuron.

An axon ends in a bunch of small fibers. Each small fiber delivers its impulse to the synapse. Most end-fibers in mammals have a **synaptic knob** at the tip. Sacs within the knob contain chemicals that transmit the nerve impulse. These chemicals are called **transmitter substances**. When an impulse reaches the

A synapse can be as narrow as 20 millionths of a millimetre.

13–9 The synapse: **a.** numerous synaptic knobs contact the cell body of a neuron; **b.** structure of a synaptic knob and synapse.

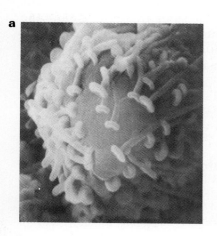

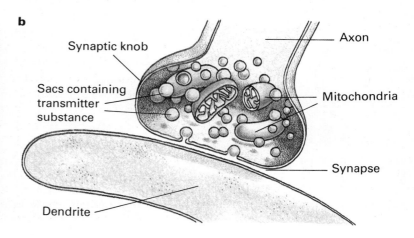

13–10 Transmission of a nerve impulse: **1.** A nerve impulse stimulates the release of transmitter. **2.** The transmitter combines with receptor molecules in the membrane of the dendrite or cell body, affecting the membrane's permeability to ions. The impulse either continues or is blocked. **3.** The transmitter is broken down.

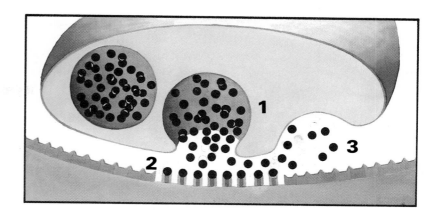

synaptic knob, it stimulates the sacs to release their transmitter substance into the synapse. The transmitter substance in the synapse combines with receptor molecules in the cell membrane of the dendrite or cell body of the next neuron. The union of receptor molecules and transmitter substance affects the permeability of the membrane of the neuron to ions. Whether the membrane becomes more permeable or less permeable depends on the type of neuron and the type of transmitter substance.

The Action of Transmitter Substances A large number of an animal's neurons produce **acetylcholine** (as-uh-tul-KŌ-leen) as their transmitter substance. Acetylcholine stimulates the transmission of impulses across synapses by making the membrane of the neuron more permeable to sodium ions. After the impulse is transmitted across the synapse, the enzyme **cholinesterase** (KŌ-luh-NES-tuh-RAYS), breaks down the excess transmitter substance. This prevents continuous stimulation of the neuron at the synapse. Other transmitter substances include *noradrenalin, dopamine,* and *seratonin.*

Some neurons produce inhibitory transmitter substances, which block the transmission of an impulse across the synapse. Many of these act by increasing the permeability of the membrane to chlorine ions. This allows more chlorine ions to enter the cell, making it more negatively charged.

Impulse Transmission Numerous axon end-fibers form synapses with the dendrites and cell body of a single neuron. This means that one neuron can receive impulses from a large number of other neurons. Some of these impulses stimulate the

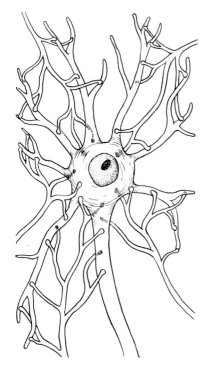

13–11 A single neuron can receive impulses from many different neurons, resulting in complexity in impulse transmission.

Section 13.1 Characteristics of Nervous Systems

neuron to "fire" an impulse; others inhibit it from firing. Whether or not the neuron fires depends on the signals received from all the neurons. If the combined stimuli of the other neurons are great enough, the neuron will fire an impulse.

Building a Science Vocabulary

receptor	nerve
effector	sensory nerve
impulse	motor nerve
central nervous system	mixed nerve
neuron	all-or-nothing response
cell body	resting state
dendrite	polarized
axon	sodium pump
glial cell	synapse
sensory neuron	synaptic knob
motor neuron	transmitter substance
interneuron	acetylcholine
ganglion	cholinesterase

Questions for Review

1. Replace each of the underlined words in the sentence below with one of the following terms: stimulus, receptor, central nervous system, effector, impulse.

    The student's <u>salivary glands</u> produced saliva when the <u>sound</u> of the classroom bell entered her <u>ear</u> and the <u>message</u> reached her <u>brain</u>.

2. Name the three main parts of a neuron. State the function of each part.

3. What is a nerve? How do the three types of nerves differ?

4. Briefly describe the resting state of a neuron. How is the resting state of a neuron membrane restored after an impulse passes?

5. How is an impulse transmitted across a synapse?

13.2 AN OVERVIEW OF INVERTEBRATE AND VERTEBRATE NERVOUS CONTROL

Invertebrate Nervous Systems

The nervous systems of most invertebrates are relatively simple. In general, invertebrates have a low ratio of neurons to other body cells. In addition, their neurons are quite large. They can be easily seen under a microscope and isolated for study.

The complexity of the nervous system varies greatly among the invertebrates. As you would expect, the larger and more complex species tend to have better developed nervous systems. Simple invertebrates, such as the hydra, have relatively few neurons and no central nervous system. In addition, their neurons conduct impulses in any direction. Complex invertebrates, such as grasshoppers and squids, have many more neurons than simple invertebrates. They also have a central nervous system, and their neurons carry impulses only in one direction.

To help you better understand the differences among the nervous systems of the invertebrates, nervous control in the hydra, earthworm, and grasshopper is considered below.

Nervous Control in the Hydra A hydra's behavior is slow and quite limited because it lacks a central nervous system that coordinates impulses. The hydra's nervous system consists of a **nerve net** that contains neurons scattered throughout the body. Impulses can move in any direction in the nerve net. When certain sensory receptor cells are stimulated, impulses spread

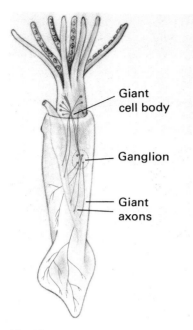

13–12 The giant axons of a squid may be as large as 1 mm in diameter.

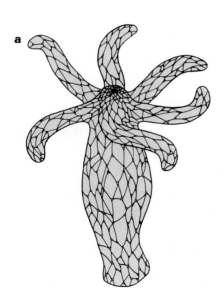

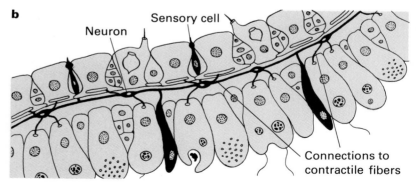

13–13 Nerve net of a hydra (**a**). Detail of the body wall shows receptor cells connecting to neurons between the cell layers (**b**). The neurons connect to contractile fibers in body cells.

Stinging cells and some cells with contractile fibers pick up stimuli and respond directly.

out from the point of stimulation over the nerve net. The hydra responds by contracting its body in the stimulated area. The impulses decrease in number as they spread from the point of stimulation. The stronger the initial stimulus, the farther the impulses spread and the more the hydra contracts.

Nervous Control in the Earthworm The earthworm's nervous system is more complex than that of the hydra. A pair of nerves in each segment carries impulses into and out of the central nervous system, which consists of the **ventral nerve cord** and ganglia. The ventral nerve cord is a double cord that runs along the underside of the earthworm's body from its head to its tail. It separates into two parts at the head end, forming a ring around the pharynx. A large pair of ganglia are located on the nerve cord above the pharynx. These ganglia are sometimes called the earthworm's "brain." Smaller ganglia are located on the nerve cord in each segment of the earthworm's body.

venter = belly

The earthworm has various types of receptor cells in its skin. These detect touch, light, chemicals, and moisture. Sensory neurons in each segment are stimulated by receptor cells. They transmit impulses to the ganglia, which coordinate the sensory impulses with motor impulses to the earthworm's muscles and glands.

Remember that the earthworm does not really have a head. A more accurate term is anterior end.

During locomotion, the contraction of muscles in one segment of the earthworm stimulates sensory neurons in the next segment. The sensory neurons send impulses to the ganglia in their segment. From the ganglia, impulses on motor neurons stimulate the muscles of the segment to contract. Muscle contraction in this segment then stimulates the neurons in the next segment, and the process is repeated.

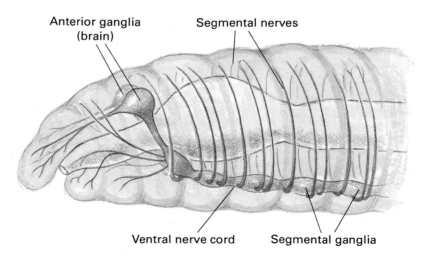

13–14 The nervous system of an earthworm. The nerves relay sensory impulses from receptor cells and send motor impulses to muscles.

An earthworm with its brain removed can still move but is not capable of stopping.

Impulses travel from segment to segment on the ventral nerve cord, which contains fast-conducting axons. This makes it possible for the earthworm to contract its whole body and move quickly. The ventral nerve cord also transmits impulses into and out of the brain. The coordination of muscle action occurs mainly in the smaller ganglia, but the brain has some control.

Nervous Control in the Grasshopper Like the earthworm, the grasshopper has a central nervous system consisting of a double ventral nerve cord and ganglia. The ventral nerve cord extends from the tail to the head. In the head, it divides into two cords. These cords encircle the esophagus and rejoin, forming two large ganglia, which are considered to be the grasshopper's brain. Pairs of ganglia also occur along the length of the nerve cord. Nerves from the brain carry impulses to and from the head organs. Nerves from the ventral nerve cord carry impulses to and from the segments, legs, wings, and body organs.

The grasshopper has specialized receptors, or sense organs, that pick up stimuli from the environment. These specialized receptors are much more complex and efficient than the earthworm's receptor cells. The grasshopper has two sets of eyes. Two large eyes see objects, particularly when they move. Three small eyes detect light and motion. Specialized sense organs also detect touch, taste, odor, and sound. In addition, the grasshopper has receptors that detect its body movement.

When the nerves in the grasshopper's sense organs are stimulated, nerve impulses begin. These impulses travel to the central nervous system, which coordinates the incoming impulses with impulses going to effectors. For example, the quick motion of a bird or other predator is detected by the grasshopper's eyes. The eyes send impulses to the central nervous system, which interprets the motion. The central nervous system then sends impulses to the grasshopper's leg and wing muscles. The grasshopper jumps or flies away, avoiding the predator.

The grasshopper is aided by rapid automatic movements in avoiding predators. Its movements—walking, jumping, and flying—require coordination. As in the earthworm, coordination of muscle movements in the grasshopper occurs mainly in the ganglia. Nevertheless, the brain exerts some control. Without its brain, a grasshopper can still walk, jump, and fly; however, it does these things continually in response to any stimulus.

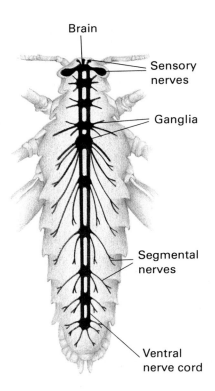

13–15 The nervous system of a grasshopper. A grasshopper has specialized sense organs, which are more complex than an earthworm's receptor cells.

Vertebrate Nervous Systems

The nervous systems of vertebrates are considerably more developed than those of invertebrates. For example, invertebrates have neurons usually numbering in the thousands, while many species of vertebrates have billions of neurons. Vertebrates also have many more interconnections between neurons. In addition, instead of a double ventral nerve cord, vertebrates have a single, hollow spinal cord, which is located dorsally (in the back). Vertebrates also have much larger brains, and control of body functions is more centralized in the brain and spinal cord.

Vertebrate Neurons Special glial cells called **Schwann cells** form a covering around the axons of many vertebrate neurons. These cells aid in the production of a white, fatty material called **myelin** (MĪ-uh-LIN) that forms a sheath between the Schwann cells and the axon. The myelin sheath pinches at regular points along the axon. When an impulse travels along the axon, it jumps from one pinched-in area to the next. This leaping greatly increases the speed of transmission of an impulse.

An axon with a myelin sheath can carry impulses 50 times faster than the same axon without myelin. Whereas some myelin-sheathed axons carry impulses at the rate of 200 metres per second, small axons without myelin carry impulses only a few millimetres per second.

Like complex invertebrate neurons, vertebrate neurons carry impulses in only one direction.

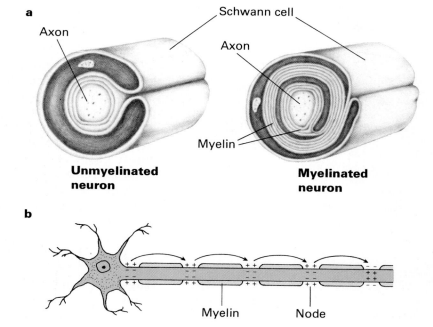

13–16 Unmyelinated axons are simply buried in Schwann cells; myelinated axons are surrounded by many coils of the Schwann cell membrane (**a**). The myelin sheath consists of specialized Schwann cell membrane tissue. In myelinated axons, impulses jump from node to node (**b**).

Chapter 13 Nervous Control

Nerve Pathways The route traveled by a nerve impulse is called a **nerve pathway**. Impulses in vertebrates can travel over a tremendous number of different nerve pathways. One common and relatively simple nerve pathway is the **reflex arc**. A reflex arc usually consists of a receptor, a sensory neuron, one or more interneurons in the central nervous system, a motor neuron, and an effector.

Figure 13–17 shows the path of an impulse through a reflex arc. The stimulus in the example is heat. A person touches a hot object. Receptors in the skin of the hand pick up the stimulus, and an impulse is generated along a sensory neuron. The neuron carries the impulse to the spinal cord. In the spinal cord, one or more interneurons transmit the impulse to a motor neuron. The motor neuron carries the impulse to muscles in the hand. They respond to the impulse by moving the hand away from the hot object.

The reflex arc allows the body to respond rapidly and automatically to certain kinds of stimuli. Even though reflex arcs are simple, they play an important role in many body responses. In addition, the more interneurons in a reflex arc, the more complex the pathway becomes. For example, other interneurons stimulated by the sensory neuron send impulses to various parts of the brain. The brain processes this information, producing the feeling of pain, memory of the incident, and other sensations.

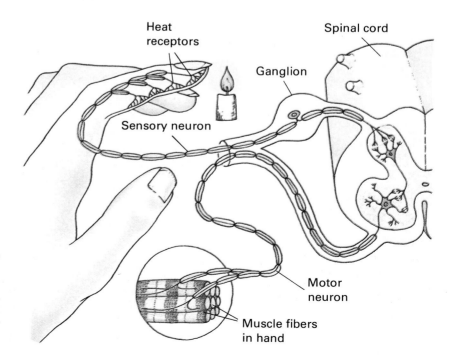

13–17 Reflex arc. When stimulated by heat from a flame, heat receptors in the finger transmit impulses to the spinal cord. As a result, a motor neuron is stimulated and muscles in the hand contract, pulling the finger away from the flame.

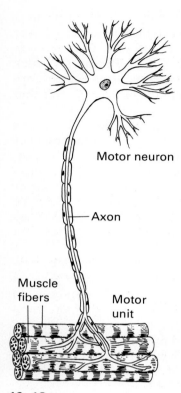

13–18 A motor unit consists of a motor axon and all the muscle fibers the axon branches contact. A motor unit may contain more than 500 muscle fibers.

The Stimulation of Skeletal Muscle Contraction When a skeletal muscle is stimulated by a motor nerve impulse, it contracts. The axon of each motor neuron branches as it reaches the muscle. Each of these branches ends at a muscle fiber. The motor neuron axon and all the muscle fibers its branches contact are known as a **motor unit**.

The end of each axon branch is embedded in a groove on the surface of the muscle fiber. Transmission of the nerve impulse to the muscle fiber is similar to transmission of the impulses between neurons. When an impulse reaches the end of a branch, a transmitter substance pours into the gap between the axon and muscle fiber. The transmitter substance stimulates the release of calcium and ATP in the muscle fibers, and the fibers contract.

Building a Science Vocabulary

nerve net   myelin   motor unit
ventral nerve cord nerve pathway
Schwann cell   reflex arc

Questions for Review

1. Name three major differences between invertebrate and vertebrate nervous systems.
2. How does the hydra pick up stimuli from the environment and transmit impulses throughout its body?
3. Name two structures found in the nervous system of the earthworm that are lacking in the hydra. What is the function of each of these structures?
4. How do impulses travel from segment to segment during locomotion in the earthworm?
5. What structures are considered to be a brain in the grasshopper? What function does the brain have during locomotion?
6. What is myelin? What function does it perform?
7. How do nerve impulses on motor neurons result in the contraction of skeletal muscle?

13.3 NERVOUS CONTROL IN HUMANS

The human nervous system is usually divided into two parts: the **central nervous system** and the **peripheral nervous system**. As mentioned before, the central nervous system consists of the brain and spinal cord. The peripheral nervous system is made up of all the nerves and ganglia that exist outside of the brain and spinal cord.

The Central Nervous System

The organs of the central nervous system are well protected. The brain is encased by the bones of the skull, and the spinal cord is surrounded by the bony vertebrae. Three layers of tissue, the **meninges** (meh-NIN-jeez), surround the brain and spinal cord. A space between two of the meninges contains **cerebrospinal** (suh-REE-brō-SPĪ-nul) **fluid**, a liquid substance that cushions the central nervous system against shock.

In spite of their protection, about 10 000 neurons disintegrate each day owing to aging, disease, injury, and alcohol and other harmful drugs. Neurons do not reproduce; yet, they do have some ability to repair themselves and to grow new parts.

The brain and spinal cord are composed of gray and white areas. The outside of the brain is gray and the inside is white. The reverse is true in the spinal cord. The gray areas, or **gray matter**, contain masses of nerve cell bodies and glial cells. The white areas, or **white matter**, contain myelin-covered axons. Impulses are transmitted between interneurons in the gray matter. They are transmitted to and from various parts of the body in the white matter.

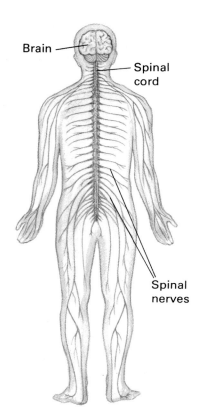

13–19 The central nervous system consists of the brain and spinal cord; the peripheral nervous system includes all the nerves and ganglia outside of the brain and spinal cord.

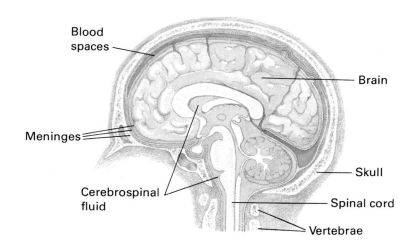

13–20 The brain and spinal cord are protected by the skull and vertebrae, the meninges, and the cerebrospinal fluid.

The Brain

The human brain weighs only about 1.4 kg. Small as it is, the brain contains nearly ten billion nerve cells and is capable of controlling and coordinating thoughts, emotions, behavior, and body functions. Many people compare the brain to a telephone switchboard, because it coordinates incoming and outgoing signals. No switchboard exists, however, that has the billions of connections found in the human brain.

The transport system supplies the brain (and spinal cord) with nutrients and removes wastes. Water, oxygen, and carbon dioxide pass through brain cell membranes easily, but ions, proteins, and other substances are transferred slowly. Even glucose, the brain's major source of food, enters brain tissue slowly. The slow exchange of materials helps preserve the constancy of the brain's internal environment. The brain's neurons cannot withstand rapid changes in the composition of the fluid around brain tissue.

The brain takes in about 49 mL of oxygen each minute and is very sensitive to an oxygen deficiency. If the total blood supply to a person's brain is cut off, the person becomes unconscious from lack of oxygen in ten seconds.

The supply of glucose to the brain is also of great importance. If a person does not have enough sugar in the blood, the brain cannot function properly. The result is confusion, convulsions, and eventually the person goes into a coma.

Temperature is another critical factor in brain functioning. Normal body temperature averages about 37°C. A person whose body temperature exceeds 41°C for a prolonged time will suffer some permanent brain damage. Temperatures above 43°C can result in heatstroke and possible death. Humans can survive low body temperature better than high body temperature. Some people recover even after having their body temperature drop to between 22°C and 24°C. At very low temperatures, heart rate and respiration become slower, and a person loses consciousness. For this reason, surgery is sometimes performed at low temperatures.

The brain gives off electrical impulses called **brain waves**, which can be detected by electrodes attached to the scalp. A person's brain waves vary according to wakefulness. A person gives off one kind of brain waves when awake and resting. When the person's brain is active, another type of wave is detected. A third type of wave is emitted when the person sleeps.

One cubic centimetre of the human brain can contain six million cell bodies, and each neuron can connect to as many as 80 000 other neurons.

The brain stores glucose in the form of glycogen. If the blood supply is shut off, the brain uses up all its glucose and glycogen within two minutes.

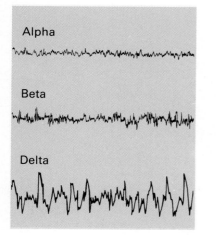

13–21 Brain waves. A person produces alpha waves while resting, beta waves while active, and delta waves while sleeping.

In the 1920s Wilder Penfield, a Canadian neurosurgeon, studied patients with epilepsy. Penfield stimulated various parts of the brain with a mild electric current, using a technique that did not harm his patients. Penfield's work led to the understanding that different parts of the brain are centers for various functions. Today the brain centers that control the senses and certain other activities are well known. Still, the location and function of the centers for many behaviors and mental abilities, such as memory, are poorly understood.

The Parts of the Brain

The brain is a complex organ, containing many structures that perform different functions. In general, the brain is divided into three main parts: the **cerebrum** (suh-REE-brum), the **cerebellum** (SAYR-uh-BEL-um), and the **brainstem**.

The Cerebrum   The cerebrum is located behind the forehead. It is covered by an area of gray matter about 2 mm thick called the **cerebral cortex**. The cerebral cortex is highly wrinkled and folded in a pattern characteristic of mammals. The inside of the cerebral cortex contains white matter. Axons in the white matter connect the various parts of the cerebral cortex and connect the cerebral cortex with the rest of the brain.

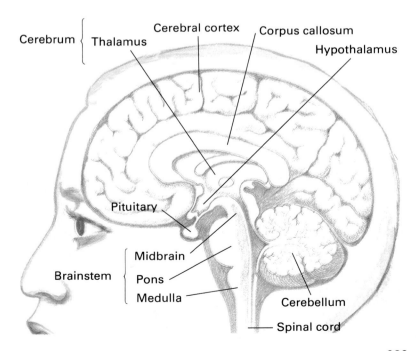

13–22 Longitudinal section of a human brain, showing the three main areas of the brain and their parts: the cerebrum, which includes the cerebral cortex, corpus callosum, thalamus, and hypothalamus; the brainstem; and the cerebellum.

Section 13.3 Nervous Control in Humans

A blow to the back of the head can cause blindness, because that is the location of the sight centers.

A deep groove divides the cerebrum into right and left hemispheres. The two hemispheres are connected by a bridge of tissue, the **corpus callosum** (KOR-pus kuh-LŌ-sum). Impulses pass between the left and right sides of the cerebrum through this bridge. Each hemisphere is further divided into four lobes: *frontal, parietal, temporal,* and *occipital.*

Certain centers of the cerebral cortex control specific functions. Parts of the cerebral cortex control voluntary muscle movements. For example, one center in your cerebral cortex controls the movement of your lips when you speak, and another center controls your hand when you write. The lobes of the cerebrum contain centers for the senses. The centers for hearing are in the temporal lobes, and those for sight are in the occipital lobes. Taste and touch are centered in the parietal lobes.

The cerebral cortex is also involved with thinking, memory, judgment, responses, and actions; however, the centers of these activities are unknown. Some information about these processes has come through studies of people in whom the corpus callosum has been cut. Without the connecting bridge, these people lose coordination between the left and right sides of the cerebrum. This has made it possible to determine what each side of the brain does alone. Such studies show that the left side of the brain functions more strongly in speech, logic, writing, and mathematics. The right side functions more strongly in discriminating shapes and forms, music, and nonverbal communication.

The people in these "split brain" experiments had the corpus callosum severed as a treatment for epilepsy. Epilepsy causes uncontrollable seizures or convulsions. Some medicines help prevent or control the seizures, but no cure for epilepsy is known.

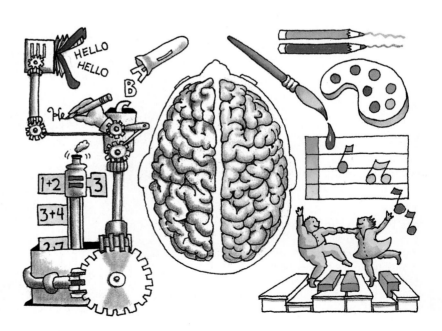

13–23 The left side of the brain functions mainly in speech, logic, writing, and arithmetic. The right side of the brain is more involved in imagination, as well as spatial, artistic, and symbolic functions. In addition, motor control of the right side of the body originates in the left side of the brain. Motor control of the left side of the body originates in the right side of the brain.

A structure at the bottom of the cerebrum, the **thalamus**, transmits impulses between the cerebrum and the brainstem below. It is also the center for pain. The **hypothalamus** is located below the thalamus. It performs many important functions, including controlling hunger, thirst, body temperature, and the pituitary gland (see Chapter 12). The hypothalamus is also the center for the sex drive and emotions such as anger and pleasure.

The Cerebellum The cerebellum is located behind and below the cerebrum. It helps maintain balance and coordinates muscle activity. The cerebellum receives information from the eyes and ears about the position of the body relative to its surroundings. It also receives information from internal receptors in joints and muscles about the physical state of the body. In addition it receives information from the cerebral cortex about its control of individual muscle actions. The cerebellum processes all this information to coordinate the body's movement with its position. It then sends impulses to the cerebrum and other parts of the brain. Finally, impulses that integrate muscle activity in a unified way leave the brain. Suppose you are walking off balance. Information from the eyes, ears, and internal receptors is processed by the cerebellum, which then directs the cerebrum to "correct" its impulses to the muscles involved in walking. Impulses from the cerebrum stimulate a correction in your movements and balance is restored.

Large quantities of alcohol reduce the ability of the cerebellum to coordinate muscle activity. This results in distorted speech and unbalanced walking.

13–24 Running hurdles quickly and without losing balance requires the coordination of numerous nerve impulses by the cerebellum.

Section 13.3 Nervous Control in Humans

The Brainstem   The brainstem is the part of the brain that attaches to the spinal cord. It consists of three sections: the *medulla*, *pons*, and *midbrain*. The brainstem contains nerve fibers that connect the brain and the spinal cord. In addition, it contains centers for the control of vital functions, such as heartbeat and respiration.

The respiratory centers are located in the brainstem.

A complex nerve pathway known as the **reticular activating system (RAS)** is located in the brainstem. Stimulation of the RAS allows us to be conscious and alert. The brainstem is also the center for reflexes that occur in the head. A reflex is an inborn, automatic action that does not have to be learned. For example, you automatically blink your eyes when something comes toward you. Other head reflexes are coughing, sneezing, and producing saliva.

Most invertebrate reactions are inborn. Many vertebrate reactions depend on learning.

The Spinal Cord

The spinal cord carries impulses to and from the brain. In addition, it is the center for body reflexes, such as withdrawing the hand from a hot object. During a medical examination, the physician might tap your kneecap with a rubber hammer. The resulting involuntary movement of your leg is a body reflex called the "knee jerk." The doctor learns two things from this test. It reveals whether the sensory and motor nerve connections are working between the muscles and the spinal cord. It also shows how responsive the spinal cord is.

Persons with severe damage to the spinal cord may lose sensation and muscle function in the lower portions of the body (paralysis). Such individuals are called **paraplegics** (PAYR-uh-PLEE-jiks). Paralysis occurs because impulses to and from the lower body muscles cannot be transmitted across the injury and therefore do not contact the brain. As a result, voluntary control of skeletal muscle is lost. Involuntary reflexes with centers in the spinal cord still work. For example, paraplegics can empty their urinary bladders, but not voluntarily.

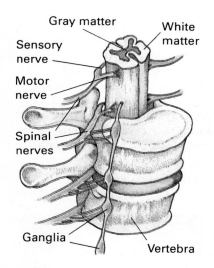

13–25   Spinal nerves enter the spinal cord between the vertebrae. Each spinal nerve carries sensory and motor neurons.

The Peripheral Nervous System

The peripheral nervous system is made up of the neurons, ganglia, and nerves that occur outside the central nervous system. It transmits impulses between receptors, effectors, and the central nervous system. The peripheral nervous system consists of two main parts. One part, known as the **somatic nervous system**, transmits impulses for voluntary actions, such as walking and singing. The other part, known as the **autonomic nervous sys-**

Twelve pairs of cranial nerves, which connect to the brain, and 31 pairs of spinal nerves, which connect to the spinal cord, are included in the peripheral nervous system.

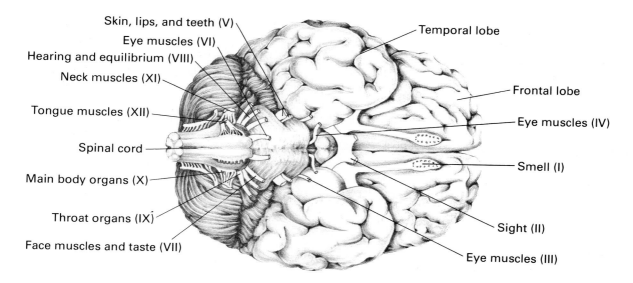

13–26 Ventral surface of the brain. The cranial nerves are labeled by function and number. Number II is the optic nerve; X is the vagus nerve.

tem, controls the involuntary actions of vital organs, such as the heart, intestines, lungs, and glands.

The nerves that control vital organs come from two different divisions of the autonomic nervous system: the **sympathetic nervous system** and the **parasympathetic nervous system**. Sympathetic nerves connect with the spinal cord at the thoracic area and lower part of the spine. Most sympathetic nerve impulses pass through ganglia just outside the spinal cord. Sympathetic nerves produce noradrenaline at their synapses. Parasympathetic nerves connect with the central nervous system at the base of the brain and the lower back. Their impulses pass through ganglia in or near the organ that they help control. Parasympathetic nerves produce acetylcholine at their synapses.

13–27 Organization of the human nervous system.

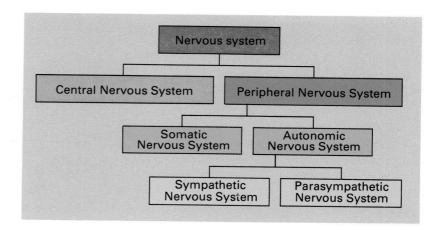

Section 13.3 Nervous Control in Humans

13–28 The autonomic nervous system controls vital organs. Four cranial nerves and about half of the spinal nerves contain neurons of the autonomic nervous system. Sympathetic and parasympathetic nerves have opposite effects on organs.

Sympathetic and parasympathetic nerves have opposite effects on organs. One nerve speeds up the activity of the organ; the other slows it down. You can see examples of this in the control of the heart and the intestine. The vagus, a parasympathetic nerve, slows the heartbeat, while the accelerator, a sympathetic nerve, speeds it up. Parasympathetic nerves do not always slow down organs; nor do sympathetic nerves always speed up organs. Their effects depend on the organ. For example, a parasympathetic nerve speeds up muscle action in the intestine wall, while a sympathetic nerve slows it down.

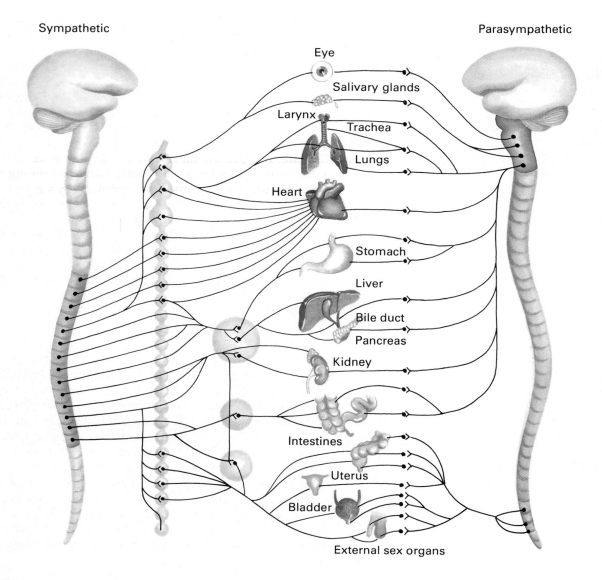

Nervous System Disorders

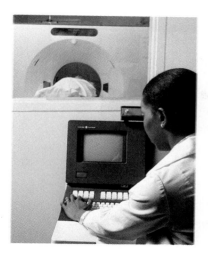

13-29 A technician operates a computerized axial tomography (CAT) scanner from a remote computer terminal. The CAT scanner produces detailed images of body parts and is a valuable tool in the diagnosis of tumors and other conditions.

Several disorders are associated with the impairment of nervous system tissues. **Meningitis** (MEN-in-JĪT-is) is an inflammation of the meninges caused by bacteria. This disease causes a high fever with a severe headache and stiffness of the neck. In severe cases it may cause coma and death. To help diagnose meningitis, a sample of cerebrospinal fluid is withdrawn and analyzed. The disease is treated with medications that stop bacterial growth.

The abnormal growth of tissue, resulting in a **tumor**, can also cause damage to the nervous system. In the brain, a tumor exerts pressure on brain tissue, frequently resulting in loss of vision or other functions or changes in personality and emotions. Brain tumors are often successfully removed by surgery.

Certain nervous system diseases involve blood vessels that supply brain tissue. For example, in arteriosclerosis a weak-walled artery can form a bubblelike swelling called an **aneurism** (AN-yoor-ism). An aneurism in an artery supplying blood to the brain can break. This can decrease the amount of blood that reaches brain tissues and result in a stroke. A person who suffers a stroke may die. If a person survives a stroke, the area of the brain near the stroke may be permanently damaged owing to lack of oxygen. Aneurisms are sometimes removed by surgery. A blood clot in a brain artery can also result in a stroke. Some clots of this kind can be removed by surgery.

Tissues degenerate in diseases such as **multiple sclerosis (MS)**. In MS, myelin breaks down leaving axons with patchy areas that lack myelin. Victims of MS can suffer from loss of coordination and control of muscles. The cause and cure of MS is still unknown.

The Effects of Drugs and Poisons

The disease lockjaw is caused by tetanus bacteria.

Many drugs and poisons affect synapses. **Strychnine** (STRIK-nīn), a plant poison, blocks inhibitory transmitter substances. So does the poison produced by **tetanus** bacteria. These poisons cause continuous stimulation of neurons, resulting in convulsions and muscle spasms. Anesthetic drugs, such as ether, opiates, and barbituates, block transmitter substances that stimulate neurons, resulting in loss of consciousness. Amphetamines cause more transmitter substance to be produced in the brain. The additional transmitter substance stimulates neurons and results in increased activity and alertness, which can be harmful to the body. Caffeine and nicotine imitate transmitter substances and cause specific nerves and effectors to be stimulated.

Section 13.3 Nervous Control in Humans

Building a Science Vocabulary

central nervous system	hypothalamus
peripheral nervous system	reticular activating system (RAS)
meninges	paraplegic
cerebrospinal fluid	somatic nervous system
gray matter	autonomic nervous system
white matter	sympathetic nervous system
brain wave	parasympathetic nervous system
cerebrum	meningitis
cerebellum	tumor
brainstem	aneurism
cerebral cortex	multiple sclerosis (MS)
corpus callosum	strychnine
thalamus	tetanus

Questions for Review

1. Name the part of the central nervous system that controls each of the following functions.
   a. thinking
   b. body reflexes
   c. balance and coordination
   d. heart rate and breathing

2. The following questions refer to Figure 13–30.
   a. If *b* is the vagus nerve and *a* is heart muscle tissue, what substance is produced where *a* and *b* meet?
   b. How does stimulation of *b* affect the heart rate?
   c. What division of the autonomic nervous system does the vagus nerve belong to?
   d. What nerve opposes the action of the vagus nerve in controlling the heartbeat?

3. The following questions refer to your ability to see this book.
   a. What receptor picks up stimuli from the book?
   b. What part of the brain is the center for sight?
   c. What parts of the nervous system transmit impulses between your brain and your hand as you hold the book?

4. Why does brain tissue often die when a stroke occurs?

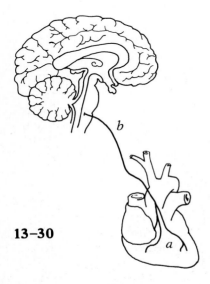

13–30

Perspective: A New View
Tomorrow's Dinner

The menu at Joe's Diner does not seem very exciting, so you decide to try the fare at El Futuro, the restaurant of the future. El Futuro is famous for serving very unusual food.

To begin, there is soup produced by bacteria. For your salad, El Futuro offers sea lettuce prepared with a delicate dressing. There is a choice for the main course—manatee steak, dugong stew, or broiled filet of capybara. The bread being served is derived from green and brown algae, and dessert is a delicious cake made of amaranth flour.

Although the food is quite unusual, you are surprised to see that the meal is only half as costly as a typical dinner at Joe's Diner. On the back of the menu you find descriptions of all these unusual dinner items. As you read, you begin to understand how the price can be so low.

Bacteria reproduce very rapidly and can grow on waste products such as sawdust and oil spills. As a result, they are much cheaper to obtain than animals that grow on expensive foods such as wheat or corn. The nutritious soup is made from the breakdown products of the bacteria. The sea lettuce used in the salad grows abundantly in sea water and is easily harvested, so the costs of farming are low. The manatee is a freshwater mammal, the dugong a marine one, and the capybara is a large terrestrial rodent. The muscle from each of these animals makes delicious meat. Since the animals are all herbivores and not at all picky about the plants or algae they eat, they are inexpensive to raise.

Amaranth is a fast-growing, highly productive weed. It served as the major part of the diet of the Aztec civilization five centuries ago and has been cultivated for at least 8000 years. Like corn and wheat, it produces seeds in abundance. Amaranth seeds are high in protein, and they contain an amino acid lacking in most cereals (lysine). The seeds have a nutty flavor and can be popped like corn seeds. The nutritious leaves have a mild spinach-like flavor.

Eating at El Futuro may be a bit too much today. But as the food needs of an ever expanding world population increase, people might have to look for new sources of food. Such a menu may not be so strange in a few years.

Seconds, anyone?

Chapter 13
Summary and Review

Summary

1. Receptors detect stimuli in the environment and transmit impulses on neurons to effectors. In most animals, a central nervous system coordinates incoming and outgoing impulses.

2. The neuron is the basic unit of structure and function in the nervous system. A neuron consists of a cell body, dendrites, and an axon.

3. Neurons are grouped according to the direction that the impulse travels. There are sensory neurons, motor neurons, and interneurons. Bundles of neuron cell bodies make up the brain, spinal cord, and ganglia. Bundles of axons make up nerves.

4. In its resting state, the neuron membrane is polarized. During a nerve impulse, the membrane changes charge as ions move across it. The movement of ions in one part of the membrane stimulates the movement of ions in nearby parts of the axon.

5. When the impulse reaches the end of the axon, sacs in the synaptic knobs release a transmitter substance into the synapse. Transmitter substances stimulate or inhibit the continuation of the impulse on the dendrites or cell body membranes of the next neuron.

6. Invertebrates have fewer neurons than vertebrates. They also have less well-developed central nervous systems.

7. The hydra's nervous system consists of a nerve net that lacks central control.

8. Nervous control in the earthworm and grasshopper depends on nerves and ganglia. The grasshopper has well-developed sense organs.

9. Neurons are organized into nerve pathways, the simplest of which is the reflex arc.

10. Skeletal muscles contract when stimulated by transmitter substances released by axon fibers carrying a nerve impulse.

11. In humans, the central nervous system consists of the brain and spinal cord. The peripheral nervous system consists of neurons, ganglia, and nerves outside the central nervous system.

12. The main parts of the brain include the cerebrum, cerebellum, and brainstem. Each of these parts contains structures that function in the coordination and control of body parts.

13. The peripheral nervous system is made up of the somatic and autonomic nervous systems. The autonomic nervous system controls vital organs by a system of opposing nerves, the sympathetic and parasympathetic nervous systems.

14. Disorders that harm nervous system tissue include meningitis, tumors, aneurisms, strokes, and multiple sclerosis.

Question Summary

Application

1. Compare the following nervous system characteristics of an ant with those of a bear.
    a. position of the nerve cord
    b. number of neurons
    c. number of interconnections
    d. reliance on reflexes

2. List all the steps in the "knee jerk." Begin with a stimulus and end with a response. Name all the structures through which the impulse passes.

3. How does ether "put you to sleep" during an operation?

4. State one function that would be lost by damage to each of the following.
   a. the earthworm's nerve cord
   b. the thalamus in humans
   c. the myelin on axons
   d. the cerebellum in humans

Interpretation

1. Why is the hydra unable to make a rapid response to a sudden stimulus?

2. Why is neuron transmission considered an all-or-nothing response, while nerve transmission is not?

3. Is impulse transmission at the synapse an all-or-nothing response? Explain.

4. How would an animal be affected if its hypothalamus were stimulated by a mild electric current?

Extra Research

1. Make a clay model of the human brain. Use paper flags on toothpicks to label structures and functions.

2. Look up the structure of the brain in each of the five classes of vertebrates. Make a drawing of the brain of one animal from each class. Develop a list of similarities and differences among vertebrate brains.

3. Look up the effects of alcohol and various drugs on the structure and functioning of the nervous system. Present your findings in an article aimed for the general public.

4. Prepare an index of disorders of the nervous system. Include the name of the disorder, structures affected, functions affected, cause (if known), treatment, and tools for diagnosis. Include at least four disorders in addition to the ones mentioned in the chapter.

5. Prepare a chart showing the chemical changes that occur during a nerve impulse and during impulse transmission across a synapse. Use different colors to represent the change in charge of the neuron membrane.

Career Opportunities

Neurosurgeons are highly trained medical doctors who perform delicate operations on the brain and other parts of the nervous system. They often work as part of a team in operations on accident victims. Several years of training beyond medical school are required. For information, write the Canadian Medical Association, Box 8650, Ottawa, Ontario K1G 0G8.

Electroencephalographic (EEG) technicians operate equipment that is used to record the electrical activity of the brain. They adjust EEG equipment to give accurate recordings and write reports for each patient. Most of their work is in hospitals and clinics. A year or more of training beyond high school is recommended. Contact the Canadian Association of Electroencephalograph Technologists Inc., 160 Wellesley St. E., Toronto, Ontario M4Y 1J3.

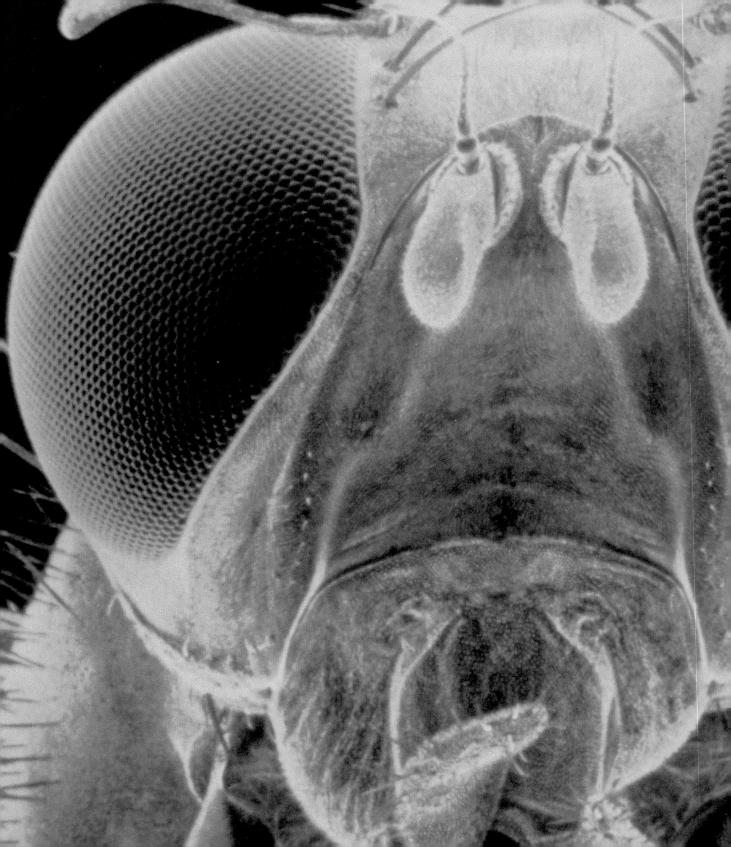

Chapter 14

The Senses

Focus *A radio receiver picks up signals from the atmosphere and converts them to sound. A television antenna gathers signals and changes them to the picture on a television screen. These mechanical devices pick up signals from the environment and change them to information that is meaningful to humans. In a way, they are extensions of our own natural senses. All living things are able to detect signals from the environment and integrate them into useful information. This ability is most highly refined in animals. It helps them detect danger, find food, sense the weather, locate mates, and communicate with one another.*

14.1 RECEPTORS GATHER INFORMATION

In 1975 a devastating earthquake rumbled through north-central China. It leveled buildings and tore apart streets, dams, and other structures. Before the earthquake began, researchers predicted that it was coming. They alerted authorities, who were able to evacuate tens of thousands of people before the earthquake hit. The researchers based their prediction on the strange behavior they noticed in certain animals. Dogs started to yelp, chickens failed to roost at sunset, rats left their burrows, and snakes came out of hibernation. The researchers knew from

14–1   A Mediterranean fruit fly's head is dominated by its eyes.

14-2 The highly developed sense organs in a cat's head—its eyes, ears, whiskers, nose, and tongue—transmit sensory information to the brain, where it is interpreted.

past experience that these unusual behaviors precede an earthquake. No one knows exactly how certain animals sense that an earthquake is coming; however, they must be sensitive to certain stimuli that we cannot detect.

Detection and Perception

Animals use specialized receptors and the nervous system in coordinating detection and response. Animals detect different kinds of stimuli, including light, heat, sound, chemicals, pressure, and body position.

A receptor may simply be part of a cell or neuron. It may also be a sense organ that contains many cells, such as the eye or ear. In complex animals, information from receptors is transmitted by nerve impulses to the brain. The brain interprets the information, and sight, hearing, touch, taste, smell, and other perceptions result.

Each receptor is primarily sensitive to one kind of stimulus. An insect's antenna, for example, is not sensitive to light, but it readily detects chemicals in the air.

Kinds of Receptors

Receptors are classified according to the type of stimulus they detect. Pressure and vibrations are detected by **mechanoreceptors** (muh-KAN-ō-rih-SEP-turz), such as the sensory cells in the ears and some skin receptors. Their action brings about the perception of hearing and touch. Light is detected by **photoreceptors**, including eyespots and eyes. In complex animals, photoreception results in sight. Chemicals are detected by **chemoreceptors** (KEE-mō-rih-SEP-turz). For example, the nose and tongue have chemoreceptors that enable animals to perceive smell and taste. **Thermoreceptors**, which are usually lo-

Vibrations are back and forth movements.

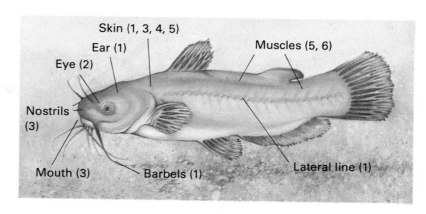

14-3 Kinds of receptors in a catfish: 1—mechanoreceptor; 2—photoreceptor; 3—chemoreceptor; 4—thermoreceptor; 5—pain receptor; 6—proprioceptor.

cated in the skin, lead to the perception of warmth and cold. **Pain receptors**, also located in the skin, bring about the perception of pain. **Proprioceptors** (PRO-pree-uh-SEP-turz), located in the ears, joints, tendons, and muscles detect body movements. They enable animals to perceive body position, balance, location of body parts, movement, and locomotion.

Stimuli Not Detected by Humans

We do not know very much about the stimuli that signal the coming of an earthquake; however, we do know about several kinds of stimuli that only certain animals detect.

Most animals are able to detect light. The light we see is called **visible light**. Visible light, however, makes up only part of the light spectrum. Other kinds of light are invisible to us. For example, we cannot see **ultraviolet light**. Bees, on the other hand, are capable of detecting ultraviolet light. Ultraviolet rays from the sun pass more easily through clouds than visible light does. Since bees can detect ultraviolet light, they are able to locate the sun when it is hidden behind clouds.

Some caterpillars also detect ultraviolet light.

Bees can also locate the sun when it is blocked by trees or buildings, as long as they can see a patch of blue sky. The light coming from blue sky is mainly **polarized light** (it vibrates in

14-4 Vision in a bee. A bee can detect light in the ultraviolet and visible light regions. They do see in color, as shown in the figure; however, how they perceive color appears to differ somewhat from how humans perceive it.

Section 14.1 Receptors Gather Information

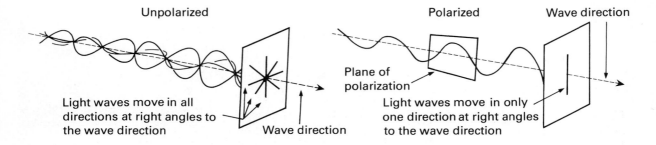

14–5 Unpolarized light waves move through many planes at right angles to the wave direction. Polarized light waves move in only one plane.

one plane), which bees can detect. Bees determine the sun's location from the direction that polarized light hits them.

The ability to perceive the location of the sun when it is blocked by clouds, mountains, or trees is an important adaptation in bees. They use the location of the sun as a reference point in communicating the location of food sources to other bees. Efficient communication is vital to the survival of members of the hive.

Boa constrictor-type snakes include the boa constrictor, python, and anaconda. Pit vipers include the rattlesnake, copperhead, and water moccasin.

Boa constrictors and pit vipers are types of snakes that prey on small mammals and birds. Mammals and birds are warm-blooded and give off heat. Some of the heat is in the form of **infrared radiation**, a kind of heat wave. Both types of snakes have thermoreceptors that detect infrared rays. The thermoreceptors are located in pits along the jaws of the snake. Heat perception and response are extremely efficient in these snakes. A boa constrictor can attack its prey 35 ms after its thermoreceptors are stimulated. Heat perception also helps snakes find a location in which to stay. Although snakes are cold-blooded, they function best in a narrow temperature range. Thermoreceptors enable snakes to sample the amount of heat in an area before settling down.

The sidewinder snake functions best between 31°C and 32°C.

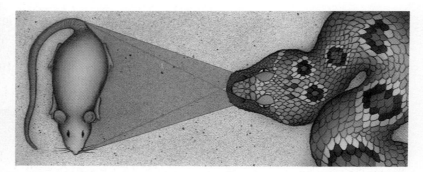

14–6 Pit vipers such as the rattlesnake are named for the heat-sensing pit beneath each eye. The position of the pits helps the snake determine the location of its warm-blooded prey.

14–7 Sound. The pitch of a sound is determined by the frequency of the sound waves (measured in hertz). The greater the frequency, the higher the pitch.

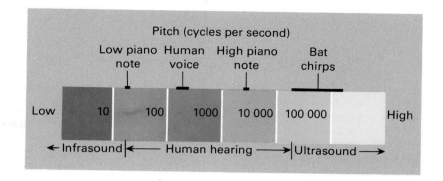

Some animals detect **ultrasound**, which is made up of sounds higher than humans can hear. Dolphins, for example, hear sounds that are two octaves higher than we can detect. Dolphins send out ultrasound waves as they swim. The sound waves bounce off objects in the water and return to the dolphins. A dolphin perceives an object's location by the pitch and loudness of the reflected sound (echo). The ability to hear ultrasound helps dolphins distinguish between the sound echoing off objects and the noise of the ocean waves.

Dogs and rodents also perceive sounds we cannot hear.

Locating objects using ultrasound echoes is called echolocation.

A fish perceives the motion of water currents using its **lateral line**, a mechanoreceptor that land animals lack. The lateral line is a series of grooves on the fish's head and sides. Receptor cells in the grooves detect water currents, enabling a fish to perceive whether something is disturbing the water. The strength of the stimulus helps the fish determine how far away it is from a predator or prey.

14–8 Electricity is another stimulus that only a few animals can detect. The African electric fish produces an electric field. Objects in the electric field distort the lines of current flow. Sensory pores in the head detect the distortion, informing the fish about the object.

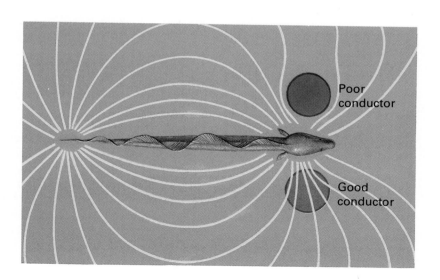

Section 14.1 Receptors Gather Information

Action of Receptors

A stimulus causes a response in a receptor that triggers a nerve impulse in a nearby neuron. The type of response brought about by a stimulus depends on the type of receptor. Pressure causes a mechanoreceptor to stretch. Heat and sound cause vibrations in parts of their receptors. Light causes chemical changes in pigments located within a photoreceptor. Food and odor molecules combine with chemical receptor molecules located within chemoreceptors.

The transmission of an impulse from a receptor to the brain is essential for perception. If the nerves connecting a receptor and the brain are cut, the animal loses sensation from that receptor.

14–9 Detection of a stimulus by a receptor results in either movement or a chemical reaction in the receptor. The response depends on the type of stimulus and receptor. All receptors transmit the detection of stimuli to the brain by nerve impulses.

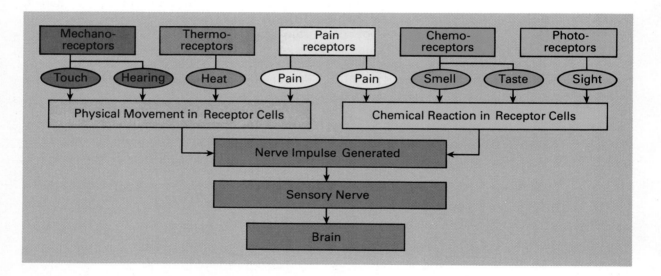

Interpretation of Sensory Impulses

For an animal to perceive sight or another sense, the brain must interpret the impulses it receives from receptors. The sensory impulses entering the brain are all similar; however, they are transmitted to different centers of the brain. Each brain center produces a specific type of perception. For example, each part of the hand has a different center in the brain. When you touch this page with your finger tip, impulses from sense organs in the skin are transmitted to the brain center for the finger. The brain center interprets the impulses as touch, and you feel the paper.

The idea that the brain is used to perceive senses is supported by experiments done on patients during brain surgery. When portions of the occipital lobes (sight center) are stimulated with a mild electric current, the patient sees light. When other portions of the sight center are stimulated, the patient sees objects. Stimulation of different centers evokes other specific senses. In people where the brain center for a given sense is damaged, the sense is reduced or lost. Damage to the sight center, for example, can result in blindness even if a person's eyes and optic nerves are normal.

You may wonder if other people perceive things in the same way that you do. You could get some idea of how others perceive stimuli by comparing your perceptions with theirs. Differences in brain or sense organ structure would account for differences in perception. For example, a nearsighted person would see a distant object differently from a normally sighted person, because their eyeballs have different shapes.

In general, members of the same species perceive things in similar ways. Members of different species, however, can perceive things quite differently. This is often due to differences in their sense organs. For example, an arthropod's eyes do not resemble human eyes; similarly, its antennae, which are organs for smell, do not resemble the human nose. Brain differences also exist among species. In fish, the center for smell is a dominant part of the cerebrum and smell is a leading sense. In humans, apes, and monkeys, the center for smell is overshadowed by the rest of the cerebrum, and smell is not a leading sense.

The human brain, which is far more developed than the brain of any other species, does more than organize sensory impulses into meaningful perceptions; it also enables us to be aware of our senses.

14–10 Locations of smell and vision centers in two vertebrate brains. Note the differences in brain shape and size.

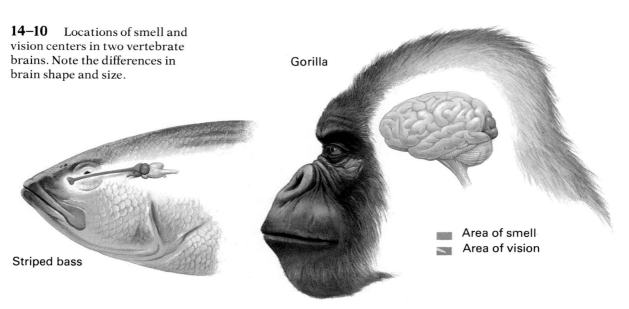

Gorilla

■ Area of smell
◪ Area of vision

Striped bass

14-11 Owls hunt primarily at night. An owl's survival depends on its highly developed sight and hearing, which enable it to locate rodents and other small prey in the dark.

Perception Aids Survival

Detecting and responding to stimuli help organisms and species survive. Sight, hearing, smell, taste, and touch help animals locate food and determine whether their food is edible. Sight helps most animals find shelter. Perceiving changes in the weather or seasons helps animals protect themselves against severe conditions. Using their senses, animals identify dangers and differentiate between predator and prey. Sight, hearing, smell, and touch help mothers and offspring recognize each other and help animals find mates. Perception of body position prevents an animal from falling, and perception of pain informs an animal that it has been injured.

Building a Science Vocabulary

mechanoreceptor   pain receptor   polarized light
photoreceptor   proprioceptor   infrared radiation
chemoreceptor   visible light   ultrasound
thermoreceptor   ultraviolet light lateral line

Questions for Review

1. Give an example of a receptor that detects each of the following.
   a. chemicals
   b. body position
   c. light
   d. movement or vibration
2. How does the ability to detect ultraviolet light adapt bees to their environment?
3. What evidence supports the idea that the brain is necessary for perception?

14.2 PERCEPTION OF SOUND, BODY POSITION, TOUCH, TEMPERATURE, AND PAIN

On land, a moving object displaces the air and possibly the ground around it. In water, a moving object displaces the surrounding water. The motion of objects causes vibrations, which are detected by an animal's mechanoreceptors. The perception of sound, pressure, pain, temperature, and body position all depend on the detection of motion.

Hearing

When an object vibrates in air, it causes the air molecules around it to vibrate also. Some parts of the air are squeezed together, forming *compressions*. Other parts of the air are pushed further apart, forming *rarefactions*. Compressions and rarefactions occur in regular patterns called **sound waves**. The vibrations of sound waves are picked up by an animal's ear, which changes them to nerve impulses. In the brain, these impulses are interpreted as sound.

Hearing in Bats and Moths Bats produce high-pitched cries that humans cannot hear. These ultrasound waves are reflected off objects, and the echoes are picked up by the bat's ears. The vibrations stimulate nerve impulses on the **auditory nerves**, which carry the impulses to the brain.

A bat determines whether the object it is approaching is a branch or a moth by the kind of echo it detects. If the object is a branch, the bat turns and avoids the crash. If it is a moth, it continues to cry and pick up echoes until it reaches the moth. As

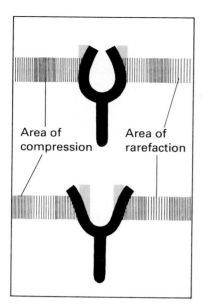

14–12 After being struck, the arms of a tuning fork vibrate rapidly back and forth, creating areas of compression and rarefaction.

Many species of bats catch moths and other insects in their tail membranes and then bend down to transfer the insects to their mouths.

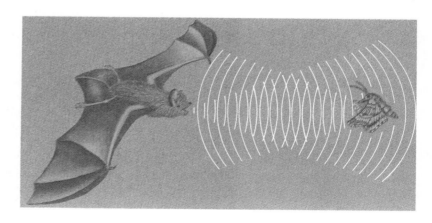

14–13 A bat produces ultrasound signals, which bounce off objects and return to the bat. The ultrasound signals travel only about 6 m before they fade away.

14–14 A tympanic membrane is located on each side of a moth's thorax. These "ears" pick up ultrasound signals emitted by bats, helping the moth avoid its predator.

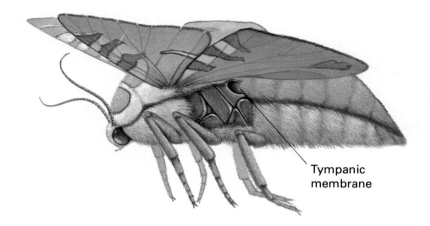

The tympanic membrane is also called the eardrum.

In some animals, the ear flaps move like radar antennas to pick up sound.

You can feel the equalizing of pressures in your ears when they "pop" as you go up in an airplane or in a high-speed elevator.

the bat approaches its prey, the cries increase in frequency from about 30 per second to about 60 per second. Echolocation makes it possible for a bat to capture moths in total darkness.

Certain moths have ears that detect the bat's cries. A moth's ears are located on the sides of its thorax. Ultrasound emitted by a bat causes a thin sheet of cells, the **tympanic** (tim-PAN-ic) **membrane**, to vibrate. This, in turn, causes vibrations of the air in air sacs inside the head. These vibrations stimulate nerve impulses to the brain, and the moth dives, drops down, or turns away.

Hearing in the Grasshopper The grasshopper has two ears, one on each side of the body in the first segment of the abdomen. Each ear has an oval-shaped, thin, tight, tympanic membrane, which detects vibrations on the ground or in the air. Unlike moths, grasshoppers cannot detect ultrasound.

Hearing in Humans The human ear does not hear all the sound waves that many other animals hear. Even though our ears do not pick up all sound waves, we have good hearing. The human ear can withstand the roaring of a jet plane in the air and then concentrate on the sound of ice tinkling in a glass of water being shaken.

The human ear is divided into three parts: the *outer ear*, *middle ear*, and *inner ear*. The outer ear consists of the **ear flap** and the **auditory canal**, which leads to the **tympanic membrane**. The membrane vibrates in response to sound waves that enter the ear.

The middle ear is connected to the pharynx by the **Eustachian** (yoo-STAY-shun) **tube**. This connection keeps the air pressure inside the ear equal to the air pressure outside. Equalizing the internal and external pressures prevents the tympanic

14–15 Structure of the human ear. Sound waves entering the ear vibrate the tympanic membrane. The bones in the middle ear conduct the vibrations to the oval window.

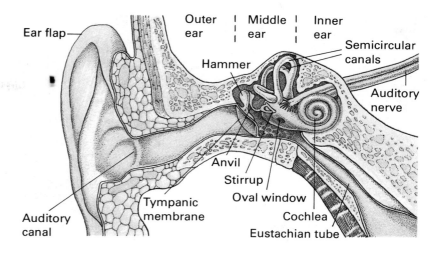

Decibel Scale and Level of Common Sounds		
Danger area	140 Physical damage	Pain threshold / Pneumatic riveter / Jet taking off
	120 Painful	Discotheque / Power mower / Rock band / Riveting machine
	100 Very loud	Subway train / Farm tractor / Motorcycle / Food blender
	80 Loud	Heavy traffic / Garbage disposal / Vacuum cleaner / Noisy office
	60 Moderate	Conversation / Light traffic / Quiet office
	40 Quiet	Library / Soft whisper
	20 Faint	Broadcasting studio / Leaves rustling / Hearing threshold

membrane from breaking. The middle ear contains three tiny bones: the **hammer, anvil,** and **stirrup.** The terms hammer, anvil, and stirrup are used because these bones look like the tools found in a blacksmith's shop. These three bones pick up the vibrations from the eardrum and conduct them to a membrane in the inner ear called the **oval window.**

The oval window is part of a coiled tube called the **cochlea** (KAHK-lee-uh), which is one of the structures located in the inner ear. Vibration of the oval window causes fluid inside the cochlea to vibrate in waves. The waves of fluid strike pressure receptor cells, called **hair cells,** located within the cochlea. The pressure waves cause the hairs in the hair cells to bend. The bending of the hairs stimulates neurons in the cochlea, which carry impulses to the brain by way of the **auditory nerve.** The brain processes the impulses from the ear, and hearing results.

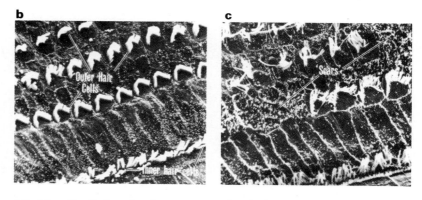

14–16 Decibel scale, showing noise level of common sounds (**a**). Micrographs of normal hair cells in the cochlea of a guinea pig (**b**) and destroyed cells after exposure to 24 hours of loud rock music (**c**).

Section 14.2 Sound, Body Position, Touch, Temperature, and Pain

Body Movement and Balance

Several kinds of receptors pick up stimuli that give the perception of movement and balance. These receptors include the eyes, proprioceptors, and touch receptors. They also provide information about the position of the body in space and where the parts of the body are.

Perhaps the most important receptors involved in movement and balance are the **semicircular canals**. These three proprioceptors are positioned perpendicular to each other in the inner ear. The canals are bony on the outside, lined with a membrane on the inside, and filled with liquid. Each canal contains a receptor that is made, in part, of hair cells. A small, doorlike structure is located at the end of the receptor. When the head moves, the liquid inside the semicircular canals begins to move. The moving liquid pushes the "door," causing it to swing. The swinging door bends the hairs in the hair cells. This stimulates nearby neurons to fire nerve impulses to the brain.

You can sense the movement of your head even if your eyes are closed. Usually, however, the brain coordinates the impulses from the semicircular canals, the eyes, and other receptors in developing a perception of motion and balance.

The semicircular canals are connected to the cochlea by a relatively large chamber, or vestibule.

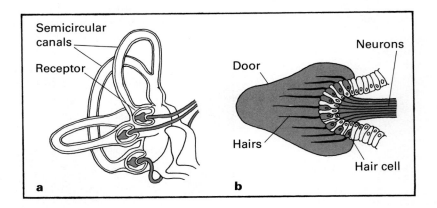

14–17 Structure of the semicircular canals showing balance receptors (**a**). Each receptor contains a door and numerous hair cells (**b**).

Other Proprioceptors You know when your arm is raised or your head is bent, because impulses from your eyes, touch receptors, and semicircular canals are transmitted to the cerebrum. In addition, proprioceptors located in and near joints detect skeletal movements and transmit impulses to the cerebrum.

You are not consciously aware of all the information about position that is transmitted to the brain. The movement of muscles and tendons stimulates their proprioceptors to send

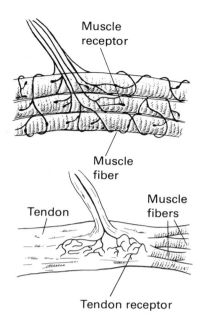

14–18 Proprioceptors in muscles and tendons inform the body about movement and position of body parts.

impulses to the cerebellum. Without your being aware of it, the cerebellum coordinates these impulses with impulses from other sense organs. As a result, you are able to maintain your balance and move in a coordinated fashion.

Touch, Temperature, and Pain

The skin contains receptors for three different kinds of stimuli: touch (including pressure), temperature, and pain. These receptors are composed of nerve endings. When these receptors are stimulated, impulses are conducted to the brain, and the appropriate perception results.

Touch receptors detect physical contact. Brief contact is perceived as touch; prolonged contact is perceived as pressure. Touch receptors are more numerous in the skin of the fingers and lips than in the skin of the chest and back. Some are located close to the surface of the skin; others are in deeper skin layers.

There are many touch receptors around the bases of hairs. When a hair is moved, it stimulates these touch receptors. This makes it possible to feel even slight pressure against a hair.

There are two types of thermoreceptors in the skin. **Heat receptors** detect the amount of heat present when the temperature is above body temperature. **Cold receptors** detect the

The presence of touch receptors at the base of a cat's whiskers helps guide it in the dark.

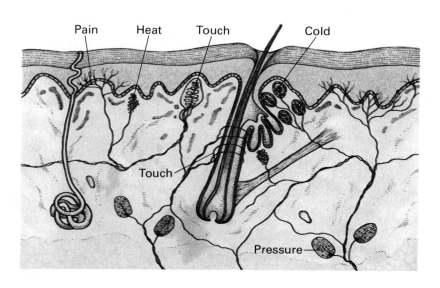

14–19 Cross section of the skin showing certain skin receptors and the senses believed to be associated with them.

From "What is pain?" by W. K. Livingston. Copyright © March, 1953 by Scientific American, Inc. All rights reserved.

Section 14.2 Sound, Body Position, Touch, Temperature, and Pain

amount of heat present when the temperature is below body temperature. There are distinct cold-detection and warm-detection areas in the skin. Four to ten times more spots on the skin detect cold than detect warmth. At extreme temperatures, skin tissue is damaged and perception of warmth and cold stops. If your skin is burned or frozen, the perception of pain overshadows the perception of heat or cold.

Pain receptors are located in the skin and in almost every body tissue. They detect many different stimuli, including tissue damage, harmful chemicals, excessive pressure (including high blood pressure), insufficient blood supply to a muscle, and certain emotions. Two types of pain can be distinguished: **fast pain** and **slow pain**. Fast pain is sharp pain in a given area of the body caused by a painful stimulus. Slow pain often follows fast pain and consists of a dull ache covering a large area.

Building a Science Vocabulary

sound wave	hammer	semicircular canal
auditory nerve	anvil	heat receptor
tympanic membrane	stirrup	cold receptor
ear flap	oval window	fast pain
auditory canal	cochlea	slow pain
Eustachian tube	hair cell	

Questions for Review

1. Why is the ear considered a mechanoreceptor?
2. Explain how the inner ear changes vibrations into nerve impulses.
3. How does a bat find a moth in total darkness?
4. How do the semicircular canals change a head movement into nerve impulses?
5. What four types of stimuli are detected by receptors in the skin?
6. Why do your ears "pop" in a fast-moving elevator?

Perspective: A New View
Call of the Wild

Rat-a-tat-tat, rat-a-tat-tat. Is that the sound of a jackhammer breaking up pavement? If you were in the woods, that sound would probably be a woodpecker making a hole in a tree in search of food.

Sound is a means of communication. The song of a sparrow, the buzz of a bee, and the bellow of a bull are different ways of communicating. The sparrow's song is not one of delightful glee, but a warning to other sparrows that "This territory is occupied." The buzz of a bee lets it be identified by others of its own kind, and thus it avoids a fight. The bellow of a bull may be a mating signal.

Sound is also a by-product of our civilization. Whining motors, jet engines, the roar of a nearby freeway, the hum of heating and air-conditioning systems that support our indoor environments—all these things can become so familiar that we cease to hear them.

You might consider such sounds to be noise. Noise is sound that disturbs or annoys us. Even the lovely songs of a nightingale or wood thrush can be noises if they disturb our sleep. The attitude of the listener determines which sounds are pleasing and which are not.

The variety of sounds and the mechanisms for producing them strain the imagination. Crickets and grasshoppers make sounds by strumming the hind leg against a rasplike plate, almost like strumming a guitar. Birds, bats, and insects produce sound by flapping their wings. Shifting the angle and speed of the motion enables some insects to change the tone and loudness of their sounds.

Many animals make sounds through some kind of vocalization. Humans do so by means of the voice box, or larynx. As air rushes by the two thin strands of tissue in the larynx, sound is produced. Tightening and loosening these strands changes the pitch. Forcing more air through the larynx increases the loudness.

Living things produce a myriad of sounds: birds sing, snakes hiss, frogs grump, horses neigh, dogs bark, and lions roar. The sounds and noises of life represent activity, interaction, and communication. Without them the earth would be a very strange place indeed.

14.3 PERCEPTION OF SIGHT, SMELL, AND TASTE

Light, odor-bearing molecules, and taste-bearing molecules cause chemical reactions in their receptors. The chemical reactions trigger nerve impulses to the brain, which interprets them as sight, smell, and taste.

Light Reception and Sight

Almost all protists and animals detect and respond to light. Depending on the organism, light energy is picked up by individual cells, eyespots, or eyes.

Light Reception by Individual Cells Little is known about how individual cells detect light energy; however, the effects of light can be easily seen. The paramecium responds to light by turning away. The earthworm, which does not have eyes, detects light in specialized cells in its skin. When these cells are stimulated, the earthworm quickly burrows underground.

Eyespots Some organisms have light-sensitive areas called **eyespots**, which react differently in light and darkness. Eyespots are groups of pigment molecules that change their chemical form in the presence of light. The change is a signal to the organism that light is present. Eyespots do not form images, nor do they detect movement.

The flatworm *Planaria* has two eyespots that look like eyes, but they are not eyes. The eyespots contain pigment in cuplike structures. Special light receptors cells are also located near the pigment cups. These cells detect light that comes from above and in front of the animal. As a result, a planarian detects both the presence of light and the direction of the light source.

Simple Eyes A **simple eye** has a lens which gathers light and concentrates it on light receptor cells. Simple eyes can detect movement and can form crude images at close range. For these reasons, they are more efficient than eyespots and can be considered true eyes.

Simple eyes are found in many arthropods. Most spiders have eight simple eyes in a cluster on the back of the head. The arrangement of the eyes enables a spider to detect movement in sequence from one eye to the next. Crustaceans and insects also have simple eyes. Some crustaceans have one simple eye between two more complex, compound eyes. Insects have three simple eyes between two compound eyes.

14–20 The eyespot at the tip of the head of a copepod detects the presence or absence of light.

True eyes detect movement and form images.

14–21 Spiders have a cluster of simple eyes. The single lens in each eye is made of cuticle secreted by the epidermis. Light-sensitive cells collect visual information and relay it to the brain.

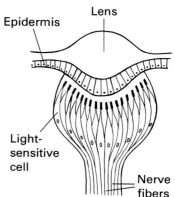

Compound eyes are found in crustaceans and insects.

A dragonfly's compound eye contains about 30 000 ommatidia.

Impulses from each ommatidium are transmitted separately to the brain.

Compound Eyes A **compound eye** is a hemispherical structure made up of many small regular surfaces called **facets** (FAS-its). It senses movement and forms images and therefore is a true eye. Each of the facets in a compound eye is the lens of a tubelike structure called an **ommatidium** (AHM-uh-TID-ee-um). A compound eye can contain hundreds or thousands of ommatidia bundled together.

Pigment cells surround the tube of the ommatidium and prevent the passage of light from one tube to the next. The lens concentrates and focuses entering light toward the center of the tube. Light-sensitive receptor cells within the tube detect the incoming light and fire nerve impulses, which are transmitted to the brain.

An insect sees a world quite different from the one we know. An insect has a broad field of vision. The rounded surface of its eyes allows it to see more than what is straight ahead. An insect does not see objects in sharp focus. For example, a wasp sees no

14–22 Insects such as the horsefly (left), have two large compound eyes. Each compound eye is made up of numerous ommatidia, which contain light-sensitive cells (right).

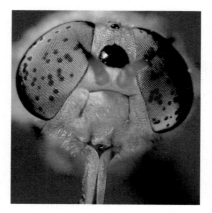
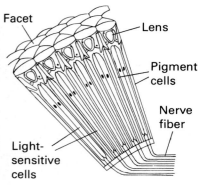

Section 14.3 Perception of Sight, Smell, and Taste

14–23 Artist's conception of vision in the bee. A bee sees a wide field of vision and some color; however, images appear blurred.

difference between a nail and a fly on the wall. The reason for its blurred focus is that each of its many ommatidia forms its own image. Although its vision is not sharp, an insect can detect movement instantly. Each ommatidium detects light independently. Therefore, when the object an insect is watching moves, the image perceived by many ommatidia changes. These changes are quickly relayed to the insect's brain. For example, a honeybee flies directly to a flower that moves only slightly in the wind.

Camera-type eyes are found in mollusks as well as vertebrates.

Vertebrate Eyes The vertebrate eye is often referred to as a camera-type eye. Like a camera, it uses a lens to focus light on a light-sensitive surface. Since they detect movement and form sharp images, vertebrate eyes are true eyes.

The eyes of all vertebrates are not exactly alike, but a basic pattern exists for all camera-type eyes. Light passes into the eye and is focused onto a layer of light receptor cells by an adjustable lens. A sharp, inverted image forms on the layer of light receptor cells and is transmitted to the brain by nerve impulses. The image is inverted because light travels in straight lines to the eye and then is bent during focusing.

14–24 Vertebrate eyes and cameras work according to the same principle. Both use a lens to focus an inverted and upside-down image on a light-sensitive surface.

In most vertebrate eyes, the lens changes shape during focusing. In fish and amphibia, however, the lens does not change shape. Instead, focusing occurs when muscles move the lens toward and away from the retina. This is the same type of focusing that is used in cameras.

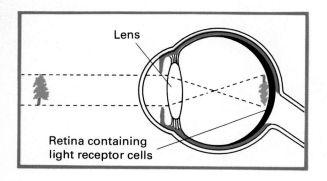
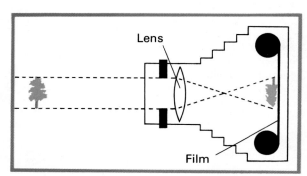

Chapter 14 The Senses

Sight in Humans

Structure of the Eye The eye is composed of three layers. The outside layer is the **sclera** (SKLIR-uh). The part of the sclera that we see is the white of the eye. The sclera is tough and protects the eye. Muscles that move the eyeball are attached to the sclera. The rounded transparent part of the sclera at the front of the eye is called the **cornea** (KOR-nee-uh). Light first passes through the cornea as it enters the eye. The cornea begins to focus incoming light by bending it as it enters.

The middle layer of the eye is the **choroid** (KOR-oyd). It contains blood vessels and black pigment. The black pigment prevents reflection of light rays within the eye. The colorful part of the choroid at the front of the eye is called the **iris**. The iris can be blue, brown, green, gray, or hazel. Light passes through an opening in the center of the iris called the **pupil**. Circular muscles in the iris control the amount of light entering the pupil. In bright light, the muscles in the iris contract, and the pupil becomes smaller. This allows less light to enter. In dim light, the muscles relax, and the pupil becomes larger, permitting more light to enter.

The **lens** is located immediately behind the iris. It focuses light on the inner layer of the eye by changing shape. The inner layer of the eye, or **retina**, covers about 65 percent of the inner surface and is the site of all the light receptor cells in the eye. The small depression in the middle of the retina, known as the **fovea** (FO-vee-uh), is the site of the eye's sharpest vision.

Blood vessels in the choroid and in the iris transport food and oxygen to the eye and remove wastes. Nourishment is also supplied by a watery fluid, called the **aqueous humor**, that fills

Sight is a very well-developed sense in humans. Humans are able to change focus or react to changes in light intensity quickly. Humans can also perceive many different colors and shades.

iris = a rainbow

The diaphragm in a camera functions similarly to the iris of the eye. It regulates the amount of light that reaches the lens.

14-25 Structure of the human eye (cross section in three-quarters view).

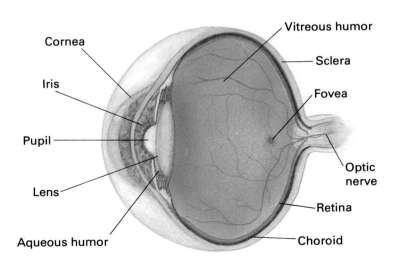

Section 14.3 Perception of Sight, Smell, and Taste

14–26 The axons in the optic nerve from the right half of each eye go to the right side of the brain. Those from the left half go to the left side of the brain.

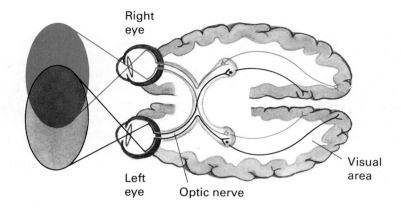

the space between the cornea and the lens. The **vitreous** (VIT-ree-us) **humor**, a transparent jelly that helps maintain the shape of the eye, fills the area between the lens and the retina.

The optic nerves connect the eyes and the brain. Notice that half the axons from each eye cross each other and enter the opposite side of the brain (see Fig. 14–26).

Light Reception and Vision The retina contains two types of light receptor cells. **Rods** are spread throughout the retina, but they are not found in the fovea. They are responsible for black and white vision. **Cones** are concentrated at the center of the retina and are clumped in the fovea. They provide color vision. Whereas rods work in dim light, cones only work in bright light. For that reason, we cannot see color at night. There are about 120 million rods and 6 million cones in the retina.

The terms rod *and* cone *refer to the shapes of these cells.*

14–27 Structure of the retina (**a**); scanning electron micrographs of rods (**b**) and cones (**c**) from amphibian retinas.

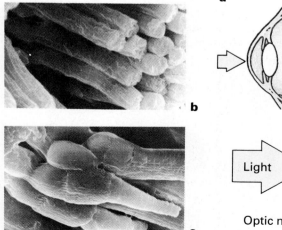

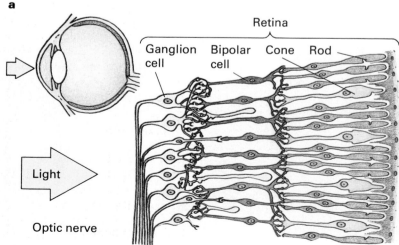

Chapter 14 The Senses

14–28 Color vision. When looking at a green object, for example, green light enters the retina. Green rhodopsin in cones that are sensitive to green light breaks down into green opsin (gO) and retinene (R). These events generate a nerve impulse.

The remaking of rhodopsin takes several minutes.

Cones can also detect ultraviolet rays, but these are normally filtered out by the lens. Cones cannot detect infrared light.

People with a vitamin A deficiency do not produce enough rhodopsin to see well in the dark.

The sense of depth perception also requires learning by experience.

Rods and cones contain light-sensitive pigments. These pigments contain a protein called an **opsin** and **retinene**, a substance made from vitamin A. The light-sensitive substance in rods is called **rhodopsin** (rō-DOP-sin). When rhodopsin absorbs light, it breaks down into retinene and an opsin. This stimulates a nerve impulse. After the impulse passes, an enzyme recombines retinene and the opsin into rhodopsin.

There are three kinds of cones. Each kind contains a different opsin: one is sensitive to blue light, the second to green light, and the third to red light. The colors we see stimulate different combinations of the three types of cones. As in rods, when the pigment in cones absorbs light, it breaks down, stimulating a nerve impulse. The pigment is later reformed.

Nerve impulses from the rods and cones are transmitted to the brain on the optic nerve. The cerebral cortex processes impulses from the rods, and we perceive black, white, and shades of gray. It also processes the impulses from cones, resulting in the perception of color. No rods or cones are at the point where the optic nerve leaves the eye. This area is called the **blind spot**, since no light reception occurs there.

Adjusting to Dim and Bright Light   It is hard to see when you enter a darkened room after being in bright light. During the period of adjustment, your rods are rebuilding the rhodopsin that was broken down in the light. As more rhodopsin is made, the retina becomes more sensitive to the small amount of available light. When you reenter the light, its glare is uncomfortable and you are often partially blinded. Eventually, as rhodopsin breaks down, the retina becomes less sensitive to light and normal vision returns.

Depth Perception   When you look at objects in front of you, you see depth as well as height and width. Your perception of depth is aided somewhat by using two eyes. Each eye gets a slightly different perspective, and your brain interprets the different views as depth. Nevertheless, most depth perception occurs because an eye looks at different parts of an object at the

Section 14.3 Perception of Sight, Smell, and Taste

14–29 To locate your blind spot, hold the page in front of your right eye, keeping your left eye closed. While staring steadily at the target bring the page closer until the X disappears. When this occurs the image of the X has fallen on the blind spot, where no light reception occurs.

same time. It can also detect shadows and the relative size of objects (nearby objects appear larger than distant ones). Further, depth perception occurs when one object moves in front of another. In all these situations, the images that are focused on the retina reveal the relative sizes and locations of objects. The brain interprets these relationships as depth.

Sight Defects Imperfections or flaws in the structure of the eye can prevent proper focusing of an image on the retina. A condition known as **astigmatism** (uh-STIG-muh-tizm) may be the result of an improperly shaped cornea. In astigmatism, light rays do not enter the eye evenly. As a result, the eye cannot focus a perfect image on the retina, and vision is distorted. People with astigmatism are helped by wearing eyeglasses. The lenses of the eyeglasses even out the entering light rays.

Sometimes the shape of a person's eyeball prevents proper focusing. If the eyeball is not perfectly round, the lens may not be able to bring the image to focus on the retina. If the eye is too long, distant images fall in front of the retina. This conditon is known as **nearsightedness**. If the eye is too short, close images fall behind the retina and **farsightedness** results. Both nearsightedness and farsightedness are alleviated by wearing glasses with corrective lenses.

Aging also affects focusing. The lens becomes less able to change shape easily as a person gets older. An older person has trouble focusing on small close objects. Special reading glasses or glasses with bifocal lenses help improve focusing.

Smell and Taste

The sense of smell in most animals is highly developed. For example, we can smell the substance that gives garlic its characteristic odor if there is as little of it as one-millionth of a milligram in a litre of air. Some people think that the sense of smell is more sensitive than the sense of taste. This is not always true; the sense of smell is more sensitive to certain molecules, while the sense of taste is more sensitive to others. Taste and smell differ mainly in the way that molecules contact the receptor. In order to taste something, it must be in the mouth. A substance can be smelled, however, at a distance from its source.

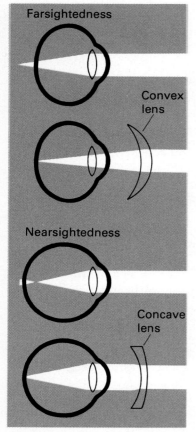

14–30 Correcting vision. Focusing a sharp image on the retina may require the use of a convex lens for farsightedness (above) or a concave lens for nearsightedness (below).

14–31 Organs of smell. In addition to using its nostrils, a snake flicks its tongue to collect samples for testing in its nasal cavity. A moth's feathery antennae and a snail's tentacles also contain chemoreceptors.

Odors are carried both in water and in air.

Smell

Many animals have a keen sense of smell. A bloodhound is able to track down a person, through a mixture of odors, by detecting molecules that the person gives off. Some animals have receptors that pick up chemical signals over great distances. The male silkworm moth has smell receptors on its antennae that can detect the chemical attractant released by females many kilometres away. Fish have chemoreceptors that work under water. Salmon smell molecules from their home streams when they are in the ocean hundreds of kilometres away.

Smell in Humans Chemoreceptors for smell are located in the nose, in an area at the top of the nasal cavity. When a person inhales, chemicals in the inhaled air dissolve in the moisture of the nasal membrane. Certain of these chemicals stimulate smell receptor cells, which send nerve impulses to the smell center of the brain.

14–32 Smell receptors are concentrated in one spot at the roof of the nasal cavity. Long, narrow sensory cells end in cilia, which are believed to be the actual odor receptors.

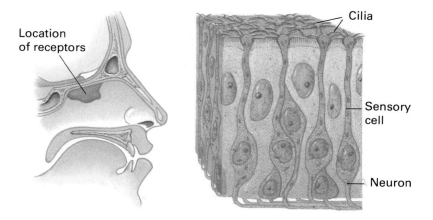

Section 14.3 Perception of Sight, Smell, and Taste

People who work in the manufacturing of perfumes use about 4000 ingredients on the job. These people can usually identify about half of the ingredients by smell alone.

Sniffing aids smell by bringing incoming molecules closer to the smell receptors.

Although humans do not have as keen a sense of smell as other animals, humans can distinguish between 2000 and 4000 odors. No one knows how we are able to tell the differences among so many odors. The shapes of molecules may play a part in the process. Odor molecules may fit into receptors like substrates fit into enzymes. When you keep smelling the same odor, you lose the power to detect it after a while. It is possible that receptor molecules undergo changes in shape when they detect stimulus molecules. When all the receptors have changed shape, you can no longer smell the odor. The perception returns when receptor molecules are restored to their original state.

Taste

Chemoreceptors for taste help organisms survive by providing information about food. Different animals have different taste preferences. Most animals avoid bitter flavor; birds avoid bitter-tasting insects, and meat-eaters avoid the bitter-tasting opposum. Most animals will not eat skunk meat that contains the skunk's repellent fluid. Yet, oddly enough, the great horned owl eats it readily.

Taste in Humans The sense of taste is affected by temperature, texture, and especially smell. Humans can distinguish between many degrees of heat and cold and can tell the difference between a tasteless oil and water, because the oil feels slippery and the water does not. Smell and taste work closely together when you perceive taste. You may get a first impression of food from a distance through the sense of smell. The impression is made more complete by taste when the food actually enters your mouth. The effect of your sense of smell on taste is obvious when you have a cold. You cannot taste as well when your nasal passages are clogged.

The receptors for taste are called **taste buds**. Most taste buds are located on the upper surface of the tongue. Others are on the epiglottis, palate, and pharynx. Each taste bud contains taste receptor cells. The taste receptor cells have hairlike structures that project into the opening or pore of the taste bud (see Fig. 14–33). Molecules of food dissolve in saliva and come in contact with the taste receptor cells. They appear to fit into their recep-

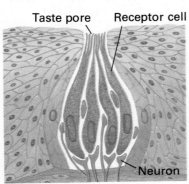

14–33 Taste buds: **a.** scanning electron micrograph of a taste bud; **b.** taste bud in longitudinal section. When receptor cells contact food molecules, they stimulate nearby neurons.
From TISSUES AND ORGANS: A TEXT ATLAS OF SCANNING ELECTRON MICROSCOPY by Richard G. Kessel and Randy H. Kardon. W. H. Freeman and Company. Copyright © 1979

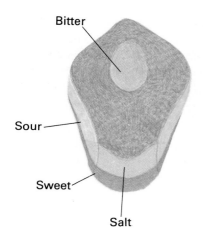

14–34 Taste buds for the four basic tastes are located mainly in the areas shown.

tors in the same way that odor molecules are believed to fit into smell receptor cells. When receptor cells contact food molecules, they stimulate nearby neurons, which carry impulses to the brain. Processing of the information by the brain results in the perception of taste.

Without the sense of smell, we can recognize four basic tastes: sweet, sour, bitter, and salt. Taste buds for each of these tastes are found clumped in specific areas of the tongue. Bitter substances are tasted on the back of the tongue, sour ones along the edges, sweet ones at the tip, and salty ones at the front. Some simple substances stimulate only one type of taste bud. Complex substances stimulate two or more types; however, no one knows how we distinguish among so many tastes.

Building a Science Vocabulary

eyespot	pupil	retinene
simple eye	lens	rhodopsin
compound eye	retina	blind spot
facet	fovea	astigmatism
ommatidium	aqueous humor	nearsightedness
sclera	vitreous humor	farsightedness
cornea	rod	taste bud
choroid	cone	
iris	opsin	

Questions for Review

1. Why are a planarian's eyespots not considered eyes?
2. How are an insect's compound eyes adapted to detect moving objects?
3. State the function of each of the following parts of the eye.
   a. cornea   b. choroid pigment   c. iris   d. lens
4. How do cones detect color?
5. Why is vision blurred in a farsighted person?
6. What are the four basic tastes that we perceive?
7. How is the vertebrate eye like a camera?

Chapter 14
Summary and Review

Summary

1. In complex animals, stimuli are detected by receptors, which trigger nerve impulses that are transmitted to the brain. The brain interprets the incoming impulses, resulting in perception.

2. Different types of receptors detect pressure and vibrations, light, chemicals, heat, cold, pain, and body position.

3. Some animals detect stimuli not detected by humans, including ultraviolet and infrared light, polarized light, and ultrasound.

4. Receptors for a given sense transmit impulses to a specific brain center.

5. Differences in perception among species can be due to differences in receptors or brain differences.

6. An ear detects sound waves in vibrating air or water. In the grasshopper, the ear is a tympanic membrane. Bats use ultrasound echolocation to detect objects in the environment.

7. In humans, sound waves are conducted through the outer ear, middle ear, and inner ear. In the cochlea, the bending of hairs in receptor cells stimulates nerve impulses, which are transmitted to the brain on the auditory nerve.

8. The semicircular canals in the inner ear and receptors in various parts of the skeleton and muscles receive information about body position and balance.

9. Nerve endings in the skin are receptors for touch, pain, cold, and warmth.

10. Pigments in photoreceptors detect light. Photoreceptors include eyespots, simple eyes, compound eyes, and camera-type eyes. True eyes detect movement and form images.

11. An insect's compound eyes form hundreds to thousands of separate images and instantly detect a moving object.

12. Humans and other vertebrates have camera-type eyes that use a lens to focus an inverted image on a light-sensitive retina.

13. Rods and cones are receptor cells in the retina. Rods detect black and white; cones detect color.

14. Taste and smell are forms of chemoreception. Taste buds detect chemicals that contact them directly; smell receptors detect chemicals that arrive from a distance. Odor and taste molecules are believed to fit into molecules in their receptors.

Review Questions

Application

1. Explain how each of the following helps the animal described survive in its environment.
   a. The eyes of owls contain many rods.
   b. Smell receptors are located in a butterfly's long antennae.
   c. Certain sand crabs have eyes on stalks.

2. Why would infrared receptors be an adaptation in a pit viper and not be an adaptation in a cow?

3. Spiders have hairy bodies. What sense is aided by body hairs? How would this sense help a spider sitting in its web?

4. What part of the eye corresponds to each of the following camera parts?
   a. diaphragm
   b. lens
   c. film

5. How does movement trigger a nerve response in each of the following?
   a. the ear
   b. semicircular canals
   c. the skin

6. How does chemical activity trigger nerve impulses in rods in the retina? in taste buds?

Interpretation

1. Some people are unable to detect the odor of skunk. Give two possible explanations for this lack in smell perception.

2. What two advantages does the bat have by using ultrasound rather than ordinary sound during echolocation?

Extra Research

1. Use a bunch of drinking straws to make a model of a compound eye. Each straw will represent one ommatidium. Cover one end of each straw with plastic wrap to represent the facet. Bundle the straws together with a string. Lace colored thread or wool through the ends on the open end to represent the nerve fibers.

2. Construct a pinhole camera. Punch a hole in the side of a small box that has no top. Place the box into a darkened room and shine a light through the hole from the outside of the box. Move the light so that it shines into the hole from above, below, and straight ahead. Note how the light focuses on the wall of the box from each of these positions. Write a paragraph comparing this simple camera with the human eye.

3. Make a map of the cold receptors in your skin by touching different parts of your hand and arm with a thin sliver of ice.

4. Devise an experiment that would help determine whether a parakeet can distinguish different colors.

5. Design and construct a demonstration that a blind student could use to learn about a given topic in biology. Design the demonstration so that the student uses as many senses as possible.

6. The next time you visit an art museum, try to figure out how artists create the illusion of depth in paintings.

Career Opportunities

Optometrists provide and adjust eyeglasses and contact lenses. Optometrists also refer people whose eyes are injured or diseased to a physician who specializes in eye care (an ophthalmologist). Usually, six or seven years of training are required to become an optometrist. For information, contact the Canadian Association of Optometrists, 210 Gladstone Ave., #2001, Ottawa, Ontario K2P 0Y6.

Speech therapists work in schools and clinics to help people improve their speech. Some also work with people who are deaf or hard of hearing. Most of their work is with children or older adults. Specialized training in speech therapy and a thorough knowledge of the biology of speech and hearing are required. Write the Canadian Speech and Hearing Association, #308, Corbett Hall, University of Alberta, Edmonton, Alberta T6G 2G4, or the Association of Canadian Educators of the Hearing-Impaired, 500 Shaftesbury Blvd., Winnipeg, Manitoba R3P 0M1.

Unit IV
Suggested Readings

Arnold, Caroline, *Sex Hormones: Why Males and Females Are Different*, New York: William Morrow & Co., 1981. *Hormone-influenced sexual differences in animals and humans are explained. Includes a lucid discussion of the role of hormones in many aspects of animal behavior.*

Arehart-Treichel, Joan, "Gut Hormones," *Science News*, January 19, 1980. *Describes hormones that are manufactured in the digestive system. Some may play a role in disease.*

Burgess, Robert F., *Secret Languages of the Sea*, New York: Dodd, Mead & Co., 1981. *Communication among marine organisms includes sound, especially in whales and dolphins, visual signals such as bioluminescence, and pressure-sensitive lateral-line systems.*

Cousteau, Jacques, *Window in the Sea*, New York: Henry M. Abrams, Inc., 1973. *Spectacular photography and easy text explore eyes and light underwater. Positions of eyes are discussed, as are camouflaged eyes.*

Galston, Arthur W., "Sex and the Soybean," *Natural History*, October, 1978. *Flowering, grafting, and the role of auxins are explored.*

Love, Milton, "With a Little Help from My Friends," *Natural History*, November, 1981. *Unusual forms of protection described in this beautifully illustrated article include use of toad toxins by hedgehogs and the ability of sea slugs to store hydras' stinging cells.*

Moffatt, Anne, "Beware That 'Innocent' Plant's Guile, Trickery, Aggression!," *Science Digest*, March, 1979. *Describes numerous chemical and physical defenses plants use against predators.*

Murchie, Guy, "Love at First Sniff," *Science Digest*, July, 1979. *Describes seven primary smells produced by differently shaped molecules.*

Patent, Dorothy Hinshaw, *Butterflies and Moths: How They Function*, New York: Holiday House, Inc., 1979. *The author relates the series of events from egg-laying through maturation of the butterfly or moth. Provides lively reading about the anatomy, physiology, and ecology of their life cycles.*

Rahn, Joan Elma, *Eyes and Seeing*, New York: Atheneum Publishers, 1981. *Concerned with a variety of animal eyes and their visual capabilities. Features simplified explanations of optics and of sensory perception of various forms of imagery.*

Reaka, Marjorie L., "The Hole Shrimp Story," *Natural History*, July, 1981. *Beautifully illustrated article describes how the usually aggressive behavior of a burrowing species of shrimp changes during molting.*

Silverstein, Alvin, and Virginia B. Silverstein, *The Sugar Disease: Diabetes*, New York: Lippincott & Crowell Publishers, 1980. *Explains glucose metabolism, the different types of diabetes, current research on this disease, and theories about its causes.*

Simpson, Lance L., "Deadly Botulism," *Natural History*, January, 1980. *Effect of botulism toxin on nerve endings is clearly explained and illustrated. Good overall discussion of the nervous system and toxin-induced paralysis.*

Ward, Brian R., *The Brain and Nervous System*, New York: F. Franklin Watts, Inc., 1981. *Modern, eye-catching diagrams accompany short, readable text covering topics ranging from protective membranes of the brain to learning and memory.*

Witzmann, Rupert F., *Steroids: Keys to Life*, New York: Van Nostrand-Reinhold Co., 1981. *An enjoyable and informative book on the role of steroids in life and the key scientists involved in steroid research.*

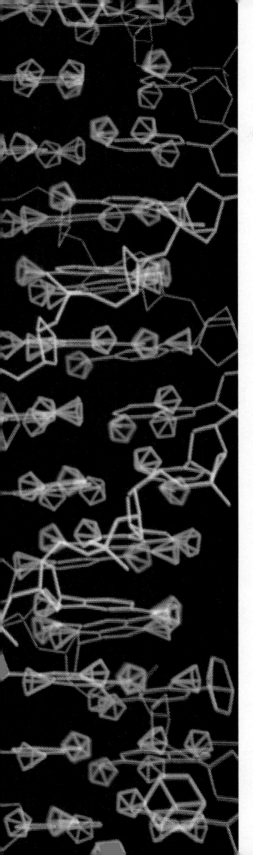

Unit V

Reproduction

How do species survive from generation to generation? The answer lies in the process of reproduction. Species span time by producing offspring. Reproduction is an efficient and well regulated process. During reproduction, information about the species is transmitted from parent to offspring on coded molecules. The coded molecules specify the traits of the species: basic body features, such as eye color; basic body size and shape; and body function. The coded molecules offspring get from their parents guide their development and growth.

Reproduction takes on many forms. In the simplest form, individual organisms divide in two, making exact copies of themselves. In the most complex form, parents of different sexes contribute to the formation of the next generation. As a result, the offspring are unique, having characteristics of both parents. Regardless of how reproduction occurs, it is the thread that connects the life of a species past, present, and future.

A computer model of the molecular structure of DNA reveals its regular spiral pattern.

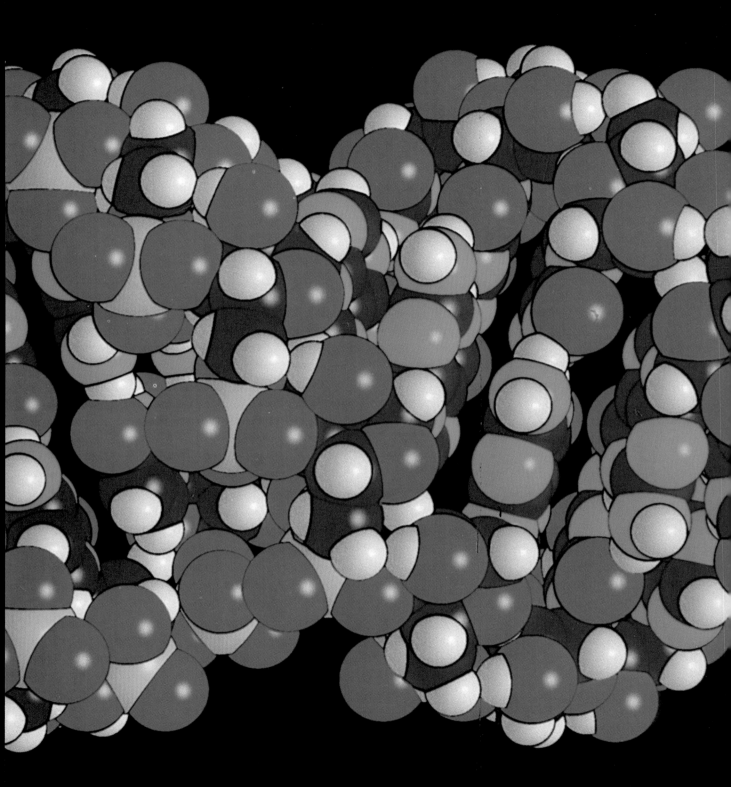

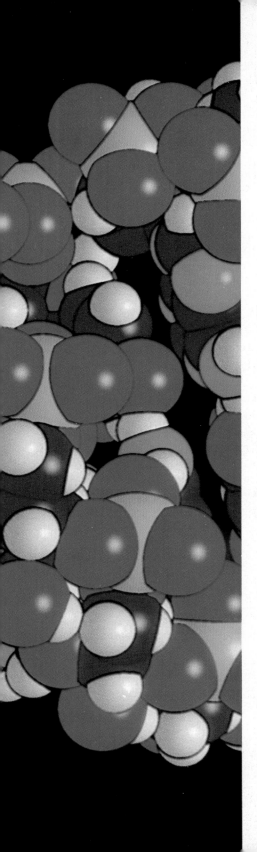

Chapter 15

Reproduction of Molecules and Cells

Focus *We live in an age of computers that can accurately store and reproduce large volumes of information. Cells, too, have the ability to store and reproduce information; however, they are far more powerful than even the most sophisticated computers. A cell contains all of the information for its structure and function in one type of molecule. This molecule controls the cell's growth and reproduction and ensures that offspring and parents resemble each other.*

15.1 THE MOLECULAR BASIS OF REPRODUCTION

Human beings have lived on the earth for many generations. As a species well adapted to a variety of environments, we expect to survive. Species survive from generation to generation by producing offspring. The process of producing offspring is called **reproduction**. Organisms of every species, from the smallest to the largest, have a process for reproducing their own kind.

In all organisms reproduction and growth involve making new molecules and cells. One molecule found in cells is capable of reproducing itself. This unique molecule is **deoxyribonucleic acid** or **DNA**. The reproduction of DNA is often called **replication**. The ability of DNA to make copies of itself is the basis for the continuation of life from one generation to the next.

15–1 Computer model of the molecular structure of DNA, which carries hereditary information in every living cell.

15–2 Offspring often closely resemble one or both parents.

Transmission of Traits from Parents to Offspring

When an organism reproduces, it transmits certain traits or characteristics to its offspring. We are aware of the transmission of traits in our own species. Each of us inherits features such as hair color and eye color from our parents. As a result, we may tend to look like them. We also resemble our parents in less noticeable ways by inheriting characteristics such as blood type and bone structure from them.

The transmission of traits or characteristics from parents to offspring is known as **heredity**. Many years ago, people did not understand how characteristics are inherited. They were aware that parents of all living things produced offspring of their own species but did not understand how this happened. Today, we know that the information that determines an organism's heredity is found in all cells, where it is stored in chemically-coded form in DNA.

A Model of DNA

Scientists worked for many years to understand how DNA stores hereditary information. A major breakthrough in DNA research came in 1953 when James Watson, an American biologist, and Francis Crick, a British biophysicist, developed a model of the structure of DNA. Their model also helped explain how DNA reproduces. To picture the DNA molecule, imagine that you have a rope ladder with rigid rungs and flexible sides. If you twist this ladder from both ends, each side forms a spiral or

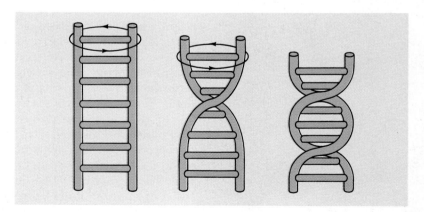

15–3 Rope ladder model of DNA. When the ladder is twisted the sides form two intertwined spirals. The double-spiral shape is known as a double helix.

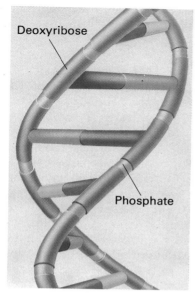

15–4 Geometric representation of DNA. The rungs are paired nitrogen bases, the sides phosphate groups and sugars.

helix. Since both sides are twisted, the ladder forms a set of spirals with rungs between them. The double-spiral shape is known as a **double helix**. The DNA molecule has the shape of a double helix.

The DNA molecule is made up of nucleotides linked together. Recall that a nucleotide consists of a simple sugar (in this case deoxyribose), a phosphate group, and a nitrogen base. The Watson-Crick model of DNA describes how the nucleotide units are arranged in the DNA ladder. In this model, the sides of the ladder are made of sugars and phosphate groups. The rungs are made of pairs of nitrogen bases.

Nitrogen Base Pairing in DNA   A DNA nucleotide can contain any of four nitrogen bases: **adenine** (AD-un-EEN), **thymine** (THĪ-meen), **guanine** (GWAH-neen), and **cytosine** (SĪT-uh-SEEN). The nitrogen bases are often referred to by their initial letters: A, T, G, and C. Each rung of the DNA ladder is made up of a pair of nitrogen bases. Each base pair is specific, meaning that each base matches up, or pairs with, only one other base. Adenine pairs only with thymine; guanine pairs only with cytosine.

Because each nitrogen base pairs with only one other base, only four base combinations are found in DNA. These base pairs are A-T, T-A, G-C, and C-G. The position of base pairs in the rung

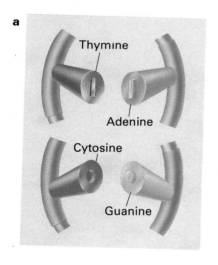

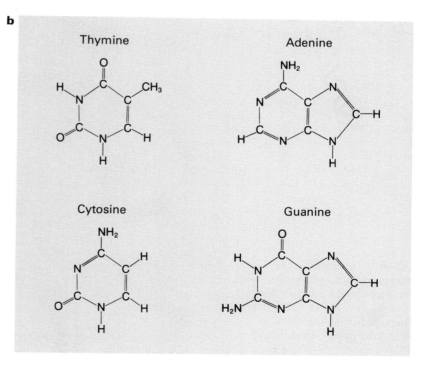

15–5 DNA nucleotides: **a.** geometric representations; **b.** structural formulas. In forming the rungs of the DNA ladder, adenine pairs with thymine, and cytosine pairs with guanine.

Section 15.1 The Molecular Basis of Reproduction

is very important, because the bases carry hereditary information in coded form. The code is built into the arrangement of the bases in DNA.

Replication of DNA In order for DNA to make a copy of itself, nucleotides, as well as the proper enzymes, must be present in the cell. Cells normally contain the necessary enzymes and nucleotides for replication. DNA replicates in three steps. First, the nitrogen base pairs separate. In other words, the rungs of the DNA ladder come apart in the middle. You can compare this to a zipper unzipping. The DNA molecule separates like a zipper. Second, nucleotides in the cell join with their matching nucleotides, making pairs along the rungs of the unzipped halves. Unpaired nucleotides with thymine bases bond with nucleotides with adenine bases. Similarly, unpaired cytosines pair with guanines. Last, the sugars and phosphates of the new nucleotides bond to each other, forming the sides of the ladder.

DNA makes a copy of itself when it replicates.

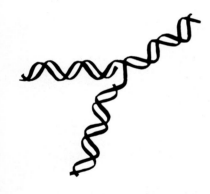

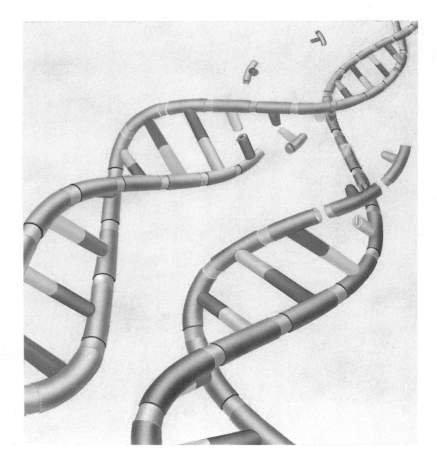

15–6 During DNA replication the paired nitrogen bases separate, opening up the double helix. The bases of free nucleotides pair with the exposed bases, forming two identical molecules of DNA from one.

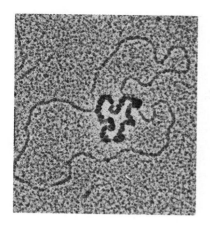

15-7 DNA replication in a bacterial plasmid. The inner loop is the newly replicated DNA.

The result of these steps is the formation of two complete molecules of DNA. Each molecule is an identical copy of the original.

Recall that enzymes are needed for each activity performed within a cell. This includes the process of DNA replication. For example, during DNA replication, an enzyme called **DNA polymerase** (puh-LIM-er-ays) is necessary to bring about the unzipping of DNA into two separate chains.

The replication of the DNA molecule is an extremely accurate process. This is remarkable, especially if you consider two facts. First, if all the pieces of DNA in one human cell were joined end to end, the resulting strand would be about a metre long. This strand would contain about six million pairs of bases. Each time a cell reproduces, all six million pairs of bases separate and re-pair with new nucleotides. In addition, this process occurs over and over again. Every day, billions of new cells are produced in our bodies. Each time a cell reproduces, its DNA replicates with very few changes.

Building a Science Vocabulary

reproduction	double helix	cytosine
deoxyribonucleic acid (DNA)	adenine	DNA polymerase
replication	thymine	
heredity	guanine	

Questions for Review

1. What function does reproduction perform for a species?

2. What is heredity? Where is information about heredity stored in a cell?

3. Suppose that the sequence of bases on one side of a DNA molecule is thymine, adenine, cytosine, guanine. What is the sequence of bases that matches it?

4. Place the following events in DNA replication in the correct order of occurrence.
    a. nucleotide bases pair
    b. DNA unzips
    c. sugars and phosphates bond

Section 15.1 The Molecular Basis of Reproduction

15.2 THE CELLULAR BASIS OF REPRODUCTION

Up to now, you have been learning about the reproduction of the DNA molecule. But cells also reproduce when an organism grows and when it replaces worn-out or injured cells. The reproduction of every cell is controlled by DNA. In eukaryotic cells, DNA is located in the nucleus, where it binds with large proteins. The resulting molecules can be seen under a microscope when they are stained. In cells that are not dividing, these molecules look like long pieces of twisted thread. These threads are called **chromosomes**.

chrōma = color
sōma = body

Every plant, animal, and protist species that we know of has its own characteristic number and type of chromosomes. The same number of chromosomes are found in nearly every cell of an organism's body. The chromosomes in body cells occur in pairs. Each member of the pair has basically the same structure as the other member. Both members carry chemical information for the same traits, such as eye color or hair color. The two chromosomes making up a pair are called **homologous chromosomes**. Humans have 23 pairs of homologous chromosomes—a total of 46 chromosomes. Fruit flies have four pairs of homologous chromosomes (eight altogether), and dogs have 24 pairs (48 altogether).

homologos = agreeing

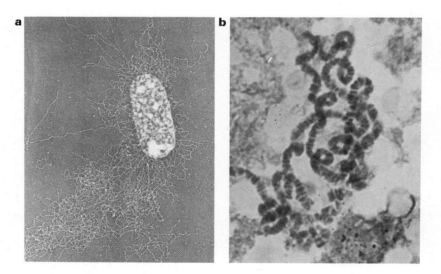

15-8 Chromosomes: **a.** electron micrograph of *E. coli* and its long, circular chromosome, which "spilled out" after the bacterium's cell wall was gently disrupted; **b.** micrograph of the salivary gland cell of a fruit fly, showing the tightly coiled giant chromosome.

Cell Division

A parent cell produces two offspring cells that have the same number and type of chromosomes as the parent.

mitos = thread

Cells reproduce by a process called **cell division**. During cell division an original, or parent, cell divides to produce two offspring cells. The chromosomes of each offspring cell are identical to those of the parent cell. Before a cell divides, the DNA of the chromosomes replicates, providing two complete sets of chromosomes—one for each offspring.

Cell division occurs in two stages. During the first stage, known as **mitosis** (mī-TŌ-sis), replicated chromosomes are separated into two nuclei in the cell. In the second stage, **cytoplasmic division**, the cell's cytoplasm divides, separating the nuclei and their chromosomes into two separate cells.

Mitosis

In the 1880s, a biologist named Walther Flemming studied cell reproduction. He observed that during cell division, the chromosomes look like pieces of thread. Because of this, Flemming named the cell division process *mitosis*.

Flemming and other biologists who observed cell growth and division saw it as an ongoing and dynamic process. To aid in the study of the changes they observed, they identified and named certain steps, or phases, of the process. A cell undergoing mitosis goes through four stages: **prophase** (PRŌ-fayz), **metaphase** (MET-uh-FAYZ), **anaphase** (AN-uh-FAYZ), and **telophase** (TEL-uh-FAYZ). A cell that is not dividing is said to be in **interphase**. Mitosis is actually continuous; biologists talk of "phases" only as an aid in studying this process.

pro- = before
meta- = between
ana- = toward
telo- = end
inter- = between

Interphase A cell in interphase is active, producing materials such as nucleic acids and proteins that are used in the growth and repair of cell parts. Replication of DNA occurs near the end of interphase. As a result of this replication, the number of chromosomes is doubled. During interphase, the chromosomes spread out like a tangled net of threads in the nucleus.

In general, a cell spends about 90 percent of its lifetime in interphase. The four other phases of mitosis are short in comparison. For example, a connective tissue cell in the human body spends about 17 hours in interphase and less than 1 hour in mitosis.

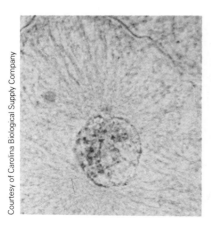

15-9 Interphase of a whitefish cell. During interphase the chromosomes appear loosely arranged within the nucleus.

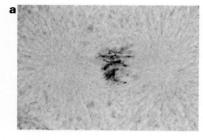
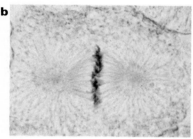
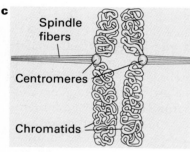
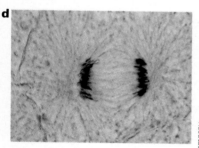
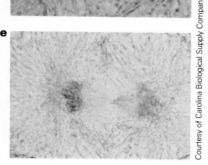

Prophase During prophase, the chromosomes become shorter and thicker. The individual chromosomes become visible as double chromosomes that are joined together. Each chromosome of a double chromosome is called a **chromatid** (KRŌ-muh-tid). They are attached to each other at a point called the **centromere** (SEN-truh-mir).

During prophase, the nucleoli disintegrate and disappear, and the nuclear membrane breaks apart. When this happens, the nucleus is no longer separated from the cytoplasm. In animal cells, the centrioles in the cytoplasm then move to opposite sides of the nuclear region. As they move, a network of fibers forms between them. These fibers are made of microtubules and are called **spindle fibers**. By the end of prophase, the spindle fibers stretch from centriole to centriole across the entire cell. Together, they make up a football-shaped structure called the **spindle**. Although plant cells generally lack centrioles, spindles do form in plant cells in much the same way that they do in animal cells. The only major difference is that the spindles in most plant cells are not organized between centrioles.

Metaphase During metaphase, spindle fibers attach to each pair of chromatids at the centromere. The chromatids line up along the center of the cell at the middle of the spindle. Then the centromeres divide. As a result, each chromatid in a pair becomes free of its partner.

Anaphase During early anaphase, the chromatids begin to move apart from one another and toward the centrioles on the opposite sides of the cell. The chromatids appear to be pulled apart by the spindle fibers; however, how this occurs is not understood. Once the chromatids are separated, they are again called chromosomes. At the end of anaphase, a complete set of chromosomes is at each end of the spindle.

Telophase During telophase, the spindle disintegrates, and a new nuclear membrane forms around each complete set of chromosomes, creating two nuclei. The chromosomes spread apart and assume the netlike appearance typical of interphase. In addition, nucleoli reappear. The two nuclei each have the same number and type of chromosomes that the parent nucleus had. In addition to the formation of the nuclei, the division of the cytoplasm usually occurs during telophase.

15–10 Mitosis in a whitefish cell: **a.** prophase; **b.** metaphase; **c.** close-up of chromatids during metaphase, showing attachment of spindle fibers to the centromeres; **d.** anaphase; **e.** telophase.

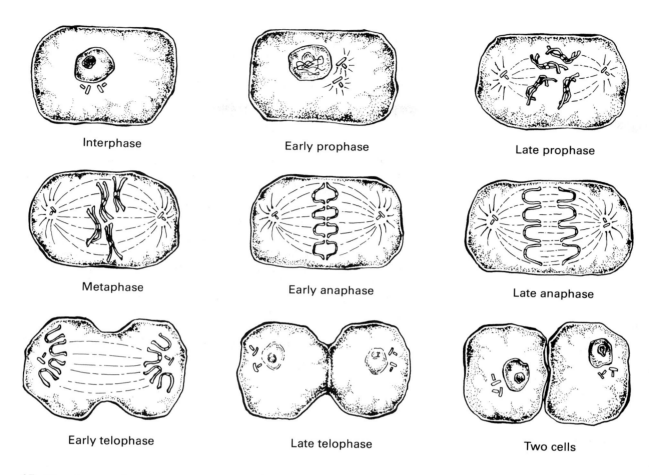

15–11 Stages of mitosis in an animal cell.

Table 15–1 Summary of Cell Division

Phase	Major Events
Interphase	Chromosomes replicate
Prophase	Nuclear membrane and nucleoli disintegrate Chromosomes shorten and thicken Two chromatids are connected at the centromere Spindle forms
Metaphase	Chromatids line up at the center of the spindle Centromeres split Free chromatids are now called chromosomes
Anaphase	Chromosomes move to opposite ends of the spindle
Telophase	Spindle dissolves Nuclear membrane and nucleoli form

Cytoplasmic Division

Cytoplasmic division results in the separation of the two nuclei into separate cells. Other organelles in the cell are also distributed between the two offspring cells. In animal cells, division of the cytoplasm between the two new cells takes place as a groove or furrow forms across the middle of the parent cell, pinching it in two. In plants, a structure called the **cell plate** forms between the two nuclei. New cell walls are formed along the cell plate.

When cell division is complete, the two offspring cells—each having the same number and type of chromosomes—enter interphase.

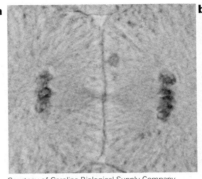 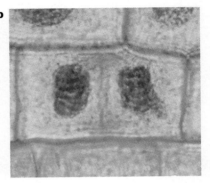

15–12 Cytoplasmic division: **a.** In an animal cell a groove forms, eventually pinching the cell in two. **b.** In a plant cell, a cell plate forms between the two nuclei.

Courtesy of Carolina Biological Supply Company

Life Span

A cell does not live forever; it lives for only a certain period of time, known as its **life span**. During a cell's life span, it grows and changes. At the end of its life span, a cell may reproduce or it may die. Different kinds of cells have different life spans. For example, a cell from the lining of the intestine lives about one and one-half days. Intestinal cells that die of "old age" are replaced by new cells produced by other intestinal cells. In general, plant cells live 10 to 30 hours and animal cells live 18 to 24 hours. Nerve cells may live for a very long time. Since nerve cells do not reproduce after birth, a seventy year old person has nerve cells that are also seventy years old.

Reproduction of cells is a continuous process. It is normal for cells to wear out. Cell reproduction provides new cells to take the place of the old ones. Many cells reproduce at a rapid rate when an organism is young. As an organism ages, cellular reproduction and growth slow down, and cellular processes begin to deteriorate. An organism's life ends when it ceases to have the ability to replace its cells and repair its body parts.

15–13 Many people remain active as they grow old.

Chapter 15 Reproduction of Molecules and Cells

Abnormal Cell Division

The human body is composed of approximately 100 trillion cells. Millions of them die every second; new ones take their place. The continual growth, death, and replacement of cells in the body is a controlled process that enables a person to grow, heal wounds, and even stay alive. Sometimes the mechanisms that control cell reproduction become disrupted. One major type of disease in which cell reproduction is disorganized and uncontrolled is **cancer**. The uncontrolled growth of cancer cells upsets the normal structure and functioning of body tissues. The damaging effects of cancer on the body contrast sharply with the regulated cycle of normal cell division.

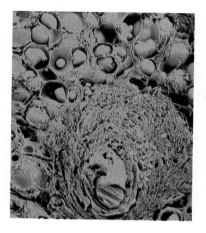

15–14 Cancerous tissue in a human ovary appears very dense (lower right).

Building a Science Vocabulary

chromosome	metaphase	spindle fiber
homologous chromosome	anaphase	spindle
cell division	telophase	cell plate
mitosis	interphase	life span
cytoplasmic division	chromatid	cancer
prophase	centromere	

Questions for Review

1. What are the two stages of cell division? State the final result of each stage.

2. Place the following events of mitosis in their proper order.
   a. DNA replicates
   b. chromatids line up at middle of spindle
   c. spindle forms
   d. nuclear membranes form
   e. chromosomes move to ends of spindle
   f. nuclear membrane disintegrates

3. What structure takes part in animal cell mitosis but not in plant cell mitosis? What structure is unique to plant cell division?

4. What are the two ways that a cell's life can end at the completion of its life span?

Section 15.2 The Cellular Basis of Reproduction

15.3 PROTEIN SYNTHESIS

After cell division, offspring cells grow by manufacturing new parts. These activities require organic molecules, especially proteins, which are made in a process called **protein synthesis**. In certain ways, protein synthesis resembles DNA replication. Before discussing protein synthesis, it will be helpful to review the structure and function of protein.

A cell contains more protein than any other substance except water. Proteins have several important functions. They are the building blocks for cell structures such as the cell membrane and ribosomes. They are found in hormones, which help regulate body functions. They help store food in eggs and seeds. They also make up enzymes, which control the many thousands of chemical reactions that take place in a cell every day.

Proteins are chains of amino acids. The proteins in all living organisms are made up of about 20 amino acids. Unicellular bacteria contain 600 to 800 different kinds of protein. Complex organisms contain up to several thousand different kinds of protein. Each type of protein molecule is determined by the number of amino acids in the chain and by the order, or arrangement, of the amino acids. The information specifying the number and arrangement of amino acids in every type of protein made by a cell is stored in the coded message in DNA.

Protein synthesis takes place in ribosomes located in the cytoplasm. The transfer of the chemical information stored in DNA, as well as the joining together of amino acids at the ribosomes, is accomplished by another type of molecule, **ribonucleic acid (RNA)**.

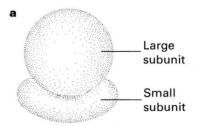
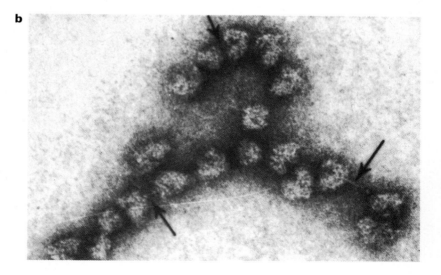

15–15 Ribosomes. **a.** A ribosome has two subunits both made up of ribosomal RNA and proteins. **b.** In areas of active protein synthesis, ribosomes are connected by a strand of messenger RNA (arrows), forming a polysome.

Similarities and Differences Between DNA and RNA

The difference between deoxyribose and ribose is that deoxyribose lacks one oxygen atom, hence the prefix deoxy

Because of uracil's similarity in structure to thymine, it is able to form a base pair with adenine.

Both DNA and RNA are nucleic acids made up of long strands of bonded nucleotides. The nucleotides in both are made up of a simple sugar, a phosphate group, and a nitrogen base. The simple sugar found in DNA differs from that found in RNA. In DNA, the sugar is deoxyribose; in RNA, the sugar is ribose. As you can see, DNA (*deoxyribo*nucleic acid) and RNA (*ribo*nucleic acid) are named for the kind of sugar each contains.

Both DNA and RNA nucleotides contain four possible nitrogen bases. Adenine (A), guanine (G), and cytosine (C) are found in both DNA and RNA. The fourth base is different: whereas DNA contains thymine (T), RNA contains **uracil** (YUR-uh-sil), (U).

Another difference between DNA and RNA is that RNA is usually a single strand of nucleotides, and DNA is characteristically a double helix. Most DNA is found in the nucleus of eukaryotic cells. On the other hand, RNA is found in both the nucleus and the cytoplasm.

In addition to differing in structure, DNA and RNA also differ in function. Molecules of DNA direct all cell processes and carry the code for heredity. They do this by holding the code for protein synthesis. The information coded in DNA is "read" by RNA. The RNA then transfers the information to the ribosomes and directs the synthesis of proteins. Three types of RNA are

15–16 Comparison of DNA and RNA. Both are made up of nucleotides, but the nucleotides differ in their sugars and in one nitrogen base. RNA base pairs with DNA but cannot replicate itself.

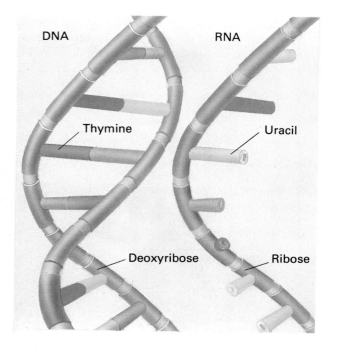

Differences Between DNA and RNA	
DNA	RNA
Double-stranded	Usually single-stranded
Deoxyribose	Ribose
Thymine	Uracil
Found primarily in nucleus	Found primarily in cytoplasm

Section 15.3 Protein Synthesis

found in cells: **messenger RNA (mRNA)**, **ribosomal RNA (rRNA)**, and **transfer RNA (tRNA).** Each of these types of RNA has a different function in protein synthesis. Messenger RNA transfers the information stored in DNA from the nucleus to the ribosomes. Ribosomal RNA is the main structural component of ribosomes. Transfer RNA decodes the information carried by the mRNA and ensures that the proper amino acids are built into the protein.

Reading of DNA

During the process of protein synthesis, DNA unzips, exposing the nitrogen bases, which act as a pattern, or **template**, for the formation of mRNA. The RNA nucleotides in the nucleus bond with the unpaired bases on the unzipped DNA molecule. The pairing of bases during the formation of mRNA is much the same as that found in DNA replication. The RNA nucleotide guanine bonds with the DNA nucleotide cytosine; RNA cytosine bonds with DNA guanine; and RNA adenine bonds with DNA thymine. The only difference between base pairing in DNA replication and the formation of mRNA is that the RNA nucleotide uracil bonds with the DNA nucleotide adenine. After RNA nucleotides have matched up with the DNA template, they bond with each other, forming a long, single-stranded molecule of mRNA. The mRNA carries the message for protein synthesis in the arrangement of its bases. As mRNA is formed, the DNA template zips back together. The strand of mRNA leaves the nucleus and enters the cytoplasm.

Messenger RNA takes the code for protein synthesis from the nucleus to the ribosomes.

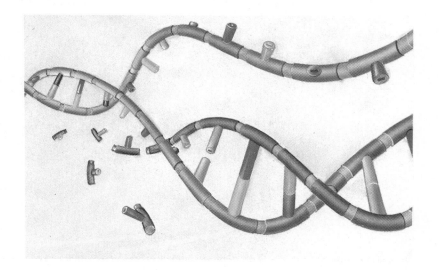

15–17 Formation of RNA from DNA. The double-strand of DNA separates, exposing the bases. The bases of RNA nucleotides then pair with the DNA bases. The RNA nucleotides bond together, forming a single strand of RNA (brown strand). The RNA eventually pulls away from the DNA template. All three types of RNA—mRNA, rRNA, and tRNA—are made in this way.

Decoding of the mRNA Message

Once the mRNA reaches the cytoplasm it attaches to a ribosome. There transfer RNA molecules decode the information on the mRNA and provide the appropriate amino acids from the cytoplasm. Each molecule of tRNA contains three "active" bases in a specific place on the molecule. The order of the three bases in tRNA makes up a triplet code, which corresponds to three consecutive bases on mRNA. The triplet code on mRNA is known as the **genetic code**. Each molecule of tRNA bonds with a specific amino acid. For example, a tRNA molecule with the bases GGU bonds with the amino acid glycine. Several kinds of tRNA molecules can carry the same amino acid. Each of four different tRNAs carries glycine. Table 15-2 shows the amino acids that correspond to each possible base triplet on mRNA.

15-18 Messenger RNA is decoded by transfer RNA. The three active bases on tRNA match up with corresponding bases on mRNA. Each tRNA carries a specific amino acid.

Table 15-2 The Genetic Code: Base Triplets on Messenger RNA and Their Corresponding Amino Acids

First Base in the Triplet	Second Base in the Triplet				Third Base in the Triplet
	U	C	A	G	
U	Phenylalanine	Serine	Tyrosine	Cysteine	U
	Phenylalanine	Serine	Tyrosine	Cysteine	C
	Leucine	Serine	End chain	End chain	A
	Leucine	Serine	End chain	Tryptophan	G
C	Leucine	Proline	Histidine	Arginine	U
	Leucine	Proline	Histidine	Arginine	C
	Leucine	Proline	Glutamine	Arginine	A
	Leucine	Proline	Glutamine	Arginine	G
A	Isoleucine	Threonine	Asparagine	Serine	U
	Isoleucine	Threonine	Asparagine	Serine	C
	Isoleucine	Threonine	Lysine	Arginine	A
	Methionine	Threonine	Lysine	Arginine	G
G	Valine	Alanine	Aspartic acid	Glycine	U
	Valine	Alanine	Aspartic acid	Glycine	C
	Valine	Alanine	Glutamic acid	Glycine	A
	Valine	Alanine	Glutamic acid	Glycine	G

Protein Formation

As the ribosome moves along the messenger RNA, the mRNA bonds with molecules of transfer RNA that are attached to amino acids. The bonding of mRNA with tRNA molecules occurs by base pairing: the three active bases on tRNA molecules match up with three bases on mRNA. As the ribosome moves along the mRNA, tRNA molecules and their amino acids are added one by one in a long chain. As an amino acid comes into position, it bonds to the other amino acids in the chain. Then, the bond between the amino acid and its transfer RNA molecule is broken. The result is the formation of a protein chain. The protein molecule is complete when the ribosome crosses the end of the messenger RNA molecule. The newly formed protein molecule leaves the ribosome and is ready for use by the cell.

Certain triplets on mRNA specify where to begin and end the protein chain.

Summary of Protein Synthesis

The formation of a protein is determined by a segment of DNA in the nucleus. The sequence of nitrogen bases in the DNA determines the sequence of amino acids in the protein. The steps by which the coded information from DNA is translated into a finished protein are as follows.

1. DNA unzips.
2. RNA nucleotides bond with bases on an unzipped DNA segment, forming messenger RNA.
3. Messenger RNA leaves the nucleus and attaches to a ribosome in the cytoplasm. In this manner, DNA's message is transferred to the site of protein synthesis.
4. Transfer RNA molecules, carrying amino acids, move into position along the messenger RNA. The three active bases on tRNA molecules match up with bases on mRNA. In this way, the original DNA code is translated into specific amino acids.
5. The amino acids attached to the transfer RNA molecules bond, forming a protein.

Protein synthesis can take less than a second from the formation of mRNA to the finishing of a protein.

15–19 *Opposite page:* protein synthesis. The DNA "message" read by mRNA in the nucleus is carried to ribosomes in the cytoplasm. There, tRNA molecules bonded to amino acids decode the message on mRNA by base pairing. The amino acids bond, forming a protein.

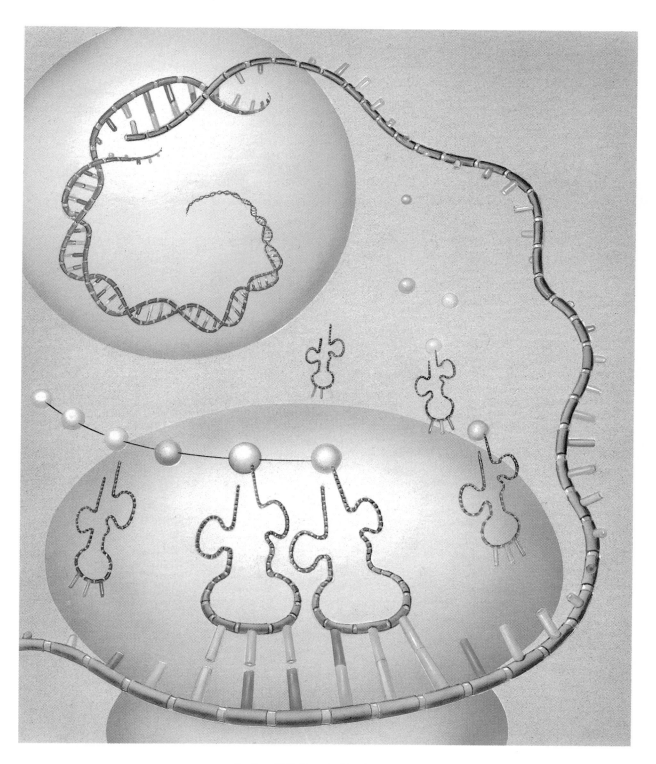

Section 15.3 Protein Synthesis

DNA Controls Structure and Function in Organisms

As you read in Section 15.2, DNA replicates before a cell divides. The replication of DNA ensures that each offspring cell receives a complete set of instructions for its heredity. These instructions contain the blueprints for protein synthesis within the offspring cell. Some proteins are used to build the cell's organelles and other structures. Other proteins become part of enzymes, which control all of the cell's functions. They do this by regulating the chemical reactions that occur in the cell.

15-20 The hereditary information of a species is contained in its DNA. By holding the code for protein synthesis, DNA directs the development, growth, and reproduction of an organism.

Building a Science Vocabulary

protein synthesis
ribonucleic acid (RNA)
uracil
messenger RNA (mRNA)
ribosomal RNA (rRNA)
transfer RNA (tRNA)
template
genetic code

Questions for Review

1. How does DNA differ from RNA in each of the following characteristics?
   a. kind of sugar in the molecule
   b. usual location in the cell
   c. nitrogen bases present in the molecule
   d. usual number of strands
   e. functions in the cell

2. Place the following steps of protein synthesis in the proper order.
   a. transfer RNA bonds to messenger RNA
   b. RNA nucleotides bond to DNA template
   c. DNA unzips
   d. amino acids bond, forming a protein
   e. messenger RNA leaves nucleus and goes to ribosome

3. What part does each of the following play during protein synthesis?
   a. DNA
   b. messenger RNA
   c. transfer RNA
   d. ribosomes

4. How does protein synthesis result in control of a cell?

Perspective: Discovery
Competition and Cooperation in Science

Progress in science requires cooperation. Scientists cooperate by publishing their findings and forming teams to work together on the same problem. Scientists loan one another equipment and sometimes even trained personnel. As a result, major scientific problems that cannot be solved by an individual scientist working alone can be resolved.

But, competition also takes place in science, especially when the stakes are high. Competition was extremely intense to determine the structure and mechanism of replication of DNA. Because of DNA's central importance to all of biology, the people involved in solving the DNA puzzle would gain worldwide fame and recognition.

This competition, however, was still very dependent on cooperation and building on the works of others. For example, Linus Pauling, an American scientist, had discovered a method to determine the molecular structure of molecules. Using a technique known as x-ray crystallography, Pauling had figured out the spiral, or helical, structure of proteins. Using the same approach, Pauling began to turn his attention to the DNA molecule.

At the same time in England, two different teams were working on the same problem. Francis Crick, a Cambridge University scientist, and James Watson, an American who had gone to England to work with Crick were one team. Maurice Wilkins and Rosalind Franklin were the other team, working about 80 km from Crick and Watson. The English sense of fair play did not permit competition between these two teams, but both raced to beat Pauling to the winner's table.

Crick and Watson used Pauling's technique of building molecular models and analyzed the results of x-ray crystallography of DNA done by Franklin. They also factored in the biochemical studies of Erwin Chargaff, an Australian, who showed that there are equal amounts of adenine and thymine and of guanine and cytosine in DNA. Crick and Watson proposed a helical structure for DNA in which complementary base pairs of adenine-thymine and of guanine-cytosine formed the rungs of a ladderlike structure, the backbone being of deoxyribose and phosphate. They further proposed this ladderlike structure existed as a double helix.

Thus was "born" the widely known molecular model of DNA, the result of extremely intense competition. This discovery later brought to Crick and Watson the most prestigious award in all science, the Nobel Prize. The stakes were indeed high.

James Watson and Francis Crick are shown here with their model of DNA.

Chapter 15
Summary and Review

Summary

1. Reproduction is the process of producing offspring. DNA molecules, cells, and organisms can reproduce.

2. Information about the characteristics of an organism is transmitted from generation to generation in code on DNA molecules.

3. The Watson-Crick model of DNA describes the molecule as a ladder twisted into a double helix. The ladder is composed of two chains of nucleotides bonded together by their nitrogen bases. The nucleotides found in DNA are adenine, thymine, guanine, and cytosine.

4. DNA replication starts with the unzipping of the double helix. Each half then builds a new half by pairing with nucleotides in the cell. In this way, two identical molecules of DNA are formed from one.

5. Cells reproduce by cell division. The two stages of cell division are mitosis and cytoplasmic division. The phases of mitosis are prophase, anaphase, metaphase, and telophase. A cell is in interphase when it is not dividing.

6. Cells as well as organisms have definite life spans. As cells grow old, they either reproduce or die.

7. The code for making proteins is stored in DNA. The code is read and translated by RNA, which is made up of a single strand of the nucleotides adenine, cytosine, guanine, and uracil.

8. During protein synthesis, the DNA code is transferred to messenger RNA, which leaves the nucleus and travels to the ribosomes. Transfer RNA molecules are bonded to specific amino acids. At the ribosomes, transfer RNAs deliver amino acids in the proper order, according to the triplet code on messenger RNA. The amino acids bond, producing a protein.

9. Protein synthesis results in the production of proteins, which make up a cell's structure and, as enzymes, govern its functions.

Review Questions

Application

1. When a paramecium reproduces by cell division, reproduction occurs at the levels of molecule, cell, and organism. Explain this process.

2. Inside certain cells, the disaccharide maltose is broken down to the monosaccharide glucose. How does the DNA of such a cell control this process?

3. Why is DNA replication necessary for the reproduction of a cell?

4. If a protein has 400 amino acids, how many nucleotides are in the section of DNA that codes for its amino acid sequence?

5. What sequences of nitrogen bases in mRNA and tRNA correspond to the following DNA base sequence: AATGCAGGT? What amino acids correspond to the DNA code? (Refer to Table 15–2.)

Interpretation

1. What problem would arise with a DNA model that had nitrogen bases as the sides of the ladder and sugars and phosphates as the rungs?

2. Cells can be kept alive and reproducing in test tubes. Is it possible for these cells to live forever? Explain.

3. How can a complex organism have 800 different kinds of proteins when there are only 20 different amino acids in cells?

4. Would a genetic code made up of four bases in combinations of two be sufficient to code for the making of proteins from the 20 amino acids normally found in cells? Why or why not?

Extra Research

1. Build a model of mitosis. Use wires for the spindle, rubber balls for the centrioles, and halves of ping-pong balls for the chromosomes. Support the spindle on two dowel rods standing in wooden bases. String the halves of the ping-pong balls on the wires.

2. The following poem was written by Payson Stevens. The lines of the poem are shown divided into six groups. Interpret each group of lines in your own words according to what you have learned about DNA replication and protein synthesis.

Molecular Rhythms

Imagine being
   a strand of DNA
unwinding in the
   fluid maze   1
uncoiling like a breeze
   while the cloud enzymes
envelope your backbone   2
and the cell wisdom
   replicates its patterns   3
drawn by the needs of bonds
your mosaic
   a vision of the past
is locked in chemical messages   4
which
   etch the face of life
and energy,
   energy, energy   5
is lowered and raised
while
   the economy of balance
telescopes
   the cytoplasmic ballet   6
into
   the forms of the future.

Payson R. Stevens

Try to write your own poem about the structure and function of DNA or the processes of mitosis or protein synthesis.

3. Research the story of how one of the following was discovered: Watson-Crick model of DNA; steps in mitosis; protein synthesis. Write the story of the discovery in the form of a short feature for a magazine.

Career Opportunities

Microbiologists study the life cycle and chemical composition of organisms such as microscopic fungi, bacteria, and viruses, which are important in agriculture, medicine, food processing, and drug manufacturing. They develop new medicines and food-production methods. Microbiologists require four or more years of specialized education, and they usually work for universities, private companies, or governments. Write the Canadian Association of Medical Microbiologists, Henderson General Hospital, Concession St., Hamilton, Ontario L8V 1C3.

Biology editors work for magazine or book publishers in the production of biology-related articles or books. Their work can involve planning and rewriting material, as well as working with designers in laying out pages and selecting artwork and photographs. They must have a strong science background and be able to write clearly. A university degree is necessary. Contact the Canadian Science Writers Association, 160 Wellesley St. E., Toronto, Ontario M4Y 1J3.

Chapter 16

Reproduction of Organisms

Focus How would you define a successful organism? Would it be big? Strong? Swift? Some successful organisms have these characteristics, but many do not. A species is considered successful if it is able to stay alive and reproduce. Offspring are essential for a species to survive over time. In this chapter you will learn about the different ways organisms reproduce.

16.1 ASEXUAL REPRODUCTION

Living things reproduce in two basic ways. In **asexual reproduction** one cell, or a group of cells, from a single parent develops into an offspring like the parent. **Sexual reproduction** requires the joining, or fusion, of the nuclei of two types of specialized reproductive cells—usually one from each of two parents. The fusion of the two reproductive cells results in a cell that grows and develops into an offspring.

Some species, amebas for example, reproduce only by asexual reproduction. Other species, including grasshoppers, mice, and humans, reproduce solely by sexual reproduction. Many species, such as paramecia, hydras, bacteria, and most plants, can reproduce both sexually and asexually.

16-1 In their reproductive cycle, sockeye salmon make a long, difficult journey from the sea upriver back to the freshwater streams in which they were spawned.

Types of Asexual Reproduction

In asexual reproduction, a cell or group of cells from one organism gives rise to offspring. Geranium plants can reproduce in this way. When a small piece of a geranium stem is cut off and placed in wet sand, it grows new roots. Eventually a new plant develops. Asexual reproduction is usually accomplished by a cell or cells that are not specialized reproductive cells.

Four types of asexual reproduction exist: **binary fission, budding, spore formation,** and **fragmentation.** Each of these types of reproduction makes use of the process of mitosis in producing new cells.

Binary Fission Some bacteria, unicellular algae, and protozoa reproduce asexually by the process of binary fission. During binary fission one parent cell divides into two offspring cells of equal size. The chromosomes within the organism replicate before cytoplasmic division occurs. Bacteria and blue-green algae do not have organized nuclei. When these cells divide, their replicated chromosomes are divided equally between the offspring cells. Other unicellular organisms, such as amebas and paramecia, have organized nuclei. The nuclei within these organisms divide by mitosis, which is followed by equal cytoplasmic division.

Binary fission results in two offspring cells that contain identical DNA to that of the parent cell. Each offspring cell grows and in turn undergoes binary fission, producing two new offspring cells with the same DNA.

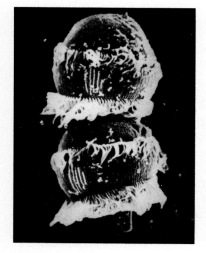

16–2 *Didinium*, a protozoan, undergoing binary fission.

Budding Yeast, hydras, and some kinds of plants reproduce by budding. During the process of budding, a cell or cells in the parent's body produce a miniature version of the parent, called a **bud.** When the bud is first formed, it is attached to the parent. The bud grows and eventually breaks off the parent as a new organism.

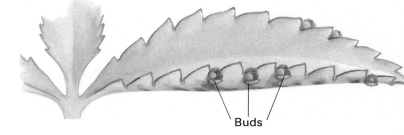

16–3 A *Bryophyllum* reproduces by budding. Each bud is an individual identical to the parent plant.

Chapter 16 Reproduction of Organisms

A hydra can have several buds developing simultaneously.

In yeast, mitosis in the nucleus is followed by unequal division of the cytoplasm. This results in a large parent cell and a small offspring cell, which is the bud. In hydras, a bud appears on the side of the parent's body. The bud grows, elongates, and sprouts tentacles. In two to three days it looks like an adult hydra—only smaller. Soon the offspring breaks away from the parent and lives as an independent organism. The plant kalanchoe (also known as *Bryophyllum*) also reproduces by budding. Tiny buds form in notches along the edges of leaves of the parent plant. The buds grow, drop from the parent, and begin their independent life.

Budding produces organisms with identical DNA.

Budding, like binary fission, produces offspring from only one parent. All the buds from the parent organism have DNA identical to the parent. Budding results in the formation of new organisms that look and function like their parent.

Spore Formation Many plants and protists reproduce by means of **spores**. A spore is a reproductive cell that contains a nucleus and a small amount of cytoplasm. A spore usually has a hard protective covering that resists drying. This enables it to survive during unfavorable environmental conditions. Spores are usually produced in large numbers by mitosis and are very small and lightweight. They are easily dispersed by wind or water. Under favorable growing conditions, a spore can develop into a new organism. Molds, mushrooms, mosses, and ferns reproduce by spore formation. As in other forms of asexual reproduction, new organisms formed from spores have the same DNA as their parent.

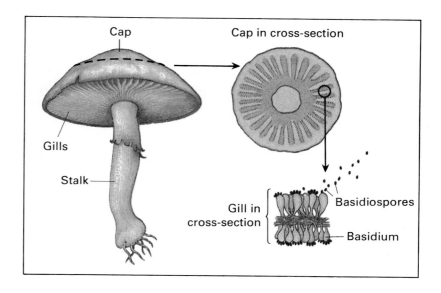

16–4 Spore formation. The visible part of a mushroom is the fruiting body. The gills are lined with basidia that produce asexual basidiospores during one part of the mushroom's life cycle.

Section 16.1 Asexual Reproduction

Fragmentation In organisms that reproduce by fragmentation, small pieces of the organism's body grow into new organisms. For example, the ocean's waves can tear sponges into many pieces, each of which can grow into a new sponge.

Starfish also reproduce by fragmentation. Oyster fishermen sometimes try to kill starfish that they catch in their nets because the starfish eat oysters. If the fishermen tear up the starfish and throw the pieces back into the ocean in an attempt to kill them, they end up helping produce more starfish. Each starfish ray can grow into a whole new starfish, as long as it has a piece of the central disc.

Many plants reproduce by fragmentation. Recall that a geranium plant can grow from a piece of the parent plant. In plants, fragmentation is called **vegetative propagation.**

As in binary fission, budding, and spore formation, all the new organisms produced by fragmentation have the same DNA their parent.

16–5 Each part of a starfish that has been broken apart can grow new arms if it includes a piece of the central disc.

Cloning

All the organisms produced from one parent during asexual reproduction make up a **clone**. For example, all the little hydras that bud off one parent hydra represent a clone, as do all the offspring starfish that result from tearing up one starfish. The members of a clone and their parent organism all have the same kind of DNA. Since members of a clone inherit identical chromosomes from their parent, all the members of a clone are alike and like their parent.

Artificial Cloning A specialized body cell, such as a nerve or muscle cell of a complex organism, does not give rise to a whole organism. Normally, a cell in a maple leaf does not develop into a maple tree; neither does a cell in a tiger's toe produce a baby tiger. Scientists have wondered whether specialized body cells hold the information to reproduce the whole organism. The processes of DNA replication and mitosis suggest that every cell in an organism contains this information.

J. B. Gurdon, an American biologist, hypothesized that specialized vertebrate body cells do contain the information needed to reproduce an entire organism. His experiments on frog cloning support this idea. In 1962, he developed a clone of frogs from the intestinal cells of one parent frog. Figure 16–6 shows the steps in Gurdon's experiment. Gurdon's artificially cloned frog looks no different than a frog produced in the normal way. Gurdon's experiments prove that specialized body cells do contain the information to produce a new organism.

16–6 J. B. Gurdon obtained an intestinal cell from a tadpole (1) and a mature egg cell in which the nucleus had been destroyed (2). He implanted the nucleus from the intestinal cell into the egg cell (3). The egg developed into a frog with the same characteristics of the tadpole that donated the nucleus (4). This experiment indicated that the intestinal cell nucleus contained all the information needed by all the cells of the organism.

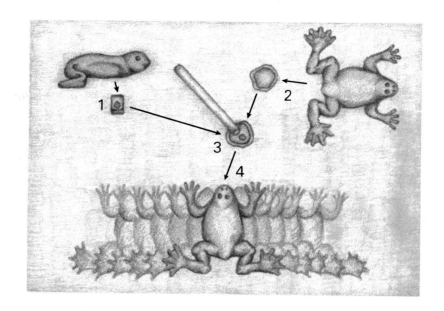

Asexual Reproduction: Absence of Variations

Asexual reproduction is an efficient method of producing identical offspring. The offspring have the same DNA as their parent organism. Such offspring can differ from their parent only if an accidental change in the DNA occurs. Such an accidental change in DNA is called a **mutation** (myoo-TAY-shun), which can be caused by radiation or certain chemicals. Since mutations do not occur often, asexual reproduction usually results in a new generation with few **variations**, which are inherited differences that exist among the members of a species. An easily seen variation in humans, for example, is eye color. A person's eyes can be green, blue, hazel, brown, or multicolored.

Lack of variations can sometimes be a disadvantage to long-term survival. Consider a field of pineapples in which all the plants are exactly alike. Suppose a disease-causing virus were to attack them. Since all the pineapples are alike, the chances that they would all be affected by the virus and die are greater than if variations existed. Variation in their ability to resist disease might enable some of them to survive the attack.

There are advantages to asexual reproduction. The organism does not have to find a mate; it is fast; and the result is an exact copy of a successful organism.

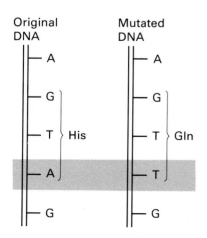

16–7 Mutation. In this illustration, radiation or a mutation-causing chemical has changed the nitrogen base of one nucleotide of DNA from adenine to thymine. The new triplet codes for the amino acid glutamine (gln), instead of histidine (his).

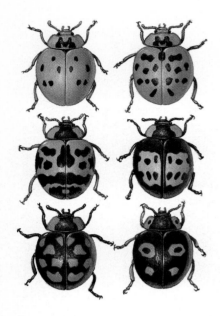

16–8 Variation can be seen in the wide range of color patterns on the wing covers of ladybugs. Each of these types of ladybug inhabits a different geographical area. Variations can help species survive when changes in the environment occur.

From "The Mechanisms of Evolution" by Francisco Ayala. Copyright © September, 1978 by Scientific American, Inc. All rights reserved.

Variations can be an advantage to a species by helping it to succeed. When the environment changes, variations give some organisms a better chance of surviving. One way a species gets variations among its members is by mutations. Another way is by sexual reproduction. As you will see in the second section of this chapter, sexual reproduction results in variations because it combines the characteristics of two parents.

Building a Science Vocabulary

asexual reproduction
sexual reproduction
binary fission
budding
spore formation
fragmentation

bud
spore
vegetative propagation
clone
mutation
variation

Questions for Review

1. What is the difference between asexual reproduction and sexual reproduction?
2. Name and briefly describe the four types of asexual reproduction. Give examples of organisms that reproduce in each of these ways.
3. What is a clone? In what ways are the members of a clone like their parents?
4. How did Gurdon's experiment on frog cloning help support the idea that all body cells contain the information to reproduce an organism?
5. How can variations be an advantage to a species?

Perspective: Technology
Myself All Over Again

Imagine the surprise of that early ancestor of ours who stuck a willow branch into the ground, only to find that it grew into a tree. Since then we have become quite used to the idea that we can grow exact copies of certain plants and trees. When we plant leaves of African violets and coleus, or the stems of potato tubers and onion bulbs, we take advantage of a natural type of asexual reproduction that occurs in many plants and animals.

Asexual reproduction results in an exact genetic duplicate of the parent. It occurs by simple mitotic division as in algae and protozoa, or by vegetative reproduction as in the case of the willow.

The term that has become quite common for such reproduction is "cloning." And while cloning has been known for a long time in plants, only recently has there been an increased interest in the possibility of cloning higher animals, even humans.

A few years ago, a popular novel described the cloning of a wealthy person. Scientists regarded the book as science fiction. They did so because no true cloning of any mammal had yet occurred in the laboratory. The highest animals in which successful cloning had taken place were amphibians. The cloning technique in amphibians involves transplanting the nucleus of a cell into an egg cell from which its own nucleus has been removed. If the transplant is successful an organism exactly like the organism that "donated" the nucleus develops, because the transplant nucleus carries the organism's full load of genetic information.

If cloning becomes possible in higher animals, it will be possible to breed livestock asexually. Scientists will be able to select the kinds of genetic traits they want and reproduce them exactly. When it comes to cloning humans, we really aren't too far from being successful. If it should become possible there will be many difficult questions to answer. Who is to be cloned? Who decides who is to be cloned? How many of the cloned individuals shall there be? Who thinks enough of himself or herself to be cloned?

How would you like a whole room or house—or city—filled with people exactly like yourself? The effect of having a group of identical willow trees can be very pleasant. But a group of identical people might prove to be rather monotonous.

16.2 THE BASIS OF SEXUAL REPRODUCTION

Sexual reproduction occurs in most species of living things. During sexual reproduction, two specialized reproductive cells called **gametes** (GAM-eets) fuse to form a one-celled **zygote** (ZĪ-gōt). The zygote grows and develops into a new organism.

Every species has a specific and characteristic number of chromosomes. This number is known as the **diploid** (DIP-loyd) **number** and is represented as **2n**. In human cells, there are two each of 23 different types of chromosomes, making a total of 46 chromosomes. Therefore, the diploid number in humans is 46. The process of mitosis ensures that every body cell as it is formed contains the diploid number of chromosomes.

Offspring resulting from sexual reproduction have the same diploid number as their parents. But if gametes were formed using mitosis, the diploid number of the offspring would be different from the diploid number of the parents. For example, if a human female gamete with 46 chromosomes fused with a human male gamete also with 46 chromosomes, the result would be a zygote with 92 chromosomes—twice the normal diploid number. If the zygote could develop and grow, each cell of its body cells would have 92 chromosomes. Problems would arise if this continued, because the number of chromosomes in the species would double every generation.

Cell division by mitosis would not be effective in forming gametes for sexual reproduction.

16–9 Sexual life cycle. The male and female adults produce haploid gametes—the egg and sperm—which combine during fertilization to form a new zygote.

The situation described in the example does not occur because gametes are not formed by mitosis. Instead, they are formed by a special type of cell division called **meiosis** (mī-Ō-sis). During meiosis only *one* replication of chromosomes occurs, while *two* cell divisions take place. As a result, four cells or gametes each containing half the diploid number of chromosomes are formed. Half of the diploid number of chromosomes is known as the **haploid** (HAP-loyd) **number** (**n**). Instead of having *two* of each type of chromosome, each gamete has only *one* of each type of chromosome. During sexual reproduction, the haploid gametes fuse, forming a zygote with the diploid number of chromosomes. The zygote then divides by mitosis, which maintains the diploid number in all the body cells during the growth and development of the offspring.

Meiosis Maintains All Characteristics

Chromosomes contain the information for inherited characteristics. Because of this, the number of chromosomes in a species is important and must be maintained. During meiosis gametes are formed with the haploid number of chromosomes. But, meiosis does not result in the loss of traits. Offspring develop all the characteristics of their species. This is true because the diploid number of chromosomes for an organism is made up of homologous pairs of chromosomes. The members of a homologous pair of chromosomes are identical in size and shape and carry information for the same traits. The segment of DNA in a chromosome that carries the code for a specific protein is called a **gene**. Both chromosomes in a homologous pair have genes for the same traits.

Consider one trait, the length of eyelashes, as an example of how homologous chromosomes work. On one of your chromosomes, inherited from your mother, you have a gene that determines the length of your eyelashes. In the same location on the corresponding homologous chromosome from your father is another gene with instructions for the length of your eyelashes. Both genes may carry instructions for long lashes. Both genes may carry instructions for short lashes. Or, one may carry instructions for long lashes and the other for short lashes. Together, this pair of genes determines how long your lashes will be.

The gene pair for eyelash length is only one example of all of the gene pairs that produce an organism's characteristics. Meiosis ensures that offspring receive one copy of each type of chromosome from each parent.

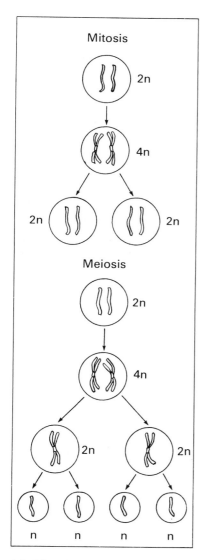

16–10 Meiosis and mitosis compared. During mitosis, one replication of chromosomes and one division take place, forming two diploid cells. During meiosis, one replication of chromosomes and two divisions take place, forming four haploid cells.

Steps in Meiosis

Meiosis separates the members of a homologous pair of chromosomes, forming cells that have only one copy of each chromosome.

Meiosis is a special case of cell division, much like mitosis. The major difference between them is that whereas mitosis results in two diploid cells, meiosis results in four haploid cells. The basic pattern in the phases of each cell division of meiosis are similar to those in mitosis and are identified by the same names as the phases of mitosis. Before a cell begins meiosis, it is in interphase, during which the cell's chromosomes are replicated. At the end of interphase, a cell has twice the diploid number of chromosomes. The cell then enters the first of its two divisions, known as the **first meiotic division**.

First Meiotic Division During Prophase I of meiosis, the replicated chromosomes shorten and thicken and become visible through the microscope. They appear as chromatids joined at their centromeres. A spindle appears and the nuclear membrane and nucleoli disintegrate.

Also during Prophase I, the homologous chromosomes form pairs. This pairing of homologous chromosomes is called **synapsis** (sih-NAP-sis). A homologous pair of chromosomes consists of four chromatids and two centromeres. Because it is made up of four chromatids, this structure is called a **tetrad**. During synapsis, the chromatids in a tetrad twist and wind around each other.

16–11 Meiosis. The first meiotic division results in separation of homologous chromosomes. The chromatids become separated during the second meiotic division. Four haploid cells are formed.

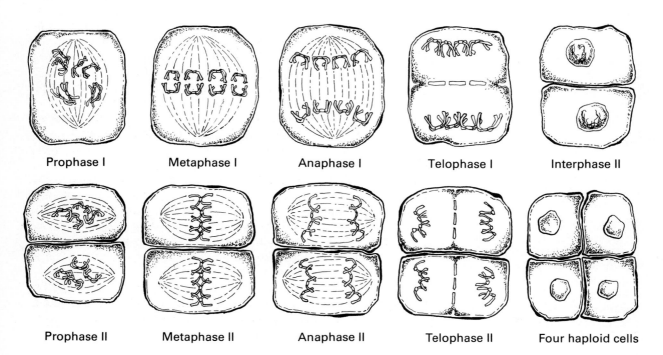

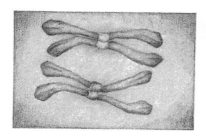 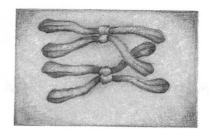 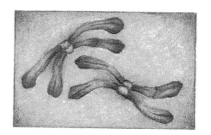

16–12 Crossing-over occurs during Prophase I. Pieces of chromatids of homologous chromosomes are exchanged, resulting in a new combination of genetic material.

As Prophase I continues, the two centromeres of the tetrad move apart slightly. The chromatids that are twisted around each other stay in contact. The intertwined chromatids often exchange pieces during this part of prophase. The exchange of parts of two homologous chromatids is called **crossing-over**. Crossing-over is an important source of variations in species because it results in new and different combinations of genes on a particular chromosome. Crossing-over is also called **recombination**. Toward the end of Prophase I, the tetrads move to the center of the spindle.

During Metaphase I, the tetrads line up along the center of the spindle. The homologous chromosomes are positioned in pairs. During Anaphase I, the homologous pairs of chromosomes separate. The paired chromatids move to opposite ends of the cell; however, the centromeres between the chromatids do *not* divide. During Telophase I, division of the cytoplasm occurs, forming two cells. In some organisms Telophase I is very distinct. In other organisms it is only a very short phase before the second division. In either case, during Telophase I, the chromatids are still joined together at the centromere.

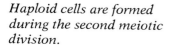

Homologous chromosomes separate during the first meiotic division.

As the first meiotic division comes to an end, the homologous chromosomes have been separated into two cells. There is no further replication of DNA in these cells, which then undergo the second division, known as the **second meiotic division**.

Second Meiotic Division Prophase II occurs after the first meiotic division is complete. It occurs in both cells formed by the first meiotic division. Centrioles in animal cells replicate and move to opposite ends of the cell, and a spindle is formed. In Metaphase II, the chromatids, still attached to each other by the centromere, line up in the center of the spindle. In Anaphase II, the centromeres divide, and the chromatids separate and are drawn to opposite poles.

Haploid cells are formed during the second meiotic division.

During Telophase II, division of the cytoplasm occurs in both cells. As a result, four cells are formed, each of which has a haploid number of chromosomes. The chromosomes in each cell elongate, and the nuclear membrane reforms.

Summary of Meiosis

Meiosis begins with a diploid cell. During interphase, each chromosome in the diploid cell replicates. Two meiotic cell divisions then occur. During the first division, the homologous pairs of chromosomes separate and two cells are formed. During the second division the chromatids separate, and four haploid cells are formed. Each haploid cell contains one chromosome from each original tetrad.

Building a Science Vocabulary

gamete
zygote
diploid number (2n)
meiosis
haploid number (n)
gene

first meiotic division
synapsis
tetrad
crossing-over
recombination
second meiotic division

Questions for Review

1. What would happen during sexual reproduction if meiosis did not occur during gamete formation?
2. What happens to the members of homologous pairs of chromosomes during meiosis?
3. Rewrite the following steps of meiosis in their proper order.
    a. second cytoplasmic division occurs
    b. homologous pairs separate
    c. first cytoplasmic division occurs
    d. synapsis occurs
    e. chromatids separate
4. State whether each of the following is the result of meiosis or mitosis.
    a. two diploid cells
    b. four haploid cells
    c. cells different from parent cells
    d. gametes are formed
5. Why are traits not lost during meiosis?

16.3 SEXUAL REPRODUCTION

Meiosis results in the formation of haploid cells. In animals, these cells become specialized reproductive cells. In males, these cells are **sperm**; in females they are **eggs**. In plants, the haploid cells become spores, which develop into haploid plants that produce specialized reproductive cells.

Some species produce gametes that look alike in size and shape. Gametes that look alike are called **isogametes** (ī-sō-GAM-eets). Examples of organisms that produce isogametes are found among the protists. Most species produce gametes that are different in size and shape. These are called **heterogametes**. Whether gametes are isogametes or heterogametes, they differ from each other in that each gamete carries a somewhat different set of information in its chromosomes. When two gametes fuse during sexual reproduction, each contributes its own particular genes to the zygote. The result is a new combination of genes that brings about variations within the species.

isos = equal

hetero- = different

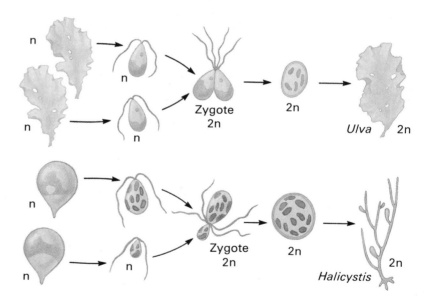

16–13 Isogametes and heterogametes. In one stage of its reproductive cycle, *Ulva*, or sea lettuce, produces isogametes, which fuse to form a diploid zygote. *Halicystis*, an alga, produces heterogametes during part of its reproductive cycle.

Egg and Sperm Production

Complex animals such as vertebrates produce gametes by a process called **gametogenesis** (guh-MEET-uh-JEN-uh-sis). Gametogenesis consists of meiosis followed by specialization, or **differentiation**, of the cells into eggs or sperm.

Section 16.3 Sexual Reproduction

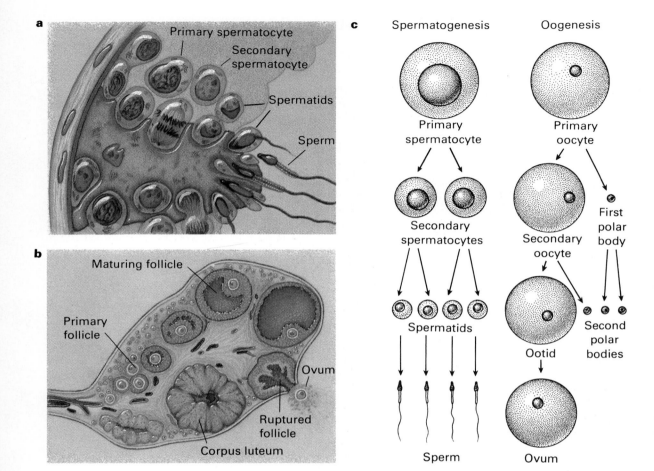

16–14 Spermatogenesis and oogenesis. **a.** In the testis, haploid sperm are formed. **b.** In the ovary, a haploid egg, or ovum, is formed. **c.** The production of ova and sperm occurs by meiosis differentiation.

ovum = egg

The process of sperm production in male animals is called **spermatogenesis** (spur-MAT-uh-JEN-uh-sis). Spermatogenesis takes place in specialized reproductive organs called **testes**. During spermatogenesis, diploid cells in the testes undergo meiosis, each forming four small haploid cells of equal size. Each cell then differentiates into a sperm. A sperm has a tail and other specialized structures that enable it to swim to the egg.

Female animals produce eggs, or **ova** (Ō-vuh), in specialized organs called **ovaries**. The process of egg formation in female animals is called **oogenesis** (ō-uh-JEN-uh-sis). Like spermatogenesis, oogenesis consists of meiosis followed by differentiation. A diploid cell in the ovary undergoes meiosis to produce four haploid cells. Unlike spermatogenesis, the resulting cells are not the same size. One large cell, the egg, and three smaller cells, or **polar bodies**, are formed. The polar bodies die and only the egg survives.

Types of Parents

The usual pattern in sexual reproduction is that males produce sperm and females produce eggs. Male and female parents are usually different in appearance and structure. In some cases, however, parents are not differentiated into male and female. In organisms that produce isogametes—bread mold, for example—the parent organisms look alike. Because they are so similar structurally, they are referred to as (+) or (−) strains.

In a few animals, such as certain parasitic worms, the same individual is able to produce both eggs and sperm. The sperm produced by these animals fuse with their own eggs. In other animals, such as earthworms, although a single organism produces both sperm and eggs, fertilization does not occur between eggs and sperm from the same worm. Instead, two individuals come together and exchange sperm. In this way, the eggs of one earthworm are fertilized by the sperm of another earthworm. In these cases it is not possible to apply the terms *male* and *female*.

An organism that has both male and female reproductive organs is known as a hermaphrodite.

16–15 Fertilization in earthworms. The pairing earthworms secrete a thick mucus, which aids in the transfer of sperm between both individuals. This process may last two hours.

Fertilization

Fertilization and crossing-over provide variation.

The joining, or fusion, of a sperm with an egg is called **fertilization**. When an egg and sperm combine, a single-celled zygote is formed. Each gamete contributes its own set of genes to the zygote. Therefore, the zygote contains a unique combination of the genes in each parent.

In some organisms, the sperm and egg fuse outside the body of the female. This type of fertilization is called **external fertilization**. External fertilization is common among animals living in the water, including most fish and amphibians. During external fertilization, the male and female mating partners release sperm and eggs into the water at the same time. Many of the sperm and eggs released into the water are washed away or eaten. Species with external fertilization survive because large

16–16 Types of fertilization. **a.** External fertilization of frog eggs depends on coordination of the release of male and female gametes. **b.** Internal fertilization in green anole lizards takes place during a mating embrace.

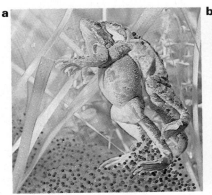 

Production of large numbers of gametes is an adaptation that increases the survival chances of species.

numbers of gametes are released by the parents. A female codfish, for example, lays over 6 million eggs at one time. The male produces many more times that number of sperm. Such large numbers of gametes increase the chances that fertilization will occur in spite of hazards in the environment.

In other organisms, the sperm and egg fuse inside the body of the female. This is known as **internal fertilization**. During internal fertilization, the male delivers sperm cells into the female's body, where a watery environment provides sperm cells with a way to survive and swim to the egg. Although some aquatic organisms practice internal fertilization, it is most common in land animals. Internal fertilization is an adaptation that increases the survival chances of species that live on land.

Internal fertilization increases the chances of a sperm meeting an egg by decreasing the likelihood of the sperm drying out or becoming lost.

Parthenogenesis

parthenos = virgin
-genesis = origination

The development of an organism from an unfertilized egg is called **parthenogenesis** (PAR-thuh-nō-JEN-uh-sis). Although it is not common, parthenogenesis does happen in nature. In bees and wasps, for example, fertilized eggs develop into females; unfertilized eggs develop into males. Most organisms that are produced by parthenogenesis have the haploid number of chromosomes in their body cells. In earthworms, on the other hand, parthenogenesis occurs in specialized diploid eggs.

The information obtained in these experiments on parthenogenesis, as well as in Gurdon's experiments on frog clones, supports the idea that a gamete contains all the information needed to reproduce an organism.

In most species, parthenogenesis does not occur naturally. Scientists are able to induce parthenogenesis in the laboratory, however, using methods that imitate the action of a sperm entering an egg. Various chemicals, temperature shocks, and electric shocks cause the unfertilized egg to act as if it had been fertilized by a sperm. The egg begins to develop and eventually forms an offspring. Artificial parthenogenesis has been accomplished in sea urchins, frogs, hens, and rabbits.

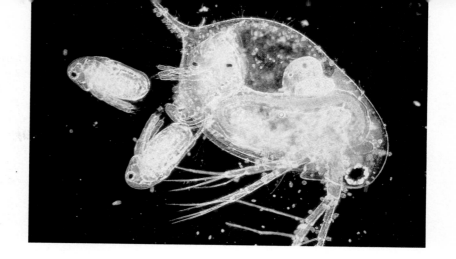

16–17 Parthenogenesis is one means of reproduction in *Daphnia*, the water flea. The brood is seen emerging from the female parent.

Building a Science Vocabulary

sperm	spermatogenesis	fertilization
egg	testis	external fertilization
isogamete	ovum	internal fertilization
heterogamete	ovary	parthenogenesis
gametogenesis	oogenesis	
differentiation	polar body	

Questions for Review

1. What is the difference between isogametes and heterogametes?

2. How many eggs are produced from one diploid cell during oogenesis? How many sperm are produced from one diploid cell during spermatogenesis? Explain the difference in number.

3. For what type of environment is internal fertilization adapted? For what type of environment is external fertilization adapted?

4. What adaptation in fish reproduction helps overcome the loss of gametes during external fertilization?

5. What is parthenogenesis? Give an example of natural parthenogenesis.

Chapter 16
Summary and Review

Summary

1. Organisms reproduce by sexual and asexual reproduction. Asexual reproduction involves one parent. Sexual reproduction involves the fusion of gametes—usually from two parents—to form a zygote.

2. Types of asexual reproduction include binary fission, budding, spore formation, and fragmentation. All the offspring produced by one parent during asexual reproduction are called a clone.

3. J. B. Gurdon's experiments with clones of frogs support the idea that all body cells contain the information to reproduce an organism.

4. Offspring produced by sexual reproduction show more variation than offspring produced by asexual reproduction.

5. Gamete production occurs by meiosis, which results in the production of cells, that have the haploid number of chromosomes.

6. Crossing-over, which occurs during the first cell division of meiosis, increases variations within a species by creating new combinations of genes on chromosomes.

7. Gametes may be isogametes, which look alike, or heterogametes, which look different. Gametes have chromosomes with unique combinations of genes.

8. The fusion of gametes to form a zygote is called fertilization and results in a unique combination of genes in the zygote.

9. In higher animals, gametes are produced by gametogenesis, which consists of meiosis and differentiation. The production of sperm in males is called spermatogenesis. Spermatogenesis produces four functioning haploid sperm from one diploid cell. The production of eggs in females is called oogenesis. Oogenesis produces one functioning haploid egg and three polar bodies from one diploid cell.

10. Fertilization may be external or internal. Loss of eggs and sperm during external fertilization is compensated by the large number of gametes produced. These problems are reduced in internal fertilization, which helps adapt organisms to life on land.

11. Parthenogenesis is the development of an organism from an unfertilized egg. It occurs naturally in a few species but can be induced artificially in several species.

Review Questions

Application

1. A cell in a certain animal's ovary has a diploid number of ten. How many chromosomes are present in that cell during each of the following stages of meiosis?
   a. Prophase I   b. end of Telophase I
   c. Prophase II   d. end of Telophase II

2. Certain plant growers clone orchids in order to sell them. What advantage does the cloning of orchids have for a plant grower?

3. Seedless grapes are reproduced asexually. Why is a crop of seedless grapes in greater danger of being wiped out by disease than a crop that reproduces sexually?

4. Draw a diagram of the cells that would be formed by each of the following in a cell with a diploid number of two.
   a. first meiotic division
   b. second meiotic division
   c. mitosis and cell division

5. What advantage does the overproduction of sperm cells have for an animal species?

6. Why are bees formed by parthenogenesis always haploid?

7. A young child found an earthworm and asked whether it was a male or a female. How would you answer this question?

Interpretation

1. A student had a fish tank containing 28 guppies. A thermostat in the tank kept the water temperature between 20°C and 30°C. Most guppies survive within that temperature range. One night, the thermostat stopped working, and the water temperature dropped below 20°C. In the morning, all but two of the fish were dead. One of the surviving fish was a male; the other was a female. In time they reproduced and replenished the population.
    a. How does this incident illustrate each of the following?
       (1) variations
       (2) advantage of variations to a species
    b. How do variations come about within a population of guppies?

2. An earthworm enters a pile of leaves and reproduces by parthenogenesis. What advantages does this form of reproduction have for the species?

Extra Research

1. Make a chart comparing mitosis and meiosis in a cell with a diploid number of four. Draw a picture of the stages in each process. Show the number and kind of chromosomes present in each stage.

2. Science fiction writers have written stories about imaginary societies in which cloning of human beings has occurred. Pretend that you are going to write such a fictional story. Make a list of the dangers that you think might happen as a result of human cloning. Can you think of any benefits? If you can, make a list of them as well.

3. Do you think that all peas in a pod are alike? Empty a pod of its peas. Remove the skin from around each pea. Cut each pea in half and measure the diameter of the cut surface with a ruler. Record your results. Is there variation in pea size? Under what circumstances might variation in pea size benefit the species of pea plant you are observing?

Career Opportunities

Pest control workers control or eliminate pests in hotels, restaurants, food stores, and private homes. Some are termite specialists. Generally, pest controllers work alone, applying pesticides, setting traps, leaving bait, or using biological controls. On-the-job training is usually provided. Write the Canadian Pest Management Society, Pesticides Division, Plant Products & Quarantine Directorate, Agriculture Canada, Ottawa, Ontario K1A 0C6.

Greenhouse operators grow seedlings, ornamental flowers, and some vegetables. They must have a good knowledge of indoor farming and be able to run their own business. Agricultural training is helpful. Contact the Canadian Nursery Trades Association, 3034 Palstan Rd., Mississauga, Ontario L4Y 2Z6.

Chapter 17

Reproduction, Growth, and Development of Plants

Focus *Plants are necessary to sustain life on earth. We would all die if the supply of plants ran out. Fortunately, plants are a renewable natural resource. Plant reproduction keeps up the supply.*

17.1 ASEXUAL REPRODUCTION IN PLANTS

It is impossible to fully appreciate the value of plants. Plants that lived in the past helped form the fossil fuels that now supply power to our civilization. Plants that are alive today give us oxygen and food. They hold down and build soil. They provide materials for our homes, clothing, buildings, tools, and machinery. They are also a source of beauty.

As the human population continues to increase in size, we need more and more plants for food. Agriculture provides millions of tons of food each year. Figure 17–2 shows the yield of 30 major crops. The demand for greater output from these crops is still increasing. Agricultural researchers try to help farmers increase the quality and amount of crops produced. They study plant reproduction, growth, and development. In this chapter, you will learn how the plant world replenishes itself by reproduction.

17–1 Rice paddies flourish on terraced hills in Indonesia, a country which produces the world's third largest rice crop.

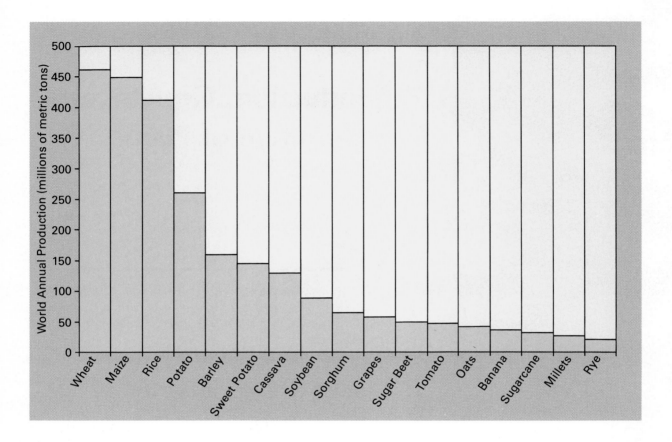

17-2 Seventeen major food crops, showing world annual production (1980) in millions of metric tons (1 metric ton equals 1000 kg).

Spider plants, which are popular house plants, also reproduce by runners.

Vegetative Propagation

Plants reproduce asexually by a process called **vegetative propagation**. In vegetative propagation, a whole new plant is reproduced from a growing portion of one parent plant. Vegetative propagation in plants is similar to asexual reproduction by fragmentation in animals. The parts of plants involved in vegetative propagation are leaves, stems, and roots. Flowers function only in the sexual reproduction of seed plants—not in vegetative propagation.

Vegetative propagation occurs as a natural reproductive process in many plants. Strawberry plants, for example, produce special stems, or **runners**, that grow, or run, along the surface of the soil. Leafy shoots and roots begin to grow at several points along the runner. A new, individual strawberry plant develops at each of the growing points. The offspring plants become separated when the runners die.

478 Chapter 17 Reproduction, Growth, and Development of Plants

17–3 Vegetative propagation in the strawberry is by runners—long stems that extend from the parent plant and take root to form new plants.

Bananas and pineapples are seedless plants that are commercially reproduced from cuttings.

Artificial vegetative propagation is widely used by houseplant owners, gardeners, and commercial plant growers. One of the most common types of artificial vegetative propagation is making a **cutting**. A new plant can be produced by cutting a piece of stem, the "cutting," from a plant. When the cutting is placed in water or moist sand, it grows roots and leaves. Geraniums, ivy, coleus, and many other plants can be reproduced from cuttings.

Grafting is another type of artificial vegetative propagation. In grafting, a stem cut from one plant, the **scion** (SĪ-un), is attached, or grafted, to a rooted growing plant, the **stock**. In a successful graft, the growing areas of the scion and the stock must come in contact. Flowers and fruit are produced much faster after grafting than after other types of reproduction. Grafting is used to advantage when a plant has desirable flowers or fruit, but a weak root system. A scion with the desired flowers or fruit can be grafted to a strong, healthy stock. Grafting is also used to propagate plants with seedless fruit, such as certain varieties of oranges and grapes. Many new varieties of roses are grafted onto older root stocks.

In grafting, the stock's branches are cut back so that the stock does not bear fruit.

17–4 Types of grafting: **a.** wedge grafting and budding; **b.** two shoots (scions) are wedge-grafted into one stock so that the vascular tissues—xylem and phloem—of scions and stock are in contact.

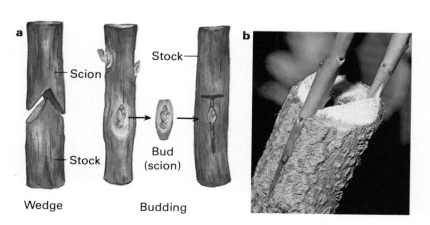

Section 17.1 Asexual Reproduction in Plants

Table 17-1 Types of Natural Vegetative Propagation

Part of Plant Involved	Description	Examples	
STEMS			
1. Unspecialized stems	Stem contacts soil, takes root, and grows new stems and leaves	Raspberry Ivy Blackberry	
2. Runners	Stem specialized for reproduction, creeps along soil, takes root, and grows stems and leaves	Strawberry Spider plant	
3. Stems that store food			
a. Rhizomes	Horizontal underground stem, offspring plants grow along stem	Iris Fern Some grasses	
b. Tubers	Thick stem with buds or eyes, offspring grow from buds	White potato	
c. Bulbs	Thick stem with food-storage leaves, stem grows roots and leaves	Tulip Onion	
ROOTS			
Roots that store food	Thick underground root, develops roots and leaves	Carrot Beet Sweet potato Radish	
LEAVES	Leaf contacts soil, offspring plant develops from leaf surface or edge	Kalanchoe African violet Begonia	

Advantages and Disadvantages of Vegetative Propagation

Artificial vegetative propagation has advantages. It is fast, convenient, and usually successful. As in other types of asexual reproduction, the offspring plants have DNA identical to the parent plant's. Plants with desirable fruit or flowers, or seedless varieties, can be quickly reproduced with little chance of variation.

Both natural and artificial vegetative propagation have a disadvantage. Asexual reproduction results in offspring with almost no variation, which, as you learned in Chapter 16, can help a species survive in a changing environment. Plants reproduced only by vegetative propagation are at a disadvantage in a changing environment. Potatoes, for example, are usually reproduced by vegetative propagation. Between 1845 and 1847, the potato crop in Ireland was attacked and destroyed by a fungus called the potato blight. Without variation in their ability to resist blight, all the potatoes died. Since potatoes were a main source of food for the people of Ireland, many people died of starvation. Others left the country, many coming to North America. If there had been variation among the potatoes, some might have been able to resist the fungus and produce offspring that also could resist the disease and survive.

17–5 A search in a stubblefield turns up a few small potatoes during the famine caused by the potato blight in Ireland. Millions of people left Ireland during the famine.

Building a Science Vocabulary

vegetative propagation	cutting	scion
runner	grafting	stock

Questions for Review

1. Why are plants referred to as a renewable natural resource?
2. What is the difference between natural and artificial vegetative propagation? Give an example of each.
3. What parts of a plant are involved in vegetative propagation?
4. Why is artificial vegetative propagation an advantage to plant growers? Why is it a disadvantage to plant species?

Section 17.1 Asexual Reproduction in Plants

17.2 SEXUAL REPRODUCTION IN PLANTS

Most plants do not reproduce solely by vegetative propagation in nature.

Plants lead a kind of double life. Reproduction in most plants involves a life cycle that consists of two stages. One stage is sexual and the other is asexual. In their sexual stage, plants produce gametes; in their asexual stage, they produce spores. Each distinct stage is called a generation. Since plants alternate between a sexual stage and an asexual stage, their life cycle is referred to as **alternation of generations**. Figure 17–6 represents the life cycle of a plant showing alternation of generations between the two stages.

Alternation of Generations

The Gametophyte A haploid spore undergoes mitosis and develops into a plant called a **gametophyte** (guh-MEET-uh-FĪT). Since the gametophyte is formed by mitosis from a haploid cell, all the cells in the gametophyte are also haploid. Gametophytes produce gametes. Certain haploid cells of the gametophyte undergo mitosis and develop into either eggs or sperm. These gametes unite during fertilization and form a diploid zygote.

The gametophyte produces gametes. The sporophyte produces spores.

The Sporophyte The zygote undergoes mitosis, resulting in the formation of a plant called a **sporophyte** (SPOR-uh-FĪT). All the cells in the sporophyte are diploid. Specialized cells in the sporophyte undergo meiosis, producing haploid spores. A haploid spore is the first cell of the gametophyte generation. The life cycle of the plant is complete and ready to begin again.

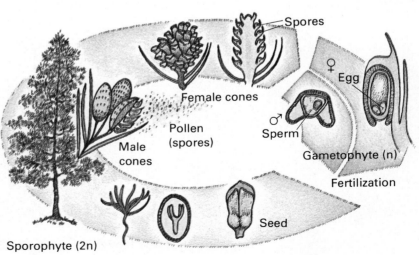

17–6 The life cycle of plants alternates between a sexual stage, the gametophyte, and an asexual stage, the sporophyte. In many plants the gametophyte is very small and almost never seen. In pines, for example, the male and female gametophytes mature inside the female cone.

Chapter 17 Reproduction, Growth, and Development of Plants

Life Cycle of Mosses

If you have ever walked through a forest, you may have observed dense patches of small green plants along the forest floor. Or, you may have noticed mats of green plants covering the banks of rocks along a stream. These soft masses of living green plants are moss. They are actually moss gametophytes. Moss gametophytes are the most noticeable form in the life cycle of a moss. Therefore, the gametophyte is referred to as the **dominant generation** in mosses.

Moss gametophytes are haploid plants that produce eggs and sperm in specialized reproductive organs. Moss gametophytes are usually separate male and female plants. Sperm from the male reproductive organ swims to the egg, which is located in the female reproductive organ. Fertilization occurs when the sperm reaches the egg. Water must be on the ground so that sperm can swim to the egg before fertilization can occur. For this reason, mosses do not survive in dry areas. Fertilization of the egg results in a diploid zygote, which stays in position at the top of the female plant and grows into the sporophyte generation.

The moss sporophyte may look like part of the gametophyte plant, but it is actually a separate organism. Inside the sporophyte, cells undergo meiosis, forming haploid spores. The sporophyte bursts open, releasing thousands of spores into the air. Each spore that lands on the moist ground is capable of growing into a haploid gametophyte moss plant. The life cycle of a moss is then ready to begin again.

Mosses live in or near water or in damp and humid environments.

Mosses build soil, help prevent soil erosion, and provide food for birds and mammals.

17-7 Moss life cycle. The sporophyte (stalk and capsule) grows out of the female gametophyte, the leafy portion of the plant. A moss sporophyte cannot live on its own.

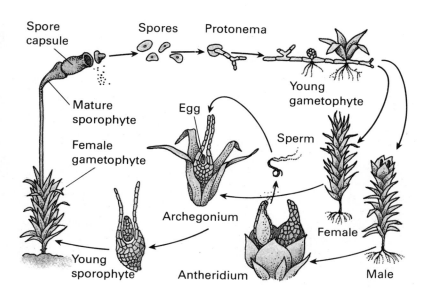

Life Cycle of Ferns

In contrast to mosses, the dominant generation in the life cycle of a fern is the diploid sporophyte. When you think of ferns you probably picture the large, leafy sporophyte plant. This familiar stage of the fern life cycle produces haploid spores in specialized structures that are usually found on the underside of fern leaves. These specialized organs are called **sporangia** (spuh-RAN-jee-uh). Cells inside the sporangia undergo meiosis, forming haploid spores. The spores disperse when the sporangia burst open and shoot the spores into the air. Each spore that lands on the ground is capable of growing into a fern gametophyte.

The fern gametophyte is very small—often smaller than a dime. It is flat, heart-shaped, and grows on moist ground. The haploid gametophyte produces male and female reproductive organs, which produce eggs or sperm. Sperm, produced by the male reproductive organ, swim to the eggs, which are contained in the female reproductive organ. The sperm swim to the eggs in water collected on the ground. After fertilization, the zygote grows and develops into the large, easily recognized sporophyte fern. As the sporophyte grows, the gametophyte gradually disintegrates.

17–8 At certain times of the year colored, beadlike structures appear on the underside of fern leaves (**a** and **b**). These structures are called sori. Each sorus contains numerous sporangia. **c.** Details of the fern life cycle.

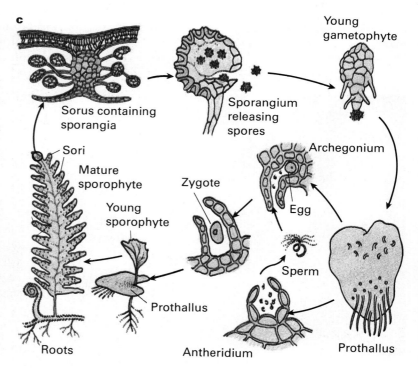

Chapter 17 Reproduction, Growth, and Development of Plants

Life Cycle of Seed Plants

Seed plants are the most successful of all the plant groups. They have three important advantages for reproduction. The first is that they do not depend on water that is on the ground for sperm to swim to the egg. The second is their remarkable reproductive structure, the seed. The third is their symbiotic relationships with animals that help in their fertilization and dispersal.

In Chapter 2 you read that there are two main groups of seed plants—gymnosperms and angiosperms. Gymnosperms do not produce flowers, and their seeds are not enclosed in fruit. In most of them, the seeds develop on scales in cones. Almost all gymnosperms are evergreens or conifers. Pine, spruce, fir, and the giant California redwood are examples of gymnosperms. Angiosperms produce flowers and have seeds enclosed in a fruit.

We will use the flowering plants, the angiosperms, as an example to explain the life cycle of a seed plant. Some differences exist between the life cycles of angiosperms and gymnosperms, but the basic plan is the same.

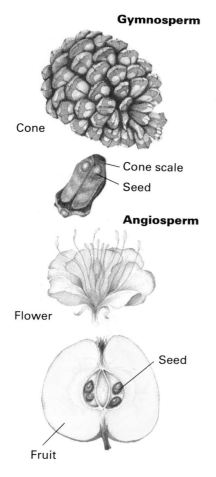

17-9 Seed types. The seeds of gymnosperms lie exposed on cone scales. The seeds of angiosperms are enclosed in fruits.

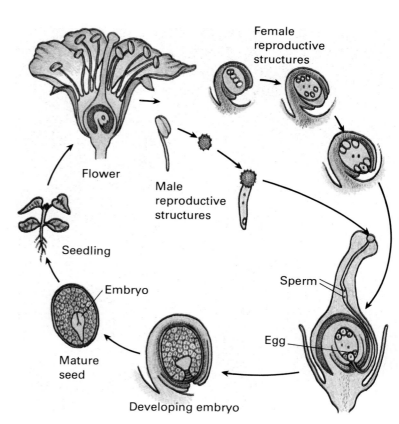

17-10 General life cycle of an angiosperm.

Section 17.2 Sexual Reproduction in Plants

In some plant species, imperfect flowers of both sexes grow on the same plant. In other species, they grow on separate plants.

17-11 Flower types: **a.** wood lily, a simple flower; **b.** primrose flowers form an umbrella-shaped cluster called an umbel; **c.** spike flowers of English laurel; **d.** structure of a perfect flower.

Structure and Function of the Flower Flowers are the specialized reproductive structures of the angiosperms. They grow on the sporophyte plant, and their primary function is to produce spores. Flowers contain male and female reproductive organs. A **perfect flower** contains both male and female reproductive organs. An **imperfect flower** has either male or female reproductive organs—but not both.

The base of a flower is called the **receptacle**. The outer ring of flower parts is the **calyx** (KAY-liks), which is composed of individual leaflike parts called **sepals** (SEE-pulz). The sepals, which are usually green, help protect the flower when it is a bud. They also support the rest of the flower when it is in bloom. Inside the calyx is the **corolla** (kuh-RŌL-uh), which is made up of the **petals**. Often the petals (and sometimes the sepals) are brightly colored and may produce special odors and nectar.

The parts of the flower that take part directly in reproduction are in the center. The male reproductive organs are the **stamens** (STAY-munz). The number of stamens varies from species to species, but all are composed of two main parts. One part is the stalk, or **filament**; the other part is the **anther**, which produces the male reproductive cells. The female reproductive organ is the flask-shaped **pistil**. The pistil has three basic parts. The top is the **stigma**, which is often sticky or hairy. Below the stigma is a slender stalk called the **style**. Below the style is the **ovary**, which produces the female reproductive cells.

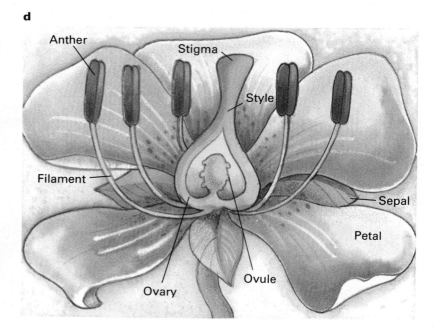

Chapter 17 Reproduction, Growth, and Development of Plants

All the trees and flowering plants that you can see are sporophytes.

The Sporophyte As in mosses and ferns, the sporophyte of a seed plant is diploid. The sporophyte is the dominant generation in the life cycle of a seed plant. In flowering plants, the sporophyte is made up of roots, stems, leaves, and flowers.

The flowers of the sporophyte produce two types of spores. The male reproductive organs, the stamens, produce haploid male spores called **pollen**. Pollen is produced from specialized diploid cells in the anther. Each diploid cell undergoes meiosis, forming four haploid spores. Each spore can grow into a male gametophyte. During growth, its nucleus divides by mitosis and forms two nuclei. A hard outer wall develops around it. The male spore with its two nuclei and protective wall is a **pollen grain**. The pollen grain is the male gametophyte.

The female reproductive organ, the pistil, produces haploid female spores. Female spores are produced inside an **ovule** within the ovary. An ovary can contain one or more ovules. A specialized diploid cell within the ovule divides by meiosis, forming four haploid spores. Three of the four spores die. The nucleus of the remaining spore divides three times by mitosis, resulting in eight haploid nuclei. Cytoplasmic division occurs and seven cells are formed. One cell has two haploid nuclei. The group of seven cells is called the **embryo sac**, which functions as the female gametophyte.

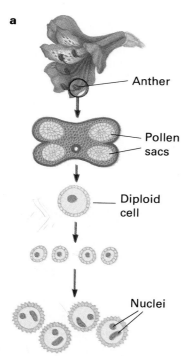

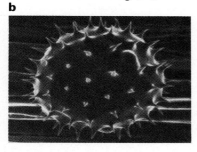

17–12 Pollen grains are produced from diploid cells within the pollen sacs in the anther (**a**). In many species, the tough outer wall of the pollen grain has spines, as in the marigold (**b**).

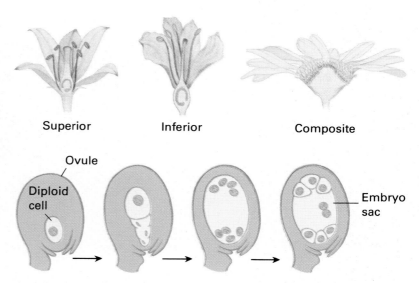

17–13 Ovary position (above) and ovule development (below). A diploid cell in the ovule develops into an embryo sac containing seven cells.

Section 17.2 Sexual Reproduction in Plants

The Gametophyte The gametophyte generation in seed plants is microscopic. Most people do not even know it exists. The male gametophyte, the pollen grain, must be released from the anther before it continues to develop. After it is released, it comes in contact with the stigma of the flower. This transfer of the pollen grain from the anther to the stigma is known as **pollination**. Once pollination occurs, one nucleus within the pollen grain divides by mitosis, forming two haploid sperm. The other nucleus within the pollen grain divides and forms the **pollen tube**. The pollen tube grows down the style of the female reproductive organ and enters the bottom of the ovary. The sperm move down the pollen tube to the egg.

The female gametophyte, the embryo sac, is made up of seven cells. One cell in the center of the embryo sac has two nuclei. The other six have one nucleus each. One of the cells at the bottom of the embryo sac becomes the egg. Fertilization occurs when one of the sperm fuses with the egg, and a diploid zygote is formed. The zygote is the first cell of the sporophyte generation. The other sperm fertilizes the two nuclei in the large cell in the center of the embryo sac. The result is one large cell with a nucleus that contains three haploid sets of chromosomes. This cell develops into **endosperm**, which is stored food that will be used by the developing zygote. The process of fertilization by two sperm is called **double fertilization** and occurs only in flowering plants.

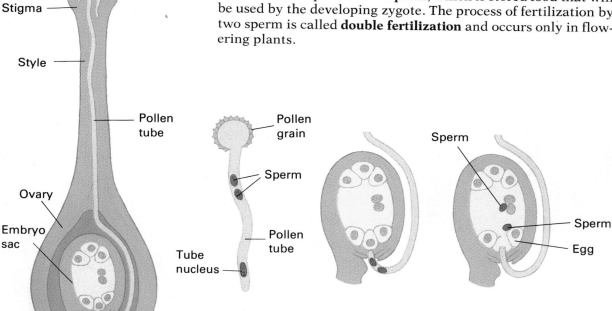

17–14 After a pollen grain is transferred to the stigma, a pollen tube grows down the style and enters the ovary, depositing the sperm. One sperm fertilizes the egg; the other combines with the two nuclei in the center of the embryo sac.

Chapter 17 Reproduction, Growth, and Development of Plants

Building a Science Vocabulary

alternation of generations	sepal	pollen
gametophyte	corolla	pollen grain
sporophyte	petal	ovule
dominant generation	stamen	embryo sac
sporangium	filament	pollination
flower	anther	pollen tube
perfect flower	pistil	endosperm
imperfect flower	stigma	double fertilization
receptacle	style	
calyx	ovary	

Questions for Review

1. Rewrite the following events in the life cycle of a plant to show the order in which they actually occur. Start with *sporophyte grows*.
    a. sporophyte grows
    b. zygote forms
    c. gametophyte grows
    d. fertilization occurs
    e. spores form
    f. gametes form
    g. meiosis occurs

2. What is the dominant generation in the life cycle of a moss? of a fern? of a flowering plant?

3. What are the three important adaptations seed plants have for reproduction?

4. What is the function of each of the following flower parts?
    a. sepals   c. stigma
    b. anther   d. ovary

5. Where are the male and female spores produced in angiosperms?

6. Describe the processes of pollination and fertilization in angiosperms.

Section 17.2 Sexual Reproduction in Plants

17.3 DEVELOPMENT OF FLOWERING PLANTS

Pollination

Pollination is the process in flowering plants whereby the sperm is transferred to the egg. **Self-pollination** occurs when pollen from the anther of one plant falls on the stigma of a flower on the same plant. **Cross-pollination** occurs when pollen from the anther of one plant lands on the stigma of a flower on another plant. Cross-pollination results in variations, which can be an advantage to a plant species.

Some flowers have adaptations that prevent self-pollination. In certain species, the stamens and pistils develop at different times. In others, the shape or position of the flower's parts make self-pollination impossible. Self-pollination is also prevented in species that have imperfect flowers on different plants.

Most cross-pollination is done by wind or insects and other animals. Some cross-pollination, however, is done by people. **Artificial pollination** occurs when a person purposely transfers pollen from anther to stigma. Plant breeders use artificial pollination to develop new varieties of plants. The sweet corn that we eat is an example of a food that was developed with the use of artificial pollination.

Pollination must take place before fertilization occurs.

17-15 Artificial pollination of corn plants is used to develop hardy hybrid varieties. In this photo, workers are covering corn ears to prevent cross-pollination.

17-16 In corn, self-pollination occurs when pollen from male flowers reaches the silk tassels of female flowers on the same plant. Cross-pollination occurs when the pollen reaches the female flowers of another plant.

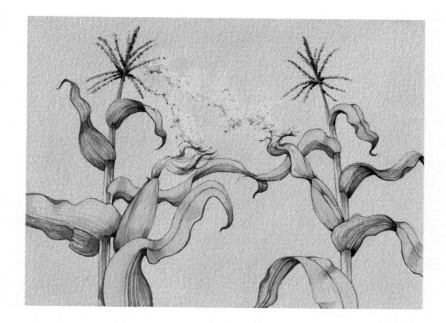

Chapter 17 Reproduction, Growth, and Development of Plants

17–17 Pollen swirls in the wind as it is released from the male cones of the bristlecone pine.

Pollination by Wind Certain flowers are adapted so that their pollen is dispersed by the wind. Their flowers are usually small and odorless and do not attract insects. Instead, long stamens produce great quantities of pollen that is so smooth and light that it travels easily in the wind. Tall pistils in wind-pollinated flowers catch pollen from the air on sticky or feathery stigmas. Corn, for example, is wind-pollinated. The tall tassels at the top of the corn plant are flowers that contain stamens. Wind shakes the pollen out of the stamens and blows it to the corn silk that sticks out of the corn husk. Each thread of corn silk is a style. Pollen lands on the stigmas at the tips of the corn silk. The tall stamens and threadlike styles adapt corn for wind pollination. Other wind-pollinated species include willow trees and walnut trees.

Pollination by Insects and Other Animals Insects pollinate roses, apples, orchids, sunflowers, and many other flowers. Insect-pollinated flowers have brightly colored petals and a sweet odor. They produce a sugary solution called **nectar**. Insects (especially bees), hummingbirds, and even bats are attracted to flowers with these features. When animals gather nectar, they spread the pollen from flower to flower. Animal-pollinated pollen has ridges and is sticky, making it cling to the pollinator. Animal pollination benefits both plant and animal: the animal obtains food; the plant is pollinated.

17–18 A honeybee collecting nectar from a cosmos also helps pollinate the flower by transferring pollen from one flower to another.

Section 17.3 Development of Flowering Plants

Results of Fertilization

Fertilization takes place after pollination occurs. The fertilization of the egg by the sperm begins a new sporophyte generation. The zygote inside the ovule undergoes mitotic divisions to become an immature plant called an **embryo** (EM-bree-ō). The embryo plant develops a stem and root region. Seed leaves, or **cotyledons**, also develop. Cotyledons are immature leaves that store food for the developing embryo. At the same time the embryo is growing, the endosperm also increases in size and becomes a nutrient-rich food supply for the embryo. The walls of the ovule harden to form a protective covering, the **seed coat**.

The **seed** consists of the plant embryo surrounded by food and wrapped in the protective seed coat. The seed coat is watertight and airtight and keeps the embryo from drying out. The seed coat also prevents the embryo from sprouting until water and temperature conditions are right. The embryo in a seed does not grow. It is in a resting or dormant condition in which it uses very little food, oxygen, and water. A seed may last for months or even years in the dormant state. When conditions are right, the seed coat splits open, and the embryo plant grows.

Seeds are usually not set free directly from the parent. They are enclosed in a **fruit**, which is a ripened ovary. For example, the edible parts of a watermelon, peach, and squash are fruits.

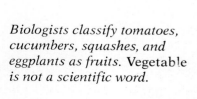

17–19 Seed structure. Cotyledons of the peanut, a dicot, contain stored food, which will nourish the embryo when the seed germinates.

Biologists classify tomatoes, cucumbers, squashes, and eggplants as fruits. Vegetable is not a scientific word.

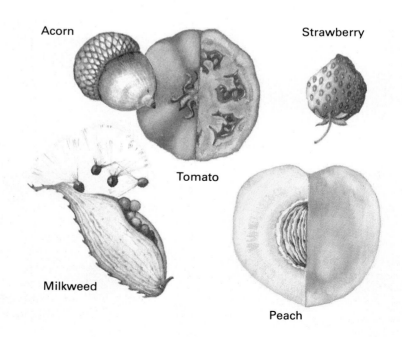

17–20 A variety of fruit types. Fruits both protect seeds and aid in their dispersal.

Chapter 17 Reproduction, Growth, and Development of Plants

Seed Dispersal

Dispersal separates the seed from its parent plant, avoiding competition for light and water between the young plant and its parent.

Seeds, like pollen, can be dispersed by wind and insects and other animals. A tiny dry fruit encloses a dandelion seed. A parachute on the fruit adapts the seed for wind dispersal. Maple seeds, enclosed in dry-winged fruits, are also carried by the wind. Nuts, such as acorns, hazelnuts, and chestnuts, may be dispersed by animals. They are also dispersed by water when they fall into a river or stream. The dry fruits of needlegrass, beggar's-ticks, cocklebur, and wild carrot have barbs or spines. Seeds of these plants are dispersed when their fruits stick to an animal's fur or a person's clothing. Raspberries, cherries, and currants are eaten by birds and other animals. The seeds pass undamaged out of the animal's digestive tract. Human beings also disperse seeds by planting crops.

17–21 Seed dispersal. Both the cedar waxwing and the box turtle are aiding in dispersal of seeds contained in berries.

Seed Germination

The sprouting of a seed is called **germination**. Three things that a seed needs for germination are water, oxygen, and the proper temperature. Germination is usually triggered by a large intake of water. In some cases, a seed may absorb so much water that it becomes 200 times larger than its normal size. When germination begins, the embryo starts actively using its stored food. It grows, develops, and bursts out of its seed coat. Eventually it develops green leaves and starts to make its own food by photosynthesis. Once an offspring plant is capable of photosynthesis, germination is considered over.

Many types of seeds stay dormant for months or years until conditions for germination are right.

Section 17.3 Development of Flowering Plants

17–22 Germination. The two cotyledons of a bean seed (top) are carried above ground with the developing shoot. They gradually wither and fall off. The single cotyledon of a corn seed (bottom) remains underground.

The endosperm holds most of the food stored in a corn seed.

Figure 17–22 shows the germination of a corn seed and a bean seed. Although some differences exist in the way these seeds germinate, their basic growth pattern is similar. Consider the bean. The bean embryo is made up of a shoot with stem and root regions and two cotyledons. The cotyledons nourish the young embryo until it develops green leaves. The root is the first structure to sprout from the seed. As it grows downward, anchoring the plant in the ground, the stem grows upward in an arch. The stem pushes up through the soil and pulls the cotyledons along with it. Green leaves develop, and when they reach the light, photosynthesis begins. The young seedling can then manufacture its own food. The cotyledons eventually dry up and fall off. At the end of germination, the offspring plant has leaves, stems, and roots. The new plant manufactures food in its leaves and obtains water and minerals from the soil through its roots.

Plant Growth and Development

Primary Growth   Growth and development in plants take place in areas of rapid cell division called **meristems** (MEHR-uh-STEMZ). The lengthwise or vertical growth of stems and roots is called **primary growth**. It occurs due to the action of **apical** (AYP-uh-kul) **meristems**, found at the tips of stems and roots. The cells immediately next to the meristem region increase in

meristos = divided

Chapter 17   Reproduction, Growth, and Development of Plants

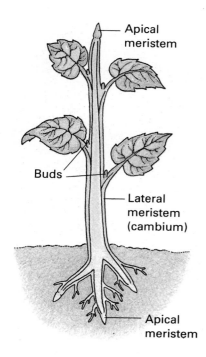

17–23 Apical meristems at the tips of stems and roots produce primary growth. Lateral meristems along stems, branches, and roots produce secondary growth.

17–24 Transverse sections of a stem tip (left) and a root tip (right). The apical meristems are visible as dense areas of small undifferentiated cells that are rapidly dividing.

Each growing season new xylem and phloem cells originate from the vascular cambium.

size by elongation, pushing the stem apical meristem up and the root apical meristem down. Differentiation of cells into various plant tissues occurs next to the area where elongation takes place. A plant grows taller and develops leaves, branches, and flowers from cells formed by the stem apical meristem. A root gets longer and develops branch roots from cells formed by the root apical meristem.

Three of the major tissues formed during primary growth are **primary xylem, primary phloem,** and the **epidermis.** Primary xylem is a system of connected tubelike cells that carry water and help support the plant. In addition to their normal cell walls, primary xylem cells have secondary walls, shaped like spirals or rings, which reinforce the cell and give it strength. When the cells reach maturity, they die. Primary phloem is also made up of tubelike cells. These are still living at maturity, though they lack nuclei. Primary phloem conducts food from the leaves to the other regions of the plant. The epidermis is a thin layer of cells that lines and protects the outside of the plant. Epidermal cells secrete the waxy cuticle that makes most of the plant surface impermeable to water. Certain epidermal cells develop into guard cells on leaves.

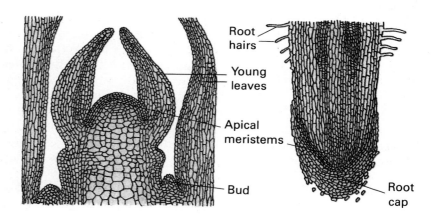

Secondary Growth Most plants that live for more than one growing season continue to grow throughout their life spans and produce woody stems. These plants grow taller by primary growth, but their stems, branches, and roots increase in width by means of **secondary growth.** Secondary growth is accomplished by **lateral meristems,** which are also referred to as **cambium** (KAM-bee-um). Two types of cambium are found: **vascular cambium,** which produces secondary growth of xylem and phloem, and **cork cambium,** which produces **cork,** the protective outer layer of woody plants.

Section 17.3 Development of Flowering Plants

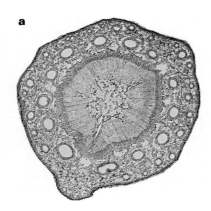
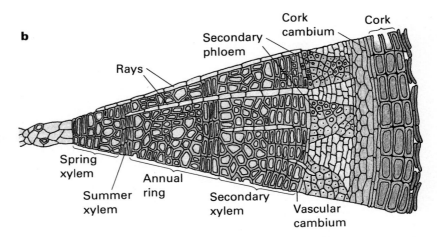

17–25 Secondary growth is produced by the vascular cambium and the cork cambium. **a.** Cross section of a one-year-old pine stem. **b.** Cross section of a three-year-old woody stem, showing the secondary growth tissues.

*A species of oak (*Quercus suber*) is the source of commercial cork. It grows in southern Europe and northern Africa.*

The vascular cambium is a thin layer of cells between the primary xylem and the primary phloem. The cells of the vascular cambium divide, specialize, and become **secondary xylem** and **secondary phloem**. The cells on the inside of the vascular cambium become secondary xylem; those on the outside of the vascular cambium become secondary phloem. More xylem is produced than phloem, pushing the vascular cambium away from the center of the stem. As the vascular cambium moves, it crushes the old layer of phloem cells, which usually have died at the end of the previous growing season. The vascular cambium also produces horizontal conducting tissues called **rays**. These radiate outward from the center of the stem and transport materials laterally from the xylem and phloem.

Water is conducted in much of the xylem. As the tree or shrub ages, however, waste products accumulate in the xylem at the center of the stem, clogging the cells so that no liquids can pass through them. In this way, xylem at the center of the stem becomes woody. Food is transported in the thin layer of phloem near the outside of the stem.

The cork cambium is located near the outside of the stem. It produces the cork, which is impermeable to water, oxygen, carbon dioxide, and other gases. Lenticels in the cork enable water and gases to be exchanged with the interior of the plant.

The tissues outside of the vascular cambium—the secondary phloem, crushed phloem, cork cambium, cork, and layers of dead cork tissue—are collectively referred to as the **bark**. The inner bark is made up of the living tissues. The outer bark is made up of the dead tissues.

Growth of Wood Wood is secondary xylem tissue. In areas with temperate climates, the growing season usually lasts from spring through early fall. In the spring and early summer, when water is plentiful, the vascular cambium produces large xylem

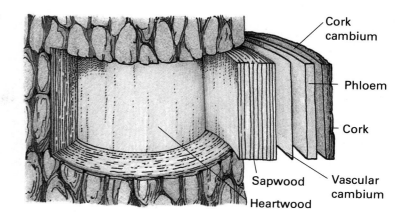

17–26 Wood. **a.** The annual rings of an acacia stem record the growth of xylem tissue. Note the contrast between the dark heartwood and the light sapwood. **b.** A dissected tree shows the arrangement of the secondary tissues.

A tree can continue to live after its heartwood has burned or rotted away, though it is weakened structurally.

Biennials develop roots, stems, and leaves during the first year. During the second year they produce flowers and seeds.

cells. In the later summer and early fall, when less water is available, fewer and smaller xylem cells are formed. As a result, spring wood appears lighter in color than summer wood, and the growth of the tree is marked by alternating light and dark rings, called **annual rings**. It is possible to determine the approximate age of a tree by counting its rings. It is also possible to learn something about the growing conditions that existed in previous years by observing the rings. A wide light band indicates plenty of spring rain and favorable growing conditions. A narrow band indicates poor conditions, perhaps caused by a drought or insect or fungal damage to the tree.

The older xylem toward the center of a tree becomes filled with resins, gums, tannins, and other waste products that give the wood a dark-reddish color. This hard, dark wood is known as **heartwood** and helps support the tree. The lighter-colored, conducting wood is called **sapwood**. With each year of new growth, some sapwood is changed into heartwood.

The Life Span of Plants

Plants that live for only one growing season, such as corn, geraniums, and sweet peas, are **annuals**. Others, such as carrots and beets, are **biennials** (bī-EN-ee-ulz). They live for two growing seasons. Still others, **perennials** (puh-REN-ee-ulz), live for more than two growing seasons. Some of them live for many years. Oak trees, irises, and rosebushes are perennials.

Large differences exist in the life spans of different plants. Some annuals live only a few weeks. On the other hand, one bristle-cone pine tree in California is the oldest tree known at 4500 years of age. Whatever its individual life span, eventually a plant ages and dies. But as you have seen, its species survives by means of reproduction.

Building a Science Vocabulary

self-pollination	secondary growth
cross-pollination	lateral meristem
artificial pollination	cambium
nectar	vascular cambium
embryo	cork cambium
cotyledon	cork
seed coat	secondary xylem
seed	secondary phloem
fruit	ray
germination	bark
meristem	annual ring
primary growth	heartwood
apical meristem	sapwood
primary xylem	annual
primary phloem	biennial
epidermis	perennial

Questions for Review

1. What happens to the ovary of a flower after double fertilization? What happens to the ovules?

2. What three conditions are necessary for seed germination to occur?

3. Describe the germination of a bean seed.

4. Compare primary growth and secondary growth in plants. Where does each occur? What tissues are formed by each?

5. What is wood composed of? Why does wood from the center of a tree look different from the wood near the outside of a tree?

6. Most plants that people grow in vegetable gardens live for only one growing season. What types of plants are these in terms of their life spans?

Perspective: A New View
What Good's A Plant?

So often we take things for granted until we stop to think about them. Plants are like that. They are all around us but most often we tend not to realize how important they are to us.

Plants are an important source of food. Carrots, radishes, and beets are roots. White potatoes, asparagus, onions, garlic, and bamboo shoots are stems. Lettuce, spinach, cabbage are leaves, as are many spices such as basil, bay leaf, and oregano. And don't forget tea. Broccoli, cauliflower, and artichokes are flowers. Apples, peaches, and all the berries and nuts are fruits, and actually, so are peas, tomatoes, beans, and cucumbers, although most people call them vegetables. The most important basic food around the world—bread—comes from grain, which is a seed, as is rice.

Plants are used to make many kinds of cloth and rope. The white fibers that surround cotton seeds are used for clothing, towels, sheets, and other products. The thick, coarse fibers in the stems of hemp plants are used in making rope.

Plants are the main source of free oxygen. If it were not for the process of photosynthesis there would not be any major amount of oxygen in the world. Without oxygen we wouldn't be here either.

Plants are a source of many medicines. In fact, many years ago it was widely believed that the shape of a plant indicated what it was to be used for. Walnut meats, which resemble the cerebrum of the brain, were used to treat mental illnesses. Liverwort, whose leaves look somewhat like the liver, were used to treat ailments of the liver. Plants with bladder shaped leaves were used to treat bladder problems. Plants with red juices, like bloodroot, a spring flower, were used to treat diseases of the blood.

While most of these uses have not proved of actual value, plants do give us many medicines. Cortisone, used in the treatment of arthritis, is now produced from Mexican yams. Cascara (a laxative), digitalis (a heart stimulant), reserpine (used in mental illnesses), and many other modern drugs are derived from plants.

So roots, leaves, and stems provide food, fiber, oxygen and medicine. Indeed, plants are too important to be taken for granted.

Chapter 17
Summary and Review

Summary

1. Vegetative propagation is asexual reproduction by fragmentation in plants. In vegetative propagation a leaf, stem, or root produces an offspring plant.

2. Runners, bulbs, and tubers are structures that take part in natural vegetative propagation. Cutting and grafting are types of artificial vegetative propagation.

3. Vegetative propagation is rapid and convenient, and the results are certain. Scions that produce desirable flowers or fruit, such as seedless varieties, can be grown on hardy stocks. Lack of variation is the major disadvantage of vegetative propagation.

4. The life cycle of a plant involves two generations and is referred to as alternation of generations. The gametophyte generation produces haploid eggs and sperm. Fertilization results in a diploid zygote, which develops into the diploid sporophyte generation. Cells in the sporophyte form haploid spores, which grow into gametophytes.

5. In mosses, the gametophyte is the dominant generation. In ferns and seed plants, the sporophyte is the dominant generation.

6. In flowering plants, the sporophyte flower produces two types of haploid spores. Pollen grows and develops into the male gametophyte, the pollen grain, which produces two sperm. The other type of spore is a haploid cell in the ovule, which develops into the female gametophyte, the embryo sac. The embryo sac produces an egg.

7. Double fertilization occurs in angiosperms. The diploid zygote is the beginning of the new sporophyte generation.

8. After fertilization, the ovule becomes a seed, which consists of an embryo, stored food, and a seed coat. The ovary usually becomes a fruit, which surrounds the seed.

9. Seed germination requires water, oxygen, and the proper temperature.

10. Plants grow and develop at special regions called meristems. All plants undergo vertical or primary growth. Plants that live more than one growing season undergo lateral or secondary growth.

11. Wood is formed by secondary growth. The pattern of growth can be observed in the annual rings.

12. Individual life spans in plants vary greatly from species to species. In general, plants are divided into annuals, biennials, and perennials.

Review Questions
Application

1. What is a disadvantage of vegetative propagation? Why do plant growers use this means of reproduction?

2. Write the names and functions of the numbered parts of the flower shown in the illustration.

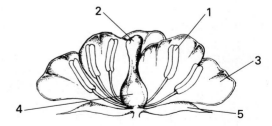

3. A scion of a red rosebush is grafted onto the rooted stock of a white rosebush. What color roses will grow on the grafted branch the following season? Explain your answer.

4. Suppose you wanted to cross a variety of corn having white kernels with a variety having yellow kernels. What steps would you follow in cross-pollinating the two types of corn?

Interpretation

1. A boy and girl carved their initials on a young beech tree trunk one metre up from the ground. They returned ten years later to see where their initials would be. Where do you think they found them? Explain.

2. Refer to the illustration of the cross section of a tree trunk in answering the following questions.

   a. What was the approximate age of the tree?

   b. What might have caused the band labeled *a* to be so much wider than the other light-colored bands?

   c. What might have caused the band labeled *b* to be so much narrower than the other light-colored bands?

Extra Research

1. See if you can determine what effect, if any, the amount of stored food has on the growth of potato buds (eyes). Grow potato buds on moist sand or vermiculite. Include different-sized pieces of potato tuber (stored food) with each bud. Try one with no stored food. Record the growth of each one over a period of time. Present your results in a graph.

2. Grow a pineapple plant from the crown of a pineapple fruit. Slice off the top of the pineapple about two inches below the leaves. Trim off the outer skin and fruit, up to the hard, stringy core. Let the fruit dry out for a few days to prevent rotting. Insert the core in moistened sand or potting mix. Keep the plant in a warm, well-lit location, and mist the leaves regularly with water. As the plant takes root, leaves will come out from the center.

3. Dissect several types of flowers. Locate and count the parts. Decide whether the flowers are monocots or dicots and whether they are adapted for cross-pollination or self-pollination. If they are cross-pollinated, decide whether the flowers are adapted for insect or wind pollination. Record your observations in sketches and notes.

4. Try to grow pollen tubes from pollen. Shake pollen from several kinds of flowers into a saucer filled with a strong sugar-water solution. Cover with plastic wrap, and keep in a warm place for several hours. Observe with a hand lens. Draw and label what you see.

Career Opportunities

Agricultural engineers need a working knowledge of both engineering and agriculture in order to design agricultural equipment and to improve on current farming methods. Many agricultural engineers work for farm equipment manufacturers and distributors or for utility companies. A degree in engineering is required. Contact the Canadian Society of Agricultural Engineering, 151 Slater St., Ottawa, Ontario K1P 5H4.

Nursery operators grow and sell plants. Some have retail stores; others specialize in raising trees, shrubs, or house plants for wholesale. Many run their own businesses. A knowledge of botany and an aptitude for business are necessary. Contact the Canadian Nursery Trades Association, 3034 Palstan Rd., Mississauga, Ontario L4Y 2Z6.

Chapter 18

Reproduction and Development of Vertebrates

Focus *Every organism alive today is part of an unbroken chain. If an organism is sterile or dies before reaching sexual maturity, the chain breaks. Through reproduction and development, organisms keep the chain of life going from one generation to the next.*

18.1 VERTEBRATE REPRODUCTION

Vertebrate reproduction, unlike plant reproduction, does not include alternation of generations. The dominant—and only—generation in vertebrates consists of diploid organisms that reproduce solely by sexual reproduction. Diploid males produce haploid sperm; diploid females produce haploid eggs. The haploid gametes fuse to form a diploid zygote, which grows and develops into an adult organism capable of reproduction.

If you compare reproduction in various vertebrate species, you will observe similarities in structure. In both male and female reproductive systems, reproductive organs are found in pairs. Animal reproductive organs produce haploid gametes and are called **gonads** (GŌ-nadz). A system of tubes connects the gonads to the outside of the body.

The general pattern of vertebrate reproduction and development is illustrated in the first two sections of this chapter using the example of reproduction and development in humans.

18–1 An orangutan infant clings to its mother's back.

Human Reproduction

The human reproductive system is controlled and regulated by hormones. Humans are not able to reproduce until they reach **puberty** (PYOO-ber-tee). At puberty, hormones cause physical changes in the human body that enable the production of gametes to begin. On the average, males reach puberty at about age 14. Once they reach puberty, males produce sperm continuously for most of the rest of their lives. Females reach puberty between the ages of 10 and 14. Once females reach puberty, they begin to develop and release eggs. Females continue to release eggs for only a portion of their life span. Sometime between the ages of 45 and 50 a female's hormone levels change, and she no longer releases eggs. This stage in the female life cycle is called **menopause** (MEN-uh-PAHZ).

The change in hormone levels at puberty also cause changes in the body called secondary sex characteristics. In males, the voice deepens, facial and body hair grows, and the general body shape changes: the shoulders widen and the hips narrow. In females, the breasts enlarge, the hips widen, and menstruation begins. In both sexes, pubic hair and underarm hair begin to grow.

Male Reproductive System   The paired gonads of the male reproductive system are the **testes** (TES-teez). The testes have two functions. They produce sperm, the haploid male gametes, and they produce male hormones, including testosterone. The

puber = adult

All of the immature eggs that a female will develop and release are produced during her own embryonic development.

18–2   Male reproductive system: **a.** human sperm cell, which measures about 60 μm in length; **b.** male reproductive organs.

From TISSUES AND ORGANS: A TEXT ATLAS OF SCANNING ELECTRON MICROSCOPY by Richard G. Kessel and Randy H. Kardon. W. H. Freeman and Company. Copyright © 1979

a

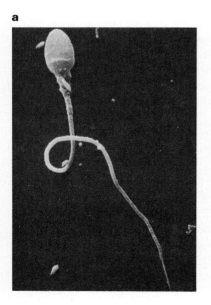

b

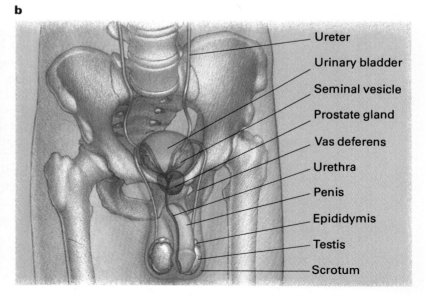

Chapter 18 Reproduction and Development of Vertebrates

The temperature within the scrotum is lower than normal body temperature (37°C), which is too warm for the production of healthy sperm.

A sperm cell is one of the smallest cells in the human body.

testes are located in a sac called the **scrotum** (SKROT-um), which hangs in front of and between the thighs. Each testis contains loops of folded tubes, the **seminiferous** (SEM-uh-NIF-er-us) **tubules**. The total length of the seminiferous tubules in both testes is about 250 m. Cells in the walls of the seminiferous tubules produce sperm.

Sperm are very small cells, each being about 60 μm long. A sperm has a head region, which contains the chromosomes, and a tail region, which helps the sperm swim to the egg. The large surface area of the seminiferous tubules adapts the testes to produce large numbers of sperm. Over 100 million sperm can be released by a human male at one time.

Sperm leave the seminiferous tubules and enter a system of tiny tubules, which lead to a larger duct called the **vas deferens**. Sperm are stored in the tubules and vas deferens until they are released from the body. Glands located near the vas deferens secrete liquids into the sperm. These secretions nourish the sperm, keep them from drying out, and help them reach the egg once inside the female. Together, the sperm and secretions are called **semen** (SEE-mun). The semen moves through the vas deferens into the urethra, which passes through the **penis** (PEE-nis) to the outside of the body.

The urethra also functions to remove urine from the body.

Female Reproductive System The paired gonads of the female are the **ovaries**. An ovary is about the size of a walnut. Each egg is produced near the surface of an ovary in a saclike structure called a **follicle**. A follicle nourishes the egg that devel-

18–3 Female reproductive system: **a.** human egg cell, which measures about 100 μm in diameter; **b.** female reproductive organs.

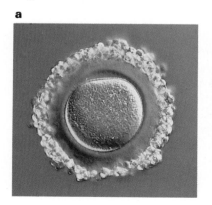

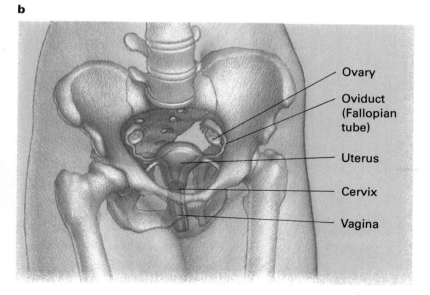

The egg is the largest cell in the human body.

ops within it and also produces the female hormone estrogen. Follicles within the ovary mature at different times. Consequently, at a given time, the follicles are at different stages of development. An ovary of a newborn female contains about 400 000 immature eggs. About 400 of these mature and are released during her lifetime.

An egg cell is much larger than a sperm cell. It is round and about 100 μm in diameter—nearly the size of the period at the end of this sentence. Approximately once a month, during **ovulation**, a mature egg is released from the ovary. The egg enters the **oviduct**, which is a tube located near the ovary. The inside walls of the oviduct are lined with moving cilia, which draw the egg into the oviduct and down toward the **uterus**. The uterus has thick muscular walls and is about the size and shape of a pear. If the egg is fertilized by a sperm, it becomes attached to the wall of the uterus, and development of the zygote into a baby occurs. The period of development lasts about nine months and is known as **pregnancy**. If the egg is not fertilized, it passes through the opening of the uterus, the **cervix** (SER-viks), and out of the body through the **vagina** (vuh-JĪ-nuh).

Refer to Chapter 12 for more information on the menstrual cycle.

In humans, and some other primates, the reproductive cycle of the female is called the **menstrual cycle**. One menstrual cycle takes about 28 days, but variation exists from individual to individual. During a menstrual cycle, an egg matures in the ovary and is released. Over the next three days the egg travels down the oviduct to the uterus. As the egg is traveling down the oviduct the inner lining of the uterus becomes thicker and spongy and filled with blood. These changes prepare the uterus to receive a fertilized egg. If the egg is not fertilized, it passes out of the body, and the thickened lining of the uterus breaks down. The thickened lining and extra blood are shed during **menstruation**, which lasts four to five days.

Fertilization in Humans

Although only one sperm fertilizes an egg, it takes as many as 130 million sperm to ensure fertilization.

Sperm pass out of the penis of the male through the urethra and into the vagina of the female. Using their long tails as flagella, the sperm swim up the female reproductive tract toward the ovary. Fertilization usually occurs in the upper portion of the oviduct. As the head of the sperm penetrates the egg, the tail separates from the rest of the sperm. The nucleus of the sperm unites with the nucleus of the egg, and the zygote is formed. After a sperm has fertilized an egg, a membrane forms around the egg. The membrane prevents any more sperm from entering and fertilizing the egg.

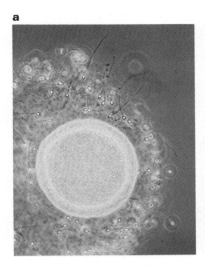
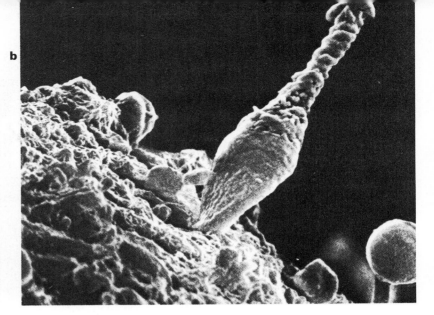

18–4 Fertilization: **a.** egg surrounded by sperm; **b.** sperm touching an egg. As the head of a sperm penetrates the egg cell membrane the tail separates. The head eventually dissolves, and the genetic material of the sperm and egg combine.

Building a Science Vocabulary

gonad	semen	pregnancy
puberty	penis	cervix
menopause	ovary	vagina
testis	follicle	menstrual cycle
scrotum	ovulation	menstruation
seminiferous tubule	oviduct	
vas deferens	uterus	

Questions for Review

1. What is puberty? When does it occur in males? in females?

2. What is the function of each of the following structures?
   a. seminiferous tubules d. oviduct
   b. urethra in males e. uterus
   c. ovaries f. follicle

3. Briefly describe the events of the menstrual cycle.

4. Where does fertilization occur? How many sperm fertilize the egg?

Section 18.1 Vertebrate Reproduction

18.2 HUMAN DEVELOPMENT

The gestation period is the length of pregnancy.

The period of time between fertilization of the egg and birth of the offspring, also known as **gestation** (jes-TAY-shun), lasts about nine months in humans. During this time, the zygote, which is 0.1 mm in diameter, develops into a baby that is approximately 50 cm in length.

Early Stages of Development

After fertilization, the zygote continues moving down the oviduct. The first mitotic cell division of the zygote happens about 24 hours after fertilization and is followed by many more cell divisions. The repeated cell division of a zygote is called **cleavage**. During cleavage the overall size of the embryo does not change; however, the cells get smaller and smaller as they divide.

In a few days, enough cells have divided to form a solid ball of cells called the **morula** (MOR-yoo-luh). The morula is still about the same size as the fertilized egg. The number of cells in the morula increases by cell division as it moves down the oviduct. About three days after the egg is fertilized, the morula enters the uterus.

18–5 Development of a frog embryo: **a.** fertilized egg; **b.** two cell stage; **c.** four cell stage; **d.** eight cell stage; **e.** early blastula; **f.** early neurula stage, formed after the gastrula.

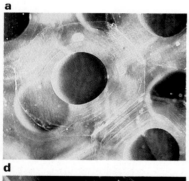
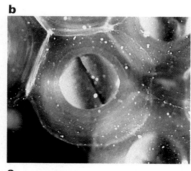

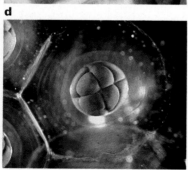
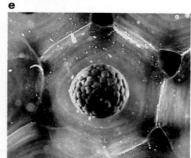

blast = bud or sprout
-ula = little

A blastula stage is found in the development of most animals.

ecto- = outside
derm = skin

endo- = inside

meso- = middle

The morula floats freely in the uterus for a few days. It continues to divide and changes into a fluid-filled ball of cells, the **blastula** (BLAS-choo-luh). The blastula is made up of a single layer of cells. Part of the blastula develops into a disc of cells, which eventually becomes the developing organism, or **embryo**. The rest of the blastula forms membranes that surround and protect the embryo. At about the sixth day, the blastula becomes implanted in the wall of the uterus.

The single-layered blastula folds over to form a pouchlike structure, the **gastrula** (GAS-troo-luh). When first formed, the gastrula is made up of two layers. The outer layer of the gastrula is the **ectoderm** (EK-tuh-DERM). During later development, the ectoderm gives rise to the brain, spinal cord, sense organs, and skin. The inner layer is the **endoderm** (EN-duh-DERM). The endoderm gives rise to the lining of the digestive system and the lungs. Eventually a third layer, the **mesoderm** (MES-uh-DERM), forms between the endoderm and ectoderm. The mesoderm gives rise to the skeleton, muscles, transport system, excretory system, and sex organs.

Membranes Around the Embryo At about eight days, three membranes begin to form around the embryo. The first membrane to form is the **chorion** (KOR-ee-AHN). The chorion is the outermost membrane surrounding the embryo. It develops fingerlike projections that push deep into the wall of the uterus, coming into close contact with the blood vessels of the mother. The fingerlike projections become the embryonic portion of the **placenta**, an organ in the uterus where the exchange of food, gases, and other materials between the embryo and the mother takes place. The second membrane to form is the **amnion** (AM-nee-un). The amnion is filled with a liquid called **amniotic fluid** that provides a watery environment in which the embryo develops. The amniotic fluid cushions the embryo and prevents injury. The third and last membrane to form is the **yolk sac**. The yolk sac provides food to the developing embryo. In humans, the yolk sac is small and usually disappears during the second month of development. At that time the function of nourishment is taken over by the placenta. In most vertebrates, however, the yolk sac is large and is a main source of food for the embryo.

The Placenta The placenta grows around the spot where the tiny embryo attaches to the uterus. The embryonic part of the placenta is the fingerlike projections of the chorion and blood vessels of the embryo. The maternal part of the placenta is made up of blood vessels in the wall of the uterus. The blood vessels of the embryo and mother are very close together; however, the bloodstream of the mother does not connect directly to the

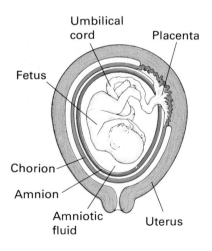

18–6 Embryonic membranes. The fetus is surrounded by the amnion and the chorion. The placenta is the site of exchange of gases, food, and wastes between the mother and the fetus.

Section 18.2 Human Development

bloodstream of the embryo. The embryo receives food and oxygen from the mother by diffusion through the placenta. Carbon dioxide and other wastes are removed by diffusion through the placenta. The embryo is attached to the placenta by the **umbilical** (um-BIL-ih-kul) **cord.** The navel is the scar that shows where the umbilical cord was attached. As the developing embryo increases in size, so does the placenta.

In 1924 Hilde Mangold and Hans Spemann discovered that embryonic induction accounts for the development of the nervous system in salamanders.

Embryonic Induction Different tissues in the embryo come in contact and interact during development. **Embryonic induction** is the process whereby one tissue in an embryo contacts another and affects its development. The process depends on the passage of a chemical substance from one tissue to another. For example, embryonic induction works to form lenses in the embryo's eyes. Two masses of tissue, known as optic cups, grow out of the brain toward the skin. Substances produced by the optic cups induce nearby skin cells to differentiate into lenses.

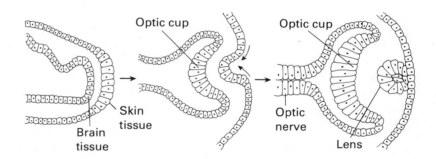

18–7 Embryonic induction, showing formation of the optic cup and lens.

Formation of the Primitive Streak At 11 to 13 days, the ectoderm develops a bulge, known as the **primitive streak.** The primitive streak is a kind of marker that separates the right side of the embryo from the left side. The primitive streak later develops into the spinal cord. Once the primitive streak forms, the embryo can be easily seen to be a vertebrate.

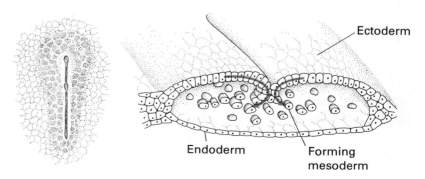

18–8 The primitive streak, seen from above (left) and from the side in cross section (right), is formed by infolding of the outer cell layer. The cells inside will form the mesoderm.

Later Development

For convenience in describing what occurs during pregnancy, it is often divided into three periods, or trimesters, each lasting three months.

The developing embryo is very sensitive to drugs, medicines, diseases, and X-rays.

First Trimester The events of the first two weeks of development, described above, mark the beginning of the **first trimester**. By the end of the first two weeks, all major parts of the body have begun to form. The developing human is considered an embryo until about the eighth week after fertilization. At that time the embryo is recognizably human, with clearly defined facial features. After the eighth week, it is called a **fetus** (FEET-us). At the end of the first trimester, the average fetus weighs about 30 g and is 9 cm long. The heart is beating, toenails and fingernails are beginning to grow, and the fetus is able to move its arms and legs slightly.

A six-month-old fetus could survive outside the mother's body with special medical attention.

Second Trimester The fetus undergoes tremendous growth during the **second trimester**. It increases in weight from about 30 g to nearly 700 g and in length from 9 cm to about 35 cm. The bony skeleton begins to form, and the heartbeat becomes strong enough to hear with a stethoscope. The fetus becomes stronger, and the mother is more able to feel the stretching movements of the arms and legs. The skin appears red and is very wrinkled.

a

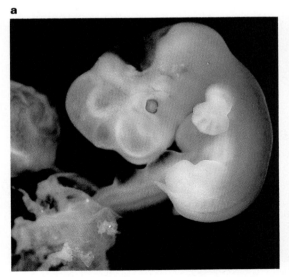

b

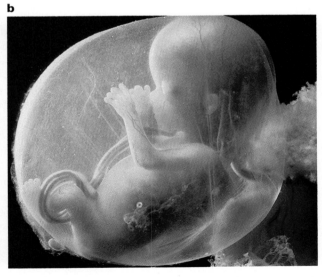

18–9 Human development: **a.** five-week-old embryo (first trimester); **b.** four-month-old fetus (second trimester).

Section 18.2 Human Development

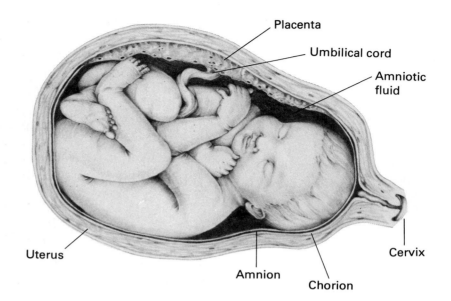

18–10 At the end of the third trimester the fetus is closely confined within the uterus. In 95 percent of all births, the head of the fetus is pointed toward the cervix.

During pregnancy a woman typically gains about 11 kg.

Throughout pregnancy the use of alcohol, tobacco, and caffeinated beverages may damage the tissues of the fetus.

The normal gestation period is 38 to 40 weeks, but it can vary from 28 to 45 weeks.

Third Trimester During the **third trimester**, the fetus increases greatly in size. Most of its gain is in weight (from 700 g to nearly 3000 g); yet, the fetus does increase in length (from 35 cm to 50 cm, on the average). Most of the weight is gained by adding fatty deposits under the skin. Further growth and development of all body parts occur during the final three months of pregnancy. The most crucial development, however, occurs in the nervous system, where nervous tissue, especially in the cerebral cortex, is formed. It is vital that the mother be well-nourished during this period so that the fetus has enough of the proteins and lipids necessary for nervous tissue formation.

Birth

The process of childbirth is known as **labor** and can last from slightly longer than two hours to more than 20 hours. Labor is usually longer with the first-born child than with later children. The onset of labor is controlled by hormones. At the beginning of labor, an increase in the estrogen level and a decrease in the progesterone level occur in the blood of the mother. Apparently, this change in hormone levels is signalled by a chemical messenger produced by the fetus.

During the first part of labor, rhythmic contractions of smooth muscles in the wall of the uterus gradually push the fetus and amniotic sac down. This causes the opening of the uterus, the cervix, to dilate or become larger. The contractions are mild at first and occur at 20 minute intervals. Toward the end of the first part of labor, they are much stronger and occur as often as every minute. The cervix dilates from 1-2 cm to about 10-11 cm. Eventually, the amniotic sac breaks, releasing the amniotic fluid. This initial stage of labor lasts the longest.

Once the cervix is fully dilated, strong and frequent contractions of the uterus and the abdominal wall force the baby out of the uterus, usually head first. The baby passes through the vagina, which is sometimes referred to as the birth canal, to the outside. The expulsion of the fetus usually takes 30 minutes to 2 hours.

As soon as the fetus is expelled from the uterus, the umbilical cord is tied and cut. The baby is now completely separate from the mother. With its first cries, the baby expels fluid that has been in its lungs since early in its development and fills its lungs with air.

The uterus continues to contract, dislodging the placenta from the uterine wall. The placenta, amniotic sac, blood and other fluids, collectively known as the **afterbirth**, are then expelled through the vagina. This process takes about 15 minutes. Further contractions of the uterus help close off ruptured blood vessels in the uterine wall.

Soon after birth, the mother's pituitary hormones stimulate her breasts to begin producing milk. The milk she produces is easily digested by the baby and provides antibodies against early childhood diseases. Formula is available for infants that are not fed mother's milk. Formula contains cow's milk or soy extract plus necessary nutrients; however, it does not contain antibodies.

18–11 The birth process involves rotation of the head and shoulders of the baby as it passes through the vagina.

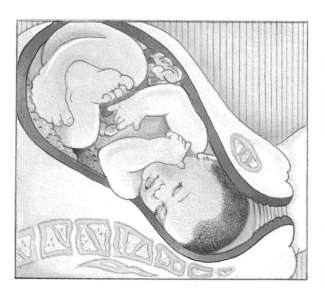
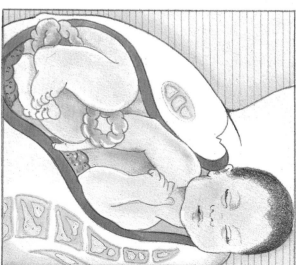

Multiple Births

Although women usually give birth to only one offspring, having more than one baby at the same time is not uncommon. Sometimes more than one egg is released from the ovaries during ovulation. If two eggs are released at the same time and fertilized by two sperm, **fraternal twins** develop. Fraternal twins have different DNA because each has a different set of genes from each egg and sperm. Fraternal twins usually look no more alike than other siblings and can even be different sexes. Fraternal twins also have different placentas. **Identical twins**, on the other hand, develop in a different way. The zygote resulting from the fertilization of one egg undergoes cell division to form two identical cells. These two cells separate, each developing into a separate organism. Since identical twins arise from a single fertilized egg, they have the same genes. Identical twins share the same placenta and are always the same sex.

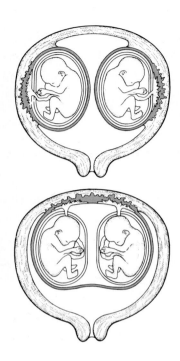

18-12 Fraternal twins (above) develop from two fertilized eggs. Identical twins develop from one fertilized egg.

Growth After Birth

The growth and development of an individual after birth is dependent upon a number of factors, especially nutrition and heredity. A general pattern can be seen, however, among most individuals. Between birth and three years of age, growth is rapid. During this time the baby is almost completely dependent on others for survival. The most significant changes in the infant are a very large increase in height and weight and the

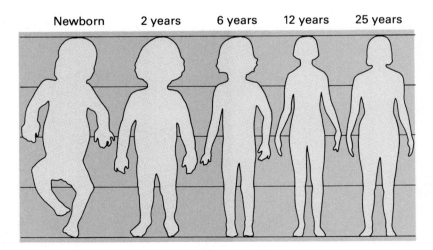

18-13 The proportions of body parts change dramatically from birth to maturity.

further development of the nervous system, enabling the infant to control its body and learn.

After the age of three, the rate of growth levels off until puberty, when there is a sudden growth spurt. Maturation of the reproductive system also normally occurs at this time, as well as the appearance of secondary sex characteristics. The growth rate drops significantly after puberty. By the age of 18, increases in height occur only gradually, if at all.

Building a Science Vocabulary

gestation	chorion	fetus
cleavage	placenta	second trimester
morula	amnion	third trimester
blastula	amniotic fluid	labor
embryo	yolk sac	afterbirth
gastrula	umbilical cord	fraternal twins
ectoderm	embryonic induction	identical twins
endoderm	primitive streak	
mesoderm	first trimester	

Questions for Review

1. Which layer in the gastrula gives rise to each of the following?
   a. skin
   b. lining of digestive system
   c. nervous system
   d. skeleton

2. What is the function of each of the three membranes that form around the embryo?

3. How does one tissue affect another during lens formation in an embryo?

4. Briefly describe a human fetus at the end of each trimester of pregnancy.

5. What is labor? How long does it usually last?

6. How do identical twins differ from fraternal twins?

7. How does a human embryo obtain food?

18.3 ADAPTATIONS IN VERTEBRATE REPRODUCTION AND DEVELOPMENT

All vertebrate species have adaptations that help them reproduce. For example, the ability to produce offspring in large numbers is a safeguard to ensure that a species survives from one generation to the next. This and other safeguards exist as adaptations in structure, function, and behavior.

Adaptations in Vertebrate Reproduction

Invertebrates also produce large numbers of gametes.

Most vertebrates produce more gametes than they need for reproduction. A female codfish lays about 6 million eggs at one time. A male horse produces about 50 million sperm at one time. A male pig releases a record number of 85 billion sperm at one time. Overabundance in the number of sperm and eggs produced is an adaptation that ensures that fertilization will occur.

Internal fertilization is another adaptation in vertebrate reproduction. Direct transfer of sperm from male to female prevents loss of sperm and keeps sperm from drying out. Other vertebrate reproductive adaptations help bring males and females together or aid in the timing of gamete release.

Adaptations that Bring Male and Female Together The behavior of a male and a female during reproduction is called **mating**. For the most part, mating behavior is inherited and not learned. For mating to occur, the male and female of a species must first find each other. Every vertebrate species has adaptations that help bring the sexes together at the right time for mating. Some of these adaptations involve the use of sight, hearing, and smell.

During the spring, which is mating season, the normally bright red color of a robin's breast becomes even brighter.

In some species of vertebrates, males and females find each other by appearance. Male birds of many species have brightly colored feathers, while females do not. These differences in appearance are secondary sex characteristics. During mating season, the plumage of some male birds becomes even more brightly colored. Bright color on a male bird is a signal to other males to stay out of his territory. It is also a sexual signal to a female of his species.

For many species of vertebrates, voice plays an important part in bringing males and females together. A female bird recognizes the special reproductive song of a male of her own species. Her response helps bring them together.

Odor is another important mating signal. For example, when certain female mammals, such as dogs, are ready to mate,

18–14 Mating signals. **a.** The male blue grouse puffs out his neck and thrusts his tail up and forward. **b.** A northern spring peeper inflates his throat sac to amplify the sounds of his mating calls.

Frogs breed early in spring.

their glands produce special substances, which are often deposited with urine. The odor of these substances serves as a chemical signal to attract the males of the species. Males that detect the odor seek out the female.

Some vertebrate species depend on one type of sensory signal more than others. In general, however, vertebrates use most of their senses together in finding mates.

Adaptations of Timing of Gamete Release In addition to finding each other, vertebrate males and females must make their gametes available at the same time. When frogs mate, males clasp females from behind as the females lay their eggs in the water. The males immediately release sperm over the eggs. The behavior of males and females is synchronized, assuring that gametes are released at the same time.

Most vertebrates mate only at certain times of the year. Special breeding periods, controlled by hormones, help synchronize gamete release. In some species, such as deer, goats, and sheep, males undergo a mating period called **rutting**. The

18–15 Rutting. Elk stags lock horns in a struggle for dominance and the first choice of mates.

rutting season for deer occurs in the fall. During the rutting season, the male deer's antlers grow large and his behavior changes. He makes sounds associated only with reproduction and behaves in special ways. Sometimes he fights with other males to determine which is the strongest, or dominant, male. The dominant males then locate females and mate with them.

In humans, the female reproductive cycle is called the menstrual cycle. In most other female mammals, the reproductive cycle is called **estrus** (ES-trus). How often estrus occurs varies from species to species. Female wolves and deer, for example, have one estrus a year; dogs have two. Cows and horses experience estrus about once a month. Rats and mice have estrus every few days. Depending on the species, estrus lasts from a few hours to four weeks. During estrus, a female produces mature eggs that are ready for fertilization. Females avoid males when they are not in estrus. During estrus, however, they search for males and try to mate. Females produce chemical signals to attract males during estrus. As a result of rutting and estrus, males and females are drawn together when reproduction is possible.

Humans do not undergo a special mating period.

Some vertebrate species have a unique set of actions that males and females perform before mating known as **courtship behavior**. Courtship behavior brings a male and female closer together and helps in the timing of gamete release. When stickleback fish breed, for instance, the male lures an egg-bearing female into the nest. He then stimulates her to lay her eggs by prodding the base of her tail. After the female's eggs are laid, the male swims over the eggs and releases sperm to fertilize them.

18–16 Touch plays an important role in courtship behavior. These albatrosses are engaged in a kind of "kiss."

Adaptations that Help Embryos Develop

Embryos cannot survive on their own. They must be supplied with food and protected from drying out. Four different patterns of development can be found among vertebrate species.

External Fertilization, Development in the Water, Food from Yolk Many fish and all amphibians lay their eggs in the water. Males deposit sperm over the eggs. The embryos develop entirely in the water, obtaining their food from yolk in the egg. They are in danger of being injured or destroyed by predators, dry conditions, and other factors.

Some species of frogs, such as the leopard frog, lay their eggs in the water in the early spring. A frog's egg has a white part and a black part. The black part develops into the tadpole. The white part is the yolk. As the eggs are fertilized, a jellylike substance that helps protect the eggs forms around them. The time it takes for a tadpole to develop varies with the temperature of the water. Development is faster in warmer water. A tadpole hatches out of its egg between 8 and 20 days after fertilization. The tadpole then begins to develop into an adult frog. It takes an additional 60 to 90 days for a tadpole to become an adult leopard frog.

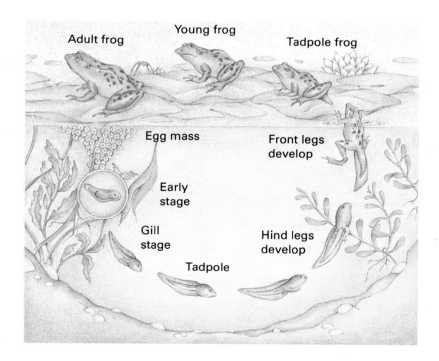

18–17 Frog development. The early stage embryo is nourished by the egg. The gill stage embryo and tadpole feed on plant material. The adult frog has lungs that enable it to live on land. Adult frogs feed on other animals, mainly insects.

Internal Fertilization, Development on Land, Food from Yolk
The females of all birds, most reptiles, and two species of mammals—the duckbill platypus and the spiny anteater—lay eggs. In all cases, internal fertilization occurs before the eggs are laid.

Consider the development of a chick embryo. The embryo develops inside the egg, where it has a supply of food. The outside of the egg, the **shell**, protects the embryo from injury and prevents it from drying out. The shell also keeps out dirt and bacteria and lets air in and out. Inside the shell, four membranes surround the embryo. The outer membrane is the **chorion**. The chorion presses against the shell and acts in the exchange of gases between the embryo and outside of the shell. Blood vessels in the chorion absorb oxygen and give off carbon dioxide.

The inner membrane that surrounds the embryo is the **amnion**. The embryo floats freely in the **amniotic fluid** within the amnion. The amniotic fluid acts as a shock absorber, protecting the delicate embryo from injury when the egg is moved. In addition, it also prevents the embryo from sticking to the shell. A third membrane, the **yolk sac**, encloses the yolk, which is the embryo's source of food while it develops in the egg.

The embryo and the four membranes that surround it develop from the zygote.

The amniotic fluid provides a watery environment for the embryo even though it is developing on land.

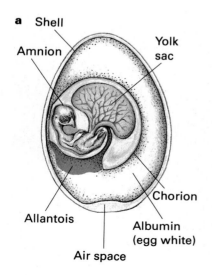
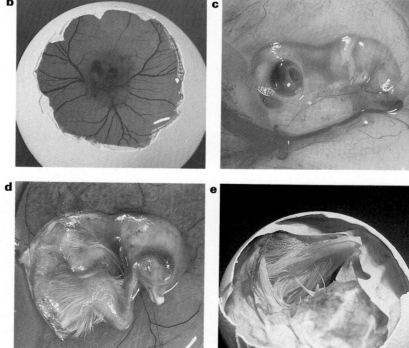

18–18 Chick development: **a.** amniotic egg; **b.** 5 days; **c.** 11 days; **d.** 14 days; **e.** hatching at 21 days. The yolk within the yolk sac nourishes the developing embryo.

Chapter 18 Reproduction and Development of Vertebrates

With the exception of the allantois, the membranes surrounding a chick embryo are the same as those that surround a human embryo.

Like all living things, the chick embryo produces nitrogen wastes, in this case uric acid. The embryo inside the eggshell cannot dispose of these wastes and collects them, instead, in a special membranous bag, the **allantois** (uh-LAN-tuh-wis). Uric acid is stored in the allantois until the embryo hatches. The allantois also functions in taking in oxygen and giving off carbon dioxide.

Internal Fertilization, Development Inside the Mother, Food from Yolk In fish and some species of reptiles, embryos develop inside the mother's body. Internal development is an adaptation that protects an embryo from the external environment. Fish and reptile embryos get food from yolk in the egg during their development.

The dogfish shark has internal fertilization and develops in an oviduct inside the mother's body. Development takes from 16 to 25 months. During that time, yolk from the egg provides the embryo with food.

Guppy offspring develop inside the mother's body, as do rattlesnake offspring.

Internal Fertilization, Development Inside the Mother, Food Through a Placenta Embryos of most mammals develop inside the mother. The mother's body provides food through a placenta. The developing embryo is enclosed in an amnion and floats in amniotic fluid. The umbilical cord attaches the embryo to the placenta. Food and oxygen from the mother's blood vessels diffuse through the placenta into the offspring's blood vessels. Carbon dioxide and nitrogen wastes diffuse out of the embryo's blood vessels through the placenta and into the mother's blood vessels. The blood of the mother and the embryo do not mix.

18–19 Ewe licking her newly born lamb. As in most mammals, the lamb develops inside the ewe and receives its food through a placenta.

Section 18.3 Adaptations in Vertebrate Reproduction and Development

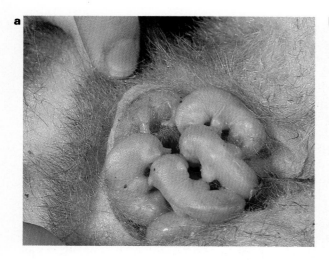 

18–20 Development of marsupials. **a.** After a brief period inside the mother's body, young opossums are born and climb blindly into the mother's pouch, where they attach to nipples. **b.** Later, the opossums develop fur and can move around; however, they remain in the pouch for two months.

The gestation period of mammals varies greatly from species to species. Table 18–1 shows the large variation in the gestation period of different mammals. The shortest gestation period occurs in marsupials, the pouched mammals. Some of them have a gestation period of as little as eight days. Marsupials are not fully developed when they are born. A newborn American opossum, for example, is about the size of a grape and weighs about 2 g. Marsupials are blind and helpless at birth. They climb into the mother's pouch and attach to a nipple. Growth and development continue while they remain inside the pouch. The gestation period of other placental mammals is longer than that of marsupials. In general, larger mammals have longer gestation periods.

Table 18–1 Gestation Periods in Mammals

Mammal	Average Gestation Period
Opossum	13 days
Gray Kangaroo	40 days
Mouse	3-4 weeks
Dog	2 months
Tiger	3 months
Human	9 months
Seal, Sea Lion, Walrus	9-12 months
Elephant	20-22 months

Adaptations that Help Offspring Survive

One adaptation for species survival is the production of large numbers of offspring at one time. Large numbers of offspring increase the chances that some offspring will survive. This adaptation is common in fish and amphibians. Toads, for example, can produce 15 000 offspring in a single year. Many of the offspring die, but enough survive to provide a new generation of toads.

Parental care is another type of adaptation that helps the young survive. This adaptation is common in birds and mammals. Parental care seems to be related to the number of offspring produced. Animals that produce small numbers of offspring show great parental care. Because of parental care, fewer offspring die. Elephants, for example, usually have one offspring at a time. A mother elephant guards and cares for the baby very carefully for the first six months of its life.

Vertebrate parents care for their young in various ways. Some prepare burrows or nests. Others carry their young on their bodies. Many parents provide food for their young. Mammalian mothers supply their young with milk from mammary glands. Some parents protect their young from predators. Some groom their offspring, keeping them free from parasites. Some parents teach their offspring skills that help them survive.

18–21 Parental care. **a.** The female wolf spider transports her young on her back. **b.** A herring gull protects and warms the developing chicks in its nest. **c.** A brown bear teaches its cub to fish.

18–22 A human baby is entirely dependent on its parents for food, shelter, and protection. Without adult care for a prolonged period during childhood, a human offspring would not survive.

Parental care is highly developed among primates, which include apes, monkeys, and humans. Primates have grasping hands, good eyesight, and a well-developed brain. Primates usually have few offspring. Consider humans, for example. A newborn baby is helpless. It depends completely on parents or other adults and takes a long time to grow. Along with the growth of the body, children learn a great deal from their parents, other adults, other children, and from the environment itself. It takes many years to grow up physically and mentally. During the prolonged period of childhood, survival depends on adult care.

Building a Science Vocabulary

mating	shell	yolk sac
rutting	chorion	allantois
estrus	amnion	
courtship behavior	amniotic fluid	

Questions for Review

1. What four major patterns of development can be found in vertebrate reproduction?

2. How does each of the following help bring the sexes together?
   a. secondary sex characteristics
   b. voice signals
   c. odor signals

3. What is estrus? What function does estrus serve in the reproduction of vertebrates?

4. What is meant by courtship behavior? What function does courtship behavior serve in vertebrate reproduction?

5. How do each of the following help an embryo survive?
   a. placenta
   b. internal development

Perspective: Careers
Agriculture — A Misunderstood Science

What is "agriculture" to you? Do you think of a giant combine sweeping over endless fields of grain? Or perhaps you think of dairy cattle being milked and sheep shorn. Actually, the science of agriculture is much broader than that. It involves both biological and physical science and, increasingly, mathematics and computer science working to meet the world's demand for food and fiber.

Animal agriculture, better known as animal science, is concerned with the production and marketing of meat, poultry, and dairy products. It includes the application of genetics to breeding stocks for a particular climate and to provide a particular quality for marketing purposes.

Animal scientists also include people trained in meat packaging, livestock marketing, feed industries, artificial insemination techniques, pesticide and disease control, as well as in basic research in physiology, developmental biology, genetics, and ecology.

Of some 350 000 species of plants, only 100 to 150 are of commercial importance as crops. Crop scientists work at improving the yield of such crops and increasing their quality in the marketplace. They are involved in breeding everything from flowers and trees to grapes and grains. They study the ways soil, water conditions, pests, and disease affect these plants.

Because the field of agricultural science is so very broad, one can enter it at many levels. A good background in high school science and mathematics enables one to perform certain limited roles. A two-year technical training program or a degree in some aspect of agricultural science permits one to do more. Research scientists in agriculture usually have degrees at the masters or doctoral level.

Given the needs of the world for enough food of the right kinds, the field of agricultural science holds one of the major keys to addressing those needs. You might want to consider it as a career.

Chapter 18
Summary and Review

Summary

1. All vertebrate reproductive systems have paired gonads with ducts connecting them to the outside of the body.

2. At puberty a male or female becomes capable of reproduction and develops secondary sex characteristics.

3. In the human male, sperm are produced by the seminiferous tubules in the testes. A system of ducts stores and carries sperm out of the body.

4. Eggs are produced in follicles in the ovaries and are carried to the uterus by the oviducts. If fertilization occurs, the zygote attaches to the wall of the uterus. If the egg is not fertilized, it passes out of the body. The reproductive cycle in females is known as the menstrual cycle.

5. Fertilization usually occurs in the oviduct and results in a diploid zygote. Only one sperm can fertilize an egg.

6. Gestation in humans lasts about nine months. During early development, cleavage of the zygote leads to the formation of the morula, blastula, and gastrula.

7. Three cell layers are found in the gastrula: the ectoderm, the mesoderm, and the endoderm. Each layer differentiates into specific parts of the embryo's body.

8. Embryos of humans and other mammals are surrounded by three membranes: the chorion, the amnion, and the yolk sac. The amnion and amniotic fluid provide protection for the embryo.

9. The placenta is the site of exchange of materials between the embryo and the mother.

10. As development proceeds, the embryo takes on human features, and it increases greatly in size and weight.

11. The onset of labor is controlled by hormones. Rhythmic contractions of the uterus and abdominal wall push the baby out of the cervix and vagina. The afterbirth is expelled shortly after the baby is born.

12. Fraternal twins develop from the fertilization of two eggs released at the same time. Identical twins develop from a single zygote.

13. Overproduction of eggs and sperm increases the probability that fertilization will occur. Internal fertilization prevents sperm from getting lost or drying out.

14. Adaptations that bring vertebrate males and females together involve sensing secondary sex characteristics, mating calls, or odors.

15. Breeding periods and courtship behavior help synchronize gamete release.

16. Some vertebrate embryos obtain nourishment from yolk in an egg and may develop in the water, on land, or inside the mother's body. Other vertebrate embryos get their food from the mother through a placenta.

17. In some species, survival of the young is related to the large numbers of offspring produced. In other species, survival of the young depends on parental care.

Review Questions

Application

1. A biology student located a mass of developing frog eggs in a pond. Suddenly she saw a group of leeches swim over and begin devouring the embryos. Her observations caused her to think of the following questions about frog reproduction. Try to answer them.

a. How do male and female frogs find each other for mating?
  b. Is fertilization internal or external?
  c. Before developing frog embryos are tadpoles, how do they get food?
  d. How does the frog species overcome the loss of embryos to predators?

2. Compare each of the following for a chick embryo in the egg and a human embryo in the uterus.
   a. source of food
   b. how oxygen is supplied
   c. how wastes are removed
   d. length of time for development

3. In an experiment, an embryologist used harmless dyes to color parts of a frog embryo. He made the ectoderm yellow, the mesoderm blue, and the endoderm green. What color most likely appeared in each of the following structures of the tadpole?
   a. muscle   c. nervous system
   b. skin   d. lining of the intestine

Interpretation

1. What does each of the following accomplish for the species described?
   a. The male peacock spreads his beautiful blue feathers into a fan and parades in front of a female.
   b. At a certain time of the year, the male and female of a species of sea lion rub necks and bring their mouths together.
   c. For the first three weeks of its life, a baby penguin is constantly guarded by its mother or father.

2. How does a chick egg compare with a seed in regard to the following characteristics?
   a. source of food for the embryo
   b. how the embryo grows
   c. adaptations that prevent the embryo from drying out
   d. length of time as an embryo

3. A sheep gave birth to identical twins. Can they be considered a clone? Explain.

Extra Research

1. Collect amphibian eggs and watch them develop. At home, place a small clump of eggs (6-12) in a large bowl of pond water or aged aquarium water. Record the date, time, and location of collecting. Monitor the water temperature, and watch for hatching. Record the date that tadpoles appear. After hatching occurs, remove the empty jelly from the bowl. The tadpoles can be fed on a daily diet of bits of crumbled dog biscuit or crumbled dry cereal. Change the water frequently with pond water or aged aquarium water. Do not let the water's temperature go over 25°C. Observe changes in the appearance of the tadpoles over a six-week period. Take photographs or make drawings of your observations.

2. Set up separate breeding tanks for guppies (live bearers) and zebra fish (egg layers). Compare reproduction and development in these two types of fish.

3. Using reference books and magazine articles in the library, research the courtship behaviors of various vertebrates. Write a short paragraph describing each one.

Career Opportunities

Fisheries workers survey the fish in streams and breed fish in artificial ponds. Most of their work is outdoors and is for provincial or federal agencies. On-the-job training is usually provided. Contact the Fisheries Council of Canada, 77 Metcalfe St., Ottawa, Ontario K1P 5L6.

Farm laborers work outdoors with their hands and farm machinery in the raising of plant crops and animals. They are usually employed year-round on livestock farms and seasonally on crop farms. Good health and stamina are essential. Farm experience or training is helpful but not required. Write the National Farmers Union, 250C 2nd Ave. S., Saskatoon, Saskatchewan S7K 2M1.

Unit V
Suggested Readings

Beim, Carol, "Mysterious Callers of the Wilderness," *Conservationist*, July-August, 1978. *The natural history of the common loon, with descriptions of male and female birds, nesting sites, calls, and courtship behavior.*

Carrington, Richard, *The Mammals*, New York: Time-Life Books, 1972. *This book has excellent, well-illustrated information on mammalian courtship, reproduction, and parental care.*

Harrison, Hal H., "Home Sweet Home," *National Wildlife*, June-July, 1979. *Discusses the role of nests in the reproduction and survival of parents and offspring.*

Howard, Richard D., "Big Bullfrogs in a Little Pond," *Natural History*, April, 1979. *Article describes the mating behavior and reproduction of bullfrogs in a Michigan pond.*

Howell, Thomas R., "Desert-nesting Sea Gulls," *Natural History*, August, 1982. *Attractively illustrated explanation of why gray gulls of the Atacama Desert of Northern Chile rear their young in one of the most inhospitable environments on earth.*

Murchie, Guy, "Compensating with Prodigality: An Exuberance in Nature," *Science Digest*, August, 1979. *This article, from the author's book* The Seven Mysteries of Life *discusses interesting adaptations for reproduction, particularly the enormous numbers of gametes produced by certain species.*

Ritzau, Fred, "Propagation: Giving Plants a New Life," *Horticulture*, October, 1978. *Describes ways to revive and replace plants using vegetative propagation.*

Robinson, Michael H., "Wondrous Ways and Means of Tropical Spiders," *Smithsonian*, October, 1978. *This article describes the remarkable adaptations, fascinating behaviors, and reproductive features of tropical spiders.*

Rugh, Roberts, and Landrum B. Shettles, *From Conception to Birth: The Drama of Life's Beginnings*, New York: Harper & Row Publishers, Inc., 1971. *Excellent book with superb illustrations, including color photographs, describing human development. Begins with the gametes and ends with the birth of a baby.*

Sayre, Anne, *Rosalind Franklin and DNA*, New York: W. W. Norton & Co., 1978. *Using x-ray diffraction techniques, Rosalind Franklin contributed much to unlocking the mystery of the structure of DNA. Sayre tells the story of Franklin both as a scientist and as a woman.*

Stokes, Donald, "Tree Flowers," *Horticulture*, March, 1979. *Fascinating descriptions of the flowers that grow on trees and their adaptations for different types of pollination.*

Todd, Frank S., *The Sea World Book of Penguins*, New York: Harcourt Brace Jovanovich, Inc., 1981. *A fact-filled book on penguins, including sections dealing with their breeding adaptations. Illustrated with excellent photographs.*

Watson, James D., *The Double Helix*, New York: Atheneum, 1968. *Watson's account of the way the structure of DNA was discovered is also a fascinating story of the personal relationships between the scientists on this project.*

Went, Fritz W., *The Plants*, New York: Time-Life Books, 1973. *Beautiful photographs and drawings illustrate flowering, pollination, seed dispersal and germination, and growth in plants.*

Witman, Jon, "The Nesting of the Lumpfish," *Sea Frontiers*, September-October, 1978. *In spite of its odd appearance, the lumpfish is a successful species of fish in temperate waters. Illustrates the lumpfish's remarkable reproductive strategy, which is responsible for its success.*

Unit VI

Continuance and Change

Everyone is interested in how traits are passed from one generation to another. People look for their parents' traits in themselves. They also seek their own traits in their children. Breeders use the principles of heredity to improve plants and animals. People try to overcome hereditary disorders by learning about inheritance.

For over a century, scientists have directed their efforts to understanding heredity. It all started late in the nineteenth century when an Austrian monk named Gregor Mendel worked out the heredity of garden peas. It was later found that his principles apply to other living things as well. In time scientists discovered many of the secrets of DNA and its role in reproduction and heredity.

While the search to understand inheritance of traits was going on, other scientists were examining the history of life on earth. Evidence obtained by carefully studying the characteristics of living species and fossils, as well as the distribution of species on earth, suggested to them that species change over time. Their theories have revolutionized the science of biology.

Fossil shell imprints in sedimentary rock record the existence of an ancient sea.

Chapter 19

Genetics: Mendel's Laws of Heredity

Focus Expectant parents always wonder what a new child will look like. Will the baby be a boy or a girl? Will its nose be shaped like its father's or its mother's? Will the baby have blue eyes or brown eyes? Will it be healthy? In the past, people could only wonder about questions like these. Today's parents-to-be have information that can help them predict what their children may be like. This information is the result of research in genetics.

19.1 GREGOR MENDEL: FIRST GENETICIST

The study of heredity—the transmission of characteristics from parents to offspring—is known as **genetics** (juh-NET-iks). In their work, geneticists are interested in learning about the similarities and differences, or variations, between parents and offspring. The scientific study of genetics began over one hundred years ago.

Gregor Mendel (1822–1884), the son of Austrian peasants, became a monk at age 21 in order to continue his education. Mendel was interested in the breeding of plants and kept a small garden. Fifteen years after he entered the monastery, Mendel conducted experiments on the heredity of garden peas. His experiments eventually led to the science of genetics.

19–1 Offspring of the same parents often show variations.

The people of Mendel's day knew very little about heredity. They believed that the characteristics of two parents somehow blended in the offspring, but they had no idea how this came to pass. Some thought that heredity had to do with something in the parents' blood. Although they were interested in plant and animal breeding, people of that time had no scientific principles to guide them in this work. Mendel conducted careful experiments, concentrating on a few hereditary characteristics. He also applied mathematical techniques to analyze his results.

In 1865, Mendel presented his work at a meeting of scientists. No one understood it or saw its importance. He published his findings a year later, but the paper went unnoticed. It was not until sixteen years after his death that scientists began to understand the significance of Mendel's work.

The Work of Gregor Mendel

Mendel decided to study the edible garden pea for several reasons. Pea plants are easy to grow and have a short life cycle. In addition, he could control their reproduction by a simple but painstaking procedure. Pea plants normally reproduce by self-pollination. The shape of their flowers prevents cross-pollination. Mendel could allow them to self-pollinate, or he could use artificial pollination to make crosses between plants. In artificial pollination, the flower's own stamens are removed, the desired pollen is applied to the pistil, and the flower is covered to prevent further pollination.

For his experiments, Mendel needed **pure-breeding plants**, that is, plants whose seeds grow into offspring with the same characteristics as the parent plants. Pure-breeding organisms are said to breed true. To make sure that his plants bred true, Mendel bred seeds of their offspring for two years.

19–2 Gregor Mendel was the first person to scientifically attempt to discover the laws at work in heredity.

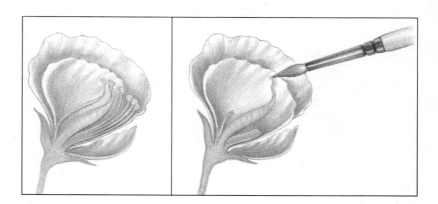

19–3 Artificial pollination in pea plants involves removing stamens before natural pollination occurs. Later, the mature pistil is hand-pollinated.

One of the reasons other researchers failed to learn the laws of heredity was that they tried to study too many traits at once.

filialis = of a son or daughter

19–4 Hybrid offspring. Mendel's cross of round-seed pea plants with wrinkled-seed pea plants produced hybrids with round seeds in the F_1 generation. When he crossed the F_1 hybrids he found both round-seed and wrinkled-seed offspring in the F_2 generation.

Unlike others investigating plant breeding, Mendel selected only a few characteristics, or **traits**, for study. Pea plants have certain traits that have different or opposite expressions in different plants. One such trait in garden peas is seed shape. Seeds are either round or wrinkled. Another trait of this sort in pea plants is stem length. Stems are either tall or short. Mendel knew that the offspring resulting from self-pollination of his pure-breeding plants were alike and like their parents. He wanted to know what the offspring would look like if he crossed two parent plants that bred true and had contrasting expressions of the same trait.

Geneticists (including Mendel) use various terms for the different generations of organisms in breeding experiments. The first generation is referred to as the **parents**. The offspring of parents are referred to as the **first filial generation**, or F_1. Similarly, the offspring of the F_1 generation are the **second filial generation (F_2)**, and the offspring of the F_2 are the **third filial generation (F_3)**.

Mendel crossed round-seed pea plants with wrinkled-seed pea plants producing *hybrid* offspring. **Hybrids** are organisms that result from crosses of parents having different expressions of the same trait. Upon examination, Mendel found that all the F_1 offspring had round seeds. What had happened to the expression of the trait for wrinkled seeds? Was it lost completely or was it somehow hidden? Mendel then crossed the F_1 pea plants with each other and found surprising results. Some of the F_2 generation had round seeds and some had wrinkled seeds. The hidden expression for wrinkled seeds had reappeared.

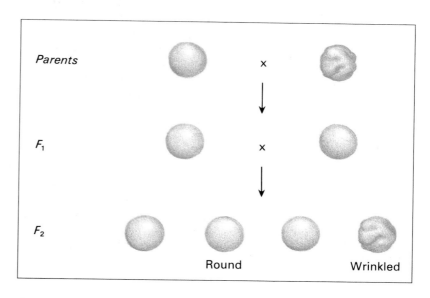

Section 19.1 Gregor Mendel: First Geneticist

Mendel knew nothing about chromosomes, genes, or DNA. However, you know that inherited traits are controlled by pieces of DNA called genes. All diploid organisms have two of each type of chromosome; consequently, they have two of each type of gene. If the two genes for a trait are alike, the organism is said to be **homozygous** (HŌ-mō-ZĪ-gus) for the trait. If the two genes are different, the organism is **heterozygous** (HET-ur-ō-ZĪ-gus) for the trait.

Mendel made hundreds of crosses using contrasting expressions of the same trait. He recorded the number and appearance of the offspring of each cross. By applying mathematical principles to his careful observations, Mendel figured out three principles of heredity that have become the basic laws of genetics.

Law of Dominance When Mendel crossed round-seed pea plants with wrinkled-seed pea plants, he was crossing homozygous parents that had contrasting expressions of the trait for seed shape. The offspring received a different gene for the trait from each parent and so were heterozygous. Yet, they all had round seeds. Mendel said that the round-seed expression of the trait is **dominant**. A dominant expression of a trait is seen in the offspring when contrasting genes for the trait are present. Mendel also stated that the wrinkled-seed expression of the trait is **recessive**, since it was hidden in the heterozygous offspring.

Mendel also experimented with stem length. He crossed homozygous tall plants with homozygous short ones. The offspring were all tall. In garden peas, tall is dominant over short. When studying genetics, biologists use symbols to represent the expressions of a given trait. The dominant expression of a trait is indicated with a capital letter and the recessive expression of a trait is indicated with a lowercase letter. In pea plants, T stands for tall and t stands for short. A homozygous tall pea plant is represented as TT. Similarly, a homozygous short plant is represented as tt, and a heterozygous pea plant is represented as Tt. Since tall is dominant over short, a pea plant that is heterozygous for stem length must be tall. The symbols TT, tt, and Tt represent an organism's gene combination, or **genotype** (JEE-nō-tīp). The expression of an organism's genotype—that is, the way the trait actually appears—is its **phenotype**. For a homozygous tall pea plant, its genotype is TT and its phenotype is tall.

pheno = appearance

Mendel crossed pea plants with contrasting expressions for seven different traits (see Table 19–1). In each case he found a dominant and a recessive expression in the offspring of the F_1 generation. These observations were the basis of the **law of dominance**. As stated by Mendel, the law of dominance says that when homozygous parents with contrasting expressions of a given trait are crossed, only one expression will be seen in the offspring. The expression that is seen is dominant. The expression that is not seen (remains hidden) is recessive. A modern way of expressing the law of dominance is:

> When homozygous parents with contrasting genes for a given trait are crossed, the effect of one gene is often seen in the offspring. The gene whose effect is seen is dominant. The gene whose effect is not seen (remains hidden) is recessive.

19-5 The seven traits in garden peas studied by Mendel. Each trait has a dominant expression and a recessive expression.

	\multicolumn{7}{c}{Traits}
	Seed shape	Endosperm color	Seed coat color	Pod shape	Pod color	Flower position	Stem length
Dominant	Round	Yellow	Colored	Smooth	Green	Axial	Tall
Recessive	Wrinkled	Green	White	Wrinkled	Yellow	Terminal	Short

Section 19.1 Gregor Mendel: First Geneticist

Law of Segregation When Mendel crossed plants that were pure-breeding tall (TT) with plants that were pure-breeding short (tt), the F_1 generation were all tall (Tt). He was anxious to see what would happen in the F_2 generation. He crossed the tall F_1 pea plants with each other (Tt × Tt). Some of the F_2 generation grew into tall pea plants; others grew into short ones. Mendel counted the offspring of each type. There were about three tall plants for every one that was short; therefore the ratio of tall plants to short plants was three to one (3:1).

During meiosis, pairs of genes separate. One gene from each pair goes into a haploid sperm or egg. This means that a sperm produced by a pea plant that is heterozygous for stem length (Tt) could carry a gene for tall (T) or a gene for short (t). Similarly, an egg produced by a Tt parent could have either gene.

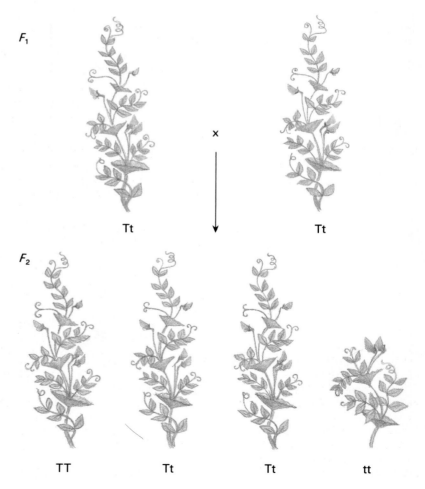

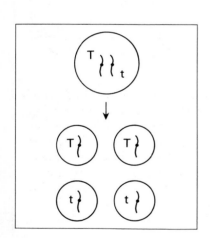

19–6 Pairs of genes separate during meiosis (above), resulting in each gamete having one gene for a given trait. The genes recombine at random during fertilization. Mendel found that when he crossed hybrid tall pea plants (right), he obtained nearly three tall plants for every short plant. These results helped him formulate the law of segregation.

When two pea plants heterozygous for stem length are crossed, each produces equal numbers of gametes containing T and t. During fertilization, a sperm combines with an egg by chance. A sperm carrying the gene for tall has an equal chance of combining with an egg that holds the gene for tall as it has of combining with an egg that holds the gene for short. The same is true for a sperm carrying the gene for short. The following four genotypes are possible in the zygote if the parents are heterozygous: TT, Tt, tT, and tt. Notice that there are two ways to get a heterozygous genotype. They are Tt and tT. Both these genotypes are usually written as Tt. The genotypes that are possible from this cross are one homozygous tall, two heterozygous, and one homozygous short. The genotypic ratio is 1:2:1. Also notice that three of the possible genotypes (TT, Tt, tT) result in tall offspring. Only one genotype (tt) results in short offspring. The phenotypic ratio is therefore three tall to one short (3:1).

Mendel explained these results in his **law of segregation**. He stated that during sexual reproduction, the individual expressions of a given trait separate, or segregate, and recombine at random. A modern way of stating Mendel's law of segregation is:

> Paired genes separate during meiosis so that each gamete possesses only one gene for a trait. Genes recombine at random during fertilization.

The law of independent assortment is also called the law of unit characters.

Law of Independent Assortment According to Mendel's **law of independent assortment**, a gene goes into a sperm or egg independently of other genes. For example, a gene for tall stem assorts independently of the gene for white flowers. A gene for wrinkled peas assorts independently of the gene for yellow peas. The law states that each gene is an independent unit that is inherited on its own. We now know that genes are located on

19–7 Independent assortment. During meiosis, homologous pairs of chromosomes separate. The genes located on one homologous pair (Rr) go into gametes independently of the genes located on another homologous pair (Yy). As a result, gametes containing different combinations of genes are produced.

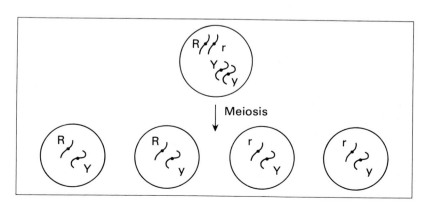

chromosomes. Consequently, it is the chromosome and not the gene that is inherited independently during reproduction. Exceptions to Mendel's law of independent assortment do exist and will be discussed later in this chapter.

Building a Science Vocabulary

genetics	heterozygous
pure-breeding plant	dominant
trait	recessive
parents	genotype
first filial generation (F_1)	phenotype
second filial generation (F_2)	law of dominance
third filial generation (F_3)	law of segregation
hybrid	law of independent assortment
homozygous	

Questions for Review

1. Explain how Mendel used artificial pollination in his experiments with pea plants.

2. State the law of dominance. What kind of offspring result when pea plants with green pods are crossed with pea plants with yellow pods? Explain.

3. If you were given a tall pea plant, you probably could not tell its genotype. Why not?

4. The law of segregation involves separation of paired genes and their random recombination. During sexual reproduction, when do paired genes separate? When do they recombine at random?

5. What types of gametes can be produced by a parent with the genotype Dd? What are the genotypes and phenotypes of offspring that can be produced when two parents bear the genotype Dd?

6. State the law of independent assortment. What structure is now known to "assort independently" during reproduction?

19.2 MENDEL'S LAWS AND PROBABILITY

Mathematics was one of the most important tools that Mendel used in his work. When he applied mathematical principles to the results of his experiments, the patterns of heredity became clear. The mathematical principles upon which Mendel's laws are based are known as the **laws of chance**.

The Laws of Chance

The laws of chance are mathematical formulas that are used to predict the chance, or **probability**, of something happening. One of the laws of chance that helped Mendel explain the law of segregation can be stated as follows:

> The chance of two independent events occurring together is equal to the chance of one event occurring alone multiplied by the chance of the other event occurring alone.

To see what this law of chance means, take out two pennies and toss one of them. The chance that the penny will turn up heads is fifty-fifty or one out of two (½). The chance that it will turn up tails is also fifty-fifty or ½. Tossing one coin is an "independent event." Now toss the two coins together. This represents "two independent events occurring together." The two events are independent because the outcome of one coin toss does not affect the outcome of the other coin toss. Now, apply the law of chance stated above:

- The chance of both coins turning up heads is ½ x ½ or ¼.
- The chance of the first coin turning up heads and the second coin turning up tails is ½ x ½ or ¼.
- The chance of the first coin turning up tails and the second coin turning up heads is ½ x ½ or ¼.
- The chance of both coins turning up tails is ½ x ½ or ¼.

Whether you actually get these results depends on how many tosses you do. If you toss two coins four times, you will probably not get these results exactly. If you toss two coins one hundred times, you will come closer to the predicted results. One thousand tosses should bring you very close to the exact outcome.

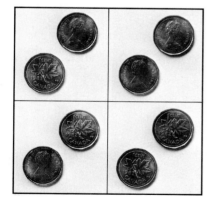

19–8 The four possible results of tossing two pennies at the same time.

Heredity and the Laws of Chance The laws of chance help figure out the probability that a particular gene combination will occur in an offspring. Recall Mendel's cross, Tt x Tt. What is the chance of an offspring being TT? To answer this question,

Section 19.2 Mendel's Laws and Probability

The laws of chance do apply even when sperm far outnumber eggs, as in human reproduction.

Mendel made two important assumptions: 1) sperm and eggs are produced in equal numbers, and 2) gametes are fertilized at random. Then he applied the law of chance.

In the cross Tt x Tt, two independent events are occurring together. A sperm carrying a T is one independent event. An egg carrying a T is another. The probability of them occurring together at fertilization is:

$$\underset{\substack{\text{(chance of T} \\ \text{in sperm)}}}{\tfrac{1}{2}} \times \underset{\substack{\text{(chance of T} \\ \text{in egg)}}}{\tfrac{1}{2}} = \underset{\substack{\text{(chance of TT} \\ \text{offspring)}}}{\tfrac{1}{4}}$$

The chance of an offspring being TT is one in four. This result does not mean that one out of every four pea plants will be tall. It means that every pea plant produced by those parents has one chance in four of being tall. The outcome is not certain; only the probability is known.

An easy way to visualize gene combinations is to use a **Punnett square**, which shows all the possible gene combinations that parents might give to their offspring. In Figure 19–9, a Punnett square is used to show the cross of two pea plants heterozygous for seed shape (Rr × Rr). Remember that in pea plants the gene for round seeds (R) is dominant over the gene for wrinkled seeds (r).

The female plant (Rr) produces eggs containing either an R gene or an r gene. These are shown on the left side of the box. The male plant (Rr) produces sperm with either R or r. These are shown across the top of the box. The gametes combine at random during fertilization. The possible gene combination for each zygote is entered in its box in the Punnett square by multiplication as if you were calculating the area of each box. You can figure out the probability of each genotype by looking over all the boxes. In this case, the chance of RR is 1 out of 4 (¼).

The Punnett square is named for R. D. Punnett, a British geneticist of the early twentieth century. He devised the square to use in his work on heredity.

19–9 Punnett square (left) with letter combinations representing genotypes resulting from cross of parents heterozygous for seed shape. To the right is a Punnett square of the phenotypes, presented in picture form.

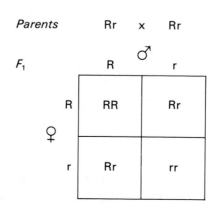
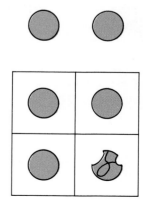

Chapter 19 Genetics: Mendel's Laws of Heredity

Predicting with Mendel's Laws

All crosses involve many more than one character, but one character may be singled out for study.

You can use Mendel's laws to predict the probability that an offspring may inherit a particular gene combination. The six problems that follow deal with such predictions. Each problem involves one trait controlled by a single pair of genes. When you consider only one trait in a cross, it is described as a **monohybrid cross**. Analyze each problem using a Punnett square. Then check your results with those shown under the problem.

1. Homozygous Dominant × Homozygous Dominant In humans, having dimples is usually a dominant trait. If two parents are homozygous for dimples (DD), what is the probability that a child of theirs will also have dimples?

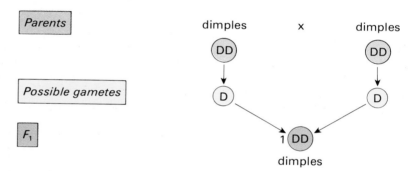

Since both parents contribute only the dominant gene, all the offspring will have dimples.

2. Homozygous Recessive × Homozygous Recessive In humans, straight hair (w) is a recessive gene. Wavy hair (W) is dominant. What is the probability that parents both having straight hair will have a child with straight hair?

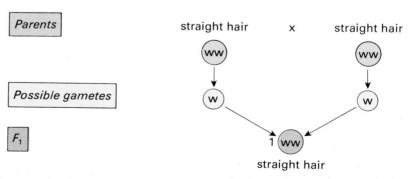

As in the first problem, both parents contribute the same gene. All the offspring will have straight hair.

In humans normal vision is recessive. Farsightedness and nearsightedness are dominant traits.

3. Homozygous Dominant × Homozygous Recessive In humans, attached earlobes is due to a recessive gene (f). Free earlobes is dominant (F). What is the probability that a parent homozygous for free earlobes and a parent with attached earlobes will have a child with free earlobes?

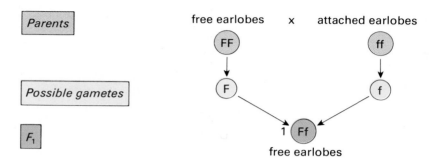

Both parents contribute only one type of gene; therefore the offspring will all be heterozygous and have free earlobes.

4. Heterozygous × Heterozygous Problems involving two heterozygous parents are similar to Mendel's crosses with bean plants, Tt × Tt and Rr × Rr. In humans, long eyelashes are dominant (L). Short eyelashes are recessive (l). What is the probability of the various genotypes and phenotypes for offspring of parents both heterozygous for eyelash length?

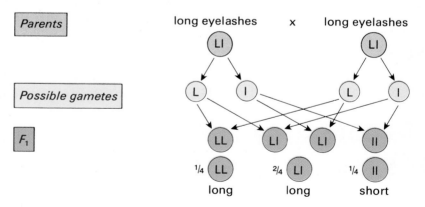

Since each parent can contribute either gene, the probability is that one fourth of the offspring will be LL, one half Ll, and one fourth ll, giving a phenotypic probability of three long to one short.

5. Heterozygous × Homozygous Dominant In squash, white fruit is dominant (W) and yellow fruit is recessive (w). What possible genotypes and phenotypes can be produced when a heterozygous squash is crossed with a homozygous white squash?

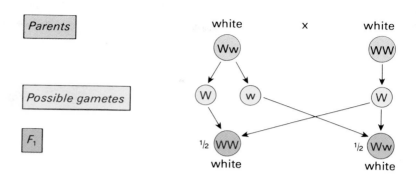

One parent can contribute either type of gene; the other parent can contribute only one type. The probability is that half the offspring will be homozygous dominant and the other half will be heterozygous. All the offspring will have white fruit.

6. Heterozygous × Homozygous Recessive In this last case, one parent is heterozygous while the other is homozygous recessive. After you analyze the problem, compare the results with the cross in the preceding case. Spotted coat is dominant in rabbits (S). Solid coat is recessive (s). What possible genotypes and phenotypes can be produced when a rabbit that is heterozygous for spotted coat is crossed with a rabbit that has a solid coat?

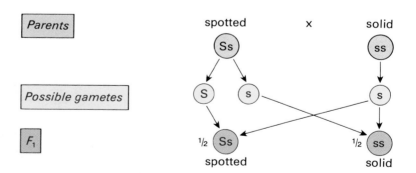

One parent can contribute either type of gene; the other can contribute only one type. The probability is that half the offspring will be homozygous recessive, half will be heterozygous. The phenotypic probability is that the offspring will be half spotted, half solid.

19-10 The offspring of a heterozygous parent and a homozygous recessive parent have a 50 percent chance of inheriting either trait.

After studying Mendel's laws and analyzing the six problems above, you should be able to work out most genetics problems where dominance exists and the trait is controlled by a single pair of genes.

Building a Science Vocabulay

laws of chance
probability
Punnett square
monohybrid cross

Questions for Review

1. According to the laws of chance, the chance of two independent events occurring together is the chance of one event occurring alone multiplied by the chance of the other event occurring alone. During sexual reproduction, what are the two independent events? When do they occur together?

2. In corn plants, albinism, the absence of pigment, is recessive to normal pigmentation. What is the probability that two normal green corn plant parents that are both heterozygous for albinism may have an albino offspring? Use a Punnett square to work out the answer.

3. In humans, wavy hair is dominant over straight hair. What possible genotypes and phenotypes can occur in the offspring where one parent is heterozygous for wavy hair and the other has straight hair? Give probabilities for each possible outcome. Use a Punnett square to work out the answer.

19.3 GOING FURTHER WITH MENDEL'S LAWS

Inheritance of Two Pairs of Genes

How Mendel's laws help explain the inheritance of one trait was discussed in Section 19.2. Mendel's laws can also be applied to the inheritance of two or more traits. To understand inheritance of more than one trait, first consider the inheritance of two heterozygous pairs of genes that are on separate chromosomes. A cross that involves two independently inherited heterozygous pairs of genes is called a **dihybrid cross**. An example of a dihybrid cross follows.

In guinea pigs, black coat color (B) is dominant over white coat color (b). Short hair (S) is dominant over long hair (s). Imagine a male and a female guinea pig both heterozygous for these two traits (BbSs). Such guinea pigs are short-haired and black. If they mated, what types of offspring could they have? A Punnett square can be used to determine all of the possible genotypes. To set up the Punnett square you must first figure out what possible types of gametes each parent can produce. Since there are two traits and each parent has both genes for each trait, there are four possible combinations of the genes in the gametes of each parent.

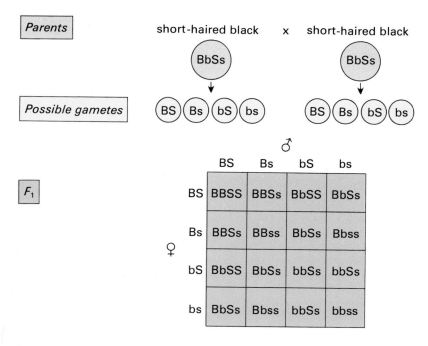

Section 19.3 Going Further with Mendel's Laws

The Punnett square gives 16 possible combinations of gametes. To determine the possible phenotypes, figure out what each of the 16 possible genotypes would look like (remembering that S is dominant over s and B is dominant over b). Then count the total number of each phenotype.

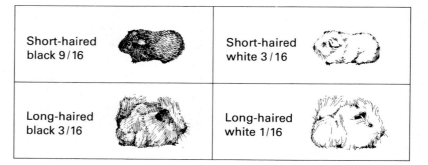

19-11 Possible phenotypes of a dihybrid cross in guinea pigs. The eight possible genotypes are expressed as four possible phenotypes in a ratio of 9:3:3:1.

As you see in the problem above, a dihybrid cross of heterozygotes results in eight possible genotypes and four possible phenotypes among the offspring. The ratio of the probability of the phenotypes is 9:3:3:1.

Exceptions to Mendel's Laws of Heredity

Although Mendel's laws are still the basis of today's genetics, a lot more information has been discovered. Some of this information indicates that there are exceptions to Mendel's laws of dominance and independent assortment.

Incomplete Dominance
Garden peas show dominance for many traits when they are crossed. The gene for red flowers is dominant in garden peas (R), and the gene for white color is recessive (r). When homozygous, red-flowered pea plants are crossed with white ones, the offspring always have red flowers.

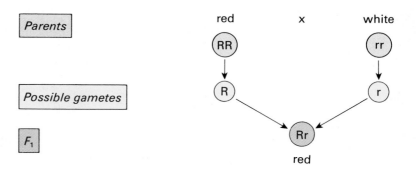

In some cases, however, dominance does not occur. For example, when red-flowered four o'clocks are crossed with white-flowered four o'clocks, the offspring plants always have pink flowers. The effects of the two genes seem to blend in the phenotype of the offspring. This type of inheritance is called **incomplete dominance**. In cases of incomplete dominance, heredity is intermediate between the two genes that are inherited for the trait.

In cases of incomplete dominance, you can always tell which offspring are heterozygous, since they have a distinct phenotype. For example, pink color in four o'clock flowers indicates that the genotype must be RR'.

When two pink four o'clocks from the F_1 generation are crossed, the genes separate and recombine at random, as Mendel's law of segregation predicts. Dominance does not occur. As a result, the 1:2:1 ratio of red to pink to white exists in the phenotype, as well as in the genotype, of the F_2 generation.

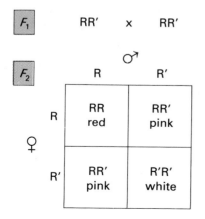

19–12 Incomplete dominance in four o'clocks. A cross between pure red and pure white parents produces pink F_1 offspring, as if the effects of the genes blend in heterozygotes.

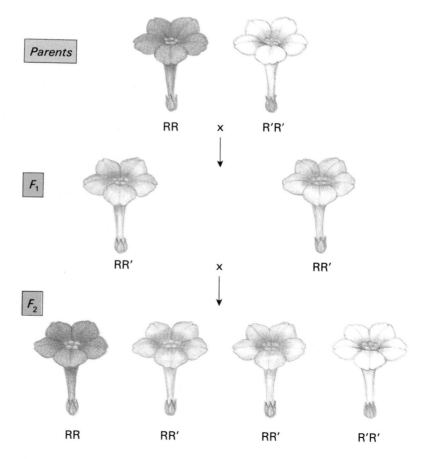

Section 19.3 Going Further with Mendel's Laws

19–13 Incomplete dominance is found in shorthorn cattle. When pure red and pure white cattle are crossed, hybrid roan offspring result. In roan cattle, red and white hairs are intermixed in the coat.

Incomplete dominance is responsible for the gradation in flower color, fur color in animals, and eye color and skin color in humans.

Incomplete dominance also occurs in snapdragons. White-flowered snapdragons crossed with red-flowered snapdragons produce pink-flowered offspring. In certain chickens with incomplete dominance, a black chicken mated with a splashed white chicken produces offspring with an intermediate type of feathers called blue Andalusian.

Linkage Mendel's law of independent assortment is based on the idea that genes are inherited independently. After watching the results of many different crosses between various organisms, however, biologists observed that a given pair of genes is not inherited independently of all other genes. Instead, groups of genes are inherited together. This occurs because several genes are located on the same chromosome. Genes on the same chromosome are said to be linked and are inherited together. The existence of two or more genes on the same chromosome is called **linkage**. Linkage is an exception to Mendel's law of independent assortment. The genes for the traits that Mendel studied in garden peas were not linked. They were located on different chromosomes.

Linked genes sometimes separate during meiosis. Genes on the same chromosome separate during crossing-over (Chapter 18). Crossing-over is therefore an exception to linkage.

19–14 Linkage. Genes located on the same chromosome are linked and are inherited together. Linked genes do not separate during meiosis unless crossing-over occurs.

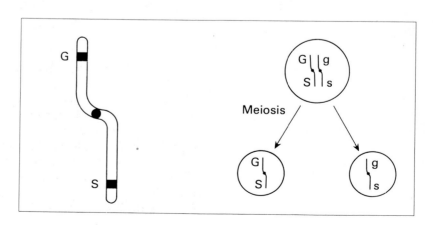

Chapter 19 Genetics: Mendel's Laws of Heredity

Applying the Laws of Heredity

The laws of heredity have practical applications. Some people may want to know more about heredity because they are concerned about a particular hereditary disorder. They may seek the advice of a geneticist, who can supply useful information by applying the laws of human heredity. In another application of hereditary laws, people benefit from the work of animal and plant breeders.

Genetic Counseling Geneticists that advise people about heredity are called **genetic counselors**. Working with physicians, they study a person's history and help determine whether the person has an inherited disorder or may be carrying a gene for one. By applying the laws of heredity, they can predict the chances of inheriting certain genetic disorders. They provide a better understanding of the nature and transmission of inherited traits.

Consider an example. A husband and wife are planning to have a baby. They both come from families with a history of **cystic fibrosis**, which is a hereditary disorder affecting the bronchi and lungs. People with cystic fibrosis often die young. Neither the man nor his wife have the disorder. They fear, however, that they both may carry the recessive gene for cystic fibrosis. If they do, would their child have the disorder? They seek the advice of a genetic counselor.

A genetic counselor *cannot* tell them whether their child *will* have cystic fibrosis. The counselor can only tell them the probability that their child *might* have the disorder. Recall that geneticists deal in probabilities. The couple may be told that if they are both heterozygous for cystic fibrosis, the probability of them having a child with cystic fibrosis is one in four. This means that every child born to them has one chance in four of inheriting the disorder. With this information the couple would then be better informed to make a decision about having a baby.

Genetic counselors trace the occurrence of inherited disorders throughout a family's history in trying to determine whether a person may be carrying a gene for one.

Plant and Animal Breeding

Plant and animal breeders have been very helpful in developing an increased and better food supply. Their work has benefited people all over the world. The development of hybrid corn is a significant example of the work done by breeders. In 1908, George M. Schull, an American geneticist, began work on the breeding of corn. Up until that time, farmers merely saved some of their best ears of corn to use as seed for the next year's crop. This method resulted in crops of mixed quality and low yield.

19–15 Hybrid vigor. When rye (genus *Secale*) is crossed with wheat (genus *Triticum*), a new grain called triticale results. Triticale has the high yield of wheat combined with the disease resistance of rye.

An example of hybrid vigor is the natural resistance to malaria found in people in tropical parts of Africa who are heterozygous for sickle cell trait. The sickle cell trait remains in the population even though the homozygous recessive produces sickle cell anemia, a disease that is usually fatal.

Schull used techniques similar to Mendel's. He used pure-breeding strains and artificial pollination. The hybrid corn he developed was sturdier and more resistant to disease than either of its homozygous parents. The term **hybrid vigor** is used to describe this condition.

Today, hybrid varieties of corn are still produced by artificial pollination. Farmers cannot use the seed of hybrid corn for the next year's planting, because it does not breed true. The original pure-breeding strains must be crossed every year to produce seeds for the next year. So, farmers buy hybrid corn seeds each year. The hybrid plants have large ears, strong stalks, large root systems, and broad leaves. They yield over five times as much corn per acre than pure-breeding varieties.

Much of the excellent corn available today is the result of crossing four pure-breeding parents twice. Each pure-breeding parent is picked for a desirable trait. Resistance to disease and sweet taste are examples of desirable traits in corn. Breeders have even developed a strain of corn that contains protein with an increased amount of the essential amino acid lysine. This work is important, because it makes corn a better source of protein in the diet.

Techniques Used by Breeders

In addition to corn, breeders have improved tremendous numbers of plant and animal types. These include wheat, rice, cattle, sheep, hogs, poultry, and many others. Breeders use special techniques that are based on Mendel's laws of heredity.

Artificial Selection The choosing of particular organisms to be parents is known as **artificial selection**. The breeder selects organisms that have characteristics that would be desirable in the offspring. The breeder may want to improve the strain or to achieve consistency among the offspring.

19-16 Artificial selection was used to produce the Brangus breed (**c**), which combines many of the desirable traits of its parents, the Brahman (**a**) and the Angus (**b**).

Inbreeding Breeders obtain consistency in a strain by **inbreeding**, or mating close relatives. Inbreeding happens easily in plants in which self-pollination occurs. In animals, inbreeding is accomplished by mating closely-related animals. Inbreeding increases the number of organisms in a stock that are homozygous for the same traits. Such organisms tend to reproduce organisms more like themselves, and their offspring tend to resemble each other to a greater extent than if they mated randomly. Pedigreed racehorses and dogs are examples of organisms developed for special traits by inbreeding.

Inbreeding results in consistency, but it brings about certain problems. The breeding of closely-related organisms tends to increase the chances of harmful recessive genes occurring together in the offspring. If a plant or animal strain has a history of a disease-causing recessive gene, inbreeding increases the chances that the disease will show up. For example, deafness in Dalmatian dogs and dwarfism in Hereford cattle are recessive traits that occur frequently as a result of inbreeding.

Inbreeding also leads to changes in some important qualities in the offspring. Compared to hybrids, inbred organisms tend to produce fewer offspring and grow less.

19-17 Inbreeding. **a.** The Morgan horse breed, known for its strength and endurance, was developed from a stallion obtained by Justin Morgan in the 1700s as payment of a debt. **b.** German shepherds, inbred for desirable traits, have also inherited a tendency to develop hip dysplasia, which is faulty development of the bone structure of the hip.

Section 19.3 Going Further with Mendel's Laws

Turkeys with large amounts of tender white meat are the result of careful breeding.

Outbreeding When breeders want organisms to have desirable traits from two different strains, they use a process called **outbreeding**. Outbreeding is the mating of different strains or families of organisms. You saw how outbreeding is used in the breeding of corn. In sheep, better wool and better meat were obtained by the outbreeding of two different inbred strains. Breeders were also able to improve upon the sheep's ability to resist disease and withstand hot weather. A disadvantage of outbreeding for commercial breeders is variation among offspring. To obtain a strain that breeds true, outbreeding must be followed by a program of inbreeding.

Test Cross Homozygous individuals breed true and heterozygous ones do not. Where dominance exists, however, it is impossible to tell whether an organism is homozygous dominant or heterozygous for a given trait. For example, in tomatoes, red fruit (R) is dominant over yellow (r). If you look at a red tomato you cannot tell whether it is RR or Rr since both genotypes appear red.

How does a breeder know whether a potential parent is homozygous dominant or heterozygous? A procedure that can help is the **test cross**. In a test cross, the breeder mates the organism whose genotype is in question with an individual that is homozygous for the recessive trait. In the case of the tomatoes, the red-fruited tomato plant would be crossed with a yellow-fruited plant. If the red-fruited plant is heterozygous, the F_1 generation has a 50 percent chance of being yellow.

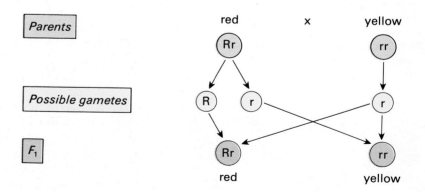

On the other hand, crossing a homozygous dominant parent with a recessive parent results in offspring that show only the dominant trait. In this case, the cross of red-fruited tomato plants with yellow-fruited plants would produce only heterozygous, red-fruited plants.

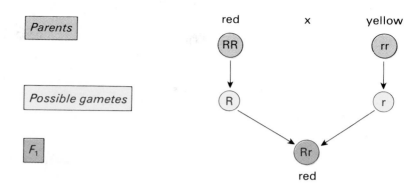

If all the offspring of a test cross show the dominant trait, the result suggests that the parent in question is homozygous. Unfortunately for the breeder, the same result *can* occur by chance when a heterozygous parent is crossed with a recessive one. In general, a breeder can assume that the test organism is homozygous if a large number of offspring are produced and they all show the dominant trait.

Interaction of Heredity and Environment

The genes that an organism inherits are a kind of *potential*. The expression of a gene depends to a large extent on the external environment, the conditions inside the body, and the action of other genes.

For example, heredity and environment interact to produce a person's height. Height is controlled by many genes. Some carry the code for growth hormone, some carry the code for digestive enzymes, and some carry the code for how the body deposits calcium in bones. Regardless of the influence of these genes, if a person doesn't get the proper foods throughout the growing years, that person's adult height is lessened. Remember that no single factor is responsible for human height.

Another example of the interaction of heredity and environment is the development of coat color in the Himalayan rabbit. The Himalayan rabbit is a pure-breeding variety that has a pattern of black paws, ears, and tail and white body. This pattern occurs because the gene for black fur can only be expressed at cool temperatures. As a result, the fur near the rabbit's warm body stays light, and the fur at the cooler extremities

Temperature affects coat color in Siamese cats in the same way it affects the Himalayan rabbit.

19–18 Heredity and environment. A patch of black fur can be induced to grow in a Himalayan rabbit by placing an ice pack over one spot. The gene for black fur is only expressed at cool temperatures.

turns black. In an experiment, a small patch of white hair on the rabbit's back was removed. An ice pack was kept in position while the hair grew back. The new hair that grew under cold conditions came in black. At first it seems that the fur pattern in the Himalayan rabbit is inherited. Instead, the experiment shows that the potential for the pattern is inherited. Its expression depends on the environment.

Building a Science Vocabulary

dihybrid cross	cystic fibrosis	outbreeding
incomplete dominance	hybrid vigor	test cross
linkage	artificial selection	
genetic counselor	inbreeding	

Questions for Review

1. Pea plants that are heterozygous for both stem length (Tt) and pod shape (Rr) are crossed with each other. What are the phenotypes of these parents? What four possible phenotypes can appear in the offspring? What is the probability of each type?

2. When round squash are crossed with long squash, the offspring are all bell-shaped. What type of inheritance is illustrated by this cross? To which of Mendel's laws is this an exception?

3. Why do farmers prefer to grow hybrid corn over pure-breeding corn?

4. Name one purpose for each of the following breeding practices:
   a. artificial selection   b. inbreeding   c. outbreeding

5. What experiment could you perform to see the effect of light (an environmental factor) on chlorophyll production (an inherited trait) in newly sprouted seeds?

Perspective: Technology
Designing an Organism

Sometime in the dim past, some of our early ancestors began breeding animals and plants. The wide variety of horses, cats, dogs, and pigeons today is evidence of those efforts, as is the variety of our modern corn and wheat. The process was relatively simple: those animals or plants that had certain traits were allowed to reproduce; the others were not. Then somewhere along the line came the idea of breeding for combinations of traits—long legs from one breeding line and short tails from another.

We have become quite good at combining desirable traits. We have dairy cattle that can withstand drought, tomatoes that are resistant to disease, corn that resists fungus infections, horses that run faster, pigs that are meatier, hens that lay more eggs. We have, for thousands of years, been designing organisms, making them fit whatever need we defined.

Recently we learned how to put new DNA directly into an organism. We call the technique "gene splicing," since it is much like splicing rope. A bacterium's DNA is split apart by a special technique and a "foreign" section of DNA is inserted. By splicing in particular DNA pieces, the scientist can have the original bacterium do things it couldn't do before, such as produce insulin, fix nitrogen, or digest oil.

Since bacteria reproduce rapidly by asexual means, large numbers of the "new" organism can be produced in a short time. Essentially these become bacterial factories, turning out identical selves by the thousands and millions. Thus, a person with diabetes can have insulin that is much purer than what we now get from pigs and cattle. Giant oil spills can be "cleaned up" by releasing barrels of oil-digesting bacteria. Nitrogen-poor soil can be fertilized by loading the soil with nitrogen-fixing bacteria, rather than by pouring on chemicals.

It is now legal to patent organisms made by gene splicing. Suppose somebody designs the "wrong kind" of new organism, one that causes disease instead of curing it, or one that interferes with some important biological process.

The ability to make organisms puts an enormous responsibility on scientists. It also means that the general population must be aware of the possibilities, so that the right decisions are made.

Chapter 19
Summary and Review

Summary

1. In his experiments with garden peas, Gregor Mendel discovered three basic laws that govern heredity: dominance, segregation, and independent assortment.

2. Mendel discovered the law of dominance when he crossed pea plants that were homozygous for contrasting expressions of a given trait. The dominant expression of the trait appeared in the heterozygous offspring; the recessive expression did not. Today, we know that a dominant gene is expressed in the offspring, while a recessive gene's expression remains hidden.

3. In modern terms, the law of segregation states that paired genes separate during meiosis and recombine at random during fertilization.

4. The law of independent assortment states that each gene of a pair is inherited independently. We now know that chromosomes and not genes are inherited independently.

5. An organism that has two like genes for a given trait is homozygous for the trait and therefore is pure-breeding. An organism with unlike genes for a given trait is heterozygous, or hybrid. Hybrids do not breed true.

6. An organism's genotype can be symbolized by letters. A capital letter stands for the dominant trait. A small letter stands for the recessive trait. Three possible genotypes are homozygous dominant (DD), heterozygous (Dd), and homozygous recessive (dd).

7. An organism's phenotype is the result of gene expression. The dominant phenotype appears in organisms that are homozygous dominant or heterozygous for a trait. The recessive phenotype appears in organisms that are homozygous recessive.

8. Mendel's laws of heredity are based on the laws of chance. They can be used to predict the probability that an organism will inherit a particular genotype and phenotype.

9. Mendel's laws apply to the inheritance of one or more pairs of genes that are inherited independently. Exceptions include incomplete dominance and linkage.

10. Mendel's laws of heredity are put to practical use by genetic counselors and by plant and animal breeders.

11. Both heredity and environment interact in the development of the characteristics of an individual.

Review Questions
Application

1. In sheep, the gene for white wool is dominant; the gene for black wool is recessive. What three crosses could result in a black sheep in the family?

2. According to the law of chance, what is the probability that a sperm carrying a gene for blue eyes will fertilize an egg carrying a gene for brown eyes where both parents are heterozygous? (Assume that eye color is controlled by one pair of genes.) What is the probability of a heterozygous offspring? Explain your answer for each case.

3. When shorthorn red cattle are bred to shorthorn white cattle, they produce roan (red and white mixed) offspring. What type of inheritance is this? If two roan shorthorns are crossed, what is the probability of red, white, and roan coat color in their offspring? Use a Punnett square.

4. The gene for the hereditary disorder cystic fibrosis is recessive. A woman with cystic fibrosis is married to a normal man who has no history of the disorder in his family. They decide to consult a genetic counselor to find out the probability of their having a child with cystic fibrosis. What do you think the counselor will tell them?

5. Which of Mendel's laws of heredity is best illustrated by each of the following?
   a. Our inheritance of eye color is not affected by our inheritance of right-handedness or left-handedness.
   b. Two red-flowered garden peas have some offspring with white flowers.
   c. When striped silk moths are bred to nonstriped silk moths, all their offspring are striped.
   d. All the children in a family have free earlobes. One parent has attached earlobes; the other parent has free earlobes.

6. In an experiment, the tip of a Himalayan rabbit's tail was kept warm from the time of its birth. What color do you think the hair at the tip of the tail became as it grew in? Explain.

Interpretation

1. In cattle, hornless (or polled), (P) is dominant over horned (p). A hornless bull is bred to three cows. First, the bull is bred to cow 1, which is horned, and a horned calf is produced. The bull is then bred to cow 2, which is also horned, and a hornless calf is produced. Finally, the bull is bred to cow 3, which is hornless, and a horned calf is produced. What are the genotypes of all the offspring and parents?

2. About 7 percent of Caucasians are anosmic (incapable of smelling) to the odor of musk. Children of two anosmic parents are also anosmic. Children of normal parents are usually normal, but a few of them are anosmic. Is the gene described above dominant or recessive? Explain your reasoning.

3. We now know that all the genes on a chromosome are inherited together. Why did it seem to Mendel that all genes were inherited separately?

Extra Research

1. Purchase garden pea seeds from a nursery or garden store. Choose two varieties that differ in height. (The Freezonia and Greenshaft varieties grow about 11 cm. The Little Marvel variety grows about 7 cm.) Follow the instructions for planting printed on the envelope. The plants may be grown indoors in window boxes or outdoors. Study heredity in garden peas using Mendel's methods.

2. Grow seeds in the light and in the dark to determine the effect of light on chlorophyll production.

3. Grow seeds under different colored lights to determine the effect of wavelength of light on stem height.

4. Research the life and work of Luther Burbank. Write a short article describing his major accomplishments in plant breeding.

Career Opportunities

Seed growers are farmers who specialize in growing plants for seeds to be planted the following year. They must sometimes carry out special procedures to produce hybrid seeds. Experience and training in agriculture are required. Contact the Canadian Seed Growers' Association, Box 8455, Ottawa, Ontario K1G 3T1.

Animal breeders must have a knowledge of genetics in order to develop new breeds of animals that are more productive and disease-resistant. Breeders also conduct tests and maintain records on new breeds developed by others. A degree in genetics or livestock production is required. Write to the Genetics Society of Canada, 151 Slater St., Ottawa, Ontario K1P 5H4, the Canadian Society of Animal Science, Centralia College, Huron Park, Ontario N0M 1Y0, or one of the many breeder's associations.

Chapter 20

Heredity: Chromosomes and Genes

Focus "*The multiplicity of shape, colour, and behavior in individuals and in species is produced by the coupling of genes, as Mendel guessed. As a matter of mechanics, the genes are strung out along the chromosomes, which become visible only when the cell is dividing. But the question is not how the genes are arranged; the modern question is, How do they act? The genes are made of nucleic acids.* That *is where the action is."*

Jacob Bronowski, The Ascent of Man

20.1 RESEARCH IN CHROMOSOMAL GENETICS

Mendel studied heredity by looking at outcomes. He was like a person learning about the automobile industry by watching cars drive out of the factory. Improved microscopes helped scientists look into the factory windows. Scientists learned that a cell's chromosomes act in accordance with Mendel's laws.

In 1902, Walter S. Sutton and Theodor Boveri each realized independently that the units of heredity, the genes, are contained in chromosomes. They identified the chromosome as the hereditary structure transmitted from parents to offspring. Spurred on by their ideas, other scientists also focused on the

20–1 Squash come in all shapes, colors, sizes, and textures. These visible traits are evidence of the great potential for variation carried in genes.

559

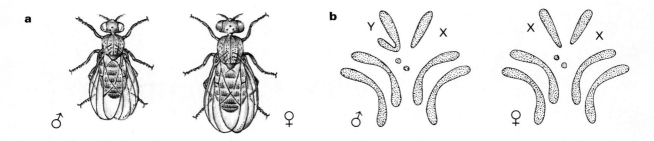

20–2 The common fruit fly, *Drosophila melanogaster*: **a.** adult male (left) and adult female; **b.** chromosomes of male (left) and female. The top pairs are the sex chromosomes.

Morgan established an American school of genetics. He and his students made many major contributions to the study of genetics.

cell's chromosomes. One of the pioneers in the new field of chromosomal genetics was Thomas Hunt Morgan. Morgan, and many of his colleagues, studied fruit flies the way Mendel studied garden peas to learn about heredity.

The fruit fly, or vinegar fly, *Drosophila melanogaster*, has a lot of advantages for genetics research. Fruit flies are easy to culture. Their life cycle is only 10 to 14 days. They produce many offspring. They have a diploid number of 8. They have very large (giant) chromosomes in their salivary glands, which are easy to isolate and study. Finally, they have several inherited variations that can be seen easily. By crossing fruit flies, Morgan and other researchers learned many things about heredity.

Sex Determination

Morgan's investigations on fruit flies showed that an individual's sex is determined by the type of chromosomes it inherits. He found a difference between the chromosomes of male and female fruit flies. What Morgan found in fruit flies we now know is also true in humans.

Inheritance of Sex in Humans In 1956, Dr. Joe Hin Tijio and Dr. Albert Levan developed a new method of separating, staining, and photographing human chromosomes. This method enabled them to observe individual chromosomes. They found that human cells contain 46 chromosomes (23 pairs). In females, all 23 pairs are alike or homologous. In males, however, only 22 pairs are homologous. The chromosomes in the other pair differ in size and shape (see Fig. 20–3). In females, the twenty-third homologous pair is referred to as XX. In males, the unmatched pair is referred to as XY. The male X chromosome is like the female X chromosome. The Y chromosome is different. The X and Y chromosomes are called **sex chromosomes**. All the other pairs of chromosomes are known as **autosomes** (AHT-uh-sōmz). A human has 22 pairs of autosomes and 1 pair of sex chromosomes.

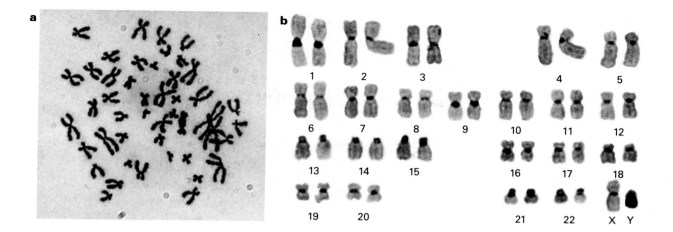

20–3 Human chromosomes: **a.** micrograph of female chromosomes; **b.** karyotype of male chromosomes. A karyotype is made by cutting an enlarged micrograph of chromosomes (taken at metaphase of mitosis) into pieces and rearranging the chromosomes into homologous pairs. The autosomes are lined up in order of decreasing size.

When eggs are formed, each egg carries an X chromosome. When sperm are formed, half the sperm carry a Y chromosome; the other half carry an X chromosome. At fertilization an egg has an equal chance of being fertilized by a sperm carrying a Y chromosome or by one carrying an X chromosome.

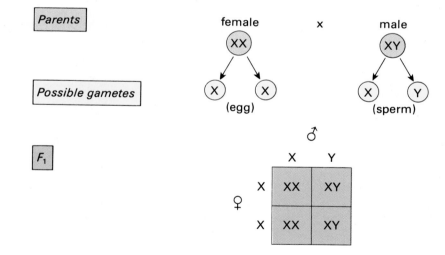

This analysis of sex inheritance deals with chromosomes and not genes.

At each fertilization the offspring has an equal chance of becoming male or female.

In humans, a Y chromosome must be present in the zygote for an offspring to develop into a male. In the absence of the Y chromosome, an offspring develops into a female. An egg can only carry an X chromosome. A sperm can carry either an X or Y chromosome. The sperm, therefore, determines whether the offspring will be male (XY) or female (XX).

Section 20.1 Research in Chromosomal Genetics

Sex Linked Inheritance

Recall from Chapter 16 that a mutation is a change in a gene. A mutated gene that is carried in a gamete is inherited.

In the fruit fly normal eye color is red. White eye color is caused by a mutation. In one of Morgan's experiments, he crossed a white-eyed male fly with a red-eyed female. All the F_1 offspring had red eyes. These results were not unusual since red eyes are dominant over white eyes. The surprising results came when Morgan crossed an F_1 red-eyed female with an F_1 red-eyed male. In one way, the results were predicted by Mendel's law of dominance. Red eyes were dominant over white eyes by a ratio of 3 to 1. But, the flies with white eyes were *all males*. Morgan realized that this did not happen by chance alone. Morgan had identified a trait (white eyes) that was directly related to the sex of the organism. Such traits are called **sex-linked**. The genes for sex-linked traits are carried on the X chromosome.

The symbol for sex-linked traits is X with the sex-linked gene shown as a superscript.

Homologous chromosomes have genes for the same trait in corresponding positions. In human females, the sex chromosomes, XX, are homologous. In human males, the sex chromosomes, XY, are not homologous. The genes on the X chromosome do not carry information for the same traits as the genes on the Y chromosome. For that reason, any gene on the X chromosome of a male is expressed in the phenotype of the offspring. Even a recessive gene is expressed if it appears on a male's X chromosome, because it is the only gene the individual has for that trait.

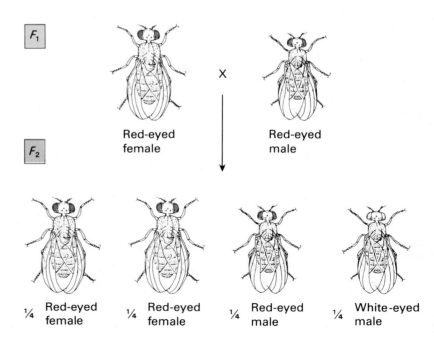

20–4 Morgan's experiment on sex-linked inheritance of eye color in fruit flies.

In the case of Morgan's white-eyed male flies, the gene for eye color is carried on the X chromosome. Half the F_2 males received the gene for red from the F_1 females; the other half received the gene for white. Since they had only one copy of the X chromosome, half the males had red eyes; the other half had white eyes. The F_2 females all had red eyes because they all received the gene for red on the X chromosome from the F_1 males.

Hemophilia: A Sex-Linked Trait You can understand sex-linked inheritance more easily if you work with a specific example. In humans, **hemophilia** (HEE-muh-FIL-yuh) is an inherited disorder that is sex-linked. The blood of a hemophiliac does not clot properly. If cut, hemophiliacs bleed uncontrollably. Hemophilia is caused by a recessive gene that is located on the X chromosome (X^h). The gene for normal blood is (X^H). The disease is rare, but it did occur in the British royal family. Queen Victoria had ten descendents with hemophilia, even though neither she nor her husband, Prince Albert, had the disease.

As in the example of the British royal family, sons often inherit sex-linked diseases from mothers with normal phenotypes. Females who exhibit the normal phenotype but pass the trait to their offspring are called **carriers**. Carriers are hetero-

20–5 Queen Victoria's lineage provides a study of a human sex-linked trait, hemophilia. Sex-linked traits show up more often in males than in females because males receive only one copy of the X chromosome.

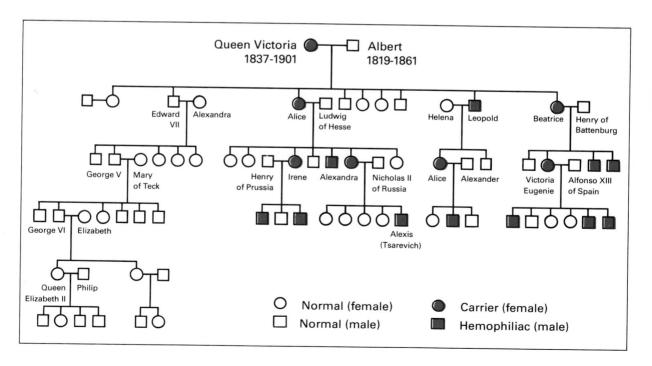

Section 20.1 Research in Chromosomal Genetics

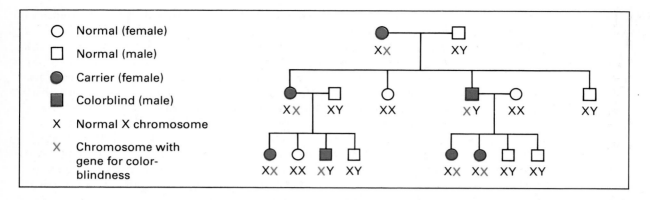

20–6 Colorblindness in humans is a recessive trait carried on the X chromosome. The diagram shows how an original carrier female (top) passes the trait to subsequent generations. On the average, half the sons of a carrier will be colorblind.

zygous for the sex-linked trait. They carry the recessive gene on one of their X chromosomes and the dominant gene on the other X chromosome. Queen Victoria's genotype was $X^H X^h$. Although she did not have hemophilia, her sons had one chance out of two to inherit the disorder.

All of the hemophiliacs in the British royal family were male. Recessive traits on the X chromosome show up more often in males than in females, since males receive only one copy of the X chromosome. In order for a female to have a recessive trait, such as hemophilia, her father must have the trait, and her mother must have at least one recessive gene for the trait.

Nondisjunction

Calvin Bridges started his career in genetics as one of Morgan's students. In the course of his career, he discovered a genetic event known as **nondisjunction**. Nondisjunction is the failure of chromosomes to separate during meiosis. Based on Morgan's explanation of sex-linked inheritance, Bridges predicted that white-eyed female flies crossed with red-eyed males should always result in an F_1 generation of red-eyed daughters and white-eyed sons. He found that the expected results happened *almost* all the time, but not all the time. About one fly in 2000 to 3000 of the F_1 offspring had an eye color that was not predicted. Occasionally, males had red eyes, and females had white eyes.

Bridges figured out how these exceptional flies with the "wrong-colored" eyes appeared. He said this happened because the mother's X chromosomes did not separate during meiosis. Therefore, the F_1 flies inherited the wrong set of sex chromosomes. Bridges said that the pair of X chromosomes in the mother failed to "disjoin," so he called the condition *nondisjunction*. Because of nondisjunction, some eggs contain two X chromosomes and other eggs contain no X chromosomes.

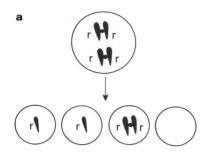

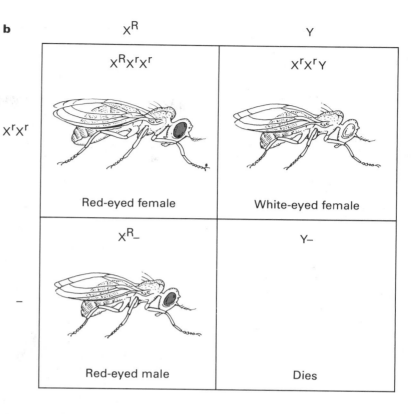

20–7 Nondisjunction. **a.** Meiosis occasionally results in chromosomes not separating so that a gamete may receive too many or too few of a chromosome, shown here for the X chromosome of white-eyed female flies. **b.** Diagram of Bridges' unexpected results of red-eyed males and white-eyed females due to nondisjunction. The red-eyed male received no sex chromosome from the female parent.

Sex in *Drosophila* is determined by the number of X chromosomes. (This differs from humans and other mammals where the Y chromosome determines the sex.) In *Drosophila*, a zygote with two or more X chromosomes develops into a female. A zygote with one X chromosome develops into a male. The unusual white-eyed female that appeared in Bridges' experiments resulted from the fertilization of an X^rX^r egg by a Y-carrying sperm. The male flies with red eyes received no sex chromosome from the females and an X^R from the males during fertilization. The unusual eggs were the result of nondisjunction.

Other Chromosomal Disorders Owing to Nondisjunction

Too many chromosomes or too few chromosomes owing to nondisjunction can result in serious disorders. By observing and counting chromosomes, scientists discovered that there was an extra chromosome in the cells of people suffering from a hereditary disorder known as **Down's syndrome**. A person with this disorder has characteristic facial features, a smaller head, a short stocky body, and is mentally retarded. Chromosome studies show that Down's syndrome is caused by nondis-

Down's syndrome was named for Langdon Down who first described the symptoms in 1866. Down's syndrome is also called trisomy 21 syndrome, which refers to the three copies of chromosome 21 in the cells of people with this disorder.

Section 20.1 Research in Chromosomal Genetics

20-8 A Down's syndrome child, though mentally retarded, is able to participate in and enjoy many activities.

junction in chromosome number 21. There are three copies of chromosome 21 instead of two. Physicians can identify the extra chromosome in Down's syndrome in a developing embryo. This is done by studying the offspring cells that appear in a sample of the mother's amniotic fluid. This type of test is used to diagnose Down's syndrome and other hereditary disorders before a baby is born.

Building a Science Vocabulary

sex chromosome   hemophilia   Down's syndrome
autosome   carrier
sex-linked   nondisjunction

Questions for Review

1. Briefly state one major accomplishment of each of the following scientists in the field of genetics.
   a. Walter Sutton and Theodor Boveri
   b. Thomas Hunt Morgan
   c. Calvin Bridges

2. How is sex inherited in humans?

3. In *Drosophila*, the gene for white eye is sex-linked and recessive (X^r). Use a Punnett square to analyze a cross between a white-eyed male ($X^r Y$) and a homozygous red-eyed female ($X^R X^R$).

4. Use a Punnett square to analyze a cross between a red-eyed male *Drosophila* ($X^R Y$) and a white-eyed female *Drosophila* ($X^r X^r$). Compare the results with those from the cross in question 3.

5. What is nondisjunction?

6. The cells of people with Down's syndrome contain an extra chromosome. Explain how this condition occurs.

20.2 CHROMOSOMES CONTAIN GENES

Thomas Hunt Morgan associated genes with chromosomes. He explained that genes were lined up on the chromosome like beads on a string. Neither he nor the scientists of his day had any idea of the structure or working of DNA. Nevertheless, they continued to make discoveries about genes.

A chromosome contains a helix of DNA 1 m long plus other nucleic acids plus protein all packaged into a structure only 1 µm long.

Gene Mapping

When Morgan suggested that genes were located in a single line along the length of a chromosome, scientists in his laboratory went to work to construct a **gene map** for fruit flies. A gene map shows the location of the genes on a chromosome. Morgan's co-workers studied linkage and crossing-over to help determine the order of the genes.

Since genes are linked on chromosomes, offspring tend to inherit somewhat the same sets or combinations of traits that their parents have. When crossing-over occurs during meiosis, the linked genes break apart and reattach to the genes on the homologous chromosome. Offspring then get a different combination of traits than is seen in the parents. Morgan reasoned that crossing-over has a greater chance of occurring between genes that are far apart on a chromosome than between genes that are close. Geneticists in his laboratory watched the results of thousands of *Drosophila* crosses. Each time they watched for new gene combinations in the offspring. The rate of new combinations gave them the information to construct gene maps for *Drosophila* chromosomes.

0.0	Yellow body
1.5	Scute bristles
3.0	White eyes
5.5	Facet eyes
7.5	Echinus eyes
	Ruby eyes
13.7	Crossveinless wings
20.0	Cut wings
21.0	Singed bristles
27.7	Lozenge eyes
33.0	Vermilion eyes
36.1	Miniature wings
43.0	Sable body
44.0	Garnet eyes
56.7	Forked bristles
57.0	Bar eyes
59.5	Fused veins
62.5	Carnation eyes
66.0	Bobbed hairs

20–9 Gene mapping in *Drosophila*. Genes for some of the traits on the X chromosome are shown. The numbers on the left represent the relative positions of the genes along the chromosome.

20–10 More crossovers can occur between genes that are far apart on a chromosome (B and C) than between genes that are close (A and B). This principle underlies the making of gene maps.

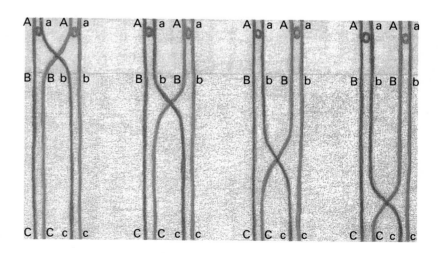

Mutations

Mutations that occur in body cells cannot be inherited.

Crossing-over is one cause of variation from parent to offspring; mutation is another. A gene or chromosome that has undergone mutation may be present in a gamete. Then the mutation is inherited. Some mutations, **chromosomal mutations**, may involve addition of part of, or a whole, chromosome, loss of part of, or a whole, chromosome, a broken chromosome, or rearrangement of a section of a chromosome. Down's syndrome is an example of a chromosomal mutation involving the addition of a whole chromosome. Other mutations, **gene mutations**, result from changes in DNA. Hemophilia is caused by a gene mutation. Most mutations are recessive.

20-11 Normal and albino corn seedlings. Albinism in plants is a lethal mutation since albino plants cannot photosynthesize.

Some mutations help a species adapt to a changing environment; however, many mutations are harmful. Harmful mutations reduce an individual's chances of survival. Some mutations are killing, or **lethal**. When a lethal mutation is expressed in an organism's phenotype, the organism dies. Albinism in green plants is an example of a trait caused by a lethal gene. Since albino plants are unable to produce chlorophyll, they cannot carry out photosynthesis and die of starvation. Mutations occur rarely. If they occurred often, we would not be able to predict heredity with Mendel's laws.

Causes of Mutations Mutations may occur from radiation or chemicals in nature; for example, the ultraviolet radiation the earth receives from the sun or the chemicals dissolved from rocks by water. They may also occur due to the effects of human activities. In 1927, Herman J. Muller was the first person to artificially cause mutations in the laboratory. He irradiated fruit flies with X-rays, producing many different types of flies. Many of them were quite strange. Most of the mutations produced would be lethal if the flies lived in nature. As a result of his work, Muller tried to alert people to the dangers of radiation in the environment.

Substances that cause mutations are known as **mutagens**. Nitrogen-mustard gas, which was used in World War I, was the first substance to be identified as a mutagen. Since then, many chemicals and types of radiation have been tested, and a long list of mutagens has been compiled. People are now much more aware of the genetic effects of radiation and chemicals. As a result, scientists are attempting to identify and control the use of these materials.

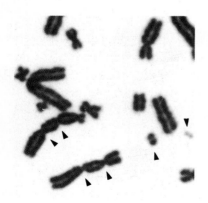

20-12 In addition to causing mutations, radiation can cause physical damage to chromosomes, as shown in this micrograph.

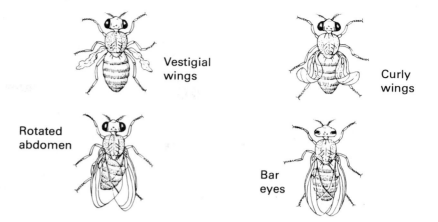

20–13 Common mutations in fruit flies as a result of treatment with radiation.

Breeders of plants and animals often use naturally occurring mutations to their advantage. For example, a short-legged breed of sheep was developed from a lamb that had unusually short legs. The advantage to the breeder is that the sheep cannot jump over fences. Delicious apples, navel oranges, seedless grapes, double flowers, and cut-leaves are examples of mutations that have been developed by plant breeders. Breeders also deliberately use irradiation and chemicals to cause mutations, hoping that a beneficial one will appear among the many bad ones. The late-ripening peach is an example of a crop that was improved in this way.

Polyploidy **Polyploid** cells have more than two sets of chromosomes and represent a kind of mutation. Scientists can induce polyploidy by applying the chemical **colchicine** (KAHL-chuh-SEEN) to cells during mitosis. Colchicine breaks up microtubules, without which a spindle does not form. As a result, the replicated chromosomes do not move apart, and cell division does not occur. The result is a polyploid cell. Polyploidy occurs more often in plants than it does in animals.

Seeds treated with colchicine produce polyploid plants, which often have large stems, leaves, flowers, and fruits. The plants are also generally larger than the parent plants. Breeders use colchicine to develop commercially important polyploid varieties, such as the large watermelons, marigolds, and snapdragons available today.

20–14 The banana grown commercially today, which is triploid, is a naturally occurring polyploid plant. Polyploid plants and their fruits are usually larger than they would be if they were diploid.

Alleles: Alternate Versions of a Gene

Genes are composed of DNA. When a gene mutation occurs, the structure of the DNA in the gene changes. Because of mutations, alternate versions of genes exist in a population. In garden peas, for example, tall and short are alternate versions of the gene for height. Alternate versions of a gene for a trait are called **alleles** (uh-LEELZ).

In the cell, alleles are found at the same position, or **locus**, in homologous chromosomes. If the allele for a given trait is the same on both homologous chromosomes, the individual is homozygous. If the allele for a given trait is different on both homologous chromosomes, the individual is heterozygous.

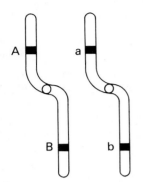

20-15 In any pair of homologous chromosomes there can be only two alleles for a given trait. Alleles are found at the same position, or locus, on each chromosome of the pair.

Multiple Alleles

Mutations can occur any number of times in a given gene. Some genes have two, three, or more alleles in a population. Genes that have more than two alleles in a population are called **multiple alleles**. Note, however, that only two alleles for any single trait can exist on a pair of homologous chromosomes.

For example, multiple alleles in humans determine whether a person's blood type is A, B, AB, or O. Three alleles for blood type exist among humans. They are designated A, B, and O. An individual inherits one of these alleles from the father and the other allele from the mother. The following combinations for blood type are possible: AA, AO, BB, BO, AB, and OO. Alleles A and B are dominant over O, but are both expressed when they occur together (alleles A and B are said to be *codominant*). As a result, four phenotypes are found: A, B, AB, and O.

Consider this situation. Jeff has blood type A. His wife, Jean, also has blood type A. Their child, Joey, has blood type O. What is the genotype of each member of this family? You may recognize this problem as an example of Mendel's law of segregation. Joey's genotype must be OO. Therefore both parents have one A allele and one O allele (AO).

The alleles for blood type are located at a specific locus on the homologous chromosome pair number 2.

Blood Types and Blood Transfusions

The blood type of a blood donor must be compatible with that of a recipient. If a person receives the wrong blood type, the incoming blood cells clump together and can cause the death of the recipient. This reaction is part of the body's defense against foreign substances.

Table 20-1
Inheritance of Blood Type

Genotype	Phenotype
AA	A
AO	A
BB	B
BO	B
AB	AB
OO	O

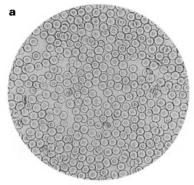
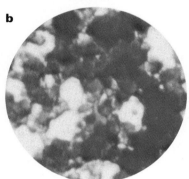

20-16 Clumping reaction in blood: **a.** normal red blood cells, **b.** cells of type B blood after mixing with serum containing anti-B antibodies. Once clumping occurs it is irreversible and the blood cells die.

Antigen-Antibody Reaction The clumping reaction of blood is due to a reaction between antigens and antibodies. An **antigen** is a foreign substance that comes in contact with the body. Bacteria and viruses are examples of antigens. Chemicals may also be antigens. The surfaces of red blood cells of blood types A, B, and AB have certain proteins, which act as antigens. The second column of Table 20-2 lists the antigens that occur in the red blood cells of each blood type. Notice that type O has no antigens on the red blood cells. Type AB has both antigen A and antigen B.

Antibodies are the body's response to antigens. Antibodies are special protein substances that "fight off" specific antigens. The antibodies that fight off antigens A and B are located in the blood plasma. The center column of Table 20-2 lists the antibodies found in the plasma of each blood type. Notice that type AB has no antibodies in the plasma. This makes sense, since antigens A and B are not foreign substances in type AB blood. Type O has antibodies anti-A and anti-B. This also makes sense, because antigens A and B are foreign to type O blood.

To see how the system works, suppose the blood of a type A donor is transfused into a type O recipient by mistake. The donor's red blood cells carry antigen A. The recipient's plasma contains antibodies anti-A and anti-B. Immediately upon contact, antibody anti-A reacts with antigen A, and clumping occurs.

Table 20-2 Antigen-Antibody Reactions in Humans

Blood Type	Antigens	Antibodies	Reaction with Anti-A	Reaction with Anti-B
A	A	Anti-B	Clumping	No clumping
B	B	Anti-A	No clumping	Clumping
AB	A and B	None	Clumping	Clumping
O	None	Anti-A and anti-B	No clumping	No clumping

Section 20.2 Chromosomes Contain Genes

20–17 Blood donors lie down while giving blood to avoid dizziness or fainting. The parts of the blood are separated and stored for future use.

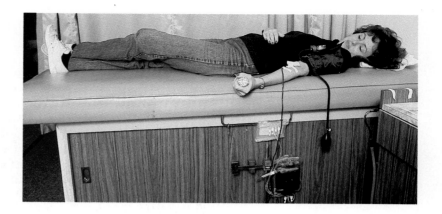

Refer again to Table 20–2. The two right-hand columns show all the possible reactions that occur between the A-B-AB-O blood types. Since type O has no antigens, it can be safely transfused into any of the other blood types. People with type O blood are therefore known as **universal donors**. However, type O individuals can receive only type O blood, since they have both anti-A and anti-B antibodies. Type AB blood has both antigens A and B and no antibodies. Type AB can receive blood from any of the other blood types. People with type AB blood are therefore known as **universal recipients**. Type AB individuals can only donate blood to others with type AB, however, since they have both A and B antigens.

Antigen-antibody reactions are the basis for determining what type blood a person has. Try the following sample problem to be sure that you understand how the system works.

You are a technician in a biology laboratory. You are given a sample of blood and told to determine its A-B-AB-O type. First, you put a few drops of antibody anti-A on a little of the blood. The blood does not clump. Then you apply some antibody anti-B to another small amount of blood. With anti-B, clotting occurs. What is the person's blood type?

Matching Blood for Transfusions A, B, AB, and O blood types are one system of inherited differences found in human blood. Other inherited blood types include the Rh factor, a system of M, N blood groups, a P system, and an HLA system. All of these blood systems must be tested and matched for transfusions and tissue grafting. The Rh factor is particularly important in pregnant women. In cases where Rh-positive embryos develop inside Rh-negative mothers, treatment is sometimes necessary to prevent an antigen-antibody reaction that would destroy the offspring's blood.

Multiple-Gene Inheritance

When you look at a population, you see a gradation of expression for many traits. People are not either tall or short. Mice are not either fat or thin. Ears of corn are not either long or short. Instead, there are many sizes in between. Even in cases of inherited diseases people do not simply have the disease or not. There are many differences in how sick a person gets.

Geneticists explain the existence of many gradations of a trait by **multiple-gene inheritance** (also known as multiple factors). In multiple-gene inheritance two or more pairs of genes control a single trait. Each gene pair works according to the laws of dominance and segregation. The organism's phenotype depends on the combined effect.

This type of inheritance is difficult to study. Geneticists are not sure how many gene pairs are involved for most traits of this kind. Furthermore, it is very difficult to separate the effects of environment and heredity in these cases. Nevertheless, the concept of multiple-gene inheritance does help explain how many traits in a population are inherited.

20–18 Multiple-gene inheritance is responsible for the wide range of corn ear length. Since more than one pair of genes affects ear length many different phenotypes are possible. The phenotype observed depends on the combined influence of the genes involved and the effects of the environment.

Building a Science Vocabulary

gene map
chromosomal mutation
gene mutation
lethal
mutagen
polyploid
colchicine
allele

locus
multiple alleles
antigen
antibody
universal donor
universal recipient
multiple-gene inheritance

Questions for Review

1. How did Morgan figure out whether two linked genes were close to each other or far away from each other on a chromosome when he developed gene maps for *Drosophila*?

2. State one difference and one similarity between a gene mutation and a chromosomal mutation.

3. What is the danger of radiation and chemical mutagens in the environment? How did Morgan's experiments on *Drosophila* make him aware of this danger?

4. How do breeders make use of mutations in their work?

5. The gene for red flowers and the gene for white flowers are alleles in garden peas. Explain what this means. Assume that the gene for white flowers was the original flower-color gene in peas. How might the gene for red color have come into being?

6. Why is a person with type O blood considered a universal donor? Why is a person with type AB blood considered a universal recipient?

7. In a certain type of wheat, kernels are red, white, or a shade of pink. When pink-seeded varieties are crossed, the offspring show a continuous gradation of color from white to red. What theory would explain this genetic result? How would this theory explain the large number of phenotypes for this trait?

Perspective: Discovery
Simultaneous Discovery

Gregor Mendel's discovery of the laws of inheritance was not recognized by the scientific world at the time his results were published. It wasn't until years later that Mendel's conclusions occurred to another scientist. In fact, rediscovery of the laws of inheritance occurred simultaneously by three scientists working independently in three different countries. Hugo de Vries, a Dutchman, Carl Correns, a German, and Erich Tschermak von Seysenegg, an Austrian, each discovered Mendel's results within days of each other.

"Simultaneous discovery" has occurred numerous times in science. In an earlier chapter you learned about the development of the cell theory by Scheliden and Schwann. They were working independently at the time. Later you will learn that Alfred Russel Wallace, who was conducting research in Malaysia, came to conclusions about evolution that were essentially the same as Charles Darwin's and at the same time. At the turn of the century, Walter S. Sutton, an American, and Theodore Boveri, a Bavarian, simultaneously and independently recognized the striking similarities in the behavior of chromosomes in nuclear division and the behavior of Mendel's factors (genes) in inheritance. Both independently proposed the chromosomal basis of inheritance.

Closer to our own time there have been discoveries about the role, structure, and action of DNA and of vaccines against poliomyelitis (infantile paralysis). Within a few months, two different approaches to preventing polio were discovered, the Salk vaccine and the Sabine vaccine. And, only by a hair's breadth did Watson and Crick beat out Linus Pauling in the discovery of the chemistry of the gene.

As science and time move on, certain problems and questions assume greater importance and thus receive greater emphasis. The time ripens for discovery as efforts are concentrated on these questions. Today, simultaneous discovery is even more likely since the speed of communicating research results is so much faster. Also today, government agencies and private foundations make funds available for particular kinds of research. As a result it is likely that there will be even more examples of simultaneous discovery in the near future.

20.3 THE NATURE OF GENES

In 1953, James Watson and Francis Crick were credited with creating the double helix model of DNA. The work they and others performed provided a vital clue to figuring out the puzzle of heredity. It showed the structure of the gene and helped explain how genes work. Geneticists no longer had to limit their work to watching the results of crosses and looking at chromosomes through the microscope. Knowing the structure of DNA brought them to the center of the hereditary process.

Research That Led to the Model of DNA

The model of DNA was a giant step forward in understanding heredity. It was preceded, however, by many smaller, yet important, steps performed by a number of brilliant researchers. Some experimenters pointed to DNA as the genetic material, and others suggested that genes control the production of enzymes. The work of some of the most important of these pioneers is described below.

Friedrich Miescher Friedrich Miescher, a Swiss biochemist, applied protein-dissolving chemicals to cells in 1869. Then he observed the nuclei as the protein began to disappear. The nuclei got smaller, but they did not completely disappear. Miescher decided that the nucleus of a cell must contain something other than protein. Further tests showed that the nucleus of a cell contains a substance that is made of carbon, oxygen, hydrogen, nitrogen, and phosphorus. Miescher discovered what we call nucleic acid in the nucleus of a cell. Today we know that the nucleic acid in a cell's nucleus is DNA.

Fred Griffith In 1928, Fred Griffith, an English bacteriologist, was working with two separate strains of bacteria. One strain, the S-strain, caused pneumonia when it was injected into mice. The other strain, the R-strain, did not cause pneumonia. Griffith killed some of the S-strain bacteria with heat. He then mixed the heat-killed S-strain bacteria with live R-strain bacteria. To his surprise, the live bacteria caused pneumonia when injected into mice. Griffith decided that some material from the dead bacteria changed the live bacteria. He said that his material *transformed* the live R-strain bacteria into live S-strain bacteria. He did not know what the substance was that caused the transformation of R-strain bacteria into S-strain bacteria, but he knew that it acted like a gene.

Griffith was searching, without success, for a vaccine. The results of his experiments were published; however, at the time he did not realize the significance of his findings.

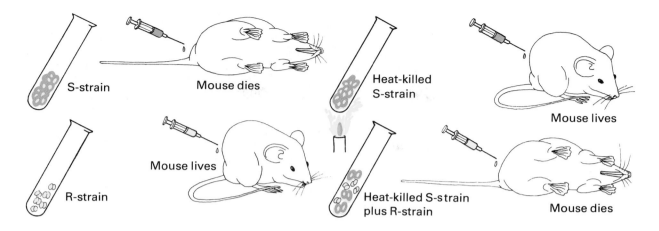

20-19 Griffith's experiment using two strains of pneumonia bacteria demonstrated that one strain of bacteria is able to be transformed into another strain.

Avery, McCarty, and MacLeod In 1944, scientists discovered the substance that caused the bacterial transformation observed in Griffith's experiment. Working at the Rockefeller Institute, O. T. Avery, Maclyn McCarty, and Colin MacLeod showed that the transforming substance was DNA. They proved for the first time that the chemical substance DNA could change an organism's genotype. They also showed that DNA is able to replicate itself precisely. Their experiments provided strong evidence that genes are made of DNA.

Hershey and Chase Alfred D. Hershey and Martha Chase studied a special type of virus that attacks bacteria. Such a virus is called a **bacteriophage**, or **phage**.

Hershey and Chase "labeled" phages with radioactive materials, which enabled them to follow the phages as they attacked bacterial cells. Hershey and Chase found that when a virus attacks a cell, it injects its nucleic acid core into the host cell. The empty protein coat stays outside. The phages in the Hershey-Chase experiment had nucleic acid cores made of DNA. Inside the bacterial host, the viral DNA replicated. Then the replicated DNA cores used host materials to form new protein coats. Finally, the newly formed viruses burst out of the host cell. The experiment showed that viral DNA carries the information to reproduce more viruses. This chemical information is the kind of message that genes carry. Here was more evidence that genes are made of DNA.

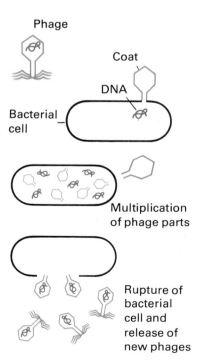

20-20 Hershey and Chase labeled phages with radioactive materials and discovered that a phage's DNA alone enters a bacterial cell, replicates, and then uses host materials to build new protein coats.

Section 20.3 The Nature of Genes

Beadle's and Tatum's hypothesis is now referred to as the "one gene–one polypeptide chain" hypothesis.

Beadle and Tatum In 1941, twelve years before Watson and Crick developed their model of DNA, George W. Beadle and Edward L. Tatum proposed a hypothesis about how genes work. They called it the **one gene - one enzyme hypothesis**. These scientists were doing research in genetics using the bread mold *Neurospora* at Stanford University. From the results of their experiments, they suggested that a gene controls the production of an enzyme.

Neurospora requires only water, sugar, inorganic salts, and one vitamin to grow and reproduce. Scientists call *Neurospora's* limited diet **minimal medium**. *Neurospora* synthesizes everything its cells need, including all 20 amino acids, from the substances in minimal medium.

Beadle and Tatum exposed *Neurospora* to X-rays, and mutations occurred. The mutations did not make the bread mold look different. Instead, some of the mutations affected the mold's ability to synthesize amino acids. Beadle and Tatum placed bits of the mutated *Neurospora* in 20 kinds of media, each containing a different amino acid. The mutated molds survived only where the medium supplied the amino acid that the mold could no longer synthesize for itself. It seemed to Beadle and Tatum that a single mutation in *Neurospora* caused the loss of its ability to synthesize one amino acid. They reasoned as follows that a single gene controls the production of an enzyme:

1. A mutation causes a change in a gene.
2. When a mutation occurs, synthesis of a specific amino acid is stopped.
3. Each synthesis in a cell requires an enzyme.
4. A mutation must stop the production of an enzyme.
5. Therefore, a gene must give rise to an enzyme.

In this way, Beadle and Tatum figured out for the first time what a gene actually does. Their one gene–one enzyme hypothesis stated that a gene controls the production of an enzyme.

After the development of the model of DNA, scientists showed that a gene actually controls the production of one polypeptide chain. The polypeptide chain may then be used as an enzyme, or part of an enzyme, that controls a chemical reaction. The reaction may be a simple one. Or, it may be one reaction in a complex series of enzyme-controlled reactions that take place in a cell.

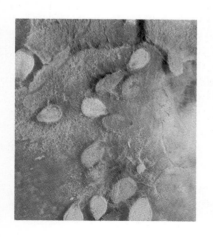

20–21 Bread mold (*Neurospora*) growing on wheat bread. *Neurospora* is easy to culture in the lab using a simple synthetic medium.

20–22 Beadle's and Tatum's experiment. 1. Expose *Neurospora* to X-rays to induce mutations. 2. Grow individual spores in enriched medium. 3. Transfer subculture of mold from enriched medium to minimal medium. No growth may indicate lack of ability to make an amino acid. 4. Grow subcultures on 20 minimal media each supplemented with a different amino acid. Medium on which growth occurs indicates which amino acid cannot be made by the mold.

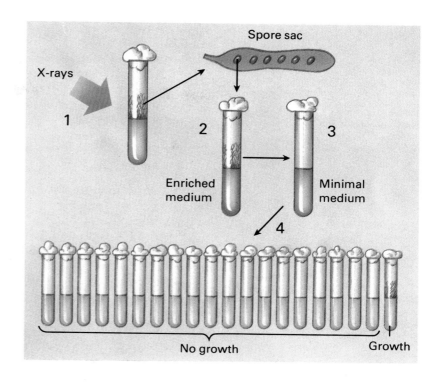

The Effect of a Gene on an Enzyme

The relationship between a gene and an enzyme can be seen in a hereditary human disorder called **phenylketonuria** (FEN-il-KEET-uh-NOOR-ee-uh), **(PKU)**. Most children inherit at least one normal gene that controls the production of an enzyme that changes phenylalanine, an amino acid common in food, into the amino acid tyrosine. A child with PKU inherits a mutant gene and is not able to produce the enzyme that changes phenylalanine to tyrosine. Without the enzyme, phenylalanine accumulates and can result in mental retardation and brain damage. Phenylketonuria affects about one child in ten thousand.

The urine of newborn babies is tested for the presence of phenylalanine. Phenylalanine in the urine identifies PKU. If an infant is found to have the disorder, a special diet without phenylalanine is prescribed, and the child develops normally. Doctors can help many people with genetic disorders, in part because they now know what certain genes do.

20–23 A white gorilla is a mutant of the normal black gorilla. The condition in which an organism lacks pigment is called albinism. It results from mutation of the gene responsible for making the enzyme needed to produce pigment.

Section 20.3 The Nature of Genes

How Genes Work

A gene works through the process of protein synthesis (see Chapter 15). It controls the number, kind, and order of amino acids in a protein. When a gene's DNA unzips during protein synthesis, a certain number and kind of DNA bases are exposed. The order of the bases in DNA is a message that determines the order of amino acids in a protein. For this reason, scientists call the order of DNA bases the **genetic code**.

The term genetic code *also refers to the sequence of bases on mRNA. Similarly,* codon *also refers to a triplet of mRNA bases.*

The Genetic Code Like other codes, the genetic code has a language all its own. The genetic code is made up of code words called **codons**. Each codon is made up of three nucleotide bases, as in AAA, CCA, GCA, and ACT. The genetic code has a triplet for each of the 20 amino acids. For example, CCA is the DNA triplet for the amino acid glycine. Some amino acids are encoded by only one triplet (tryptophan), while others are encoded by as many as six (leucine). The genetic code also has triplets that start and stop protein synthesis.

Table 15–2 on page 449 shows all the mRNA triplets and their corresponding amino acids.

Transmission of the Genetic Code The genetic code is sent from DNA in the nucleus to amino acids in the ribosomes. This transmission occurs during protein synthesis. The pathway that the genetic code follows is:

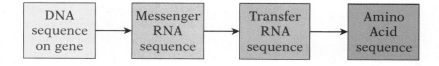

The sequence of base triplets in the DNA of a gene controls the building of a particular protein. The number, kind, and order of amino acids in the protein are determined by the number, kind, and order of triplets in the gene.

The Genetic Code and Heredity The sequence of nucleotides on a gene carries the code for the proteins in an organism. Some of these proteins make up part of the organism's structure. Other proteins make up its enzymes, which control the chemical reactions that occur in the organism. A significant part of what you are is due to your individual set of proteins. The blueprint for your proteins is the DNA that you inherited from your parents.

Proteins vary from person to person. As a result, protein analysis of blood or other tissues can solve a crime by helping to identify the criminal.

How Mutations Affect Genes

A gene mutation is a change in the DNA. If such a change is carried in a gamete, the gene mutation is inherited. The DNA in a cell can be changed in many ways. Pieces of DNA can be added, lost, or switched around. The simplest type of mutation is a change from one base to another. Such a mutation changes a DNA triplet. The changed DNA triplet results in a protein with a changed amino acid.

Sickle-Cell Disease **Sickle-cell disease** is a genetic blood disorder that affects the hemoglobin in a person's blood. People with sickle-cell disease have abnormal hemoglobin, which causes the red blood cells to be misshaped. Instead of round discs, they look like crescents or sickles. The distorted cells sometimes stick together and block small blood vessels. They also tend to break down too fast, resulting in anemia, jaundice, severe pain in the abdomen and joints, and poor resistance to infection. In severe cases, sickle-cell disease can cause death.

Sickle-cell disease results from a gene mutation in which one base in a triplet is changed. Hemoglobin is composed of four protein chains. Normal hemoglobin has the amino acid glutamic acid in the sixth position of two of its chains. A person with sickle-cell disease has valine in this position instead of

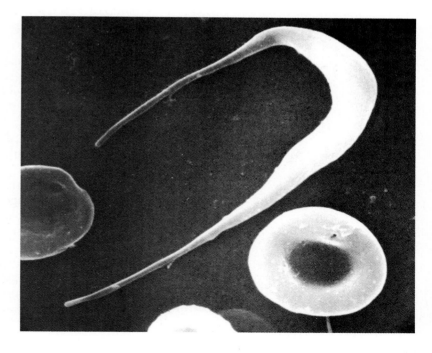

20–24 A sickled red blood cell among normal, disc-shaped red blood cells. Sickle cells contain abnormal hemoglobin.

20–25 Sickle cell anemia results from a mutation in DNA that causes the substitution of valine for glutamic acid in two of the protein chains of hemoglobin.

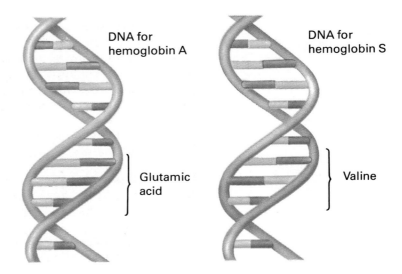

glutamic acid. Two DNA triplets code for glutamic acid: CTT and CTC. Valine, on the other hand, has four possible triplet codes: CAA, CAG, CAT, or CAC. A mutation occurring in the middle of the glutamic acid triplet that changes T to A would cause valine to appear in the protein rather than glutamic acid. This seemingly minor difference in DNA causes sickle-cell disease.

Modern Genetic Research

One of the aims of today's research in genetics is to find ways to benefit people. For example, geneticists are trying to learn how to identify, prevent, and treat chromosomal abnormalities. They use several specialized techniques in this work.

Genetic Recombination Some geneticists hope to benefit people through **genetic recombination**. Genetic recombination is the changing of the gene combination on DNA. Crossing-over is an example of natural recombination. DNA with a gene combination that has been changed by genetic recombination is called **recombinant DNA**. An organism that inherits recombinant DNA develops a combination of characteristics different from its parents.

Most research on recombinant DNA is done on bacteria, which are good organisms for genetics research because they are easy to culture, and they reproduce rapidly. A question that would take a month to answer with *Drosophila* takes a day to answer with bacteria.

Scientists have techniques that allow them to cause genetic recombination artificially. They can purposely introduce a short piece of DNA from one strain of organism into the DNA of another. This technique is called **gene splicing** or **genetic engineering**. Gene splicing can be useful. It has been used to transplant (splice) a gene for making insulin from a rat into the DNA of a bacterium. When the bacterium reproduces, its offspring inherit the insulin-producing gene. After many cell divisions, a strain of insulin-producing bacteria is produced. These bacteria could increase the insulin supply and help diabetes sufferers. Bacteria that manufacture human growth hormone have also been produced by gene splicing. Human growth hormone (HGH) is normally produced by the pituitary gland. A supply of HGH could help children afflicted with dwarfism. Bacteria that ingest oil spills and plant cells capable of fixing nitrogen have also been brought into existence by gene splicing.

Crop plants that fix their own nitrogen would not require large amounts of fertilizer.

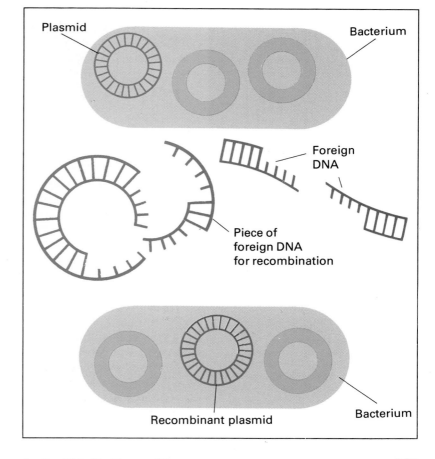

20–26 Gene splicing. In this illustration a piece of foreign DNA (natural or synthetic) is inserted by enzyme action into a plasmid. (A plasmid is a small chromosomal ring normally found in a bacterium.) The recombinant plasmid is then introduced into another bacterial cell. As the bacterium reproduces, many copies of the foreign DNA are also made. Under the appropriate conditions synthesis of the protein coded for by the foreign DNA can be turned on.

Section 20.3 The Nature of Genes

Banding A research technique that is used to help diagnose genetic disorders is called **banding**. Banding is a staining technique using a fluorescent dye and the enzyme trypsin that makes light and dark bands show up on chromosomes. This technique makes each chromosome easily identifiable and makes it possible to see any abnormality in a chromosome's size or shape. As a result, banding helps in the diagnosis of hereditary disorders and in making gene maps of chromosomes.

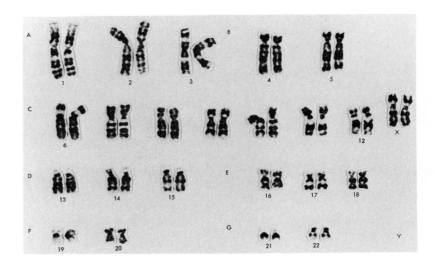

20–27 Banding is easily seen in this karyotype prepared from a micrograph of stained human female chromosomes.

In studying human chromosomes, mouse cells and human cells are fused. During their fusion, many of the human chromosomes are lost. The resulting "hybrid cells" contain one or a few human chromosomes. These hybrid cells enable scientists to concentrate on one human chromosome at a time. The procedure helps in gene mapping and studying gene expression.

Gene Mapping of Human Chromosomes Scientists are now mapping the genes on human chromosomes. They estimate that we have about 50 000 genes on our 46 chromosomes. In the past, mapping human chromosomes was very difficult. Scientists obviously cannot breed people the way they breed *Drosophila*. So they study families and watch for signs of crossing-over. Recently, banding and other new techniques have been developed for making gene maps. These techniques make it possible to locate the genes on human chromosomes more easily. Scientists believe that gene maps will help physicians in their efforts to predict and treat hereditary disorders.

Genetics Research Continues Many questions about how genes function remain unanswered. For example, how do genes control the body's resistance to disease? How do they control growth and development? What turns genes on and off? These are difficult questions; yet, researchers hope that the answers will soon be available to help people. In the meantime, the search for knowledge about heredity continues.

Building a Science Vocabulary

bacteriophage
phage
one gene–one enzyme hypothesis
minimal medium
phenylketonuria (PKU)
genetic code
codon

sickle-cell disease
genetic recombination
recombinant DNA
gene splicing
genetic engineering
banding

Questions for Review

1. The following series of reactions occurs in normal cells.

   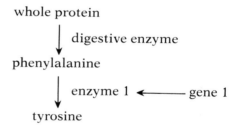

   a. What will not be formed by the cells if gene 1 is mutated?
   b. What substance will accumulate in the cell if gene 1 is mutated? What are the effects of this substance?
   c. Mutation of gene 1 causes PKU. How is PKU diagnosed in infants?
   d. What treatment prevents the harmful effects of PKU?

2. How does the Watson-Crick model of DNA explain the mutations that X-rays caused in Beadle and Tatum's studies with *Neurospora*?

3. Where is the genetic code for building protein coats located in a bacteriophage virus?

4. Why is sickle-cell disease considered to be due to a gene mutation?

5. How is banding used to identify hereditary disorders?

Chapter 20
Summary and Review

Summary

1. Much research in chromosomal genetics has been done on the fruit fly, *Drosophila melanogaster*.

2. Humans have 22 pairs of autosomes and one pair of sex chromosomes. Females have two X chromosomes. Males have an X chromosome and a Y chromosome. The genes on the X chromosome are different from the genes on the Y chromosome.

3. Genes on the X chromosome are sex-linked. Sex-linked traits occur more frequently in human males than in females, since they have only one X chromosome. Normal females may act as carriers for recessive sex-linked traits.

4. Nondisjunction occurs when the members of a pair of homologous chromosomes do not separate during meiosis. This results in extra or missing chromosomes in the offspring.

5. The order of genes on a chromosome is determined from the rate of crossing-over. Genes that are close to each other cross over more often than genes that are far apart.

6. Mutations are sudden changes in genes or chromosomes. Mutations carried in gametes are transmitted to the next generation. Radiation and certain chemicals cause mutations. Harmful mutations hinder an organism's survival. Helpful ones cause variations that may aid an organism's survival. Breeders make use of mutations in their work.

7. Alleles are alternative versions of a gene for a trait and are located at the same locus in homologous chromosomes.

8. Clumping between different A-B-AB-O blood types is due to an antigen-antibody reaction. The antigen is on the surface of the red blood cells. The antibody is in the blood plasma.

9. Multiple gene inheritance is responsible for the gradation of expression for many traits seen in a population.

10. Much information about genes and DNA was discovered by researchers before Watson and Crick made their DNA model.

11. A gene controls the production of a protein, which may then be converted into an enzyme.

12. The genetic code is determined by the order of nucleotide bases in DNA. A triplet (three bases) codes for one amino acid. The genetic code is decoded during protein synthesis.

13. A gene mutation is a change in DNA. The sickle-cell gene is an example of a gene mutation. Individuals who are homozygous for the sickle-cell gene suffer from sickle-cell disease.

14. In genetic recombination the combination of genes in a strand of DNA is changed. This occurs naturally in crossing-over and can also be done artificially.

15. Banding is a staining technique used to study chromosomes. It aids in identifying abnormal chromosomes and mapping genes.

Review Questions

Application

1. The gene for red-green colorblindness is recessive and sex-linked. Determine the probable offspring that can be produced by a colorblind father and a carrier mother.

2. Genes *a*, *b*, and *c* are located on the same chromosome. The chances of crossing-over between them are: *a* and *b*—one out of five; *b* and *c*—three out of five; *a* and *c*—two out of five. What is the order of the genes on the chromosome?

3. Females with a certain hereditary disorder have only one X chromosome in their cells. Explain how this can happen.

4. The reaction below occurs in *Neurospora*. What two things will be affected if gene 1 is mutated?

$$\text{minimal medium} \xrightarrow[\text{enzyme 1}]{\text{gene 1} \downarrow} \text{glycine}$$

5. An abandoned baby was picked up by the police. Later two different women claimed to be the mother. Blood studies revealed that woman I was type A. Woman II was type AB. The baby was type O. Which woman was possibly the mother? Explain.

6. *E. coli* bacteria live in the human intestine. Certain mutants of *E. coli* cannot produce the enzyme trypsin. How can gene splicing be used to develop a trypsin-producing strain of *E. coli* from these mutants?

Interpretation

1. Can a woman who is red-green colorblind have normal sons if she is married to a man who is not colorblind? Explain.

2. A single short-legged sheep occurred in a flock. What procedures could a breeder follow to find out whether the condition is due to a mutation or to the environment?

3. Use the Beadle and Tatum hypothesis to explain PKU.

4. Why do geneticists think that human height is controlled by several pairs of genes rather than by one pair?

Extra Research

1. Research information about the Rh factor in the library. Write a short paragraph on how this trait is inherited. Describe the problem of Rh-negative mothers who give birth to Rh-positive children.

2. About 2000 hereditary disorders have been identified. Research three disorders not described in the text. Answer the following questions about each.
   a. Is the disorder due to a gene mutation or a chromosomal mutation?
   b. What are the effects of the disorders?
   c. What tests are used to diagnose the disorder?
   d. What treatment, if any, is available?

3. Using periodicials, research the pros and cons of using genetic recombination in research. After reading what the experts say, decide where you stand on this issue.

4. Make up a code like the genetic code that can be used to send English words and sentences. Use four letters of the alphabet arranged in groups of three. Use a different group for each letter of the alphabet and for each punctuation mark. Try using your code with a friend.

5. Write a short detective story. Have your detective solve the crime by a sex chromosome test, a blood type test, or a protein identification test.

Career Opportunities

Plant breeders develop new types of crop and garden plants for seed companies or cooperative extension programs. They spend much of their time in greenhouses or outdoors. A university degree and a good knowledge of botany and genetics are essential. For information, write the Canadian Society of Plant Physiologists, Biology Dept., Dalhousie University, Halifax, Nova Scotia B3H 4J1.

Medical social workers help patients and their families deal with problems that accompany an illness or accident. They may also do genetic counseling for prospective parents. Most work in hospitals, clinics, and rehabilitation centers. A degree and specialized training are required. Contact the Canadian Association of Social Workers, 55 Parkdale Ave., #316, Ottawa, Ontario K1Y 1E5.

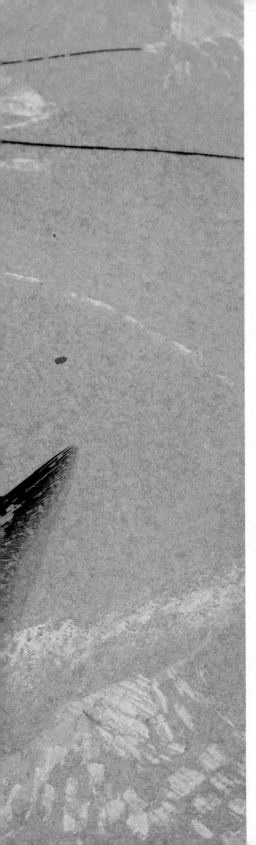

Chapter 21

Change Over Time

Focus During the past century and a half biologists have discovered many clues about the history of living things on earth. These clues suggest that all life on earth is related and that it changes over time.

21.1 THEORIES ABOUT THE ORIGIN OF LIFE

For many years, people believed that flies and many other living things grew from nonliving matter. Until the seventeenth century, people also thought that, at times, lice came from sweat, frogs and mice grew from moist earth, and flies and maggots came from decaying meat or fish. The Roman poet Virgil, living in the first century B.C., even included a kind of recipe for making bees in one of his poems. Briefly stated, the steps in Virgil's recipe for bees include:

1. Kill a bull during the first thaw of winter.
2. Build a shed.
3. Place the dead bull on branches and herbs inside the shed.
4. Wait for summer. The decaying body of the bull will produce bees.

21–1 Minerals have slowly replaced the bony skeleton of this prehistoric fish, leaving behind a fossil impression in sedimentary rock.

21–2 Spontaneous generation was once an accepted belief. Only in the last two hundred years has science shown that living things come only from other living things of the same species.

a- = without
bios = life
genesis = generation

People who believed in abiogenesis saw it as an additional means of life development. They recognized biogenesis as the major origin of new life.

Virgil said that if his instructions were followed, the bees would "swarm there and buzz, a marvel to behold." Even biologists in ancient times were sure that beetles, frogs, and salamanders formed in dust or mud.

The idea that living things arise from nonliving things under certain conditions is known as **abiogenesis** (AY-bī-ō-JEN-uh-sis). Abiogenesis is also known as **spontaneous generation**. A description of abiogenesis can be found in Aristotle's writings. He believed that living things differ from nonliving things by some "**active principle**." Living things have the active principle and nonliving things do not. Aristotle's explanation of spontaneous generation was that, under certain circumstances, the active principle enters nonliving matter and living organisms form.

Today we know that all living things arise from other living things, a process called **biogenesis**, or life from life. Beginning in the seventeenth century, scientists conducted experiments that eventually disproved abiogenesis. The two most convincing experiments to disprove abiogenesis were done by Francesco Redi and Louis Pasteur.

Redi's Experiments

Francesco Redi was an Italian physician of the middle 1600s. His work on abiogenesis began as he observed three dead snakes decay over a period of time. First, he saw some flies hovering over the dead snakes. Three days later, the snakes were covered with maggots. Redi watched the maggots gradually eat all the meat off the bones. After 19 days, he saw the maggots become pupae. Eight days later, flies emerged from the pupa cases. The flies were the same kind as those Redi had seen hovering over

The people of Redi's day and before could not see the fly eggs. So they did not see them develop into larvae. Since people believed in abiogenesis, this was the most logical explanation for the appearance of maggots (larvae).

the snakes earlier. He wondered whether the maggots were actually offspring of the flies. He suggested that flies might have the following life cycle:

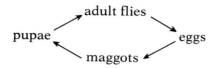

Redi formulated the problem of where the flies came from as a question: Do maggots come from meat or to it? If flies have a life cycle like the one he proposed, then maggots must come from flies and not from meat. Redi hypothesized that maggots come to the meat, not from it. He then set out to test his hypothesis.

Redi set up a controlled experiment. His experimental variable was flies getting to the meat. He placed meat and fish in open jars and in tightly sealed jars. He also placed meat and fish in jars covered with a fine screen. The purpose of the jars covered with the screen was to allow air to reach the meat and fish. During Redi's time, people believed that air was necessary for abiogenesis to take place. The sealed jars blocked off the air to the meat or fish inside. Proponents of abiogenesis claimed that keeping the air out of the jars would kill the active principle in the meat or fish, and abiogenesis could not take place. The screen-covered jars in Redi's experiment solved this problem. Air could enter the jars but flies could not.

After observing the jars for several weeks, Redi obtained the results shown in Figure 21–3. His results support the hypothesis that flies come to meat, not from it.

21–3 Francesco Redi's experiment revealed that a living agent, the fly, was responsible for the formation of maggots on meat.

Section 21.1 Theories About the Origin of Life

Redi convinced people of his day that flies do not come from decaying meat or fish. Redi's method of experimenting was used by others to repeat what he had done and to disprove abiogenesis of other organisms. For a time, it seemed that the subject of abiogenesis was settled. But the question opened up again some years later with the discovery of the microscope.

Microorganisms: Biogenesis or Abiogenesis?

With the microscope, Leeuwenhoek and other scientists discovered tiny organisms never seen before. All these scientists had to do was put broth or another kind of food in a warm place. In a day or so, the "little creatures" appeared. Many of the scientists thought that these simple forms of life were formed by the broth (abiogenesis). Others thought that they arose by reproduction (biogenesis) like other forms of life. They believed that bacteria and other microscopic organisms came to the broth and not from it. It seems odd now, but scientists were very involved in this disagreement. In fact, the Paris Academy of Science offered a prize for the scientist who could settle this question.

Scientists on both sides tried to prove their points by doing experiments. They faced one major obstacle in their experiments: they did not know how to sterilize their materials. So the broth that they studied was often contaminated with microorganisms before the experiment was started. Furthermore, some people charged that boiling and covering the broth killed the active principle. Because of these problems, the results on both sides were not very convincing. Finally, in 1862, a French chemist named Louis Pasteur disproved abiogenesis of microorganisms and claimed the prize.

Today almost all milk and beer is pasteurized, or heated, to kill microorganisms. This process is named after its discoverer, Louis Pasteur (1822-1895).

Pasteur's Experiments It had been commonly observed that when a jar of beef broth is exposed to the air, bacteria grow in the broth within a few days. This raised the problem of whether the bacteria come to the broth or are formed from it. Pasteur hypothesized that the bacteria were carried to the broth on dust particles in the air.

In order to test his hypothesis, Pasteur had to prevent dust from coming into contact with the broth. At the same time he had to allow air to contact the broth, since people believed that air was necessary for the active principle to work. Pasteur invented a new piece of equipment, the **swan-neck flask**, for his experiments. The curved neck of the flask allows air in but traps dust particles in the curve of the neck.

Pasteur filled the flask with beef broth and boiled it for several hours. Boiling killed all the bacteria in the broth and

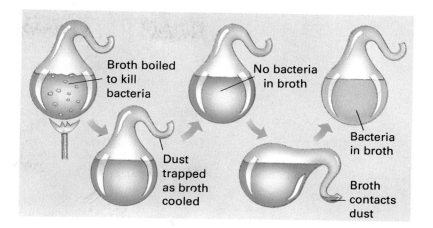

21–4 Pasteur's experiment with swan-neck flasks helped dispel renewed belief in abiogenesis by tracing the source of bacterial growth in broth to dust particles carried in air.

flask, and steam drove out the air. When fresh air entered the open end, dust particles collected in the curve of the flask's neck. The flask containing boiled beef broth was placed on a table and observed. After several months, the broth in the swan-neck flask was still free of bacteria. Bacteria did not come from the broth.

As a control, Pasteur tipped the flask and let some of the broth enter the curved neck. Then he stood the flask upright again. Dust containing bacteria that was trapped in the neck of the flask was washed into the broth. Within a few days, bacteria were growing in the broth. In another experiment, Pasteur broke off the neck of the flask as a control. This also resulted in the growth of bacteria. These controls were necessary to show that the boiled broth was capable of supporting the growth of bacteria. They showed that boiling did not destroy the active principle as some believers in abiogenesis had claimed.

Pasteur's experiments demonstrated that bacteria come to the broth and are not formed from it. These experiments finally disproved abiogenesis.

Life on Earth

The experiments on abiogenesis firmly established that life comes only from preexisting life. Only one problem remained. If all living things come from other living things, then how did life on earth begin? We may never be able to answer this question scientifically. Nevertheless, it is a fascinating subject to hypothesize about. Many scientists subscribe to a hypothesis that explains how life might have begun on earth almost 4 billion years ago. Interestingly, this hypothesis states that abiogenesis took place, forming life when the earth was young.

Formation of the Earth Most geologists and astrophysicists estimate that our planet was formed about 4.6 billion years ago from the condensation of a huge cloud of dust and gases that whirled around the sun. As the cloud condensed, heavy particles such as iron, copper, and nickel were pulled toward the center. Lighter particles such as helium and hydrogen stayed at the surface. Radioactive material and great pressure at the center caused tremendous heat to build up. The heat kept all the material at the earth's center in a molten state. Over a period of millions of years, the earth began to cool, and an outer "skin" of rock, the earth's crust, formed. The center, or core, of the earth remained hot.

As heat rose from the center of the earth the crust cracked and split. Huge amounts of light chemical compounds were sent into the space above the earth's surface, forming the atmosphere.

Some of the compounds thought to have been present in the primitive atmosphere are methane, ammonia, hydrogen, carbon dioxide, and water vapor. Oxygen gas was not present in the primitive atmosphere, since oxygen combines with other elements when heated.

Astronomers have found similar compounds in the atmospheres of some of the other planets in our solar system.

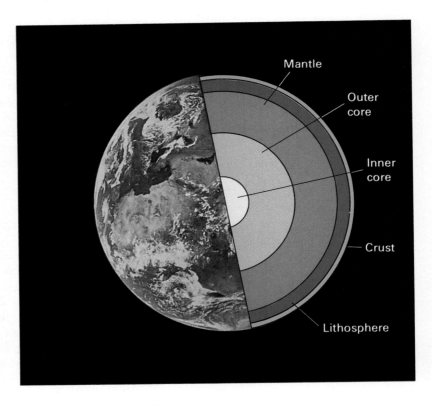

21–5 After the earth stabilized and cooled, several layers formed. The inner core consists of solid iron and nickel. Molten iron and nickel compose the outer core. The mantle consists of very hot, dense mineral matter. The lithosphere is solid and is made up of a variety of rock and mineral types.

Table 21–1 Compounds in the Primitive Atmosphere

Compound	Elements Present	Molecular Formula
Methane	Carbon, hydrogen	CH_4
Ammonia	Nitrogen, hydrogen	NH_3
Hydrogen	Hydrogen	H_2
Water	Hydrogen, oxygen	H_2O
Carbon dioxide	Carbon, oxygen	CO_2

Water vapor still erupts from inside the earth in geysers and steam fields. Naturally occurring steam is used to produce electric power in some places.

Formation of the Oceans As the earth formed, tremendous quantities of hydrogen and oxygen were trapped beneath the crust. These elements combined to form water. As it was heated, the water changed to water vapor, which steamed out through cracks in the earth's crust into the atmosphere. The water vapor condensed to rain as the earth and its atmosphere gradually cooled. For thousands of years, violent rainstorms covered the earth. Lightning flashed through the atmosphere, discharging large amounts of electrical energy.

The continuous rains wore down the rocks and mountains. Water flowed downhill, carrying salt and minerals dissolved from the rocks. It filled the depressions in the earth's surface, forming the oceans.

The Heterotroph Hypothesis

hetero- = other
trophikos = food

auto- = self

In 1924, A. I. Oparin, a Russian biochemist, published a paper describing his hypothesis for the origin of life on earth. Oparin suggested that the first living things were **heterotrophs** (HET-ur-uh-TRŌFS), which are organisms that do not make their own food. Oparin's hypothesis, known as the **heterotroph hypothesis**, states that organisms that make their own food by photosynthesis, known as **autotrophs** (AHT-uh-TRŌFS), developed after the heterotrophs. According to the heterotroph hypothesis, life on earth had its beginnings when molecules in the primitive oceans combined.

Autotrophs are producers and heterotrophs are consumers.

Molecules Combine in the Primitive Oceans The heterotroph hypothesis suggests that methane, carbon dioxide, and ammonia from the atmosphere dissolved in the rain as it fell into the seas. It also suggests that the primitive seas contained a mixture of salts and minerals, including sulfur and phosphorus. Oparin hypothesized that at first hydrogen, methane, ammonia, and water combined to form simple organic molecules, using sun and lightning as sources of energy for chemical reactions.

The sun emits heat, light, ultraviolet rays, and X-rays.

Section 21.1 Theories About the Origin of Life

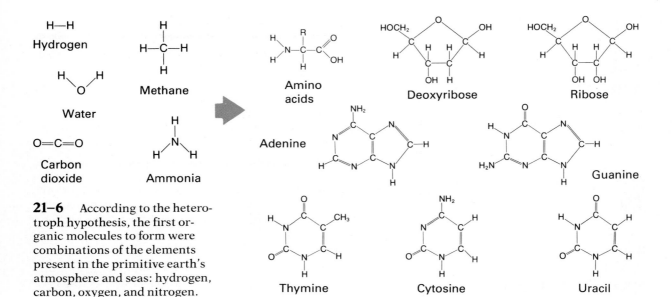

21-6 According to the heterotroph hypothesis, the first organic molecules to form were combinations of the elements present in the primitive earth's atmosphere and seas: hydrogen, carbon, oxygen, and nitrogen.

The first heterotrophs are supposed to have resembled anaerobic bacteria and were scavengers on the organic matter that formed in the sea.

An ozone molecule consists of three atoms of oxygen. It forms in the stratosphere when ultraviolet rays strike oxygen gas.

After they formed, some of the simple organic compounds clumped together and underwent further chemical reactions, forming complex organic molecules. In this way amino acids, sugars, nucleotides, and other building blocks of living things appeared. The clumps of organic molecules kept reacting and eventually gave rise to the first simple forms of life.

It is hard to say when the clumps of organic molecules could be considered alive. According to the hypothesis, certain clumps of molecules became capable of reproduction by replication. They changed by mutation and used energy to perform life functions.

Development of Autotrophs According to the hypothesis, large numbers of organic molecules formed in the primitive seas, which scientists often refer to as a "hot, dilute soup." These organic molecules were food for the first heterotrophs. As the molecules in the organic soup changed, chlorophyll was formed. Some of the heterotrophs took in this molecule and became able to produce their own food by photosynthesis. Photosynthesis provided a continuous source of food for the heterotrophs and also added oxygen to the atmosphere. Eventually, some organisms developed the ability to use oxygen to get energy from food.

Part of the oxygen that the autotrophs gave off rose high in the atmosphere and formed the **ozone layer**. Once formed, the ozone layer acted as a shield against ultraviolet radiation. To this day, it protects us from the harmful rays. Organic molecules

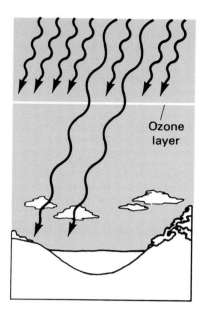

21-7 The ozone layer is unable to filter out the longer wavelengths of ultraviolet radiation. It does filter out most of the shorter wavelengths, which can damage cells in surface life (an example is sunburn).

in the seas stopped forming when ultraviolet rays from the sun were blocked and other conditions changed. As a result, abiogenesis probably could not occur again on earth.

Support for the Heterotroph Hypothesis

Although the heterotroph hypothesis may not be verifiable scientifically, scientists would like to know whether the steps that it describes are physically possible. Some scientists have tried to recreate the primitive earth's atmosphere in the laboratory. The results of their experiments support the idea that some of the steps in the heterotroph hypothesis could have happened.

Miller's and Urey's Experiment In 1953 Stanley Miller and Harold C. Urey, at the University of Chicago, tested the heterotroph hypothesis for the first time. They circulated methane, ammonia, water vapor, and hydrogen through the apparatus shown in Figure 21-8. At one place in the apparatus, electric sparks were produced. After operating the apparatus for a week, Miller and Urey analyzed the contents and found that various organic compounds, including simple amino acids, had been formed.

Other Experiments Using various setups to simulate the earth's primitive conditions, other scientists have also produced biologically important molecules from simple compounds. In some experiments, ultraviolet light is used as the energy source. In some experiments, carbon dioxide is included among the gases of the primitive atmosphere. These experiments produce sugars and other molecules found in cells.

21-8 Miller and Urey used gases and water vapor to imitate conditions on the primitive earth. The electric spark represents the effect of lightning. In the experiment organic compounds, including amino acids, were produced.

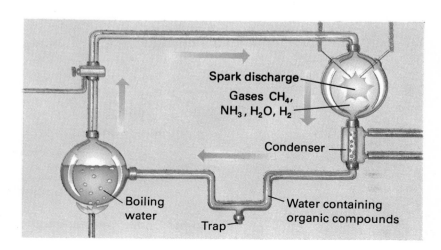

Section 21.1 Theories About the Origin of Life

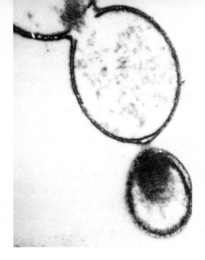

21–9 Fox found that organic molecules clump together under certain conditions to form small spheres of polypeptides which then slowly grow and bud.

The Clumping of Molecules Several explanations for how organic molecules might have clumped together in the primitive seas have been proposed. Sidney W. Fox of the University of Miami heated an almost dry mixture of amino acids together on volcanic rock. The rock was then sprinkled with water and allowed to cool. Looking at the mixture under the microscope, Fox found that small clumps of polypeptides had formed. It is possible that amino acids from the primitive seas washed onto the beach. There they dried and were heated by the sun. Then they were sprayed by the ocean's water. Theoretically, clumps might have formed. Perhaps the clumps were eventually washed back into the sea. Some scientists think that the clumps gave rise to cells.

Building a Science Vocabulary

abiogenesis
spontaneous generation
active principle
biogenesis
swan-neck flask

heterotroph
heterotroph hypothesis
autotroph
ozone layer

Questions for Review

1. What is abiogenesis?

2. What did Aristotle mean by the "active principle?" How was it supposed to work during abiogenesis?

3. The following questions refer to Redi's experiment with flies.
   a. What question was Redi trying to answer?
   b. Why were the jars with the screens necessary?
   c. What conclusion was drawn from Redi's results?

4. Briefly describe the work on abiogenesis performed by Pasteur. What effect did his experiments have on the abiogenesis controversy?

5. Briefly state the major parts of the heterotroph hypothesis.

6. How do the experiments of Miller and Urey and those of Fox support the heterotroph hypothesis?

21.2 INFORMATION ABOUT THE PAST

Evidence exists that suggests that one form of life or another has been on earth for over 3 billion years. Scientists search for clues that can tell them about the history of life on our planet. Such clues come from various sources, including body structure, embryology, development, body chemistry, organisms in different areas, and breeding studies. Much information also comes from the study of **fossils**, which are parts or traces of organisms that have been preserved over long periods of time. Fossils show the characteristics of organisms that lived in the past and can help in determining the possible relationships among living things.

Scientists who study fossil remains are called paleontologists.

The Fossil Record

How Fossils Are Formed Most dead organisms are eaten or decompose and disappear. Others are destroyed by erosion by wind or water. Some dead organisms, however, are protected from decay and erosion, enabling them to become fossils. Consider a dead animal or plant that becomes buried in mud. The minerals from the surrounding soil slowly seep into the organism's skeleton. As the minerals accumulate and harden, the bones, shells, or other hard body parts gradually become stone. Some dinosaur skeletons were preserved in this way.

Fossil formation also occurs in other ways. When an insect or other small organism is trapped in the resin, or sap, of a tree,

21–10 Fossils: **a.** imprint of a fern left in carboniferous rock; **b.** fossilized skeletons, showing an early deerlike mammal in the foreground; **c.** an insect preserved in amber, which is fossilized tree resin.

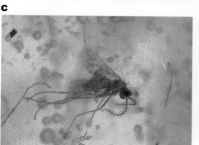

The ancient Italian city of Pompeii was discovered well preserved under ashes, cinders, and stones hundreds of years after Mt. Vesuvius erupted in 79 A.D.

and the resin hardens into amber, the organism is preserved. Some animals, such as the saber-toothed tiger, fell into natural tar pits and were preserved. Other organisms were preserved by a protective cover of volcanic ash. In other cases, organisms, such as the woolly mammoth, were preserved by being quickly frozen.

Imprints, such as footprints, tail prints, shell prints, or leaf prints made in soft ground, may harden into rock. The imprints of ancient ferns are found as fossils in the coal they helped form. These tracks and traces also become part of the fossil record.

Where Fossils Are Found   Over time, the remains of organisms are buried under layers of rock particles, or **sediment**. Layers of sediment are usually deposited by water. For example, sand sinks to the bottom of the ocean, and new layers of sand settle over the older ones. **Sedimentary rock** forms when a layer of sediment, including skeletal remains and imprints, sticks together and slowly hardens. The shells, bones, and imprints trapped in a layer of sedimentary rock become fossils.

Most of the fossils that have been found were preserved in sedimentary rock.

Fossils found in sedimentary rock may remain hidden for millions of years. Sometimes, earthquakes and other geological activity relocate large masses of rock. For example, several

21–11   Grand Canyon: **a.** sequence of strata, showing fossils typically found; **b.** south rim.

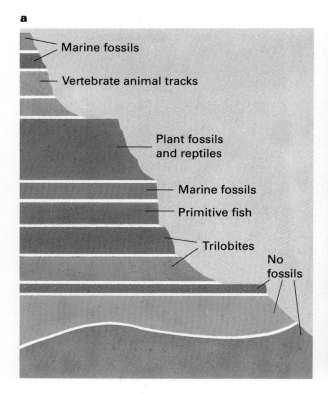

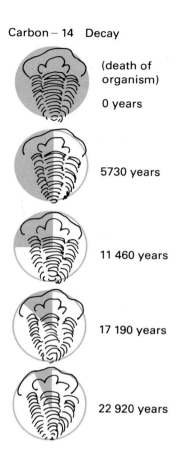

21–12 After an organism dies radioactive substances give off radiation at a fixed rate. The *half life* refers to the number of years it takes for half of a radioactive substance in the organism to decay. The radioactive form of carbon (C^{14}) has a half life of 5730 years.

times in the past areas of ocean bottom have risen and become part of mountains on land. In places such as the Grand Canyon, erosion over time has exposed the hidden layers of sedimentary rock that were once at the bottom of a sea. Very old fossils have been found in exposed layers of sedimentary rock.

The Age of a Fossil Several methods are used to determine the age of a fossil. The relative age of a fossil can be determined by its position in sedimentary rock. If the layers of sedimentary rock are not disturbed, the oldest fossils are found in the deepest rock layers. Newer fossils are found in the rock layers that were deposited on top of the older and deeper layers. The true age of a fossil can also be estimated. Scientists know about how many years it takes to form a given thickness of sedimentary rock. The approximate age of a fossil is calculated by measuring the thickness of the sedimentary rock in which the fossil is found.

Chemical analyses provide a more precise method for determining the age of a fossil. Organic remains up to 50 000 years old are dated directly using the **carbon-dating method**. Older fossils are dated by the age of the rock around them. The **potassium-argon method** is often used to date the rock around a fossil. The basis for both methods is that as a rock or fossil ages, its radioactive substances change to a nonradioactive form by the process of **radioactive decay**. Scientists know the rates at which different radioactive elements change. The proportion of radioactive to nonradioactive materials in a fossil or rock is determined by chemical analysis and compared to the proportion of radioactive to nonradioactive materials in living organisms or new sediments. The data provide an accurate measure of a fossil's age or the age of the rock around a fossil.

Fossil Clues and Biological History

Fossil clues help indicate relationships between organisms and suggest that changes in species have occurred over time. They also place a species in its period of earth history.

The Order of Appearance of Living Things Different species of organisms appear to have originated at different times. The oldest fossils that have been found so far are bacterialike cells embedded in rock dated to be 3.4 billion years old. Fossils of blue-green algae date back 2 to 3 billion years. Signs of green algae, protozoa, and simple invertebrates also appear early in the fossil record. On the other hand, most of the major phyla of complex invertebrates begin to appear only about 600 million years ago.

Table 21–2 Major Geological Periods and Evolutionary Events

Millions of Years Before Present	Era	Period	Epoch	Major Evolutionary Events
0	Cenozoic	Quaternary	Pleistocene	Human development
50		Tertiary	Pliocene Miocene Oligocene Eocene Paleocene	Mammals spread out
100	Mesozoic	Cretaceous		Last dinosaurs First primates First flowering plants
150		Jurassic		Dinosaurs First birds Conifers dominant
200		Triassic		First mammals Mammallike reptiles dominant (Therapsids)
250	Paleozoic	Permian		Major marine extinction Reptiles with large fins on the back dominant (Pelycosaurs)
300		Carboniferous	Pennsylvanian	First reptiles
			Mississippian	Scale trees, seed ferns
350 400		Devonian		First amphibians Jawed fishes diversify Fern forests
450		Silurian		First vascular land plants
500		Ordovician		Burst of diversification in animal groups
550		Cambrian		First fish First chordates
600	Precambrian	Ediacaran		First invertebrates with skeletons
650				First-soft bodied invertebrates Algae
700				First animal traces Blue-green algae dominant

21-13 Two species on the edge of extinction: the California condor and the tiger, have had to give up much of their natural habitats for human expansion.

Extinction of Species The fossil record also contains information about groups of organisms that disappeared completely. The dying out of a species is known as **extinction**. About 22 million years ago at least half of the families of marine invertebrates started to die out. During the next few million years, all the trilobites, ancient corals, and many other sea invertebrates died out. Another mass extinction occurred about 70 million years ago. Dinosaurs and their relatives, which were the dominant animals, disappeared during that period. No one really knows what caused the mass extinction of the dinosaurs, though many theories exist to explain what happened.

Relationships Among Species Fossils with features characteristic of existing biological groups suggest relationships among species. These fossils provide clues to ancestry, supporting the idea that the forms of life on earth change over time and give rise to new and different groups of living things. Consider the archeopteryx, an animal that lived about 150 million years ago. Since the archeopteryx had feathers, it is considered a bird. But it also had characteristics of reptiles: claws at the end of its wings, teeth in its mouth, and a long tail. The combination of traits in this fossil gives scientists the idea that birds originally evolved from a group of ancient reptilelike creatures. In addition to the archeopteryx, other fossils exist that show characteristics of different groups.

21-14 *Archeopteryx* was a land bird about the size of a large crow.

Interpreting Fossil Clues The fossil record suggests that the first living things on earth were simple cells that lacked true nuclei. These were the original ancestors of all life on earth. In time, these organisms underwent changes, and cells with true nuclei appeared. Later, multicellular forms of life arose. The

Section 21.2 Information About the Past

21–15 Fossil clues suggest that life as we know it today has traveled a path of continual change and diversification.

new organisms spread out, inhabiting different parts of the oceans. As much time passed, these organisms continued to change. The organisms that were adapted to life in the ocean survived; others died. Later, certain ocean-dwellers gradually began to inhabit the land. The land organisms spread to different areas and gave rise to new and different forms.

Clues That Suggest Relatedness

In addition to fossil clues, scientists obtain information about relationships among species by studying the anatomy and physiology of living organisms. Organisms that have the same ancestors are related. One characteristic of related organisms is that they are similar in certain ways.

Body Structure Structural similarities (homologous structures) suggest that different species have the same or a common ancestor. Scientists compare both large and microscopic features, including chromosomes, organelles, cells, and larger body parts. The vertebrates are a good example of how similar body parts suggest relatedness. The bones of vertebrate forelimbs are remarkably similar. For example, the human arm, foreleg of the horse, wing of the bat, paddle of the seal, fin of the whale, and wing of the bird have corresponding bones. These similarities suggest that all classes of vertebrates have a common ancestor.

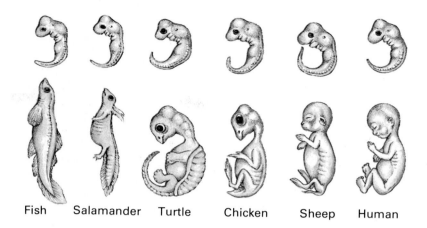

21–16 Comparison of embryos of six vertebrates shows a high degree of similarity in the early stages.

Fish   Salamander   Turtle   Chicken   Sheep   Human

Embryology The embryos of many groups of living things are sometimes similar. For example, all vertebrate embryos have a tail and a two-chambered, fishlike heart at some time during development. These similarities suggest that the vertebrates have a common ancestor. The similarity of multicellular embryos in mosses, ferns, and seed plants is evidence that these plant groups are also related.

Vestigial Organs Certain organisms have body parts that they do not use. These body parts are called **vestigial** (ves-TIJ-ee-ul) **organs**. Although a body part is vestigial in some species, it may be developed and functioning in others. For example, most vertebrates make good use of their legs. But snakes and whales have traces, or vestiges, of legs, which they never use. Many scientists believe that a species with a vestigial structure is related to species in which the structure functions.

Body Chemistry All living things are composed of similar basic molecules. The complex molecules found in many species are also similar. Chlorophyll is found in algae, mosses, ferns, and seed plants. Hemoglobin is found in all vertebrates. The similarity of nucleotides, enzymes, and hormones from species to species suggests that organisms are related.

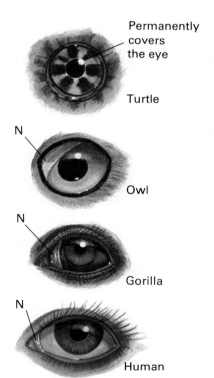

21–17 The third eyelid, or nictitating membrane (N), functions in turtles and owls. However, it is a vestigial fold of tissue in gorillas and humans.

Clues That Suggest Biological Change Over Time

Evidence from the fossil record suggests that living species have changed over the course of time. Scientists obtain evidence of change in living species from breeding experiments and studies of species in different geographical areas.

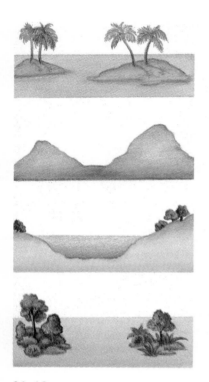

21–18 Types of barriers. Barriers separate members of a species from one another.

Organisms on Both Sides of a Barrier Suppose that at some time in the past, a species was widely distributed over a given geographic area. Then a barrier formed. Perhaps the barrier was a mountain, or maybe the ocean cut off a piece of land from the mainland. The barrier separated the species into two distinct locations. Therefore, the organisms on the opposite sides of the barrier could no longer get together to mate. According to many scientists, after much time under these conditions the separated species would probably differ somewhat in structure. If they could no longer produce offspring with each other, even if brought together, two species would have developed.

Examples exist in which species appear to have been separated by a barrier and undergone change. Many scientists believe that Australia was once attached to Asia. Now they are separated by the Indian Ocean. The pouched mammals of Australia (marsupials) appear to be related to Asian mammals. But some of the marsupials in Australia, such as the duckbill platypus and the kangaroo, look very different from Asian mammals. When the barrier formed, it prevented the Australian mammals from mating with the Asian mammals. The mammals on both continents changed in their own way. As a result, different species now exist. The evidence suggests that Australian and Asian mammals have a common ancestor.

21–19 Australian pouched mammals and their placental Asian counterparts. Over time both types of mammal developed in similar ways to adapt to similar niches.

606 Chapter 21 Change Over Time

21–20 A single wild species, *Brassica oleracea*, has diversified into various familiar cultivated species. They were selected for different traits and grown in different climates.

Experiments in Breeding Through breeding, scientists develop hybrid varieties of organisms. When scientists induce mutations in the laboratory, new and different kinds of organisms result. The production of new types of organisms by breeders shows that it is physically possible for organisms to change in the course of their biological history. If people can change species artificially, theoretically it could happen in nature as well.

Building a Science Vocabulary

fossil
sediment
sedimentary rock
carbon-dating method

potassium-argon method
radioactive decay
extinction
vestigial organ

Questions for Review

1. A student found fossils in sedimentary rock. Fossils of invertebrates were located in a lower layer of the rock. Fish fossils were found higher up.
    a. How might the fossils have gotten in the different rock layers?
    b. How could you determine which fossils are older?

2. Place the following groups of organisms in order of their appearance on earth according to fossil evidence.
    a. algae, blue-green algae, ferns, flowering plants, conifers
    b. vertebrates, invertebrates with skeletons, invertebrates without skeletons

3. Based on fossil evidence, what two groups of vertebrates have a common reptile ancestor?

4. What is a vestigial organ? Why do vestigial organs suggest that species are related?

5. A breeder developed a new breed of sheep. How does this event relate to the idea that species change over time in nature?

Section 21.2 Information About the Past

21.3 THEORIES ABOUT HOW LIVING THINGS CHANGE

Studies of living organisms and the fossil record support the idea that living and extinct species are related. In addition, the evidence suggests that species change in structure and function over time. The changing of species over time is known as **evolution**.

During the nineteenth century, a few scientists tried to explain how organisms change. The three most important of these scientists were Jean Baptiste Lamarck, Charles Darwin, and Alfred Russel Wallace. As you study their work, remember that these men had no idea how heredity works. When they first developed their theories about evolution, Mendel had not yet done his experiments on garden peas.

Lamarck's Theory

Lamarck observed that if certain individuals often used an organ or part of the body, that organ or part became larger and stronger. If a part of the body was not used, it shriveled and weakened. He called this observation **use and disuse**, implying that organs not in constant use atrophy. You can also observe use and disuse. If people exercise a lot, their muscles get larger and stronger. People confined to bed for a long time, however, find that their muscles weaken and shrink.

Lamarck used his observations of use and disuse to explain the appearance of some of the physical changes in an organism's

Lamarck claimed that acquired characteristics that were due to use and disuse were passed on to succeeding generations.

21–21 According to Lamarck's theory of inheritance of acquired characteristics, each generation of giraffes stretched to reach higher leaves. Subsequent generations inherited the longer, more stretched necks from their parents.

body within its lifetime. He then went on to explain that these changes, or **acquired characteristics**, pass from generation to generation and eventually change a species.

Lamarck's theory is not supported by observations and results of experiments. Body changes do not pass from one generation to the next. A scientist named August Weismann tested Lamarck's theory by cutting off the tails of mice for hundreds of generations. Each new generation was born with tails. Weismann concluded that body changes are not transmitted from one generation to the next. He then showed that only the material in a sperm or egg passes from one generation to the next. Weismann's experiment is evidence that acquired characteristics are not inherited. Although it may have seemed logical at the time it was written, Lamarck's theory does not explain evolution.

Darwin and Wallace: Two Naturalists with the Same Idea

Sometimes two scientists working independently have the same idea at the same time. Charles Darwin and Alfred Russel Wallace did not know each other personally and never worked together. Yet, they both came up with the same explanation for how species might change over time. Both men were English naturalists during the nineteenth century. Darwin spent five years on the *H.M.S. Beagle*, a ship chartered by the government in 1831 to map the coast of South America. During the voyage, Darwin carefully observed the abundant plant and animal spe-

21–22 Both Wallace and Darwin came to similar conclusions about how species change over time. Both men were experienced naturalists and collectors and had encountered remarkable examples of variation in their travels.

Darwin and Wallace came from different backgrounds. Darwin's father was a doctor. Darwin studied medicine at Edinburgh University for two years and then studied for the clergy at Cambridge. Wallace, whose parents were poor, left school at age fourteen and became an apprentice surveyor.

cies. His observations convinced him that species change over long periods of time. In 1838, Darwin thought of a way to explain the changes. He delayed writing his theory of evolution until twenty years later, when a letter from Wallace made him realize that they both had the same idea.

Wallace also made the observations that led to his theory about evolution during a trip to South America. The purpose of Wallace's trip, which began in 1848, was to collect rare plant and animal specimens to sell in England. Unfortunately, in 1852 the ship on which Wallace was traveling caught fire and all of his hard-sought specimens were destroyed. Nevertheless, Wallace's observations convinced him that living species are related and come from a common ancestor. He also felt certain that species change over time.

While still abroad in 1858, Wallace conceived his explanation for evolution. Having heard that Darwin was interested in the subject, Wallace sent his idea to Darwin. The unusual occurrence of two scientists coming up with the same theory had happened. Friends settled the difficulty by suggesting that both men send a paper describing their work to a scientific organization. Both papers were read at the same meeting a month later, Darwin's being read first. As a result, Darwin is credited with the theory rather than Wallace. At the end of 1858, Darwin published *The Origin of Species*, in which he describes his theory of how evolution takes place.

The Theory of Natural Selection

Both Darwin and Wallace explained the process of evolution through the **theory of natural selection**. The theory of natural selection can be stated in three steps.

1. More organisms are born than can survive. The environment cannot support all the living things that are born.

2. Organisms vary. Favorable variations aid survival.

3. Organisms with favorable variations survive and reproduce. In this way, favorable variations pass from generation to generation and collect in a population.

More Organisms Are Born Than Can Survive Most of the living things born into an environment die before they reproduce. This happens because natural resources such as food, water, and space are limited. The environment cannot supply all the needs of the millions upon millions of living things created by the process of reproduction. In addition, many organisms are eaten as food, and others are killed by either living or nonliving factors in the environment.

Favorable Variations Aid Survival Some members of a species survive. Darwin observed that organisms with favorable variations are "selected" by nature to survive. (Hence the name *natural selection*.) For example, a lightweight skeleton is a favorable variation for a bird because it helps the bird fly. A bird with heavy bones would have difficulty getting off the ground.

Favorable Variations Are Transmitted from Generation to Generation According to the theory of natural selection, members of a species with favorable variations survive and reproduce. The offspring inherit the favorable variations of their parents. Gradually, the species becomes more and more adapted to its environment.

If a species' environment changes, different variations are advantageous. Natural selection then favors these variations. Within a relatively few generations different adaptations become dominant in the species. In this way, the theory of natural selection explains both gradual and sudden changes that occur in a species during its biological history.

21–23 A variety of species of finches is found on the Galapagos Islands. Each is adapted to fill a particular niche and is not found anywhere else. It has been suggested that a few original finch ancestors reached the Galapagos Islands and, in the absence of competition, gave rise to a variety of species.

Modern View of Natural Selection

The modern view of natural selection is similar to Darwin's theory. One difference is that we now know the source of the variations among organisms. Darwin and the scientists of his day did not know about heredity. For that reason, neither Darwin nor Wallace could explain the source of variations. They merely stated that the variations occur at random and are inherited.

At the beginning of the twentieth century, a scientist named Hugo DeVries explained that mutations cause variations by causing changes in the genetic material. In addition, recombination during sexual reproduction brings about variations by introducing new combinations of genes into the zygote. Both mutations and recombination occur at random and the changes they cause are inherited.

Evolution is defined as a change in the gene frequencies of a population over a period of time. New characteristics appear in a species, and eventually new species are formed.

The Peppered Moth: Natural Selection in Action

Peppered moths have been observed in Manchester, England, for more than a century. Peppered moths vary in color. Some are light-colored with dark markings; others are black with light markings. This variation is inherited.

Before the Industrial Revolution occurred in the eighteenth century, most of the peppered moths in Manchester were light in color. The light-colored moths were practically invisible on the light-colored tree trunks in the area. The black moths, on the other hand, were very visible against the light tree trunks and became easy prey for birds. The birds, however, often could not find the camouflaged light-colored moths. As a result, more

The original variation in moth color was due to a mutation that occurred long before the Industrial Revolution.

21–24 The peppered moth. **a.** Lichen-covered trees afford camouflage protection for the light-colored moths, allowing them to thrive. **b.** A soot-covered trunk favors the black moths, which are now well camouflaged.

light-colored moths survived than black moths. The light-colored moths were more adapted to the local environment than the black moths. The light-colored moths lived, reproduced, and had mostly light-colored offspring.

During and after the Industrial Revolution, the environment in England changed. By the late nineteenth century, air pollution from the burning of coal had darkened the tree trunks. Light-colored moths, no longer protected by their color, stood out against the darkened tree trunks. Birds plucked them off the trees and ate them. Very few light-colored moths were left to reproduce. The black moths, on the other hand, were better adapted to the changed environment. Birds did not find them as easily as they found light-colored moths. Many black moths lived to reproduce, and the black peppered moth population began to increase. After a while, most of the moth population was black.

In recent years, England has adapted pollution control measures, and the trees are now becoming lighter in color. Recent studies reveal that the number of light-colored moths is increasing.

Natural Selection Is a Useful Theory The change in England's peppered moths is one example of a population that changed in the face of an environmental change. There are many examples of such changes occurring all the time. Natural selection is a useful theory because it helps us understand and predict the way that organisms change in a changing ecosystem. It provides a better understanding of ecology by relating living things to their physical environment.

Consider the example of stable flies and houseflies on Michigan's Mackinac Island. These biting, disease-carrying insects breed in manure. There is much manure on the island because transportation is by horse-drawn carriages. In the past, horse manure was piled up in alleys and along roadsides, and flies were abundant. In 1945 a program of spraying the pesticide DDT to control the flies went into effect. At first, the drastic reduction in the fly populations seemed like a miracle. But by 1949 the fly populations made a comeback. DDT-resistant flies survived and reproduced. Other chemical insecticides that were tried only helped temporarily. Chemical-resistant flies survived and thrived each time a new insecticide was used.

The theory of natural selection helps explain the shift in the fly population. More flies are born than will survive. An inherited variation exists within fly populations. For some flies, DDT is a poison. For others, DDT-resistant flies, it is harmless. When DDT was introduced into the ecosystem, resistance to DDT was a favorable variation for flies. DDT-resistant flies survived and

The tree trunks were light in color because lichens grew on them. Pollution killed the lichens, and the tree trunks darkened. Since pollution has been controlled in England, the lichens are growing back and the number of light-colored moths is increasing.

The change from light-colored to black moths took place in about 50 moth generations.

Recently the residents of Mackinac Island introduced better sanitation methods, natural fly predators, and traps to control the flies.

The increase in DDT-resistant flies was observed wherever DDT was used—not just on Mackinac Island.

Section 21.3 Theories About How Living Things Change

Species or populations that lack variations are more likely to be exterminated by pollutants, poisons, or medicines than those that have variations.

reproduced; others died. New generations contained more and more DDT-resistant flies.

Similar shifts occur in other populations as well. Where antibiotics are used, many bacteria die. But antiobiotic-resistant bacteria survive and reproduce. In ecosystems where pollution kills many organisms, pollution-resistant forms of plants and animals survive and increase in number.

The changes that can be observed in a population in a changing environment over the course of time are explained by the theory of natural selection. The theory helps us understand the ongoing history of life.

Building a Science Vocabulary

evolution
use and disuse
acquired characteristic
theory of natural selection

Questions for Review

1. Lamarck's statement of use and disuse is based on observations, but his idea of inheritance of acquired characteristics is not.
   a. What was Lamarck's observation about use and disuse?
   b. Give an example of use and disuse that is observable.
   c. What was Lamarck's idea about the inheritance of acquired characteristics?
   d. How did Weismann's experiment help disprove the inheritance of acquired characteristics?

2. What does the theory of natural selection attempt to explain?

3. What are the three main ideas in the theory of natural selection?

4. According to modern theory, what are the sources of variation described by the theory of natural selection?

5. What change in England's peppered moth population occurred after the Industrial Revolution? What change in the environment preceded this change in the moth population? Use the theory of natural selection to show how these changes are related.

Perspective: A New View
Nature's Way: Adapt or Become Extinct

As we study various plants and animals, we find that some organisms continue to change over long periods of time. Some others remain relatively unchanged and thrive, while still others change very little and become extinct.

The modern horse has an ancestry dating back some 60 million years. The earliest known horses stood only 25 cm high and had a hoof for each toe. By contrast, the opossum has survived almost unchanged for over 75 million years. The horseshoe crab has the same appearance as horseshoe crabs 500 million years ago. Nor has the Ginkgo tree changed in 200 million years. Dinosaurs are the most noted animals to have stopped changing and become extinct. For 200 million years, they were the dominant form of life. Yet, dinosaurs became so specialized that when climate and land forms changed, they could not adapt.

Changes in climate or in the elevation and the shape of the land bring about changes in the plants and animals that live on the land. Glaciers once covered much of North America; the Sahara was once fertile agricultural land. The large tree ferns that once dominated much of the landscape at the time of the dinosaurs are now gone, replaced by deciduous and coniferous forests.

Clearly, change is part of Nature's way. Only variations in the characteristics of a species enable it to adapt and avoid death and extinction in times of change. Organisms that are highly specialized to a particular environment are not likely to survive. Recognizing which organisms will survive and which will not is a constant challenge to the modern biologist.

With the human capacity to accelerate the rate of change in the environment, we are in a situation where the rate of adaptation for many organisms is not great enough for their survival. Thus, we help create more and more endangered species as we dredge harbors that would ordinarily fill, clear forests for agriculture, alter the flow of streams, flood the land by dams, and convert farm land to suburban housing. We must wonder: Are humans an endangered species? Can we change rapidly enough to adapt to our own changing environment? Only time will tell, but you and I may not be around to hear the answer.

Chapter 21
Summary and Review

Summary

1. Abiogenesis states that living things are formed from nonliving matter in the presence of an active principle.

2. Francesco Redi provided the first convincing evidence against abiogenesis by proving that flies arise only from other flies.

3. After the discovery of microorganisms, people believed that they were formed by abiogenesis. Louis Pasteur disproved abiogenesis by showing that microorganisms come to broth and not from it.

4. The heterotroph hypothesis attempts to explain the origin of life on earth. It states that under primitive conditions, simple compounds in the early atmosphere may have clumped together, combined further, and formed complex substances. These clumps may have given rise to the first forms of life.

5. The first forms of life are thought to have been heterotrophs. Autotrophs arrived later, providing a continuous source of food and adding oxygen to the atmosphere.

6. Experiments done by Miller, Urey, and others support the idea that certain parts of the heterotroph hypothesis could have occurred.

7. Information about the history of life on earth and relationships between organisms comes from fossils.

8. Fossils reveal the order of appearance of living things on earth, changes within a species, extinction, and relationships. They suggest a history of dispersal, adaptation, and change.

9. The body structures, embryos, vestigial organs, and body chemistry of living species provide information about their relatedness.

10. Studies done on organisms living on two sides of a barrier suggest that species change over time. Evidence that species can physically change over time is obtained by inducing mutations and breeding new varieties.

11. The changing of species over time is known as evolution.

12. Lamarck's theory of evolution involved use and disuse of body structures and inheritance of acquired characteristics. Lamarck's theory is not supported by observable facts.

13. Darwin and Wallace independently developed the theory of evolution by natural selection. According to the theory: the environment is unable to support all the organisms that are born, so many die before reproducing; variations exist among organisms; and organisms with favorable variations survive and reproduce, passing their favorable variations to the next generation. In time, a species becomes more adapted to its environment.

14. Neither Darwin nor Wallace explained the source of variations. Today we know that variations are due to mutations and recombination.

15. Changes in the peppered moth population over the past 150 years can be explained by natural selection. Natural selection also explains increased numbers of DDT-resistant flies and antibiotic-resistant bacteria.

Review Questions
Application

1. From each of the following pairs, select the group of organisms that is believed to have appeared on earth earlier.
   a. water plants, land plants
   b. flowering plants, conifers
   c. invertebrates, vertebrates
   d. reptiles, amphibians

2. A fossil of a fish was found high in a mountain's sedimentary rock. How could a fish fossil get to such a location? How could the age of such a fossil be determined?

3. Van Helmont, a seventeenth century scientist, believed that human sweat was the active principle in a dirty shirt. He said that if you placed kernels of wheat and a dirty shirt in an open box, mice would be formed in 21 days by abiogenesis. Design an experiment to test Van Helmont's idea in a way that would provide convincing evidence to the people of his day.

4. The largest antler span of any living animal belongs to a species of moose. The male's antlers have a span of about 300 cm. How might Lamarck have explained the enormous antlers present in this species? How might Darwin have explained them?

Interpretation

1. A small fold of tissue in the corner of the eye near the nose has no function in certain mammals. In reptiles, amphibians, and birds this structure is a third eyelid that closes and protects the eye. What type of structure is this eye tissue in mammals? How does this structure imply relationship among the vertebrates?

2. The chlorophyll molecule in algae is similar to the chlorophyll molecule in seed plants. What does this observation imply about the relatedness of autotrophs? What type of organism does fossil evidence suggest might be a common ancestor for all autotrophs?

Extra Research

1. Visit a museum and observe the fossil displays. Look for fossils that relate one group of organisms to another. Sketch such fossils and identify the characteristics of the groups they relate.

2. Many scientists tested the ideas of biogenesis and abiogenesis, but their work was not conclusive. Look up the work on this subject by the following scientists: Van Helmont, Needham, Tyndall, and Spallanzani. Explain what each scientist tried to prove and why the work was not conclusive.

3. With the heterotroph hypothesis in mind, write down your ideas about the possibility of life on other planets. How could your ideas be tested?

4. Look for sedimentary rock exposures in your area. If you have a camera, photograph the area. If possible try to dig out fossils. Determine the kinds of fossils that are characteristic of the particular rock formations that you explore.

5. Prepare plaster of Paris and make a cast of a footprint or a leaf print. Relate this process to fossil formation.

Career Opportunities

Medical and biological illustrators specialize in technical illustrations of the human body and other biology-related subjects, primarily for books and journals. Sometimes they begin with crude drawings, photographs, or descriptions. More often they work directly with their subjects in the field, in operating rooms, or in laboratories. Some work on a free-lance basis. A background in biology and an ability to draw are necessary. Contact medical schools for further information.

Paleontologists are biologists and geologists who study plant and animal fossils. The fossils are used to trace the development of past life or to indicate the presence of gas or oil deposits. Most paleontologists work for universities, energy companies, or the government. Four or more years of university are required. Write the Mineralogical Association of Canada, c/o Royal Ontario Museum, 100 Queen's Park, Toronto, Ontario M5C 2C6.

Unit VI
Suggested Readings

Bronowski, Jacob, *The Ascent of Man*, Boston: Little, Brown & Co., 1973. *Bronowski includes a summary of the ideas about the origin of life in his chapter, "The Ladder of Creation." It is beautifully presented, consistent with the high quality of both the book and related television series.*

Dunbar, Robert E., *Heredity*, New York: F. Franklin Watts, Inc., 1978. *Very short (64 pages) and easily read book that begins with Mendel and brings genetics up to date.*

Hopf, Alice L., *Nature's Pretenders*, New York: G. P. Putnam's Sons, 1979. *A collection of animal life histories that investigates the intriguing behavioral and physical adaptations, such as mimicry, that aid in survival.*

Kennedy, M. Keith, and Richard W. Merritt, "Horse and Buggy Island," *Natural History*, May, 1980. *Describes insect problems and their control on Michigan's Mackinac Island, showing natural selection at work in fly populations.*

Lamberg, Mark, *Fossils*, New York: Arco Publishing, Inc., 1979. *This accurate and current treatment of fossils includes excellent illustrations of both living forms and skeletal reconstructions of fossils.*

Milne, Lorus J. and Margery Milne, *Ecology Out of Joint*, New York: Charles Scribner's Sons, 1977. *Fascinating account of what happens when the natural rate of change is speeded up by human intervention. Explains the effects of civilization on the natural world.*

Moll, Don, "Dirty River Turtles," *Natural History*, May, 1980. *The Illinois River, once clear and pure, now is cloudy and polluted with silt and sewage. The populations of turtles living in the river are affected by the changing environment.*

Pogash, Carol, "The Left Handed," *Science Digest*, June, 1977. *A discussion of the cause and nature of handedness refers to theories on the genetic and other origins of the condition.*

Powell, Jim, "Superseeds," *Science Digest*, June, 1977. *Excellent article describes techniques and gains of plant breeding. Discusses use of artificial selection, outbreeding, and inbreeding to develop high-yield, nutritionally valuable, disease-resistant seeds.*

Ricciuti, Edward R., *Older Than the Dinosaurs: The Origin and Rise of the Mammals*, New York: Lippincott & Crowell Publishers, 1980. *This survey of mammals as they have evolved since the end of the dominance of reptiles discusses the differences between amphibiians, reptiles, and mammals and the advantages of each. Includes vivid descriptions of prehistoric environments.*

Sagan, Carl, *Cosmos*, New York: Random House, Inc., 1980. *Based on the television series of the same name, this book attempts a chronological account of great human efforts in science. Traces today's scientific knowledge and methods to their historical roots.*

Sandred, Kjell B., "Message to Moths: Trick or Perish," *Audubon*, November, 1979. *Excellent article describing life-saving adaptations that have evolved in moths. Includes peppered moths' camouflage, body shapes that resemble leaves, buds, or flowers, eyelike wing patterns, and chemical warfare.*

Sootin, Harry, *Gregor Mendel: Father of the Science of Genetics*, New York: Vanguard Press, Inc., 1958. *Well-rounded biography reveals both the gentle monk and the brilliant scientist. Includes lucid explanations of Mendel's laws.*

Tennesen, Michael, "Phantoms of the Prairie," *National Wildlife*, June-July, 1976. *Describes dominant grassland herbivores of the past, including the pronghorn antelope.*

Unit VII

The Environment

All organisms depend on the world around them for food, water, shelter, and other necessities of life. No organism lives alone, isolated from other organisms and the nonliving world.

Organisms interact in all phases of life, whether they are getting food, finding shelter, or defending territory. In any given area, numerous species live together. As one species grows in size, some species also grow; others become smaller. They all use and reuse the natural resources of the area, constantly recycling matter and funnelling energy from the sun through producers to consumers.

Living things inhabit almost every region on earth. They thrive where rainfall, temperature, availability of food, and other conditions are best for them—where they are best adapted to live. Over time, organisms change the areas in which they live. New forms thrive. New adaptations become suitable. When you observe the environment as a whole, you begin to see life as a constant interchange between organisms and between organisms and their environment.

Ladybugs cluster on a fern. Almost everywhere on earth living things interact with each other and their environment.

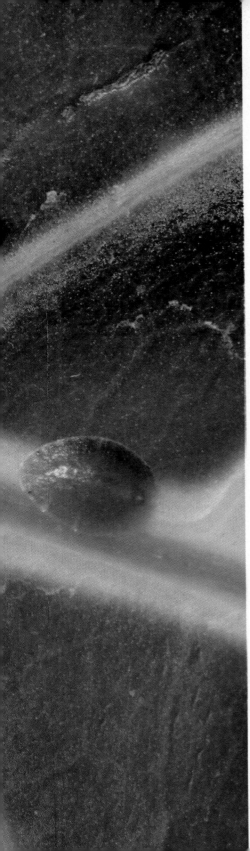

Chapter 22

Interactions in the Ecosystem

Focus *Your mental picture of a sunny meadow is probably like a painting or photograph—still and unchanging. Yet a meadow, with its birds, insects, trees, and soil, is an ecosystem supporting an unceasing round of activities and interactions. Living things interact continually with each other as well as with the nonliving environment.*

22.1 FOOD AND ENERGY IN THE ECOSYSTEM

The Role of Food

Food is a kind of currency that passes through the ecosystem. Once it is produced, it is transferred from organism to organism in a chain. In this sense, food links living things together. Organisms can be categorized into three groups according to how they obtain food. Some organisms are producers, some are consumers, and others are decomposers.

Producers Producers take in simple chemical substances from the environment and build them into food by the process of photosynthesis. Producers include producer protists, algae, and other green plants. Producers are the original source of all food on earth and are part of every ecosystem.

22–1 A honey ant tends scale insects. The ant milks the scale for honeydew, which they produce from plant juices.

22–2 Consumers. **a.** The starfish, a carnivore, feeds on a mussel. **b.** The crayfish is a scavenger. **c.** The raccoon, an omnivore, cleans its food before it eats.

Consumers Consumers depend on other living things for food, since they cannot manufacture it. They can be classified into several categories, according to their eating habits. Animals are major consumers and are classified according to the foods they eat. **Herbivores** (ERB-bih-vōrz), such as cows, horses, and deer, eat only plants; **carnivores** eat only meat. Some carnivores, including lions and hawks, are **predators**; they kill other animals and feed on them. Other carnivores are **scavengers**; they eat animals that they do not kill themselves. Vultures, hyenas, and crabs are examples of scavengers. **Omnivores**, including humans, bears, raccoons, and rats, eat both plants and animals.

Decomposers In addition to animals, **decomposers** are another important group of consumers. Most decomposers are simple forms of life, such as bacteria and fungi. Decomposers break down dead bodies and the waste products of living things. They change these materials into simple chemical substances, which they release into the environment. These chemical substances are picked up from the soil, water, and air by producers and are used to make food. For this reason, decomposers are essential to the cycling of chemicals in an ecosystem.

Food Chains and Food Webs

The food that producers make is eaten by consumers and is eventually broken down by decomposers. The sequence of organisms through which food passes in a community is called a **food chain**. In some food chains, food begins with producers and passes to one or more consumers. For example, gypsy moth

Each ecosystem has its own particular food chains.

larvae eat tree leaves. In other food chains, food begins with dead organic matter and passes to decomposers. For example, bracket fungi thrive on dead trees. Some food chains have only two or three links, such as the one shown in Figure 22–3. In the figure, the mesquite shrub is the producer link in the chain; the squirrel makes up the only consumer link. When the squirrel and mesquite die, they become food for bacteria. Some food chains have more than two or three links, such as the one shown in Figure 22–4.

Consumers are ranked according to what they eat. The consumers that eat producers (herbivores) are called **primary consumers**. Those that eat primary consumers (carnivores) are called **secondary consumers**. The ranking of consumers continues up the chain to the final consumer.

Food chains are usually interrelated with other food chains in an ecosystem. This forms a kind of feeding network called a **food web**. The food chain shown in Figure 22–4 is part of a food web. Most food webs include producers, consumers, and decomposers. When ecologists analyze the feeding relationships in all ecosystems, they find very complex food webs.

22–3 A simple, two-link food chain exists between a mesquite shrub and a ground squirrel.

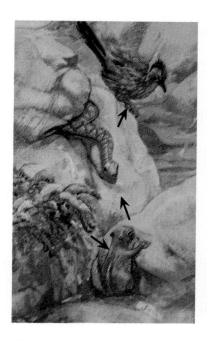

22–4 The simple food chain becomes longer with the addition of a secondary consumer (the snake) and a tertiary consumer (the roadrunner).

22–5 Numerous food chains make up a food web. Each organism in an ecosystem is part of several food chains.

Section 22.1 Food and Energy in the Ecosystem

The Transfer of Energy Through the Ecosystem

The sun is the original source of energy for every ecosystem. When the sun shines on producers in an ecosystem, part of the sun's energy is converted to chemical energy by the process of photosynthesis. The chemical energy is stored in the glucose molecule, which can be converted to starch, for long-term storage, or to other molecules. The energy stored in glucose and other molecules passes from producers to consumers and decomposers through the food chain.

On the average, only about 10 percent of the food ingested by an organism supplies materials for building and repairing its body parts, as well as food storage. Much of the food, from 30 percent to 60 percent, is not digested. The rest undergoes cellular respiration, providing energy for cell functions.

Energy must be constantly supplied to maintain life on earth, because some energy is lost at each link in the food chain. The loss occurs as heat given off during the organism's life activities. In these processes, chemical energy is changed to heat energy, which cannot be used by the organism. Heat energy leaves the ecosystem and enters the atmosphere. As a dead organism decomposes, much of the energy that is left in its body eventually goes into the atmosphere in the form of heat.

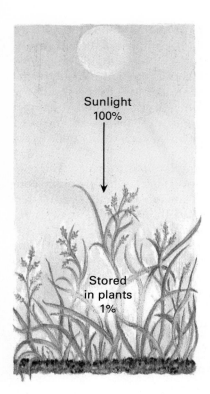

22–6 A tremendous amount of energy from the sun reaches the surface of the earth. Of this energy, only about one percent is trapped and stored as chemical energy by producers.

A Food Chain Problem

To help you understand how energy moves through the ecosystem, try this problem. Imagine that you are stranded on an island with a cow and a 25-kg bag of corn kernels. The object is to stay alive as long as possible in the hope that a rescue party will arrive. You must decide the order of eating the corn and the cow so that you get the greatest possible benefit. You have the following choices:

Eat the corn, then the cow
Eat the cow, then the corn
Feed the corn to the cow, then eat the cow
Share the corn with the cow, then eat the cow
Share the corn with the cow, then drink milk

What would you do and why? In considering this problem, remember that you want to obtain the greatest possible amount of energy from the food. Do not forget that energy is lost as heat in the body of each consumer. Also, assume there is no loss of food by spoilage.

The food chain problem has applications for human ecosystems. In countries with limited production of food crops and large populations people generally have vegetarian diets. Most of the sun's energy trapped by producers goes directly to the human consumer. There is only one transfer in which energy is lost. If cattle are fed, energy is lost in two transfers, and humans get less energy. As a result, a vegetarian diet allows more people to get food energy. By eating plants rather than meat, people take in more of the energy originally captured by the plants.

Energy Pyramids

Biologists often use a diagram called an **energy pyramid** to show the path of energy from one type of organism to another. An energy pyramid also shows the amount of energy available at each level of the ecosystem.

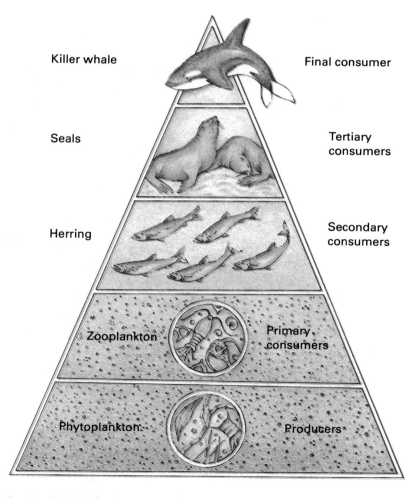

22–7 Aquatic community energy pyramid. Only about ten percent of the food energy in a given level will become incorporated into the bodies of the organisms at the next higher level. Notice that as you go from one level to the next higher level, the number of organisms decreases and the organisms' size increases.

Section 22.1 Food and Energy in the Ecosystem

Although the energy pyramid is a simple diagram, it provides much information. Producers are the first organisms to trap and convert the sun's energy to chemical energy. Therefore, the greatest amount of usable energy is located in their level. The amount of usable energy decreases from level to level up the pyramid, because at each level organisms use some energy for life activities. During these processes, energy is lost to the atmosphere as heat.

The pyramid does not show the decomposers that feed at both the producer and consumer levels. These decomposers tap off some energy for their own use. During decomposition, energy is also lost to the atmosphere as heat.

Building a Science Vocabulary

producer	scavenger	secondary consumer
consumer	omnivore	food web
herbivore	decomposer	energy pyramid
carnivore	food chain	
predator	primary consumer	

Questions for Review

1. Why are producers needed for an ecosystem to survive?

2. What is the role of decomposers in an ecosystem?

3. The following organisms might be found on a farm: mouse, corn, cat. Arrange them into a food chain. Which one is a producer? a predator?

4. Which organisms in a food chain depend directly on the sun for food? indirectly?

5. Why is a certain amount of the energy in food always lost? In what form is it lost?

6. Why is the two-link food chain (humans→ plants) of advantage to populations with limited food resources?

7. Prepare an energy pyramid for an ecosystem that contains grass, cattle, and humans. Place each type of organism at its proper level.

22.2 CHEMICAL CYCLES IN THE ECOSYSTEM

Energy flows into, through, and out of an ecosystem; it must be constantly replenished. But certain important chemical substances are used over and over again as they cycle between the living and nonliving parts of the ecosystem. In general, they begin in the nonliving environment: water, air, or soil. These chemical substances then enter the bodies of producers, pass through food chains, and are returned to the nonliving environment by decomposers. Three of the most important substances that are continually recycled are water, carbon, and nitrogen.

The Water Cycle

In the water cycle, water passes from the atmosphere to the earth and back to the atmosphere. On earth it enters bodies of water, soil, and living organisms. Water that falls from the atmosphere to the earth is called **precipitation**. Rain is liquid precipitation; snow, hail, and sleet are solid precipitation. When precipitation falls on land, water seeps underground or runs off the surface of the land into streams that flow to the ocean.

The sun heats water in lakes, streams, and oceans. As water warms, it changes into water vapor (gas). This process is known as **evaporation**. Water vapor rises into the atmosphere where it is cooled. As it cools, water vapor changes from a gas to a liquid. This process is called **condensation**. If it is cold enough, some of the liquid may freeze and become solid. Condensed water falls to the earth as precipitation and the cycle begins again.

Ninety-nine percent of the water on earth is in the oceans and polar ice caps.

Water changes from liquid to gas when sweat evaporates on the skin.

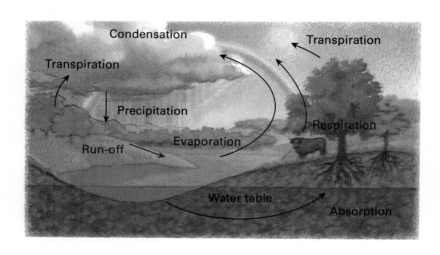

22–8 Water cycle. Water that evaporates from soil and water surfaces, or is exhaled or transpired, is returned to the earth through condensation and precipitation.

The Carbon Cycle

Carbon is one of the major components of living organisms (see Table 4–1). In the carbon cycle, carbon passes from the environment into the bodies of living things and back to the environment. Much of the carbon in the environment is in the form of carbon dioxide (CO_2), a gas that is commonly found in air and in water. During photosynthesis, green plants and producer protists use carbon dioxide, water, and light energy to build food. These producers use some of the food they make as a source of energy and raw materials for themselves; they store the rest as starch, oils, and other compounds. Animals and other consumers obtain their food from the bodies and food stores of producers.

Both producers and consumers obtain energy from food by cellular respiration. During this process, glucose and other energy-rich molecules are broken down into carbon dioxide and water, which are returned to the environment.

The waste products and dead bodies of organisms are broken down by bacteria and other decomposers. In this process, decomposers produce carbon dioxide and release it into the environment. If dead organisms fail to decompose quickly, they may slowly change to coal, oil, and gas. These fossil fuels contain large amounts of carbon. When they are burned, carbon dioxide is returned to the environment.

Carbon dioxide is the gas that makes soda bubbly.

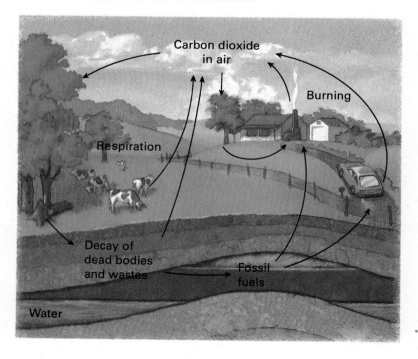

22–9 Carbon cycle. Carbon compounds are present in the atmosphere, in fossil fuels, in dead and living organisms, and in the waste products of organisms. Decomposition, burning, and respiration release carbon into the air as carbon dioxide.

The Nitrogen Cycle

The nitrogen cycle is the most complex of the nutrient cycles. It involves many important microorganisms.

Living things use nitrogen to build protein. As in the water cycle and the carbon cycle, nitrogen passes from the environment into the bodies of organisms and back to the environment. About four fifths of air is nitrogen gas (N_2); however, plants cannot use nitrogen gas to make proteins. The nitrogen they use is in the form of nitrogen-containing minerals in the soil called **nitrates**.

Green plants take in nitrates from the soil through their roots. Inside plant cells, nitrates are used to make amino acids, which serve as the building blocks of proteins. Proteins make up parts of the plant body, as well as the plant's enzymes. Consumers depend directly and indirectly on the food produced by plants to supply them with protein. Consumers use the protein that they get from food for growth and energy.

Decomposers obtain energy from the processes of ammonification and nitrification.

Decomposers break down protein in dead organisms and animal wastes. One type of bacteria decomposes protein and releases ammonia. The work done by these decomposers is called **ammonification**. A second group of bacteria decomposes ammonia, changing it to **nitrites**. A third group of bacteria changes the nitrites to nitrates. The process whereby bacteria change ammonia to nitrates is called **nitrification**. The nitrates released by nitrification are then available for plants.

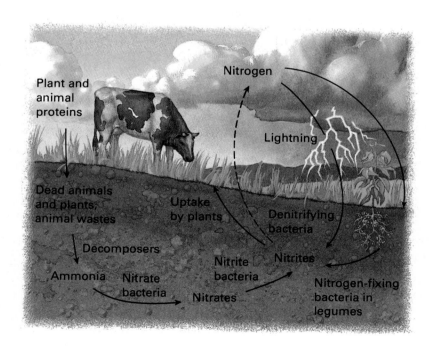

22–10 Nitrogen cycle. Lightning, nitrogen-fixing bacteria, and decomposers produce nitrogen compounds used by plants to make proteins, which recycle through the decay of body tissues and animal wastes. Free nitrogen is returned to the atmosphere by bacterial action.

Section 22.2 Chemical Cycles in the Ecosystem

Using a system of crop rotation, farmers alternate grain and other crops with pod-bearing plant crops. The nitrogen-fixing bacteria on the roots of pod-bearing plants (legumes) restore nitrates to the soil.

22–11 The roots of the soybean plant (a legume) develop nodules of root cells that harbor nitrogen-fixing bacteria in their cytoplasm.

Other Sources of Nitrates Decomposition of protein is one process that produces nitrates for plants. **Nitrogen fixation** is another. Nitrogen fixation is a process in which bacteria convert nitrogen gas to nitrates. The bacteria that carry out this process are called **nitrogen-fixing bacteria**. Some nitrogen-fixing bacteria live in the soil; others live in nodules on the roots of pod-bearing plants, such as peas, beans, clover, and alfalfa. The nitrates released by nitrogen-fixing bacteria are then available to green plants.

A small amount of nitrate is produced when lightning strikes the nitrogen in the atmosphere. The nitrates produced by this process fall to the earth with rain. Humans also add large amounts of nitrates and ammonia to the soil by the application of industrial fertilizers.

Loss of Nitrates from Ecosystems Several events remove nitrates from an ecosystem. Farm ecosystems lose nitrates when crops are harvested. Water washing over exposed soil takes nitrates with it. Water can also carry nitrates deep into the soil where they are not available for plants. Denitrifying bacteria thrive in poorly drained soil, where they convert nitrates into nitrogen gas.

Great amounts of nitrates are removed from ecosystems each year. The effects of the loss of nitrates are counterbalanced, however, by the processes that restore nitrates to the ecosystem.

Other Elements Cycle in the Ecosystem

In addition to carbon, hydrogen, oxygen, and nitrogen, about 20 other elements cycle in ecosystems. The water, carbon, and nitrogen cycles illustrate how the cycling process operates. In studying any chemical cycle, scientists try to answer four basic questions.

1. Where is the element found in the environment?
2. How does it get into the ecosystem?
3. What path does it take in the ecosystem?
4. How does it get back to the environment?

In the case of the water cycle, these questions are fairly easy to answer. In other cases, they are more difficult. For example,

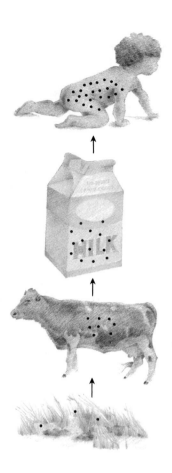

22–12 Strontium-90, introduced at the lowest level of a food chain, becomes more and more concentrated as it is passed from one level to the next. A high concentration of strontium-90 in the human body can cause cancer and other disorders.

phosphorus and sulfur may be bound up in the depths of the ocean for long periods of time, making it hard to determine how and when they will be released.

The same approach is used by scientists when they study harmful substances that occur in the ecosystem. Consider the case of strontium-90, a radioactive substance that is similar to calcium. Strontium-90 entered the soil in areas where nuclear testing occurred in the 1950s. The strontium-90 passed through the food chain, first from the soil to grass, then from the grass to cows. It ended up in the cows' milk and was consumed by children. By 1959, more strontium-90 could be found in the bones of children than in the bones of adults. By studying the way in which this substance passed through the ecosystem, scientists were able to recommend controls on nuclear testing. If left unchecked, exposure to strontium-90 could cause leukemia, bone cancer, or hereditary damage.

Building a Science Vocabulary

precipitation	nitrate	nitrification
evaporation	ammonification	nitrogen fixation
condensation	nitrite	nitrogen-fixing bacteria

Questions for Review

1. What substance is taken in by producers during the carbon cycle? the nitrogen cycle?
2. Why are nitrates important to green plants?
3. Explain what happens in the phases of the water cycle.
4. Name three processes that return carbon to the environment during the carbon cycle.
5. Explain how decomposers connect living things to the nonliving environment during the nitrogen cycle.

22.3 INTERACTIONS AMONG ORGANISMS

When you examine the links in a food chain, you can see differences in food-getting patterns. Some organisms cooperate during the process, while others compete. As a result, complex patterns of interaction exist. These interactions often involve more than the process of getting food. Consider the relationship between humans and dogs. Over 35 million dogs in North America have human owners. These dogs depend on their owners for food. The relationship between humans and dogs involves more than food, however. Dogs also depend on their owners for shelter and other essentials of life. Clearly, the dog benefits from the relationship. The owner, in turn, may receive protection, companionship, and help in hunting, herding, or some other kind of work.

22-13 A seeing-eye dog is much more than a companion to this young woman. Through her interaction with the dog, the woman is able to lead a more active life.

Symbiotic Relationships

When two species of organisms interact with each other in any way, the relationship is called **symbiosis**. In some cases of symbiosis, both parties benefit. In other cases, only one party benefits; the other party is either harmed or not affected. Five types of symbiosis can be observed.

Mutualism A symbiotic relationship in which both species benefit from the interaction is called **mutualism** (MYOO-choo-wul-ism). The relationship between a dog and its owner is an example of mutualism. Another example is the relationship between certain nitrogen-fixing bacteria and pod-bearing plants. The bacteria supply nitrates to the plants, and the plants supply the bacteria with food, water, and shelter.

A mutualistic arrangement also exists between a termite and the protozoa that live in the termite's intestine. Termites cannot digest the cellulose in wood without the protozoa. The termite benefits from the relationship by receiving glucose. The protozoa benefit by having a steady diet and the protection of the termite's intestine.

To test the termite-protozoa relationship, a scientist at Harvard University irradiated termites, killing the protozoa inside their intestines. The termites were not harmed. Soon, the intestines of the termites became clogged with wood fibers, and the termites began to starve. When the scientist put new protozoa in the termites' intestines, the termites began to digest wood normally and soon recovered.

22–14 Mutualism. The oxpecker rids the warthog of ticks and obtains food in the process.

Commensalism The kind of symbiosis in which one of the organisms benefits and the other is not affected is called **commensalism** (kuh-MEN-sul-izm). The term *commensalism* actually means "eating at the same table." The remora, a slender fish about 60 cm to 90 cm long, has a commensal relationship with the shark. The remora has a suction disk on top of its head with which it attaches itself to the body of a shark. Sharks are sloppy in their feeding habits. The remora snatches scraps of food the shark misses. The shark does not appear to benefit from the relationship, but it is not harmed either.

Certain fish, crabs, and worms find shelter inside the burrows of large sea worms and shrimp. They do not appear to harm their hosts, and they benefit by getting free shelter. In addition, they sometimes feed on bits of surplus or rejected food. One species of a small, round-bodied marine crab sometimes lives inside an oyster's shell along with the oyster. The oyster does not benefit, but the commensal crab gets housing and food. On land, Gila woodpeckers and Pygmy owls live in the desert. Gila woodpeckers peck holes in cactus plants. Later, after the woodpeckers have abandoned the holes, the Pygmy owls take them over and use them as nests.

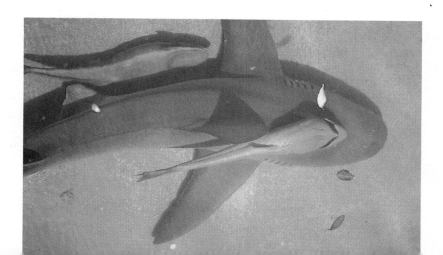

22–15 Commensalism. Remoras travel attached to a shark and share scraps of food missed by their host. The shark is apparently unaffected by this interaction.

Predation Predation is a symbiotic relationship between predators and prey. An organism that attacks a living animal to obtain food is called a **predator**. The animal that serves as food is referred to as **prey**. In a predator-prey relationship, the direct benefit to the predator is food. It appears that only harm results to the individuals that are attacked and eaten. If the predator-prey relationship is in balance, however, the prey benefits as a population, even though individual members are killed. Consider the tremendous numbers of rats, mice, insects, and other prey that would exist if they were not preyed upon by owls, cats, and other predators. Predators help control the number of individuals in the prey population.

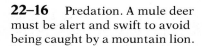

22–16 Predation. A mule deer must be alert and swift to avoid being caught by a mountain lion.

Parasites may bring about the death of their host, but most often they only weaken it.

Parasitism Parasitism is a symbiotic relationship between parasites and hosts. **Parasites** are organisms that obtain food by living on or inside the bodies of other living organisms. The organisms that parasites live on or in are called **hosts**. Host organisms are usually harmed by the parasite in some way.

Almost all living things are at some time hosts to parasites. Certain roundworms are parasites on plant hosts. The worms get food from the plants they inhabit, but they damage their hosts in the process. Each year, roundworm parasites cause millions of dollars in damage to crops. Ticks can be parasites on dogs and other animals, including humans. They feed on the host's blood, often transferring disease organisms that they carry to the host. The bacteria that cause pneumonia are one more example of parasitism. These bacteria get food and shelter in the bodies of their human hosts, but they harm humans by causing disease.

22–17 Parasitism. A tomato hornworm caterpillar has been paralyzed by a braconid wasp to become a living home and food source for the wasp's larvae.

Competition between members of the same species is called intraspecific *competition. Between populations of different species it is called* interspecific *competition.*

Competition Competition is rivalry between individuals for a specific resource and can occur between members of the same or different species. Some of the resources for which organisms compete are food, space, water, air, light, shelter, and mates. For example, tree seedlings growing in a shady forest may compete with one another for light. Field mice and house mice living together in a deserted building might compete for food.

If a needed resource is in short supply, competition can be fierce. An organism might directly attack its competitor, or it might produce a substance that is harmful to its competitor, thereby driving it away. Competition can also occur if one species reproduces more rapidly than its competitors. The more abundant species may crowd others out of the area.

22–18 Competition. Eucalyptus trees grow tall and branch as they compete for sunlight in a forest.

Section 22.3 Interactions Among Organisms

22–19 Reducing competition. Shore birds having bills and legs of different lengths are adapted to feed in various regions of the tidal zone. In this way, they avoid competing directly for food resources.

Many organisms have adaptations that reduce competition. For example, several related species of birds living in the trees of a given area compete for the same foods; however, their feeding habits reduce competition. One species of bird feeds on the lower branches of the trees; another feeds on the middle branches; still another feeds at the tree tops. In other cases where competition is reduced, competitors use the same food source at different times of the day, or they may occupy different territories. They may use slightly different food or mineral sources. Each of these adaptations reduces the competition between organisms. In general, more members of an ecosystem are adapted in ways that reduce competition than in ways that increase it.

Building a Science Vocabulary

symbiosis	predator	host
mutualism	prey	competition
commensalism	parasitism	
predation	parasite	

Questions for Review

1. Name the type of symbiosis that exists between the organisms in each of the following pairs.
   a. snake and mouse
   b. pneumonia bacterium and human
   c. shark and remora

2. State one difference and one similarity between parasites and predators.

3. What causes competition in an ecosystem? Give an example.

4. What are three ways that some organisms reduce competition?

Perspective: Environment
Can There Be Too Much Carbon Dioxide?

Millions of years ago, the remains of many dead plants and animals failed to decompose rapidly. Instead, they accumulated at the bottoms of the lakes and marshes where they grew, forming a thick muck. Over thousands of years, this muck built up into thick layers. Eventually, it was covered over by sediments from rivers or oceans. The weight of these sediments created enough pressure to convert the organic material to coal, oil, gas, and even diamonds.

This layering process trapped great quantities of carbon in the ground for millions of years. Starting just a little over 100 years ago, much of this carbon has been mined or pumped from the ground. It has been burned as fuel by our industries, homes, automobiles, and airplanes. The burning of carbon produces carbon dioxide, which is released into the atmosphere. Much carbon dioxide has been released into the atmosphere in the last hundred years or so.

The accumulation of large amounts of carbon dioxide in the atmosphere results in the "greenhouse effect." Carbon dioxide acts like

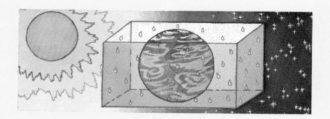

glass in a greenhouse. It keeps heat in the atmosphere in much the same way glass keeps heat in a greenhouse. If more heat is kept in the atmosphere because of an increase in the amount of carbon dioxide, the whole earth would warm up. Polar ice caps would melt, increasing the amount of ocean water. The sea level would rise, flooding the coastal areas of the world. Many cities built on or near sea coasts would be devastated.

What has happened? There has been only a slight increase in the amount of carbon dioxide in the atmosphere. Most of the carbon dioxide released in burning has been absorbed into the oceans and plants. The earth's temperature has not risen noticeably. The polar ice caps have not melted much, nor has the sea level risen significantly. The world ecosystem has been able to maintain a balance in the carbon dioxide level so far. But, if we continue to increase the flow of carbon dioxide into the atmosphere, can we be certain about the future?

Chapter 22
Summary and Review

Summary

1. In an ecosystem food passes from producers to consumers and decomposers.

2. Animal consumers include herbivores, carnivores, and omnivores. Predators and scavengers are types of carnivores.

3. Most food chains begin with producers and end with consumers, including decomposers. Upon their death, all links in a food chain become food for decomposers.

4. Food chains are connected in an ecosystem to form food webs.

5. The sun is the original source of energy on earth. Part of the sun's energy is trapped by producers and stored in food.

6. Energy passes through food chains and food webs in an ecosystem. An ecosystem requires a constant source of energy, because energy is constantly lost as heat.

7. The fewer the number of links in a food chain, the greater the amount of energy that is transferred from the producer to the final consumer.

8. Energy pyramids depict the amount of energy available at each feeding level.

9. Many chemical substances are used over and over in an ecosystem. They cycle between the environment and living things.

10. Water cycles by a continuing process of precipitation, evaporation, and condensation.

11. Producers take in carbon in the form of carbon dioxide during photosynthesis. Carbon passes through food webs and is released back to the environment by the processes of cellular respiration, decomposition, and burning.

12. Nitrogen enters food chains as green plants take in nitrates from the soil and build protein. Nitrates are restored to the soil by decomposition of protein, nitrogen-fixation, lightning, and fertilizers.

13. Symbiosis is the interaction of two species in an ecosystem. There are five types of symbiotic relationships: mutualism, commensalism, predation, parasitism, and competition.

Review Questions

Application

1. The osprey is a bird that eats fish. What type of consumer is the osprey?

The following questions are based on the diagram of a farm food web.

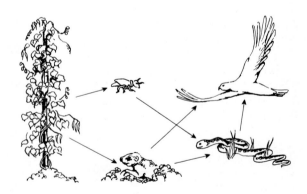

2. Arrange the organisms shown in the farm food web in an energy pyramid.

3. Make a list of all the consumers in the food web and state what type of consumer each one is.

4. If the sun stopped shining, what would happen to the farm food web? Explain.

5. Decomposers are not shown in the food web diagram. What is their role in the nitrogen cycle? in the carbon cycle?

6. How are nitrates lost from a farm ecosystem?

7. What could the farmer do to restore nitrates to the soil on the farm?

Interpretation

1. Give an example of mutualism between a consumer and a producer.

2. Why is most of the energy in an energy pyramid on the producer level?

3. A reduction in an ecosystem's sunlight could cause competition between producers and between consumers. Under conditions of less light, what would producers compete for and how might they compete? What would consumers compete for and how might they compete?

Extra Research

1. Place a well-watered plant in its planter on a flat tray or dish. Cover the plant with a glass jar or a sheet of clear plastic. Seal the jar or sheet to the tray with vaseline. Place the plant in the light. Watch for changes in the amount of moisture inside the jar or plastic over the course of a day. Diagram and label the steps in the water cycle that you observe.

2. Research and diagram some of the chemical cycles not included in this chapter. Examples are the phosphorus, calcium, and sulfur cycles.

3. Select an ecosystem in your neighborhood such as a meadow, vacant lot, pond, lawn, or garden. Find out what organisms live there and what they eat. Construct a food web based on your findings. Then arrange the organisms in the food web in an energy pyramid.

4. Try to construct a food web for a city. Based on your results, consider the question: Is a city an ecosystem? List reasons for your answer.

Career Opportunities

Soil conservationists advise farmers and ranchers on how to prevent and correct soil and water problems. They also help lumber and mining companies manage their land. Most have a degree in agriculture or a related field. Contact the Canadian Society of Soil Science, Dept. of Soil Science, University of Saskatchewan, Saskatoon, Saskatchewan S7N 0W0.

Farm operators must have a working knowledge of agricultural practices and be able to manage their time and resources well. On small farms they often do most of the physical work of caring for livestock, raising crops, maintaining machinery, and managing finances. On large farms they often supervise people who do most of the physical work. Training in agriculture is needed. Contact the Canadian Agricultural Economics Society, 151 Slater St., Ottawa, Ontario K1P 5H4.

Environmental engineers design processes and machinery that reduce the pollution of the environment. They may be hired by private firms or governments to manage solid waste disposal, waste water treatment, or air pollution control. A degree in engineering is required. For information write the Federation of Associations on the Canadian Environment, 1335 Carling Ave., #210, Ottawa, Ontario K1Z 8N8, or the Environmental Protection Service, Place Vincent Massey, Ottawa, Ontario K1A 1C8.

Chapter 23

Populations in Ecosystems

Focus *Have you ever heard anyone refer to "the straw that broke the camel's back"? This expression implies that even strong, well-developed systems have their limits. An ecosystem can support large populations of organisms, but it has its limits, too. If populations grow too large, the natural balance in an ecosystem can be upset. Natural forces are constantly at work, however, to prevent this from happening.*

23.1 CHARACTERISTICS OF POPULATION GROWTH

Imagine that you are a winner on a television game show. "For your prize you have a choice," the master of ceremonies announces. "We will give you one penny today, two pennies tomorrow, four pennies the next day, and so on for 30 days. Each day you will receive twice the amount you received the day before. Or, you can have $25 000 in cash right now. You have 10 seconds to decide." Which would you choose?

You will be much richer if you choose the first prize offered. Table 23–1 shows how much money you would receive each day for 30 days.

23–1 As food supplies dwindle in Canada at the onset of winter the monarch butterfly population migrates south to winter over in California and Mexico. They hang in large numbers from pine tree branches and on sunny days collect nectar from nearby flowers.

Table 23–1 The One-Penny Doubling Game

Day	Amount	Day	Amount	Day	Amount
1	$ 0.01	11	$ 10.24	21	$ 10 485.76
2	0.02	12	20.48	22	20 971.52
3	0.04	13	40.96	23	41 943.04
4	0.08	14	81.92	24	83 886.08
5	0.16	15	163.84	25	167 772.16
6	0.32	16	327.68	26	335 544.32
7	0.64	17	655.36	27	671 088.64
8	1.28	18	1310.72	28	1 342 177.28
9	2.56	19	2621.44	29	2 684 354.56
10	5.12	20	5242.88	30	5 368 709.12

How Populations Grow

The prize-money example illustrates the amazing results of repeated doubling. Look at another case where doubling occurs. Suppose a person is very ill with a strep throat, an extremely sore throat caused by bacteria called streptococci. A friend visits the sick person and then develops a strep throat within 24 hours.

How does the visitor become ill so fast? The answer lies partially in the way bacteria reproduce. Each bacterium produces *two* new bacteria. In other words, the number of bacteria doubles. Under favorable conditions some species of bacteria can double in number every 20 minutes. If you start with just one bacterium, how many bacteria will exist in one hour at this rate? in nine hours? If nothing stops this pattern of growth, an enormous number of bacteria will be in the body in 24 hours. The graph in Figure 23–3 shows this growth for three hours. You can see why the visitor became ill so fast.

The streptococci bacteria in the body of a sick person make up a **population**. A population is all the organisms of a certain species found in an area. Although the curve in Figure 23–3 shows growth of a bacterial population, its shape is representative of all populations whose members reproduce by doubling.

The organisms in most species do not reproduce by doubling. For example, two human parents usually produce one offspring at a time. Some adults in a population do not reproduce at all. But overall, when there are more births than deaths in any population, the population size increases.

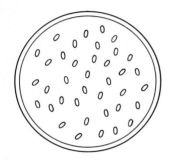
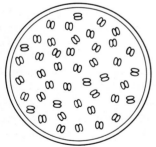
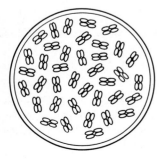

23–2 Populations of bacteria grow by doubling. Each bacterium splits in half by binary fission, producing two bacteria.

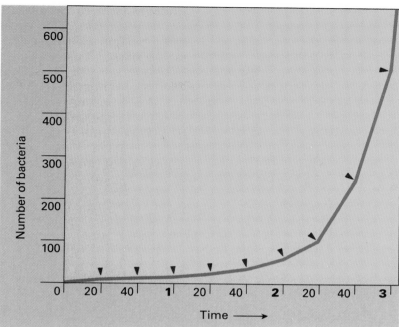

23-3 Bacterial growth by doubling every 20 minutes. Growth appears slow at first; however, it soon becomes enormous.

Population Explosions

Suppose that a family of mice is introduced to a deserted island. The mice have a large food supply and no natural enemies. The newly arrived mice thrive and reproduce, and their numbers increase rapidly. Rapid, uncontrolled population growth of this sort is called a **population explosion**.

Many factors can contribute to rapid population growth. An abundance of food, absence of predators, and lack of competition in an area can help bring about a dramatic rise in a population's size. Perhaps the area is free of parasites or pollution. Maybe unoccupied space is abundant, some necessary resource is plentiful, or natural disasters seldom occur.

One interesting example of a population explosion involved Kaibab deer living on the north side of the Grand Canyon. At the beginning of the twentieth century, the deer population was fairly stable. The number of deer was in balance from year to year. Then the government began a program to limit the number of cougars, wolves, and coyotes in the area, because they were killing livestock. Without their natural predators, the deer began to reproduce rapidly. The deer population increased from 4000 in 1907 to 100 000 in 1924.

Section 23.1 Characteristics of Population Growth

Factors that Limit Population Growth

When a population explosion occurs in an ecosystem, the natural balance is upset. The expanding population changes the ecosystem, affecting populations of other organisms. Living and nonliving factors in the changed ecosystem then operate in a way that stops the population explosion. The same factors that stimulate population growth when they are favorable also serve to retard or limit it when they are unfavorable.

Shortage of Food or Another Essential Resource Consider the population of mice on the island once again. Suppose the size of the population has grown to a large number. The mice are eating so much food that suddenly there is not enough for all. Some mice starve to death. The population explosion is limited by the food shortage, and the number of mice decreases.

In general, an exploding population uses up food or another essential resource faster than it can be replaced. The resulting shortage tends to slow population growth. In a severe food shortage, many organisms starve to death. Recall the example of the Kaibab deer. During the two winters following 1924, 60 percent of the herd died of starvation, largely because the deer had destroyed the vegetation by overgrazing.

Shortages of essential resources also affect the growth of green plants. If a plant population lacks a necessary mineral, the population size drops. For example, few nettle plants grow in soil that has a low supply of phosphorus. When phosphorus is added to the soil, more nettle plants grow.

The amount of available oxygen can be a limiting factor in aquatic communities.

23–4 In the absence of predators, the deer population on Angel Island in the San Francisco Bay grew at an alarming rate. The result was over-browsed vegetation (see left) and undernourished, disease-prone deer.

In 1982 the decision was made to sterilize the Angel Island deer in order to control their population size. The only other alternatives were slaughter or removal of the deer.

Chapter 23 Populations in Ecosystems

23–5 In 1944, when this picture was taken, the European rabbits introduced into Australia for sport hunting had prospered to the point of severe overcrowding due to the absence of their natural predators.

Crowding As the number of mice on the imaginery island grew, they filled all the places to live. Eventually they ran out of space and crowding occurred. Studies confirm that crowding causes changes in the bodies and behavior of field mice, leading them to stop reproducing.

Scientists have found that crowding limits the size of other populations as well. When crowding occurs among flour beetles, the females deposit fewer eggs. In some cases the beetles eat each other. When cockroaches are crowded, they eat each other's wings. In many cases of crowding, some organisms move out of the ecosystem, which helps to restore the original balance.

Waste Products Waste products often accumulate in crowded conditions. This happens when a growing population produces waste products faster than decomposers break them down. The buildup of waste products in an ecosystem can cause illness and death. In this way waste products limit population size.

Climate In many places, the weather changes significantly during the year, and the size of certain populations changes with the seasons. You may know when to expect a large number of mosquitoes, gypsy moths, or other insects. You may look forward to the increase in numbers of certain fish, birds, or plants. These seasonal effects on population size are due to changes in air and water temperatures and changes in the amount of light and rain. Populations also increase in size during the seasons when fruit or flowers are produced, providing food for many species.

23–6 Each fall snow geese abandon their summer grounds in Canada to find better food sources and nesting sites for the winter.

Section 23.1 Characteristics of Population Growth

During the Middle Ages the human population of Western Europe decreased sharply in size several times as a result of diseases such as the bubonic plague. The population grew between each epidemic, resulting in a fluctuating population growth pattern.

23–7 The gray ash mudflow that accompanied the Mount St. Helens eruption partially submerged trees and covered their trunks with ash. Millions of hectares of vegetation were totally destroyed.

Natural Disasters Storms, floods, earthquakes, volcanic eruptions, and fires destroy organisms and their homes. Although natural disasters are not predictable regulating factors, they do reduce population size in areas where they occur. The eruption of Mount St. Helens destroyed a lake and many square kilometres of forest land. Less dramatically, the erosion and reforestation of this land will continue to make changes in the ecosystem for many years.

Interactions Among Organisms Interactions among organisms also affect population size. Predators help regulate the population size of their prey. As the prey population grows, the amount of food available to the predator population increases. As a result, the predator population grows, decreasing the size of the prey population. The decrease in the size of the prey population decreases the amount of food available to the predator population. In turn, the predator population also decreases in size.

Parasites help limit growth in their host populations by causing disease. Disease can drastically decrease the population size in an area. The chances of disease epidemics are increased in areas of high population, because individuals are in closer contact with each other.

23–8 In the laboratory, the population growth patterns of bean weevils and parasitic wasps have regular cycles. The wasps lay eggs on weevil larvae. After hatching, the larval wasps eat the weevils.

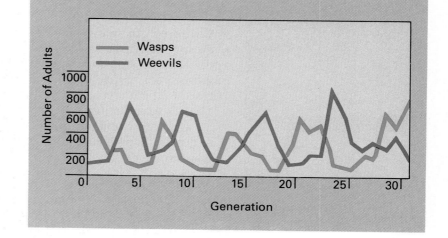

Chapter 23 Populations in Ecosystems

Patterns of Population Growth

S-shaped Growth Pattern Scientists often study the growth of populations under controlled laboratory conditions. One organism they study is yeast, which—like bacteria—reproduces by doubling. You can easily grow yeast in a test tube with food and water. In one experiment, a researcher placed a yeast cell in a tube with a known amount of food and water. He took samples from the tube every hour and determined the total number of yeast cells in the tube. The information he obtained during 20 hours of observation is shown in the graph in Figure 23–9.

Under laboratory conditions a population is considered closed; that is, individuals are not free to enter or leave. Most natural populations are open.

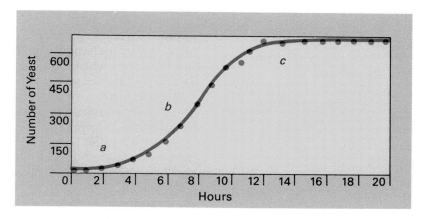

23–9 S-shaped growth curve of yeast. The population grows slowly at first, then very rapidly. Eventually, the population size levels off at the maximum number that can live in the test tube.

Scientists refer to the kind of line shown in Figure 23–9 as an **S-shaped curve**. The S-shaped curve for the yeast colony shows one type of pattern that a growing population may follow. Three stages can be seen in the S-shaped curve. During stage *a* the population is becoming established. Because the number of starting organisms is small, the results of population growth are also small. During stage *b*, as the large population doubles, growth is enormous. In part *c* population growth stops and the size of the population levels off at the maximum number of yeast that can live in the tube.

The S-shaped curve is also known as a sigmoid curve.

J-shaped Growth Pattern In another type of population growth, a population initially grows rapidly in size. Then, instead of leveling off, the population size drops rapidly, resulting in a **population crash**. A graph of rapid growth followed by a

Section 23.1 Characteristics of Population Growth

23–10 J-shaped growth curve of thrips (a winged insect) on roses. As the number of thrips peaked, the limit of the roses as a food source was reached, and a die-off resulted.

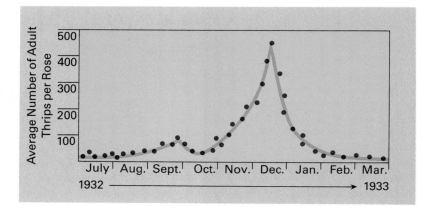

population crash appears as a **J-shaped curve** (see Fig. 23–10). In all cases of J-shaped growth, rapid growth occurs until the limits of the environment are reached. Then, some factor, such as the depletion of food, causes the population to quickly die off.

The growth of the mice population on the island is an example of J-shaped growth. At first, conditions were favorable for the growth of the mice. The population took hold and then grew in an uncontrolled way. Eventually, some factor or set of factors became unfavorable, population growth stopped, and the size of the population declined rapidly.

Building a Science Vocabulary

population   S-shaped curve   J-shaped curve
population explosion   population crash

Questions for Review

1. What factors play a role in controlling population size?
2. What is the difference between the type of population growth shown by an S-shaped curve and that shown by a J-shaped curve?
3. A small number of male and female rabbits is introduced to an island where there is plenty of food and few predators. What do you predict will happen to the population of rabbits at first? after a few years?

23.2 POPULATION SIZE IN A BALANCED ECOSYSTEM

Chemical-elements cycle in a balanced ecosystem, and energy is supplied continuously. In addition, producers make food; food passes through chains of consumers; decomposers remove and recycle waste products; and population size is regulated. The term to describe the tendency of a biological system to remain in balance is **dynamic equilibrium**. The word *dynamic* means that changes are taking place. The word *equilibrium* implies balance. As a whole, an ecosystem in dynamic equilibrium is in balance. When the size of a population changes, factors in the ecosystem work in ways to restore balance.

Ecologists study the populations of ecosystems over long periods of time. When they graph the growth of balanced populations, they find that the curves are often somewhat irregular. The reason for this lies in how populations are regulated.

Regulation of Natural Populations

To understand how population size is regulated in a balanced ecosystem, think of a thermostat that controls the temperature in a house. Imagine that you set the temperature at 20°C. When the heat goes on, the house temperature rises. When the temperature in the house goes above 20°C, the thermostat turns off the heat. When the heat is off, the temperature drops. But, if it drops below 20°C, the thermostat turns the heat back on. When the house temperature changes, regulation always occurs in the direction of the desired level of 20°C. A graph of temperature in the house would look like the one in Figure 23-12. This curve is called a **saw-toothed curve**.

23-11 A tightrope walker must make adjustments constantly to maintain balance.

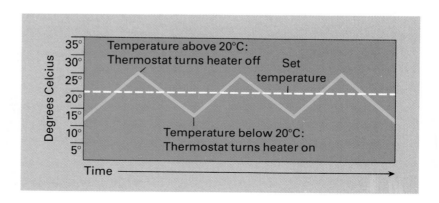

23-12 Saw-toothed curve of temperature regulation by a thermostat set at 20°C. Regulation of population size appears to work on the same principle as a thermostat.

The regulation of population size in a balanced ecosystem seems to operate on the same principle as the thermostat. Any given ecosystem can support a population of a certain size, called its **carrying capacity**. When the population size rises above the carrying capacity, factors in the ecosystem cause it to drop. When the population size drops below the carrying capacity, factors in the ecosystem cause it to rise. These changes are shown in saw-toothed curves and repeating patterns.

Different Ecosystems Have Different Carrying Capacities

In some ecosystems, conditions are favorable to living things; in others, they are unfavorable. For example, a desert has little water, its temperature is severe, and it can support only small populations. We can say that it has a low carrying capacity. A tropical forest, on the other hand, has a good water supply and favorable weather. It can support large populations and therefore has a high carrying capacity. When a population grows beyond the carrying capacity, the living and nonliving factors that you read about in the first section of this chapter help restore the balance. These factors help keep the population size within the limits of the ecosystem's carrying capacity.

S-shaped Growth and Carrying Capacity

An S-shaped growth pattern is common among organisms that are new to an area that has the proper kinds of food and other resources for their growth. The population grows slowly at first, then rapidly. Eventually, the population reaches the carrying capacity, and population growth stops. At this point, the population size is in balance with the environment.

23–13 A desert (left) and an African savannah (right) have entirely different types and amounts of vegetation to support consumer populations. As a result their carrying capacities vary markedly.

23–14 Population density.
a. A long stretch of remote coastline, occupied by few people, has a low human population density.
b. By comparison, a limited coastal beach serving a large metropolitan area has a high human population density.

Biologists are often more interested in measuring population density than total population size.

Population Density

When you think about population size, it is also important to think about the size of the area in which the population lives. Are the individuals in the population crowded? Are they far apart? **Population density** compares the number of organisms in a population with the size of the area in which they live. The word *density* describes the number of objects in a given area. High density implies more crowded conditions.

Density can be a critical factor for a population. For example, if too few organisms of a population live in a place, they may not find each other to mate, or they may be helpless if attacked. If the population density is too low, the population may die out. If, on the other hand, the population density is too high, the ecosystem cannot provide for its members, and individuals will die from starvation or other causes.

In a balanced ecosystem, population density is related to carrying capacity. Natural factors restore the balance between population density and carrying capacity when the population density is either too low or too high for a given area.

Rates that Influence Population Density

Population density is affected by the number of births and deaths in a population, as well as the migration of individuals into or out of the area. Biologists express these population changes as rates. The term *rate* compares the amount of something to the amount of something else. For example, you can say that an automobile's rate of speed is 88 km/h. In this case, you are expressing distance (88 km) in relation to time (per hour). A taxi rate of 15¢ for 0.1 km expresses cost in relation to distance. Biologists use four rates to express population size in relation to time: birth rate, death rate, emigration rate, and immigration rate.

Birth Rate The **birth rate** is the number of births in a population in a given time. If the conditions in an ecosystem are favorable and the area can support more members of a population, the birth rate often goes up. Consider this hypothetical case: A city sets aside an area for a park. Eventually, trees in the area grow tall, and squirrels in the park find food and room to nest in the trees. The abundance of nesting areas and food causes the squirrel population's birth rate to rise. As the birth rate rises, the population density rises toward the carrying capacity. If conditions become unfavorable, the birth rate may decrease. Imagine that one year the people in the community decide to chop down some of the trees to build a tennis court. The squirrels lose places to nest and their birth rate drops.

If population density is above the carrying capacity, or if conditions in the ecosystem become unfavorable, the birth rate of certain populations drops. As a result, the populations get smaller, and their population density decreases.

23–15 Changes in the birth rate and death rate of a population regulate the fluctuation of population size around the carrying capacity of an ecosystem.

Death Rate The **death rate** is the number of deaths in a population in a given time. When environmental conditions are harsh and the population density is high, the death rate goes up. If conditions improve and the population density is within the limits of the ecosystem, the death rate generally drops.

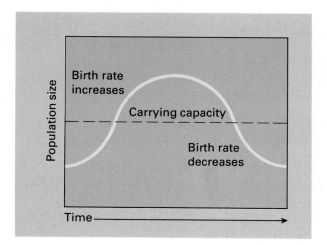
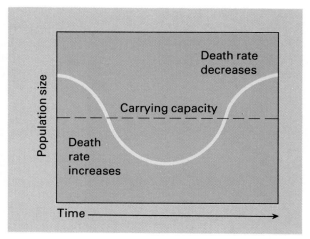

23–16 Some brown lemmings complete their life cycle without experiencing the mass emigration that is a population response to overcrowding.

e- = out
migratio = change of location

Emigration Rate The **emigration rate** refers to the number of individuals that move out of the ecosystem during a given time period. If the ecosystem has enough food and other resources and is not crowded, organisms will usually not move away. But if conditions deteriorate, the emigration rate increases. A famous example of emigration involves lemmings, which are small mouselike animals that live in cold regions. When their living space becomes too crowded, many lemmings will migrate toward the sea. In Norway, lemmings can be seen traveling in large numbers through countrysides and villages. When they get to the sea, many of them drown, drastically reducing the lemming population density.

in- = into
migratio = change of location

Immigration Rate The **immigration rate** is the number of individuals that move into an ecosystem in a given time. Imagine that conditions are favorable and the population density is low. In such a case, the population can tolerate an increase in size. More organisms move into the ecosystem, and the population density rises. If conditions are unfavorable and the population density is high, organisms will not be able to move into the ecosystem, and the immigration rate will be low.

The Combined Effect

The changes in population density in an ecosystem are the result of the combined effects of birth and death rates and immigration and emigration rates. All of these rates are affected by the living and nonliving factors in the environment. The ecosystem's supply of food and other resources, amount of space available, presence or absence of pollution, climate, occurrence of natural disasters, and interactions among organisms all operate to change the four rates. Consider, for example, an ecosystem

23–17 A change in an environmental factor, such as food supply, affects each of the rates that influence population density.

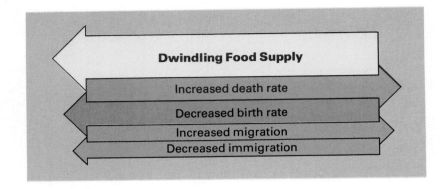

where the food supply is dwindling. What might happen to each of the four rates?

In a balanced ecosystem the living and nonliving factors in the environment operate to balance the birth and death rates and the emigration and immigration rates. Population density is thereby regulated to stay within the carrying capacity.

Building a Science Vocabulary

dynamic equilibrium   population density   emigration rate
saw-toothed curve   birth rate   immigration rate
carrying capacity   death rate

Questions for Review

1. For each of the following cases, state whether the population size will increase, decrease, or remain the same.
   a. yeast population when the food is used up in a test tube
   b. lemming population after migration has ended
   c. corn borer population if the corn crop flourishes

2. How can the amount of water in an ecosystem affect population size?

3. Explain the difference between population size and population density.

4. List four rate changes in a population that would help restore equilibrium after a population explosion.

Chapter 23 Populations in Ecosystems

23.3 HUMAN POPULATION GROWTH

Present Growth Trends

The human population has been on a sharp upward growth trend for the past thousand years. In the first century the human population size was somewhere between 200 million and 300 million. By the turn of the twentieth century the human population had reached 1.5 billion. In 1980 it reached 4.5 billion. According to one United Nations projection, by the year 2000 there will be more than 6 billion people on earth.

The human population is growing in size because the birth rate exceeds the death rate. The recent growth of the human population is a result of a decrease in the death rate, rather than an increase in the birth rate. Agricultural and medical advances are largely responsible for lowered death rates. During the twentieth century, infectious diseases such as influenza, pneumonia, tuberculosis, and diarrhea have been brought under control in many parts of the world. Improved health and better nutrition have lengthened the human life span, contributing to a lower death rate.

The world population as a whole is increasing at a rate of less than 2 percent a year. Rates of increase vary for different countries. In general, the wealthier, more highly industrialized countries have lower growth rates (1 percent or less), while the less wealthy, less industrialized countries have higher rates (often 2 percent or more). Few nations have stable populations, which remain approximately the same size from year to year.

The death rate in the population of the United States has declined steadily in the twentieth century. In 1900 the death rate per 1000 was 17.2. In 1925 the death rate was 11.7. Since 1948 it has varied between 9.0 and 9.9.

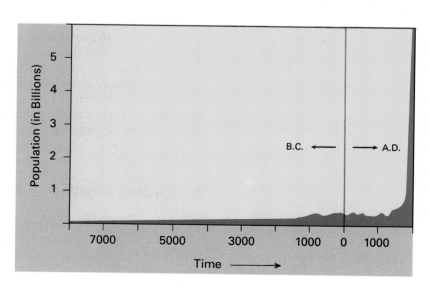

23–18 Growth of the world human population for the past ten thousand years. The last century has seen a very rapid growth rate.

Table 23-2 Population Growth by World Regions, 1750-2000 (medium estimates in millions)*

	1750	1800	1850	1900	1950	1970	1980	1990	2000
North America	2	7	26	82	166	226	240	270	290
Latin America	16	24	38	74	164	283	372	478	608
Europe (includes USSR)	167	208	284	430	572	703	749	793	832
Asia	498	630	801	925	1379	2091	2579	3078	3612
Africa	106	107	111	133	219	354	470	630	828
Oceania	2	2	2	6	13	19	23	26	30
World	791	978	1262	1650	2513	3677	4432	5276	6199

*From *Selected World Demographic Indicators by Countries, 1950-2000: Demographic Estimates and Projections as Assessed in 1978*, United Nations, New York, 1980, p. 1, Table A-1. 1980 census figures from United Nations Statistical Papers, *Population and Vital Statistics Report, Data available as of 1 April 1982*, United Nations, New York, 1982, p. 1, Table 1.

Factors that Affect Human Population Growth

The human population is subject to many of the same factors that affect the population sizes of all other organisms. Most importantly, the human population cannot exceed the carrying capacity of the environment for long before natural forces begin to restore the balance. In areas where the size of the human population has already exceeded the carrying capacity, an increase in starvation, malnutrition, and disease has led to a higher death rate.

It is difficult to estimate what the carrying capacity of the entire earth is for humans. Some experts believe it is as low as 4 billion; others think it could be as high as 30 billion. The higher limit assumes that the availability of food is the limiting factor on population growth. Moderate estimates place the carrying capacity of the environment at between 7 billion and 10 billion.

Whatever the achievable carrying capacity may be, it is necessary to bring population size into balance with the carrying capacity of the environment. Balance can be achieved by raising the carrying capacity or decreasing population growth.

Biologists are uncertain about which factor—food, energy, or space—will ultimately limit the number of people on earth.

Developing Population Balance

Raising the carrying capacity of the environment involves providing sufficient food, water, shelter, and other life necessities to an expanding population. Through **technology**, (tek-NAHL-uh-jee), which is the application of scientific knowledge and

23–19 The amaranth plant produces highly nutritious seeds and leaves. Amaranth was once a major staple in third-world countries. Now, it is being rediscovered and studied because it is a protein-rich, drought-resistant crop with a high degree of genetic variability.

methods to solving practical problems, great achievements have been made in agriculture, medicine, and other areas. Indeed, the application of technology during the nineteenth and twentieth centuries has enabled the human population to grow as it has.

Through modern agriculture, more and better food is available now than ever before. Increased and constant food supplies help support large populations. Transportation makes it possible to deliver food to areas where food supplies are low.

Many people live in areas of high population density. For example, 20 million people live in the Tokyo-Yokohama metropolitan area. In spite of the extremely high population density, people are healthy, and the life expectancy at birth is 75 years. High population density does not necessarily upset the natural balance.

People can control the indoor climate with heating, air conditioning, and artificial light, thus avoiding many of the negative effects of climate, season, and time of day. As a result, humans live and work in a wide variety of places, and work can be conducted during any time of the day or night.

We have learned how to compete fairly successfully with many of the organisms that threaten our food supplies. Insects remain our chief competitors for food, costing us billions of dollars in crop damage each year. Disease-causing parasites pose a great threat to human health; however, many diseases

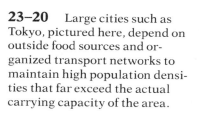

23–20 Large cities such as Tokyo, pictured here, depend on outside food sources and organized transport networks to maintain high population densities that far exceed the actual carrying capacity of the area.

have been controlled by discoveries in medicine and sanitation. Improved medical services are becoming available to more and more of the world's population.

The carrying capacity can probably be raised over the next 20 years with continued improvements in agricultural methods and other areas. However, it is unlikely that the carrying capacity can be increased indefinitely. Resources vital to modern agriculture, including oil, water, and land itself, are already becoming depleted. Greater food shortages and increased overcrowding are likely to occur.

Balance cannot be achieved if present growth rates persist. At some point, the birth rate must become equal to the death rate to achieve stability. If the birth rate and death rate soon became the same, overall population growth would still continue for some time. For example, in North America, even if each family had only two children, the population would continue to grow for another 70 years. Continued growth would occur because most of the population is young and life expectancy is long.

The human population cannot escape the fundamental rule of living populations: population size cannot indefinitely exceed carrying capacity. In order to avoid higher levels of starvation, malnutrition, and disease than exist today, it is necessary to balance human population size with the carrying capacity of the environment.

Building a Science Vocabulary

technology

Questions for Review

1. Describe human population growth during the twentieth century.

2. What is responsible for the recent growth of the human population?

3. Why is it desirable to balance human population size with the carrying capacity of the earth?

4. How can human population size and carrying capacity be brought into balance?

Perspective: Environment
Population Is Everybody's Problem

Human beings are part of the world's ecosystem. We inhabit almost every portion of the globe. We are part of complex food webs in which we compete, as well as cooperate, with other organisms. We influence mineral cycles and affect plant production. We affect the atmosphere and in turn are affected by it. We humans have a lot of control over our environment, and the environment has a lot of control over us.

Because of all these interrelationships and controls, human populations cannot grow without limit. In this chapter you learned that populations tend to level off at some kind of equilibrium level. The equilibrium level is determined by factors such as the availability of food and living space and the presence of friends or enemies. These factors—and others—affect human populations, too. There is a limit to the earth's space and ability to produce food. These limits influence the size of the human population.

Consider our food production needs. At present, each person on the earth is supported by about one hectare (two acres) of land. Half of that land is used to grow crops such as wheat and rice; the other half is used to feed livestock such as cattle and sheep. Even so, thousands of humans are starving and hundreds of thousands are without balanced diets and suffer from malnutrition. Scientists believe we cannot increase by very much the amount of land in food production except at a greatly increased expense for irrigation and fertilizers. Even these measures have environmental drawbacks.

Thus, what the maximum size of the human population can or should be depends on how much space, food, and worldly goods each of us needs, as well as on what each of us demands. If each of us wants a lot, there is room for fewer people. If each of us wants or can get by with less, there could be room for more people.

People do have a choice. Humans respond to emotional and psychological factors in making decisions. The population problem is our problem—the solution depends largely on the quality of life we want.

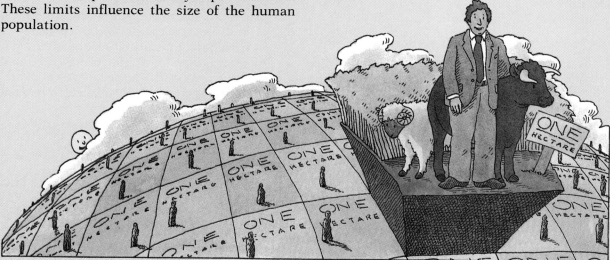

Chapter 23
Summary and Review

Summary

1. Some populations grow by doubling. Rapid and uncontrolled growth of a population results if doubling is not halted.

2. The factors within an ecosystem that affect population growth include food, space, waste products, climate, natural disasters, and interactions with other organisms.

3. Two patterns of population growth found among natural populations and in laboratory experiments are S-shaped growth and J-shaped growth. In S-shaped growth a population becomes established slowly, then grows rapidly, and finally levels off in size. In J-shaped growth a population initially grows rapidly and then crashes.

4. Within an ecosystem, population size tends to remain in balance with available resources. The property of a biological system to remain in balance is called dynamic equilibrium.

5. The regulation of population size in a balanced ecosystem can be compared to temperature control by a thermostat. Populations are regulated in the direction of optimum density, which is set by the carrying capacity of an ecosystem.

6. The living and nonliving factors that operate on population size in an ecosystem affect the birth rate, death rate, immigration rate, and emigration rate. These rates determine population density.

7. The human population is growing rapidly in size. Agricultural, medical, and other technological advances have lengthened life span and reduced the death rate so that the birth rate exceeds the death rate.

8. Like other populations, the human population cannot exceed the carrying capacity of the environment for long. Human population size must be balanced with the earth's carrying capacity.

9. Human population size and carrying capacity can be balanced by raising the carrying capacity, which is limited by the depletion of natural resources, or by limiting population growth.

Review Questions

Application

1. Each of the following descriptions of populations can be represented by a graph. From the graphs shown below, select the best graph for each description.
   a. Researchers placed five male flies and five female flies in a covered bottle with food and water. They observed the bottle for several months. The researchers counted the flies and their offspring from the day they were placed in the bottle until there were no live flies left.
   b. Ants from South America came to Alabama. Now they occupy 60 million hectares in nine of the southern states, and they are continuing to spread.
   c. A scientist introduced several males and several females of a species of salamander into a pond. The species established itself and maintained a balanced population size.

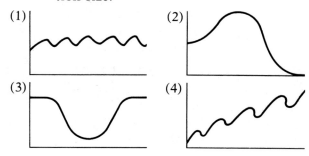

2. State whether each of the following would increase or decrease the carrying capacity of a farm ecosystem.
   a. the use of artificial fertilizers
   b. soil erosion
   c. irrigation
   d. electric power

3. How would the following changes affect the population density of an ecosystem?
   a. food supply increases
   b. number of competitors increases
   c. emigration rate rises
   d. the size of the territory decreases

Interpretation

1. Could the upward growth trend of the human population since 1000 A.D. continue indefinitely? Explain.

2. Is the increase in the human population size in the twentieth century due mainly to a high birth rate or a low death rate? Explain.

3. A farmer noticed that algae were growing on the surface of his pond. Since he wanted the pond to be clear, he told his son to remove the plants. The farmer told his son that the plants had to be removed *before* the whole surface of the pond was covered. The algae covered one square metre on the first day, two square metres on the second day, four square metres on the third day, etc. When the pond was half covered, the son announced that he would start to clean off the pond the next day. He said that he had plenty of time left. Was he right? How much time did he have?

4. Uncontrolled immigration, pollution, and a decline in business can all harm a city and its population. What is the effect of each factor on the city's carrying capacity and its population size?

Extra Research

1. Obtain an ear of corn. Count the kernels (seeds) on the ear. Assume that each kernel is capable of growing into a new corn plant and that each corn plant bears two ears with the same number of kernels that your ear of corn has. Calculate the number of corn plants that would exist in five generations if growth is not controlled.

2. With your teacher's help, buy a population of *Daphnia* (water fleas) from a biological supply house. Divide the population equally into three containers, recording the number of organisms in each. Place one container in constant light, one in constant darkness, and keep one at normal day and night conditions. Keep the temperature the same for all. After five days, count the number of *Daphnia* in each container. Determine which environment is most favorable for the growth of *Daphnia*.

3. Write a paragraph describing an imaginary city where the people are crowded together but nevertheless enjoy healthy, quality lives. What conditions would make such a city possible?

Career Opportunities

Demographers study the size, composition, and distribution of human populations using birth rates, death rates, and other information. Most demographers work for governments or research organizations. Specialized training in demography and sociology is required. Write Statistics Canada, Holland Ave. and Scott St., Tunney's Pasture, Ottawa, Ontario K1A 0T6.

Statisticians use mathematics to help describe and analyze research data. They work in many fields, including biological and psychological research. Four or more years of university are recommended. For information, write Statistics Canada, Holland Ave. and Scott St., Tunney's Pasture, Ottawa, Ontario K1A 0T6.

Chapter 24

The Geography of Ecosystems

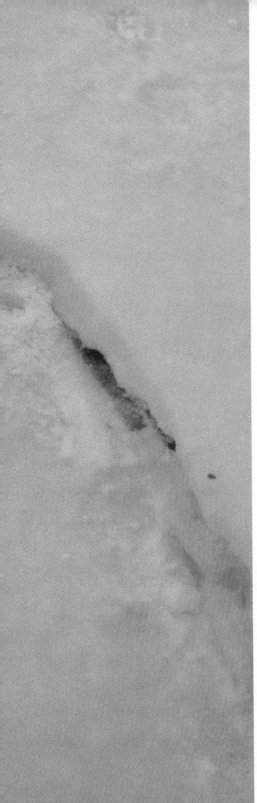

Focus *Living things spread out and inhabit almost all parts of the earth. In some areas, a few hardy pioneers survive in a harsh environment. In others, large populations interact in a wealth of resources. Wherever they live, organisms are adapted to their environment, and in large and small ways, they make changes in their world.*

24.1 THE DISPERSAL OF ORGANISMS

Plants and animals move away from their original homes in a process called **dispersal**. The reasons for dispersal vary. Food supplies may dwindle, homes may be destroyed, or climate may change. The number of organisms may become too large for the available food, or competition may drive organisms away. When such changes occur, organisms relocate to new places that may offer better chances for survival.

Adaptations for Dispersal

Most animals and plants are well adapted for dispersal. The ability to move helps many animals disperse to new locations; others are carried along by water or wind. Beavers sometimes float downstream on logs, and polar bears drift long distances

24–1 Active all winter, a polar bear has a thick, white coat, which keeps it warm and camouflages it while hunting for fish, seals, and other arctic game.

24–2 Dispersal. **a.** Wild horses explore habitats of the American Plains. The milkweed's silky tufted seeds (**b**) and the maple's winged seeds (**c**) are adapted to catch the wind.

As dispersal agents, birds may also carry seeds and fruits on the outside of their bodies, embedded in their feathers.

on chunks of ice. Young spiders spin tiny silk threads that are caught by the wind, carrying the thread and spider long distances.

Plant seeds also have adaptations that aid dispersal. Seeds may be scattered by wind or water. Wings on maple seeds and feathery projections on dandelion seeds enable them to be carried by wind. The seed of the coconut palm tree is adapted for dispersal by water. A coconut drifts in the ocean, protected by its waterproof skin from salt and water. Its thick fibrous husk keeps it afloat and protects it from the pounding of waves. When the coconut is finally washed ashore, food and liquid stored inside the seed help the young plant grow in its new location.

Some seeds are adapted for dispersal by animals. Birds eat the fruit and seeds of raspberries and cherries. The birds later drop the seeds some distance from where the fruit was eaten. Some seeds have hooks that stick to clothing or the fur of animals. Perhaps at some time you have walked through a field and had cocklebur seeds stick to your clothing. Later, you probably picked off the seeds and threw them away. By doing this you dispersed the seeds to a new location where they might take hold and grow.

Distance of Dispersal

When animals or plants disperse, each generation usually moves only a short distance. But after many years, a species can move a great distance from its place of origin. For example, look at the distribution of toads shown in Figure 24–3. Originally, toads were found only in the tropical regions of what is now Asia. Little by little they spread to Africa and the Americas. Today, toads are found in almost all parts of the world.

Table 24–1 Seed Dispersal Adaptations

Method of Dispersal	Adaptation	Examples
Wind	Wings on seeds	Maple, pine, catalpa
	Feathery projections	Dandelion, milkweed, thistle, cattail
	Tiny, light seeds (the size of dust particles)	Orchid
Water	Waterproof outer covering around seed	Coconut palm
Animals	Edible fruit surrounds undigestible seeds (seeds pass out of animal's digestive tract)	Raspberries, cherries, currants
	Seeds contain food which animals bury and store	Seeds and acorns buried by jays; hazelnuts stored by thick-billed nutcrackers; small seeds stored by ants and other insects
	Seeds fall into mud that sticks to feet of wading birds	Seeds of marsh plants carried on the muddy feet of herons and gulls
	Hooks or burrs on seeds stick to fur, feathers, and clothing	Cocklebur, burdock, beggar's tick, buffalo burr
Exploding pods	When the pods open, seeds explode into the air	Touch-me-not, witch hazel, wood sorrel, squirting cucumber

24–3 The toad, now found throughout the world, began dispersing long ago from central Asia.

Section 24.1 Dispersal of Organisms

The young and immature members of a species are usually the ones that disperse. In an experiment, adult and young house wrens were marked so that they could be identified later. After one year, only one percent of the adult birds were found outside the original nesting site. On the other hand, 15 percent of the baby birds were found far away from their original home. Some of the young birds were located as far as 1100 km away.

Barriers to Dispersal

Dispersal continues until an obstacle, or **barrier**, stops it. Sometimes the geography of an area is a barrier to dispersal. The 800 km stretch of ocean between Africa and Madagascar prevents large animals such as antelopes, zebras, or elephants from crossing. Birds, flying insects, and small organisms that cling to floating objects, on the other hand, disperse across the water. For them the ocean is not a barrier to dispersal. Table 24–2 lists some common geographic barriers.

Climate may also be a barrier to dispersal. Maple trees, polar bears, and most other organisms cannot survive the hot, dry days or cold nights of the desert. Cactus plants and tropical fish cannot tolerate frost. Green plants require a supply of sunlight. Gerbils require little water, whereas oak trees require much water. For many forms of life, climate can be a barrier to dispersal.

Competition with humans for land and resources is a major barrier to dispersal for many species of plants and animals. Organisms that live in bogs and marshes are forced to disperse when people drain such areas for building. As more wetlands are taken over by people, it becomes increasingly difficult for plants and animals there to disperse successfully. As a result, certain

24–4 The steepness of the Grand Canyon provides a natural barrier to dispersal for most organisms.

Table 24–2 Geographic Barriers to Dispersal

Type of Organism	*Barrier*
Freshwater organism	Salt water, land
Fish swimming up stream	Waterfall, land
Water dweller	Land mass
Land dweller	Ocean, canyon, desert
Lowland organism	Mountain

species of birds, ferns, butterflies, and other organisms are becoming rare.

When an organism reaches a new location, an ecological factor, such as the availability of food in the ecosystem, may present a barrier. Other ecological barriers could be the type of soil present, the amount of competition with humans and other organisms, or the types of predators or parasites in the new location. The pressure of dealing with living conditions can be as effective a barrier to dispersal as a mountain or desert.

Habitats and Niches: Addresses and Occupations

Elephants in Kenya depend on river areas for food and water. People are turning these areas into farms, ranches, and tourist sites, using water and destroying vegetation. The elephants trespass on the new farms and ranches in search of food and water and are illegally hunted. As humans take over more land, dispersal by elephants becomes more difficult.

A **habitat** is the place where an organism lives. You can think of a habitat as a biological address. A frog's habitat is a pond; a buttercup's habitat is a field or meadow. In its habitat, each organism has a role, or **niche** (NITCH). An organism's niche refers to what the organism does in the ecosystem and involves interacting with the living and nonliving factors of the environment. You can think of a niche as a kind of biological occupation or profession. A tapeworm, for example, fills a parasite's niche; a geranium fills a producer's niche.

Organisms that disperse successfully fill a niche in their new habitat. If a habitat does not change, a species can continue to fill its niche for generations. A change in the environment, however, can have an effect on a species' habitat and niche. A severe change in address or occupation can bring about the pressure to disperse all over again.

24–5 A Douglas fir provides a habitat for various warblers, each with its own niche based on nesting and food requirements.

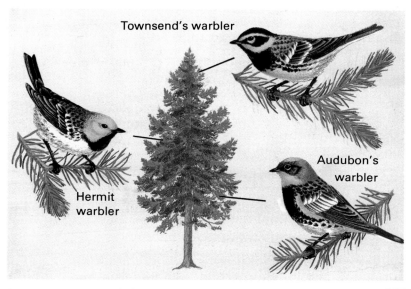

Section 24.1 Dispersal of Organisms

Changes in Ecosystems Over Time

It is impossible to predict exactly how an area will change. You cannot predict a natural disaster or a fire, but ecologists have a general idea of how an area changes if nothing unforeseen occurs. Ecosystems pass through a series of stages over time.

It is easier to understand the life stages of an ecosystem by starting with an example. Imagine a place where there are few or no living things, such as an empty lot. Seeds from plants such as ragweed, crab grass, and horseweed land on the vacant lot, take hold, and grow. As they grow, they modify the soil by loosening it. They also enrich the ground when they die and decompose at the end of the growing season. In time, seeds from other plants that are dispersed to the area take hold in the newly enriched soil. Goldenrods, asters, and tall grasses grow in the once-vacant lot. As more time passes and soil conditions improve, the lot develops into a grassy field. Eventually, tree seedlings and shrubs appear, and the area becomes a meadow.

The area in the above example underwent a series of orderly changes: vacant lot to weedy area to grassy field to meadow. During these changes the plant life changed. The animal life changed as well. Insects and mice were probably early inhabitants; birds and squirrels came later. The animal inhabitants of the area also changed the physical conditions. Their droppings and the chemicals released when they died and decomposed further enriched the soil. As conditions changed, different species entered and survived, and they in turn made changes. Communities that formed were, in time, replaced by other communities. Such a steplike series of changes continues until a balance occurs. A balanced ecosystem supports a certain number and variety of living things. As long as the conditions stay the same, the number and variety of organisms remains about the same.

The earliest plants that settle in an area are called **pioneers**. In the example, ragweed, crab grass, and horseweed were pioneers. These plants, along with insects and other animals, formed a **pioneer community**. Some organisms, such as lichens, are particularly adapted for pioneer life. Lichens survive on bare rock, producing acid that breaks down bits of the rock. Dust and dead lichens collect in small pockets, forming soil. Soil-building by lichens makes it possible for plants such as mosses to grow in the area.

In the aftermath of the Mount St. Helens eruption, pocket gophers have become major agents in incorporating the ash residue into lower soil layers, contributing to soil fertility and later growth of vegetation.

24–6 Lichens of various types cover a rock surface. They begin soil formation by breaking down bits of rock.

24–7 Ecological succession from grassland (left) to savannah (middle) to woodland (right).

succedere = to follow after

The last residents to live in an area are members of a **climax community**. A forest is an example of a climax community. A climax community is balanced. Interactions between living things and the nonliving environment occur, but major changes in populations or resources do not take place. If the environment does not change, a climax community can exist for centuries.

The process of change from a pioneer community to a climax community is called **ecological succession**. In an area undergoing ecological succession, community follows community until a climax community is reached. There are many examples of ecological succession. The development from bare land to meadow is one. The development from burned-out field to grassland is another. We will take an in-depth look at still a third example—the succession from lake to forest.

Lake to Forest Succession

The actions of humans often accelerate the aging process of lakes. An example is the case of Lake Erie, which has filled more rapidly than it would have naturally.

A lake may look as if it was always there and always will be there, but the truth is that most lakes are temporary. They have a beginning and an end. Some lakes begin when a depression or basin in the earth fills with water; others form when part of a stream gets blocked off. Some lakes last a few years; others endure for millions of years. While lakes are in existence, natural forces gradually turn them into land. The water that flows into a lake carries sediment, which consists of sand, pebbles, soil, and rocks. Sediment settles in a lake, gradually filling it in. Living organisms also play a major role in filling in the lake and changing it into a forest.

Section 24.1 Dispersal of Organisms

Pioneer Lake Community A newly formed lake usually starts out with very few inhabitants, but dispersal soon brings more. Small organisms, including bacteria, protozoa, algae, and animals, enter from the air, incoming water, and nearby soil. Birds and other animals drop seeds in and near the water. Insects, attracted to the shiny water, stay and make their homes. Frogs hop over from nearby areas, and fish enter in the streams that empty into the lake. In the lake, these organisms make up the pioneer community.

As the pioneer community develops, it changes the lake. Soil builds up on the lake's edges and bottom from dead bodies that decompose and the sediment that continues to enter. Soil-building allows different kinds of plants to grow in and near the lake. The change in the plant population makes it possible for more animals to survive. Ecological succession occurs as soil-building continues and the lake's inhabitants change.

Mature Lake Community If you return to a lake years after its formation, you can see major changes. First, the lake is smaller. Layers of sediment and decayed organisms continue to accumulate on the bottom and edges, making the lake shallower and smaller. The soil that builds up around the edges of the lake often forms a marsh where many kinds of organisms live. Among the marsh dwellers, you will find cattails, plants that resemble tall grasses, insects, snails, small crustaceans, frogs, ducks and other marsh birds, muskrats, and beavers.

Three distinct zones of life exist in the lake. One zone is at the lake's edge. If you stand on the shore, you probably can see waterlilies, pond weeds, and duckweed growing near the water's edge. You may also see fish swimming by. If you take a boat out

24-8 Lake to forest succession: **a.** pioneeer stage, a new lake; **b.** a marsh develops as the bottom and edges of the lake accumulate sediment; **c.** as soil builds up shrubs and trees follow; **d.** the climax community of animals and plants, a mixed evergreen/hardwood forest.

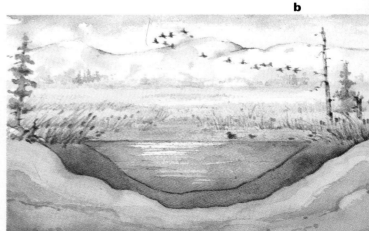

to the middle of the lake, you enter another zone of life: the open-water zone. Here you find small algae and small crustaceans floating close to the lake's surface. Fish, including certain types of bass and trout, swim in the open water of the lake. The third life zone is found on the bottom of the lake. The wavelengths of sunlight needed for photosynthesis do not penetrate to the lower levels of the lake. As a result, plant life is not found on the bottom. Dead organisms and wastes, referred to as **organic debris**, sink to the bottom. Organic debris serves as food for worms and other scavengers living on the bottom. Decomposers, such as bacteria and fungi, also live on the bottom, breaking down the organic debris that remains.

Now, go back to the shore and pass through the marsh. At the marsh's outer edge, you may see small plants beginning to take hold. You might find mosses and small shrubs. Perhaps you could even pick blueberries from the bushes growing there. These plants and the variety of insects, birds, and other animals that inhabit the land bordering the marsh will play a major role in ecological succession.

Climax Community: A Forest Soil-building continues on the bottom and edges of the lake, and the lake gets smaller. The marsh area develops closer and closer to the center of the lake; eventually, the entire lake becomes a marsh. On the outer edges of the marsh, more plants take hold and grow. When these plants die, their decomposed bodies enrich the soil. Gradually, the fertile soil at the edge hardens and is able to support the growth of trees. Willow seeds dispersed to the area may land on the soft soil and take root. The willow trees further enrich the soil; then, different trees enter and take hold. First, softwoods

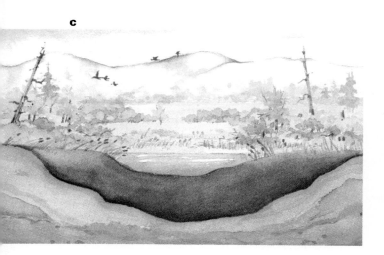

Section 24.1 Dispersal of Organisms

such as pines come in, gradually replacing the willows. Meanwhile, the process of soil-building and hardening continues. Depending on the altitude and climate, hardwood trees such as maples may come in and replace the softwood trees.

In time, the entire lake area undergoes the changes from lake to marsh to forest. Different communities of plants and animals inhabit the area during each stage. Each community changes the ecosystem in ways that prepare the way for the next community. Eventually, a climax community that includes hardwood trees and land animals is established. If conditions in the area do not change drastically, this forest will remain. It may take hundreds of years of ecological succession to establish a climax community. But once established, a climax community is balanced and able to withstand further change.

Building a Science Vocabulary

dispersal	niche	climax community
barrier	pioneer	ecological succession
habitat	pioneer community	organic debris

Questions for Review

1. List three changes in an ecosystem that would cause organisms to disperse.

2. Name three ways that seeds are transported from place to place.

3. Explain how the organisms in an existing ecosystem could be a barrier to an individual trying to move in.

4. What is the difference between a habitat and a niche?

5. What is meant by ecological succession?

6. Why are lichens considered pioneer organisms?

7. How does a climax community differ from the communities that preceded it?

8. Explain how the open-water zone of a lake gets smaller over time.

24.2 BIOMES

If you could see every part of the earth at one time, you would see large ecosystems in different areas of the world that resemble each other. For example, you would see desert ecosystems in California, Arizona, the Middle East, Australia, and Africa. Wherever they are located, deserts have similar conditions and inhabitants. Large ecosystems occurring in major land regions are known as **biomes** (BĪ-ōmz). Biomes are identified by the kinds of climax plants that grow there. You can identify a desert biome by its cacti and a forest biome by its trees. You would know a grassland biome by the widespread presence of grasses.

Because similar biomes have similar conditions, they have similar plants and animals. One reason for this is that plants live where the soil and rainfall are right for them, and many animals can live only where there are certain plants. Changes in the plant communities result in changes in the animal communities. In this sense, plants determine which animals will be found in a biome. Suppose you knew that the only plant to be seen in an area was grass. You could predict what kinds of animals live there based on that information.

Six major biomes exist on land: desert, grassland, deciduous forest, tropical rain forest, coniferous forest, and tundra. Each biome is unique because of its location, climate, and soil conditions. Also, animals and plants in each biome are adapted to the biome's unique features.

A biome can have a variety of conditions. For example, trees may occasionally be found in a grassland.

24–9 Distribution of the six major biomes of the world.

The Desert Biome

Africa's vast desert, the Sahara, is the size of the United States.

Deserts cover about one fifth of the earth's land surface and are found on all continents. Deserts are defined by lack of moisture; most usually receive less than 15 cm of rainfall in a year. Few living things can survive in these conditions.

Climate Rainfall in the desert is scanty and irregular. The Sahara gets less than 15 cm of rain per year. Some desert areas may go for years without a single drop of rain. In a desert, the temperature can go from hot to cold in a few hours. The air heats up during the day. At night, it cools rapidly because the air contains little or no water vapor to modify the temperature. Strong winds often blow in the desert, carrying large clouds of dust and sand. Winter and summer are much alike in deserts that occur near the equator. The Sahara, for example, is on the equator and is hot all year long. Deserts north and south of the equator, on the other hand, may have severe winters.

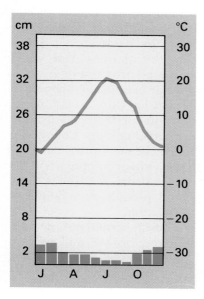

24–10 Average monthly temperature (orange) and precipitation (green) in a typical desert.

Soil Desert soil is usually sandy, dry and loose. The roots of desert plants do not hold water in the soil, nor do they hold the soil in place. Only rarely is rich soil formed from decomposing carcasses of desert organisms. More often, wind or flash floods carry away the remains. Desert soil often contains minerals like calcite or gypsum, but it does not hold enough water or organic debris to support much life.

Climax Plants and Animals The desert's climax plants are adapted to their dry environment. Some, such as cacti, have thick stems that hold water; others, such as desert shrubs, have

24–11 A desert in North Africa—miles of sand dotted with palm tree oases.

Chapter 24 The Geography of Ecosystems

24–12 Desert inhabitants: **a.** Gila woodpecker nesting inside a Saguaro cactus; **b.** kangaroo rat, named for its jumping ability; and **c.** Yarrow's spiny lizard, which uses rocks and boulders for hiding places.

thick leaves with leathery coverings that keep water in. Some plants only appear when it rains. Desert plants usually grow far apart from each other, minimizing competition for water.

Desert animals are able to withstand intense heat and lack of water. Insects and reptiles, for example, have waterproof skin that keeps them from drying out. They produce dry body wastes, another adaptation that saves water. Certain rodents, such as the kangaroo rat and the pocket mouse, even live on dry seeds. Some desert birds get water from cacti. Many desert animals sleep in burrows or shady places during the hot days and are active only at night when temperatures are cooler.

The Grassland Biome

Grasslands occur on all continents, usually in the interior. Grassland biomes in different parts of the world are called by different names. In North America, they are *plains* or *prairies*. In Argentina they are *pampas*. In the Soviet Union, they are *steppes*, and in South Africa they are *veldt*. Most grasslands cover flat land or gently rolling hills.

Climate The grassland climate is between that found in deserts and forests. Depending on its location, a grassland's average rainfall ranges from 40 cm to 80 cm per year. The grassland biome is not as dry as the desert biome, but it is drier than a forest biome. Temperatures usually range from hot to cold, and the seasons of the year are usually distinct.

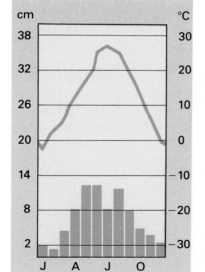

24–13 Average monthly temperature (orange) and precipitation (green) in a typical grassland.

Section 24.2 Biomes

24–14 A short grass prairie is dominated by perennial bunchgrasses less than 0.5 m tall. It receives the lowest amount of rainfall of the three types of prairie: tall, mixed, and short grass.

Soil A grassland's soil is rich in comparison to desert soil. The grass and other plants found in the grassland biome hold water in the soil. They also keep the soil from being blown away by wind or carried away by rain. Decaying plant and animal material enriches the soil. In some areas, the soil in grasslands is somewhat dry and tan in color, but in others it is moist and black.

Climax Plants and Animals The various grasses that make up the climax plants in grasslands are adapted to the conditions of their own particular area. Short grasses grow in dry grassland areas; tall grasses grow in wetter areas. In addition to grasses, plants such as clover and sunflowers are sometimes found. Grassland biomes do not have adequate water to support trees. Therefore, trees are scarce and usually grow only along streams.

Many of the animal species found in the grassland biome are herbivores. In the past, buffalo were the dominant inhabitants of North American grasslands. Today, these areas are used to raise cattle and sheep. Zebras are examples of herbivores in African grasslands. Small rodents, such as gophers and prairie dogs, are also found in grasslands. They eat grasses and roots and dig underground burrows that serve as homes. Grasslands may also be the home of large carnivores such as lions or wolves, which feed upon the large herbivores and rodents. Other rodent-eating predators of the grasslands include coyotes, badgers, snakes, and hawks. Prairie chickens, meadowlarks, and longspurs are common birds in the grasslands of North America.

24–15 Grassland inhabitants. **a.** The zebra is a common grazing animal in the South African *veldt*. **b.** In North America the coyote is a major predator of the prairie.

The Deciduous Forest Biome

A **deciduous forest** at the end of the growing season is a glorious spectacle. Deciduous forests are made up of **deciduous** trees, which lose their leaves in the fall. The green leaves turn to beautiful shades of yellow and red before they drop off. Losing their leaves adapts deciduous trees to their environment by preventing water loss.

Climate Deciduous forests have a warm growing season and a cold season when plants do not grow. In general the temperatures are moderate. The average annual rainfall in a deciduous forest biome is between 80 cm and 150 cm.

Soil The soil in a deciduous forest is rich and grayish-brown in color. It is made up largely of decomposing leaves and other organic debris. The roots of plants hold the soil down and store water.

Climax Plants and Animals The climax plants of the deciduous forest are trees with broad leaves. The type of deciduous trees found in a region depends on the amount of moisture. Maple and beech are common in moist soil. Elm and willow grow near streams. Oak and hickory grow where it is drier. A variety of wildflowers grow on the deciduous forest floor. Deciduous forests shade the ground during the nonwinter months when leaves are on the trees. The shade prevents growth of plants that require much sunlight, such as grasses. Many more kinds of plants are found in a deciduous forest than in either a desert or a grassland.

decidere = to fall off

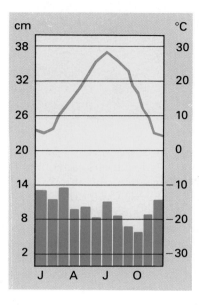

24–16 Average monthly temperature (orange) and precipitation (green) in a typical deciduous forest.

24–17 The fall color change is followed by shedding of leaves in a deciduous forest.

24–18 Several inhabitants of deciduous forests are deer (**a**), chipmunks (**b**), and cardinals (**c**). All are year-round residents and herbivores.

Deer and other animals that eat the leaves and tender twigs of trees are called browsers.

The trees and other plants found in deciduous forests provide food for many different animals. Nuts, fruits, and mushrooms help feed small rodents, including mice, chipmunks, and squirrels. Raccoons and opossums find berries and other bits of vegetation to eat. Insects also feed on plant parts. Deer browse on shrubs and small trees that grow at the forest's edge. Predators, such as hawks, owls, wolves, foxes, and mountain lions, feed on the biome's smaller animals. Turkey vultures and, sometimes, bears act as scavengers. Snakes, lizards, frogs, and various small birds also live in the deciduous forest.

The Tropical Rain Forest Biome

Forests in warm, wet areas near the equator are called **tropical rain forests**. A great number and variety of living things inhabit these forests.

Climate Some rain falls every day in most tropical rain forests. In fact, this biome receives between 200 cm and 400 cm of rain per year. Tropical rain forests receive much sunshine. The sun blazes on the treetops, but the dense foliage shades the ground, which remains moist and cool. The temperature remains fairly constant throughout the year and does not fluctuate very much between day and night.

Soil The soil of a tropical rain forest does not contain many minerals or remains of dead organisms. When an organism dies and drops to the ground, large numbers of insects, molds, and bacteria quickly attack it. Decomposition happens rapidly on the warm, wet soil, and minerals released by the decomposers

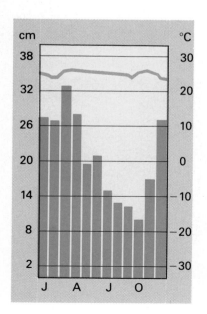

24–19 Average monthly temperature (orange) and precipitation (green) in a typical tropical rain forest.

Chapter 24 The Geography of Ecosystems

24-20 The trees of the tropical rain forest provide support and nourishment for a variety of dangling and climbing vines.

are quickly taken up by plants. As a result, the soil is not very fertile. If you let soil from a tropical rain forest dry, it becomes rocky or crusty, indicating its lack of organic material.

Climax Plants and Animals A tropical rain forest has a greater number and variety of plants and animals than any other biome. This biome actually has more kinds of plants and animals than all the other biomes combined. The climax plants are evergreen trees with broad and leathery leaves. Some trees grow as tall as 50 m to 60 m. Shorter trees branch widely at the top, their branches and leaves forming a kind of ceiling, or **canopy**, above the forest floor. The canopy prevents most of the sunlight from reaching the forest floor. Some plants, such as orchids, grow in the tree branches and take in water from the humid air. Long woody vines also grow in this biome. Few kinds of plants grow on the dimly lit forest floor. Among these is the African violet, which is widely used as a house plant.

Great numbers of different animals live in the treetops. Monkeys and other mammals are residents of the canopy, as are snakes, lizards, frogs, and endless numbers of insects. Large animals, such as leopards, spend part of their time in trees. The food webs in a tropical rain forest are very complex.

24-21 Some of the more common members of a tropical rain forest are the brilliantly colored macaw (**a**), the wild orchid (**b**), and the green tree python (**c**).

The Coniferous Forest Biome

Coniferous forests are made up of **conifers**, which are cone-bearing trees that have needlelike leaves. A great deal of our lumber comes from coniferous forests.

Climate The winters are cold and long, usually lasting at least half the year. The temperature drops below −20°C in the winter. The summer temperature in some coniferous forests reaches about 15°C. Rain and heavy snows result in an average yearly precipitation of about 60 cm.

Soil The soil in coniferous forests is not very rich. Decomposition occurs slowly in cold weather, and needlelike leaves take a long time to decompose. Fewer decomposers live in the soil of coniferous forests compared to other forest biomes. It is common to find a layer of snow covering the ground during much of the year. Under the snow, fungi decay tree needles and twigs slowly. The soil beneath the snow is grayish on top and brown below and usually lacks the minerals that many plants require.

The coniferous forest biome is sometimes called the taiga.

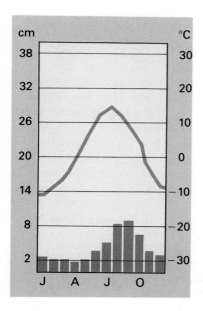

24–22 Average monthly temperature (orange) and precipitation (green) in a typical coniferous forest.

24–23 The few species of trees that grow in coniferous forests lend a uniform appearance to regions within this biome.

Climax Plants and Animals The climax trees in this biome are evergreens, such as fir, pine, and spruce. Their needlelike leaves have a protective coating that, along with their shape, helps protect the trees from drying out or freezing. Evergreens shade the forest floor to such a degree that few shrubs or flowering plants grow there. Although most of the trees in this biome are conifers, deciduous trees can be found along streams.

The trees provide food for many animals, including bears, moose, deer, porcupines, mice, hares, and grouse. Trees provide shelter for many small animals. In coniferous forests, the major predators are wolves.

24–24 Some coniferous forest inhabitants include the snowshoe hare (**a**), the porcupine, a tree dweller (**b**), and the moose (**c**).

The Tundra Biome

The **tundra** biome is found in arctic regions. The tundra is cold and lacking in water; trees cannot grow there.

Climate The tundra is characterized by severe and constant cold. Tundra temperatures drop below −25°C in winter and barely rise above the freezing point during the brief summer. The growing season lasts only about 60 days. The average precipitation in the tundra is low—about 12 cm each year, including some snowfall. Driving winter winds often pick up the snow and carry it away. In winter, the tundra is not only cold, it is dark. There are few or no hours of daylight.

The word tundra *comes from a Russian word meaning "marshy plains." Marshy describes the ground in the tundra during the summer thaw.*

Soil Little snow covers the ground. Without a thick blanket of insulating snow, the soil is permanently frozen below the surface. This layer of frozen ground is called **permafrost**. During the summer, a few centimetres of the ground thaw at the surface, becoming wet and soggy. In winter, the surface freezes again. Decomposition is very slow because of the extreme cold.

Climax Plants and Animals It is too cold for trees to grow in the tundra, and the tundra's frozen ground does not supply sufficient water. For these reasons, the climax plants are grasslike, and ecologists consider the tundra a kind of grassland. Mosses, lichens, and a few cold-resistant flowering plants grow close to the ground. An occasional shrub can be found growing in a hole that breaks the force of the wind.

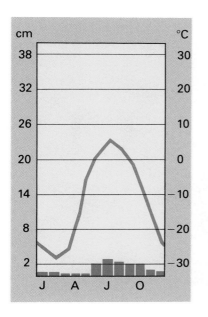

24–25 Average monthly temperature (orange) and precipitation (green) in a typical tundra.

Section 24.2 Biomes

24–26 The tundra's low vegetation leaves animals such as the grizzly bear more exposed than they would be in a coniferous or deciduous forest.

24–27 The principal mammals of the tundra are grazers such as the musk ox (**a**) and bull caribou (**b**).

Musk oxen, caribou, and reindeer are the largest animals found in the tundra. Caribou and reindeer survive by moving to new locations when they have eaten all the plants in one area. In winter they migrate to coniferous forests. Musk oxen live in the tundra all year. Their heavy coats help protect them from the winter's cold. They eat lichens, often pawing at the snow to find food. Smaller animals, such as rabbits, foxes, and ptarmigans (grouselike birds), also remain in the tundra all year. These smaller animals burrow into the snow or permafrost, thereby finding protection from storms. They feed on any bits of plant life that they can store or find. These tundra inhabitants have white coats in the winter and brown coats in the summer. The change in color of their fur or feathers adapts them to the tundra's seasons, hiding them in both winter and summer.

a

b

Summer brings more variety in tundra life forms. Insect life becomes abundant. Migrating birds, lemmings, and other small mammals arrive. Wolves, owls, weasels, foxes, and other predators thrive on the tundra's increased animal populations.

Biomes and Life Zones

The location of a particular biome can be described by its **latitude**, which is the distance north or south of the equator. Latitude is shown on a map or globe by lines parallel to the equator. As the latitude changes north and south of the equator, so do the biomes. Recall that tropical rain forests are found near the equator. Tundras are located far away from the equator. The climax plants and animals characteristic of each biome change with the latitude.

Climax plants and animals also change with elevation. The higher the elevation at any latitude, the colder it becomes. If you

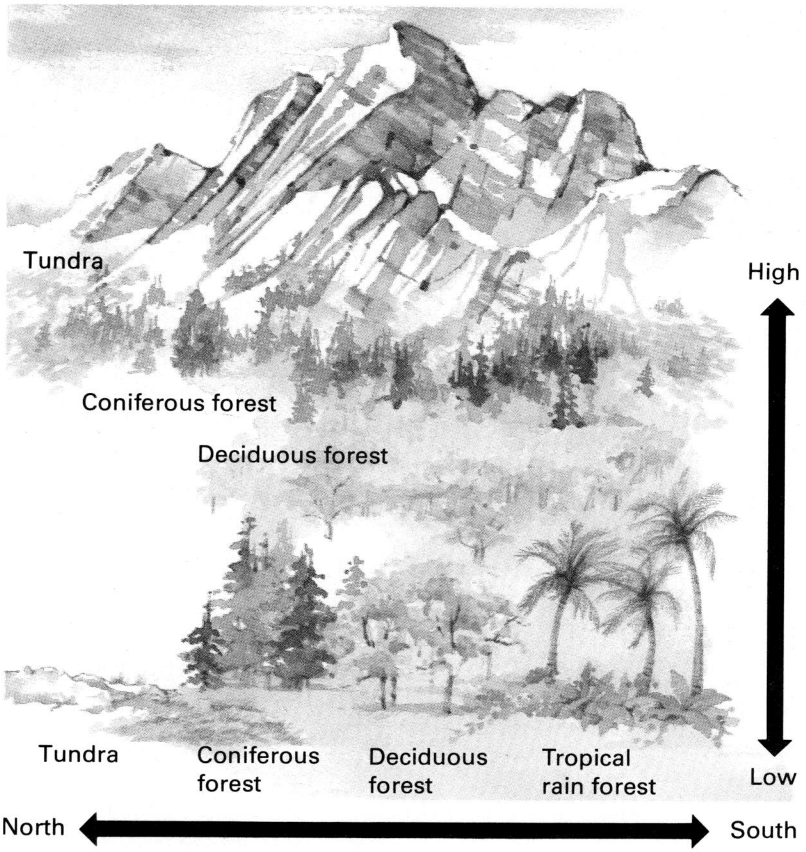

The tundra biome is also called arctic tundra and is set by latitude. Alpine tundra is a life zone set by elevation.

24–28 Land ecosystems vary according to latitude and elevation, as shown here for the Northern Hemisphere. As you move north the biomes change from tropical rain forest at the equator to tundra near the North Pole. Similarly as you climb a mountain, for example at the equator, the life zones change from tropical rain forest at the base to tundra near the top.

Section 24.2 Biomes

For every 1000-metre rise as you climb a mountain, the temperature drops about 10°C.

were to visit a mountain at the equator, you would find a tropical rain forest at the base of the mountain and a tundra at the top. Tundra conditions exist at the tops of tall mountains, as well as near the North Pole and South Pole. The zones of climax plants and animals that vary according to elevation are called **life zones**. Alpine tundra is an example of a life zone. It is the elevation on a mountain above the **timberline**—the altitude above which tall trees do not grow.

Building a Science Vocabulary

biome	tropical rain forest	permafrost
desert	canopy	latitude
grassland	coniferous forest	life zone
deciduous forest	conifer	timberline
deciduous	tundra	

Questions for Review

1. Explain the presence of cactuslike plants in the deserts of North America and Africa.

2. Select the biome from the list on the right that fits the description on the left.
   a. natural home of the buffalo
   b. a layer of permafrost occurs under the surface of the ground
   c. needlelike leaves are adaptations of climax plants against freezing and drying
   d. has the largest variety of organisms

   desert
   grassland
   tropical rain forest
   deciduous forest
   coniferous forest
   tundra

3. Give one difference and one similarity between the terms in each of the following pairs.
   a. desert and tundra
   b. biome and life zone
   c. deciduous forest and coniferous forest

4. Why are similar kinds of animals found in coniferous forests in different parts of the world?

5. Compare the average annual temperature range and precipitation in each of the six major biomes.

24.3 AQUATIC ECOSYSTEMS

Aquatic ecosystems are zones of water in which communities interact. They are classified according to the physical conditions that are present. *Lakes* and *streams* are freshwater ecosystems; *oceans* contain salt water. *Estuaries* are areas where salt water and fresh water mix. Lakes are characterized by standing water, whereas streams have running water.

As is true of biomes, a relationship exists between the physical conditions in an aquatic ecosystem and its inhabitants. Temperature, movement of water, presence or absence of salt, and amount of light help determine what kinds of organisms live in a particular aquatic ecosystem. The inhabitants of aquatic ecosystems are adapted to their environment. For example, the bodies of saltwater fish are adapted to save water and get rid of salt. The bodies of freshwater fish, on the other hand, are adapted to get rid of excess water.

The term biome *is not used for aquatic ecosystems.*

Food Webs in Aquatic Ecosystems

Producers are found as deep in the water as light can penetrate. Consumers are found throughout an aquatic ecosystem. Decomposers are usually located at the bottom.

The main producers in many aquatic ecosystems are tiny algae that live in the upper sunlit layer of the water. These algae form part of the **plankton**, which consists of microscopic algae and animals and is carried by currents in the water. The algae in plankton are called **phytoplankton** (FĪT-uh-). The tiny animals in the plankton are called **zooplankton** (ZŌ-uh-). Some zooplankton are herbivores; others are carnivores. Most zooplankton are small crustaceans. Plankton is an important food for fish and other consumers, making it a vital link in the food chains of aquatic ecosystems.

planktos = wandering
phyto- = plant
zoo- = animal

24–29 Plankton. All aquatic ecosystem food webs begin with phytoplankton, the major producer. The larger organisms in this micrograph are *Nauplius* larvae, a type of zooplankton.

Water life includes all kinds of consumers. Predators, scavengers and parasites are among the consumers that find niches in watery habitats.

Organic debris sinks to the bottom in bodies of water. There, bacteria and fungi thrive on the waste products and remains. The minerals and other nutrients released by the decomposers circulate upward, carried along by currents to the producer level.

Lakes and Ponds

A **lake** is an inland basin that contains water. Since lakes do not run off to the ocean, they are considered standing water. The term *standing water* is used even if the lake is fed by streams and has an outlet. A **pond** is simply a small lake.

Physical Conditions Lakes vary in size and depth from small ponds to masses of water such as Lake Superior, which covers 82 414 km² and is 400 m deep in parts. Newly formed lakes have little sediment on the bottom; older lakes have more. For example, Lake Baikal in Siberia, which is 20 to 30 million years old, has about 6.6 km of sediment built up on the bottom. Lake water can be salty (Great Salt Lake) or fresh (the Great Lakes). It can also range from murky to crystal clear.

Physical conditions also vary within a given lake. Water temperature varies with depth. In summer, the upper layer of lake water is warmed by the sun, and the lower layers are colder. The reverse is true in cold areas in winter: water at 0°C stays on top, and water up to 4°C remains underneath. In spring and fall, the temperatures in different parts of the lake change. The changing temperatures cause the lake water to mix. Mixing of water helps deliver supplies of oxygen and minerals to organisms living in the lake. Different parts of the lake also get different amounts of light. The wavelengths of sunlight needed for photosynthesis do not penetrate deeply into the water, and sunlight penetrates more deeply into lakes with clear water than it does into lakes with murky water.

Lake and Pond Inhabitants Although large plants may grow in a lake, phytoplankton is the main producer. Some lakes have a large supply of minerals or other nutrients. This can

24–30 The still surface of an alpine lake supports floating water plants.

24–31 Some inhabitants of lakes and ponds: **a.** backswimmer (an insect) and its minnow prey; **b.** redbreast sunfish gliding among pond weeds; and **c.** water lily, whose leaves often serve as platforms for frogs and insects.

cause an overgrowth of algae on the surface and upset the natural balance of the lake. A thick cover of algae blocks light, causing the algae below the surface to die and decompose. Decomposition uses up oxygen in the water that the lake's consumers need. The lowered oxygen level causes many of the large consumers to die. Algal overgrowth also speeds up the ecological succession from lake to land.

Recall that the three zones of life in a mature lake are the edge, open water, and bottom. Physical conditions are different in each of the three zones. As a result, the forms of life found there are also different.

Since ponds can be quite small, they often lack the open-water zone. Pond consumers include insects living on and near the water. They are part of a food chain in which frogs eat insects, and water snakes eat frogs. Other members of the pond food web are plankton, fish, crayfish, mussels, birds, mammals, worms, and bacteria.

Streams

Streams are bodies of flowing water. A stream can start with rain or melting ice and snow. It can also begin at a spring or at the outlet of a lake. A stream usually starts out as a small flowing brook and later joins up with other brooks to form a larger stream. Eventually, it can become a large river heading toward the sea. When a stream flows rapidly, it cuts into the earth, forming a bed. Sediment and other debris are carried in the current. A slow-running stream, on the other hand, deposits sediment along its banks and on the bottom.

24–32 A fast-moving stream collects runoff from the mountain slopes and scours the rocks clean of all but the most tightly clinging vegetation.

Physical Conditions Stream water is usually cool and fresh. When it flows rapidly, it mixes with air, supplying the stream water with oxygen. Streams are characterized by fast-moving water at their source and slow-moving water as they approach the sea. Sunlight generally falls on a stream's surface.

Stream Inhabitants Swift-flowing streams and slow-flowing streams present two different types of habitat. Plankton does not accumulate in fast-moving water. Instead, algae and mosses grow on rocks or on the stream's bottom. Consumers in a fast-moving stream include water insects and fish. They get food from insects and other producers as well as organisms that fall into the water. They also feed on organic debris that is washed down with rain. Decomposers do not get much to eat in a fast-moving stream, as debris does not collect on the bottom.

Plankton is abundant in the sunlit water of slow-moving streams. Plants also take root and grow in the sediment deposited on the stream's bottom. Many consumers depend on the plants and plankton in a stream. Some consumers live on the bottom, including insects, snails, mussels, and crayfish. These animals, in turn, provide food for turtles, fish, and others. Streams also provide food for birds, mammals, and large reptiles. In North America, beavers and muskrats are common consumers in river ecosystems. In the tropics, it is common to find hippopotomi eating huge amounts of plant life growing in or near streams. Crocodiles are also found feeding on fish, birds, and small mammals in the shallow water of tropical streams.

Slow-moving rivers deposit sediment, forming swamps or marshes where land ecosystems begin to take hold. Leaves, seeds, and fruits from plants growing near a stream get washed into its water when it rains. These plant parts contribute to the consumers' food supply, while decomposers break down the debris that falls to the bottom.

24–33 Stream inhabitants include: **a.** sockeye salmon; **b.** beaver; and **c.** hippopotamus, found in tropical regions.

a

b

c

Oceans

Oceans are deep basins filled with salt water that surround the earth's land masses. Oceans cover about three fourths of the earth's surface. Although some parts of oceans are deeper than the height of Mt. Everest (8.8 km), the average depth of the Atlantic and Pacific Oceans is about 3.9 km.

The deepest part of the Pacific Ocean is about 10 km.

Physical Conditions Ocean water contains various salts, the most common of which is sodium chloride, or table salt. The water temperature in an ocean depends in great part on its geographical location. The average surface water temperature in arctic oceans is less than 0°C; near the equator it is 28°C. Water temperature changes slowly. As a result, the surface temperature of ocean water does not change drastically from day to night or even from season to season. Deep water is always cold, because it is not warmed by the sun. Deep ocean water is as cold in the tropics as it is in the arctic.

Ocean water moves in two ways. One is a vertical change in sea level. This movement is connected to the cycle of evaporation and precipitation of water and to the rise and fall of the tides along the coastline. Water in an ocean basin also moves horizontally from place to place in currents. Currents are caused by the action of wind, changes in temperature, and the rotation of the earth. The movement of ocean water results in mixing, which helps deliver oxygen, carbon dioxide, and minerals to the ocean's inhabitants. The movement of water also affects water temperature.

Physical factors that influence living things in a marine ecosystem include water depth, currents, tides, light penetration, and temperature.

24–34 The featureless surface of the ocean conceals a myriad of life forms beneath, from microscopic plankton to whales, the largest mammals living on earth.

Section 24.3 Aquatic Ecosystems

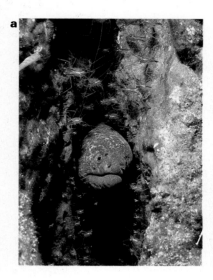  

24–35 Ocean inhabitants include the moray eel (**a**), a deep-water predator; horned puffins (**b**); and the Galapagos fur seal (**c**), whose thick, fur-covered skin insulates it against cold air and water.

The bottom of the ocean is rocky; however, much of it is covered by sediment carried by rivers to the ocean. On the average, 900 m of sediment covers the ocean's bottom.

Ocean Inhabitants The largest producers found in the ocean are seaweeds; however, they are not the main producers. Phytoplankton produces most of the food in the ocean's food webs. Ocean consumers range in size from zooplankton to whales and sharks. Some whales are herbivores; other whales and all sharks are carnivores. Other large carnivores include tuna, swordfish, dolphins, seals, and penguins. Squid, herring, mackerel, and sea birds are among the ocean's smaller carnivores.

Scavengers live on the organic debris that constantly falls toward the ocean bottom. These include mud-dwelling worms, clams, shrimps, and crabs. Other organisms living on the bottom are sponges, sea anemones, and starfish. Bacteria and fungi on the ocean bottom decompose organic matter. Rising currents bring the minerals released by decomposition to the surface, where they are taken in by the phytoplankton.

Estuaries

Estuaries are the most productive breeding grounds for fish and shellfish.

Estuaries are bodies of water that are formed where rivers meet the ocean. New York harbor, the Fraser River delta, and the mouth of the St. Lawrence River are all examples of estuaries. Estuaries are usually partially enclosed by land. They are important aquatic ecosystems because they are rich in living things.

24-36 Tidal action tears at and builds up the banks of an estuary, which supports hardy marsh grasses rooted in salt-water-saturated soil.

Physical Conditions In an estuary, salt water from an ocean mixes with fresh water from an inflowing river. Ocean tides enter the estuary, creating moderate currents that deposit sediment and organic debris on the bottom and sides. The whole system creates a kind of trap for minerals and nutrients. Estuaries occur in arctic, temperate, and tropical areas. The sun shines on the surface of estuaries but does not penetrate deeply into estuary waters. The water temperature depends on the depth of the water and the geographic location of the estuary.

Estuary Inhabitants Estuary inhabitants are adapted to the salt content of their environment. Grasses and other salt-tolerant plants grow in the marshlike ecosystems on the banks of the estuary. Decayed debris from the marsh plants enters the estuary and helps feed its inhabitants. Producers in the estuary include water plants and algae. Some of the algae grows in the mud on the bottom; other algae form part of the plankton. Many types of consumers live in estuaries. Fish, such as trout and shad, are found there. Some fish, oysters, and crabs use estuaries for breeding and nursery grounds, and salmon spend part of their lives in estuaries. Crabs, shrimps, clams, worms, and other mud dwellers live on the bottom. Decomposers on the bottom help recycle the ecosystem's nutrients.

24-37 The great blue heron, top consumer in the estuary ecosystem, spears small mammals, rodents, and fish with its sharply pointed bill.

Section 24.3 Aquatic Ecosystems

24-38 A controversy arose over the building of the Alaskan crude oil pipeline, which traverses a delicate tundra ecosystem. Much care was taken in designing and constructing the pipeline to minimize its effects on the environment.

The Human Factor in Ecosystems

In the first section of this chapter, you studied the ways that living things spread out to inhabit almost all parts of the earth, and you saw how ecosystems change during a natural process of ecological succession. In the second and third sections, you examined the main areas that organisms populate. Human beings have also spread out and presently occupy most of the land areas on the earth. There is even a possibility of establishing colonies in space some day. Aided by technical advances, we adapt to different biomes and make changes in our environment. Each change has far-reaching consequences, affecting the crucial nonliving parts of ecosystems, as well as our lives and the lives of other organisms. Understanding how living things interact with the environment should help us make intelligent decisions about living in ecosystems and using natural resources.

Building a Science Vocabulary

aquatic ecosystem	zooplankton	stream
plankton	lake	ocean
phytoplankton	pond	estuary

Questions for Review

1. What is plankton? What is its role in an aquatic food web?
2. Explain how organic debris is decomposed in the ocean food web. How are nutrients returned to producers?
3. Name three consumers found in streams.
4. Name the type of aquatic ecosystem described by each of the following.
   a. saltwater and freshwater mix
   b. running fresh water
   c. standing fresh water
5. Name two types of organisms that use estuaries for breeding grounds.

Perspective: Environment
What Good Is a Wetland?

If you have ever tried to walk through a marsh, swamp, bog, or fen—often the creepy sites favored by producers of horror movies—you probably would have supported the Swamplands Act, which the Congress of the United States passed more than a century ago. The Act supported getting rid of these wetlands by draining, diking, or filling them in. Since the Act was passed, about half of the wetlands have been filled in, leaving about 160 000 km² untouched. But, fortunately, before all the remaining ones have been obliterated, people have begun to realize how valuable and important wetlands are.

Nearly 90 percent of the wetlands are inland and contain fresh water. Freshwater wetlands serve as gigantic sponges, reducing flooding by absorbing storm and runoff water. They also serve to recharge the supply of water in underground reservoirs. The chemicals that occur naturally in wetlands neutralize sewage and other wastes and contaminants. Wetlands produce valuable timber and teem with animals and birds, which use them as refuges and breeding sites.

Even more important, wetlands are essential in fish production. Two thirds of the fish caught commercially off the Atlantic and Gulf Coasts and half of the fish caught off the Pacific Coast depend on wetlands and adjoining estuaries. Texas' wetlands are the foundation of its giant shrimp industry. Georgia's salt marshes produce 4 t of organic material per hectare every year. By comparison, a good hay field may produce up to about 1.5 t/ha in a year. One of the challenges is finding ways to use the higher productivity of wetlands effectively.

Past mistakes of filling in, diking, and draining wetlands are slowly being corrected. In recent years the Federal government of the United States has begun insisting that coastal states manage their wetlands more effectively. It has also been encouraging these states to acquire more wetlands as part of the government's property, so that they can be better protected.

People have been slow to learn the value of wetlands. What about the value of other kinds of habitats—deserts, rivers, mountains, tundras? We seem to still have much to learn.

Chapter 24
Summary and Review

Summary

1. Organisms spread out from their original habitats to new locations by the process of dispersal. Individuals disperse if physical conditions in an ecosystem change drastically or if population size increases.

2. Some animals disperse on their own. Other animals and plants are dispersed by wind, water, or humans and other animals. Some organisms have special adaptations for dispersal.

3. Climate, geography, and other organisms may be barriers to dispersal.

4. Organisms must fill a niche if they are to succeed in a habitat.

5. Ecological succession is the process of progressive change from a pioneer community to a climax community.

6. Most lakes undergo ecological succession from a lake community to a forest community.

7. Biomes are major groups of land ecosystems having similar climax plants and animals. The six major biomes are desert, grassland, tropical rain forest, deciduous forest, coniferous forest, and tundra.

8. There are similarities among the organisms within a biome because they are adapted to similar conditions. The plant life in a biome largely determines the kinds of animals that can survive there.

9. Latitude is a significant factor in determining biomes; elevation is a significant factor in determining life zones.

10. Aquatic ecosystems are characterized by their physical conditions. Lakes, streams, oceans, and estuaries are four types of aquatic ecosystems.

11. The wavelengths of sunlight that are used in photosynthesis do not penetrate to the deep layers of bodies of water. As a result, producers are not found in deep water.

12. Plankton is an important food in the food web of aquatic ecosystems. It is made up of tiny algae, the phytoplankton, and tiny crustaceans, the zooplankton. Phytoplankton is the main producer in many aquatic ecosystems.

13. Decomposers on the bottom of aquatic ecosystems break down organic debris that filters down to them.

Review Questions
Application

1. Why is a lake not considered a climax community?

2. For each of the following biomes, name one inhabitant and show how that inhabitant is adapted to conditions in the biome.
   a. tundra
   b. desert
   c. coniferous forest

3. Why is soil found in a deciduous forest richer than soil found in a tropical rain forest?

4. Why do the ecosystems change as you climb a mountain?

5. Why does plankton not grow in deep water?

6. Why is mixing of water important in aquatic ecosystems?

7. Construct an ocean food chain for the organisms listed below.

zooplankton   sharks
crabs   herring
phytoplankton

8. How can fire prevent a grassland area from undergoing ecological succession to become a forest?

9. People are clearing tropical rain forests for farming. Why are some tropical rain forests not well suited for farming? What problems can result from this practice?

Interpretation

1. Construct an energy pyramid for a grassland ecosystem. Include examples of organisms at each level.

2. Explain how a tree in a deciduous forest can be considered both an organism and a community.

3. How does the water cycle affect the up-and-down movement of ocean water?

Extra Research

1. Compare the features of the area in which you live with the general description of your biome. List the similarities and differences in climax plants, climax animals, climate, and soil.

2. Set up a terrarium or an aquarium to represent a type of biome or aquatic ecosystem. In the terrarium you can simulate a desert, marsh, or woodland biome. In the aquarium you could set up a freshwater or saltwater ecosystem.

3. Research and write a description of a biome or aquatic ecosystem not described in the text. Possibilities include chaparral or coral reef.

4. Make a collection of seeds that have different adaptations for dispersal.

5. Visit a biome or aquatic ecosystem in your area. Take along a notebook and record the kinds of plants and animals you see. Include a general description of the physical environment.

6. Human beings have made deserts out of fertile land and fertile land out of deserts. Use reference books to find specific examples of each case. Present your findings in a written or oral report.

7. Visit your local museums. In the natural history museum, observe displays of natural habitats. Study how the different habitats and biomes are displayed. In the art museum, observe paintings that depict nature. Compare the artist's view with your understanding of the ecosystem shown. Is the artist's portrayal accurate?

Career Opportunities

Foresters and their assistants develop and manage forest resources by supervising the planting, growth, and harvesting of trees. Some, but not all, of their work is done outdoors. Many foresters work for private companies, others are employed by federal or provincial governments. A degree in forestry is required. Write the Canadian Forestry Association, 185 Somerset Street W., Ottawa, Ontario K2P 0J2.

Range managers oversee the use and development of grasslands for grazing, wildlife, and recreation. Much of their time is spent outdoors evaluating the plants, soil, wildlife, and water resources of an area. Write the Prairie Farm Rehabilitation Administration, 1901 Victoria Ave., Regina, Saskatchewan S4P 0R5.

Unit VII
Suggested Readings

Anonymous, "Mammals of Sea and Ice," *Audubon*, March, 1973. *Various types of seals are well adapted to live and function in the tundra.*

Carson, Rachel, *The Sea Around Us*, rev. ed., New York: Oxford University Press, Inc., 1961. *Starts with a description of primal seas and discusses in a fascinating way currents, tides, mineral resources, and many other topics.*

Cousteau, Jacques, with Philippe Diolé, *Life and Death in a Coral Sea*, New York: Doubleday & Co., Inc., 1971. *Excellent introduction to coral reefs and islands, with beautifully illustrated discussions of geography, oceanography, and natural history.*

Graham, Ada and Frank, *Careers in Conservation*, A Sierra Club Book, New York: Charles Scribner's Sons, 1980. *Based on personal accounts of people working as conservationists, this book gives practical information on getting started in a wide range of careers.*

Graham, Ada and Frank, *The Changing Desert*, A Sierra Club Book, New York: Charles Scribner's Sons, 1981. *A short, enjoyable book on the natural history of the desert and the human inhabitants who are changing it.*

Gunderson, Harvey L., "Under and Around a Prairie Dog Town," *Natural History*, October, 1978. *Prairie dogs live in a complex system of burrows and partake in an organized society where touch, smell, sight, and voice communication are important.*

Kirk, Ruth, "Life on a Tall Cactus," *Audubon*, July, 1973. *Excellent description of adaptations of cacti and the animals living in and near them.*

Knutson, Roger M., "Flowers That Make Heat While the Sun Shines," *Natural History*, October, 1981. *Plants that bloom at low temperatures act as solar collectors, hastening the growth of pollen and seeds. This entire issue of* Natural History *is devoted to techniques for survival in cold climates.*

Leopold, Aldo, *Sand County Almanac Illustrated*, Madison, WI: Tamarack Press, 1977. *Essays, arranged by months, recount a year's changes on a Wisconsin farm. Principles of conservation are expressed with personal conviction.*

Martin, Willard K., "Seeds in Flight," *Natural History*, April, 1978. *Beautifully illustrated article on how seeds are adapted for dispersal.*

McWhinnie, Mary A., and Charles J. Denys, "The High Importance of the Lowly Krill," *Natural History*, March, 1980. *Krill are tiny shrimplike crustaceans that are essential in aquatic food webs.*

Perry, Donald, "An Arboreal Naturalist Explores the Rain Forest's Mysterious Canopy," *Smithsonian*, June, 1980. *Article and photographs describe plant and animal populations in the canopy of a rain forest.*

Quinn, John R., *The Summer Woodlands*, Old Greenwich, CT: Chatham Press, 1980. *Seasonal narrative of streams, ponds, and woods describes northern forest plants and animals. Stresses field observation techniques, manners, and ethics for budding field naturalists.*

Reiger, George, "Meanders in the Marsh," *Audubon*, September, 1977. *Excellent description of a salt marsh estuary.*

Sayre, Roxanna, "An Invasion to Remember," *Audubon*, January, 1980. *Describes the dispersal of gray owls from coniferous forests to other biomes.*

Teale, Edwin Way, "A Walk Through October," *Audubon*, September, 1978. *Fascinating day-by-day account of October in a deciduous forest.*

Unit VIII

Reacting to Each Other and the Environment

Have you ever listened to birds singing and wondered what it accomplishes? Singing is a type of communication, and communication is one aspect of behavior. In the most general sense of the word all an organism's actions and activities make up its behavior. An organism's behavior depends on its heredity and its individual life history. Certain behaviors seem to be genetically preprogrammed. Others are the result of learning through experience and interaction with others.

Many of the diseases that afflict humans are the result of interactions between parasites and human hosts. Through medical research, we have learned much about how the body's defenses fight against disease. We have made great strides in decreasing suffering and death by controlling and combatting bacteria, worms, and other parasites. Yet, many of the most treacherous diseases remain, including heart disease and many types of cancer, the causes of which are not well understood. Continued research will likely help in the diagnosis, treatment, and prevention of these diseases.

Honeybees live together in highly ordered societies. Most of their behavior appears to be inherited.

Chapter 25

Behavior

Focus *Why do geese fly in V-formation? Why do many plants grow toward light? What makes a gorilla beat its chest? No one fully understands the reasons for an organism's behavior. It appears, however, that many of the behaviors of individuals, groups, and societies of organisms help them survive.*

25.1 OVERVIEW OF BEHAVIOR

Behavior is the response of an organism to stimuli in its internal or external environment. Most behavior involves movement. Plant movement is usually slow and due to growth. Some plant movement is relatively fast; for example, the leaves and stems of *Mimosa pudica*, the sensitive plant, move quickly when touched. Most protist and animal movement involves special body structures. The paramecium swims using cilia, and most animals move using muscles.

Most behavior seems to help an organism survive in its environment. Avoiding predators or other dangers and finding food, shelter, and mates help individuals and species survive. Except for humans, organisms do not consciously behave in a certain way in order to survive. Instead, their behavior, once expressed, helps them survive.

25–1 Migrating caribou in search of more food cover the tundra for days at a time. Seasonal changes cue this instinctive behavior at the onset of winter and summer.

699

Be careful not to confuse the function of human behavior with that of other organisms. For example, if plants grow toward the light it is not appropriate to say that they "love" the sun. If two male deer butt antlers it is wrong to say that they are "angry" or "jealous." Attributing human attitudes and emotions to other organisms is known as **anthropomorphism** (AN-thruh-puh-MŌR-fizm). It should be avoided so as not to misinterpret the behavior of nonhuman organisms.

Types of Behavior

Most behaviors fall into one of two categories. Behaviors that organisms acquire through learning are called **learned behaviors**. For example, a dog learns to sit upon command, or a caged rat learns to press a lever to obtain food. Behaviors that organisms possess at birth are known as **inherited behaviors**. The characteristic behaviors of many species are inherited. For example, both the male and female stickleback fish behave in a very specific and unique way during courtship and mating. Inherited behaviors are found among all the members of a species. Newly hatched chicks walk, peck, eat, and drink. Newly hatched ducks run to the water, swim, and dive.

25–2 Stickleback fish mating behavior. After a courtship dance, the male builds a nest. The female enters the nest and deposits her eggs, which the male fertilizes.

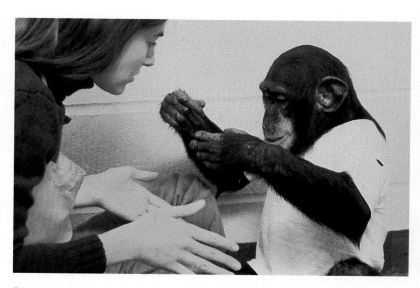

25–3 The chimpanzee is being taught how to sign the word "box." Language was once thought to be possible only in humans. Yet, some chimpanzees have learned many signs for words and ideas and use them in conversations with humans. An intriguing question is whether chimps will be able to use sign language with each other.

Inherited Behavior in Plants

The growth movements of certain plant parts toward or away from an environmental stimulus are known as **tropisms** (TRŌ-pizmz). When the movement is directed toward the stimulus, it is called a *positive* tropism. When the movement is away from the stimulus, it is called a *negative* tropism. Various environmental factors cause plant tropisms. For example, stems and leaves grow in the direction of a light source, usually the sun. This behavior is known as *positive* **phototropism**.

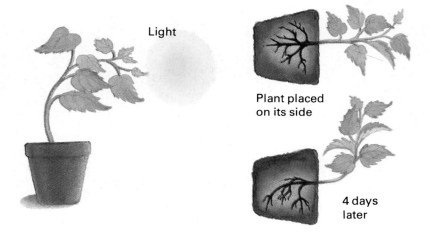

25–4 Phototropism (left) causes a plant to bend in the direction of a light source. Geotropism, or tropism in response to the earth's gravity (right), causes roots to push downward and stems to turn upward.

Table 25–1   Plant Tropisms

Environmental Stimulus	Plant Organ Affected	Tropism
Light	Stem	(+) Phototropism
	Leaf	(+) Phototropism
Gravity	Root	(+) Geotropism (growth toward the center of the earth)
	Stem	(−) Geotropism (growth away from the center of the earth)
Water	Root	(+) Hydrotropism (growth toward water)

Inherited Behavior in Protists and Animals

A fly's behavior is stereotyped. You can predict that a hungry fly will accept food. Most human behavior is not stereotyped. You cannot predict that a hungry person will accept food. For example, the person might be on a diet.

Inherited behavior in protists and animals is also called **stereotyped behavior**, because the response to a given stimulus is always the same in all members of the species. Stereotyped behavior is fixed, automatic, and predictable. Planaria always swim away from the light. When threatened, squids always propel themselves backwards with a quick spurt. When a female firefly sees a male firefly's flash, she always flashes back two seconds later. Stereotyped behavior takes on several forms.

Kineses **Kineses** (kih-NEE-seez) are very simple forms of inherited behavior that exist in animals that have simple nervous systems. When a stimulus occurs, the organism responds by moving randomly—not necessarily toward or away from the stimulus. Kineses depend on locomotion. The greater the intensity of the stimulus, the faster the organism responds.

The wood louse, a small crustacean, reacts by kinesis. Wood lice live in moist places such as under pieces of wood or loose stones on the ground. If a wood louse starts to dry out, it begins to move around randomly until it finds another moist spot. Then, the wood louse settles down again.

Negative taxes are sometimes called "avoidance behavior."

Taxes Protists and animals that do not have well-developed nervous systems often exhibit **taxes** (TAK-seez), which are movements directly toward or away from a stimulus. As with plant tropisms, movement toward a stimulus is called a *positive* taxis; movement away from a stimulus is called a *negative* taxis. Most taxes are responses to light, heat, or chemicals. Unlike tropisms, taxes are due to locomotion and not growth. A paramecium's behavior when it moves away from a harmful chemical is an example of a taxis, as is a euglena's movement toward light.

 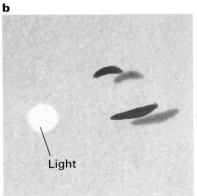

25–5 **a.** Kinesis. Wood lice, or sowbugs, move about randomly until they find a moist environment. **b.** Taxis. Planaria move away from a light source, an example of a negative taxis.

Reflexes The **reflex** is the simplest behavior of a well-developed nervous system. A particular stimulus always produces the predicted response. If you step on a tack, you immediately pull your foot away. When a foreign object approaches your eyes, you automatically blink. Many reflexes are a response to avoid or withdraw from a harmful stimulus.

Instincts A mother hen protects her chicks. She will even risk her own life to defend her offspring against aggressors. Such maternal behavior is an example of an **instinct**, which is a complex stereotyped behavior. Instincts are expressed in the same way in all individuals of a species. An animal raised away from other members of its species still shows its species' instincts.

Instinctive behavior can be triggered by internal and external stimuli. Often both types of stimuli are involved. A silkworm caterpillar's instinct to build a cocoon is first brought on by hormones inside its body. Once the cocoon is started, the cocoon itself stimulates the caterpillar to continue building.

Explaining Inherited Behavior

Scientists have devised a theory to explain inborn behavior based on the concept of **releasers**. A releaser is a specific stimulus in the environment that causes an organism's inherited response. A releaser is like a sign or special code word. When an organism senses the releaser, it responds with the appropriate behavior. Chemicals and colors are common releasers. In the stickleback fish red color on the male's belly is the releaser for fighting behavior in other males. The red belly is also the releaser for courtship behavior in the female.

25-6 A spider's behavior, including web spinning, is almost entirely guided by instinct.

25-7 A herring gull chick pecks at a red spot on the parent's beak. This behavior acts as a releaser, causing the parent to regurgitate food for the chick.

Section 25.1 Overview of Behavior

Learning

Learning is a change in behavior that is based on experience. Although we usually think of learning in humans, it occurs in many nonhuman animals as well. Scientists have taught octopuses to tell the difference between circles and squares, and snails have been trained to extend their eye stalks. However, no nonhuman animal is capable of the kind and amount of learning that human beings accomplish.

Types of Learning

Organisms learn things in different ways and at different times. Some things are learned only at certain periods in life; other things are learned throughout life. Several types of learning can be observed.

Because of habituation, you do not constantly feel the clothing that you are wearing.

Habituation An animal's environment is the source of thousands of stimuli. Some stimuli are important in the life of the animal; others have no effect. By means of **habituation** (huh-BICH-oo-WAY-shun), an animal learns not to respond to stimuli that are unimportant. By instinct, newly hatched chicks crouch when an object moves overhead. The chick remains crouched until the object passes by. If a specific object passes over a chick's head many times and no harm occurs, the chick becomes "used to" the object. Eventually, the chick no longer crouches when the now familiar object passes by. If, however, a new unfamiliar object passes over the chick's head, it will again crouch and lie still.

Imprinting During the 1930s, an Austrian zoologist named Konrad L. Lorenz observed that newly hatched geese follow the first moving thing they see. If the first thing the young geese see is their mother, they follow her. But, in experiment, Lorenz found that if the young geese saw him first, they followed him and continued following him as they grew up. Lorenz said that this was a special type of learning called **imprinting**. Imprinting occurs only during a short period in the early life of animals such as insects, birds, and mammals. In nature, one result of imprinting is that the young follow their mother, a behavior that aids in their survival.

25-8 Imprinting. Dr. Konrad Lorenz is followed by a flock of greylag goslings, which have been imprinted to respond to him as their parent.

Classical Conditioning Classical conditioning is a simple type of learning that involves a reflex. During the early 1900s, a Russian physiologist named Ivan Pavlov did some important learning experiments. In one of his experiments, Pavlov blew powdered meat into a dog's mouth and measured the amount of saliva that the dog produced. The production of saliva when food is present is an inborn reflex reaction. He then introduced a second stimulus, the ringing of a bell. Each time he gave the dog meat powder, he would ring a bell. The two stimuli, the ringing of a bell and the meat powder, were presented together five or six times. Then Pavlov rang a bell but did not give the dog meat powder. The dog salivated anyway. The dog "learned" to salivate at the ringing of the bell.

In Pavlov's experiment, the original stimulus was powdered meat. Learning occurred when a new stimulus, the bell, was paired with the meat. The dog learned to automatically produce the same reflex reaction (salivation) in response to the new stimulus. This kind of learned behavior is known as a **conditioned response**.

Almost all animals are capable of some learning by classical conditioning. For example, a dog that has been frightened by the *sound* of a gun will shrink away from the *sight* of a gun if it saw the gun when it was frightened by the sound.

Classical conditioning is not permanent. It does not last unless the original stimulus is restored periodically.

25–9 Classical conditioning. A "Pavlovian response" is created by repeated experiences in which a normal response (salivation) becomes associated with a new stimulus (ringing bell). Eventually the new stimulus causes the response in the absence of the normal stimulus.

Operant Conditioning B. F. Skinner, a Harvard psychologist, discovered many things about a type of learning called **operant** (OP-ur-unt) **conditioning**, which depends on rewards or punishments. In operant conditioning, an animal performs an action for the first time, perhaps by accident. The behavior is then rewarded, ignored, or punished. If the behavior is re-

Operant conditioning is commonly used to train animals.

Section 25.1 Overview of Behavior

25–10 Operant conditioning. **a.** Dolphins are conditioned to perform stunts in response to cues. **b.** A food reward is the reinforcer for their behavior.

warded, the animal usually repeats it, and the behavior is learned. If the behavior is either ignored or punished, the animal tends to stop the behavior. For example, a caged rat can be trained to press a lever by means of operant conditioning. As the rat moves around the cage, it accidentally presses the lever. Food is then released. The rat soon learns that it is rewarded with food when the lever is pressed. In this way, it learns to press the lever repeatedly.

A factor that makes an organism repeat a behavior is called a **reinforcer**. Food, water, praise, money, and attention are examples of reinforcers. An organism behaves in a way that brings a reward or stops a punishment. Both the reward and the stopping of a punishment are reinforcers.

Memory

When you learn something, it usually first goes into short-term memory. Then, the information is organized to fit with already stored information. Finally, it goes into long-term memory.

Memory is the storing and retrieving of learned information. The memory of recent events, or **short-term memory**, differs from memory of events that happened in the past, or **long-term memory**. Nearly everyone has been introduced to someone and remembered that person's name only for a short time. Upon seeing the same person at a later time, however, the name was forgotten. The person's name entered short-term memory, which is temporary. If a bit of information is to be retained, it must enter long-term memory. Very little is known about how information enters long-term memory. Sometimes it is relatively easy to enter information into long-term memory. Other times, it is extremely difficult. Information stored in long-term memory can be recalled for days, years, or even a lifetime.

25–11 Model of how memory works. As the mind receives sense information it selects some for short-term memory. Some of this information is then classified and filed in long-term memory. A few impressions go directly into long-term memory.

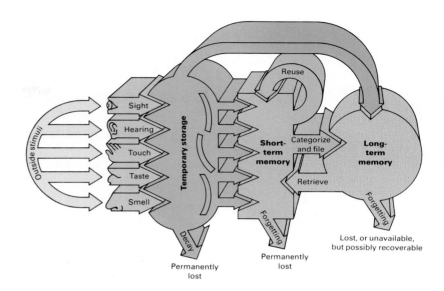

Reasoning

Reasoning is remembering one or more past experiences and using these memories to help solve a new problem. When you reason, you remember what you learned and use it in a new way. Humans seem to have the greatest capacity for reasoning; however, some reasoning occurs in all animals with a complex brain.

The term *intelligence* means "the ability to reason." It describes how well an organism understands, connects, and uses ideas. Scientists used to think that intelligence was entirely

25–12 Reasoning. A green heron has learned to drop a food pellet into the water as bait to catch its own fish. This behavior was observed at the Seaquarium in Miami, Florida.

inherited. Studies show, however, that intelligence also depends on nutrition and use. Although general mental ability is probably inherited, intelligence increases if an organism is properly nourished and gains experience in solving problems.

25–13 Chimpanzees are quite intelligent, as shown by their use of tools. Here a piece of dry grass is used to pry out insects from bark.

Building a Science Vocabulary

behavior
anthropomorphism
learned behavior
inherited behavior
tropism
phototropism
stereotyped behavior
kinesis
taxis
reflex
instinct
releaser

learning
habituation
imprinting
classical conditioning
conditioned response
operant conditioning
reinforcer
memory
short-term memory
long-term memory
reasoning

Questions for Review

1. In general, what function does behavior serve?

2. What are the differences between inherited and learned behavior?

3. State one important difference between the types of inherited behavior in each of the following pairs.
   a. tropisms and taxes
   b. kineses and taxes
   c. reflexes and instincts

4. In the laboratory, a newly hatched duck followed a moving toy train. What type of behavior is this? How does this type of behavior help baby ducks in their natural environment?

5. What is a conditioned response? Give an example of this type of behavior.

6. What is a reinforcer? How are reinforcers used to bring about certain behaviors?

25.2 CYCLIC BEHAVIOR AND AN OVERVIEW OF SOCIAL BEHAVIOR

Cyclic Behavior

Many things repeat in never-ending cycles. Day changes into night; the seasons change; the tides rise and fall. Organisms are adapted to the periodic changes that surround them. Palolo worms, for example, live on coral reefs in the South Pacific. Nearly all of these worms (99 percent) shed their eggs in the same two-hour period on one night of the year. Four environmental cycles are involved in their timed behavior. The season is spring (November). The moon is in the last quarter. The tide is very low. The time is a few hours after darkness. When all four factors coincide, the Palolo worms reproduce.

All organisms are affected by natural cycles and rhythms. Probably the most obvious cycle is the day-night, or **diurnal** (dī-YOOR-nul), cycle. The activities of most organisms are connected to the diurnal cycle. Animals that are active at night, such as bats and earthworms, are *nocturnal*. *Diurnal* animals, such as sparrows, butterflies, and most humans, are active during the day. *Crepuscular* animals, such as certain snakes and hawks, are active at dawn and dusk.

diurnalis = daily

25–14 The coyote (**a**) is diurnal. The osprey (**b**) hunts at dawn and dusk. The margay, or spotted cat (**c**), is a nocturnal animal with excellent night vision.

Biological Clocks

In many organisms, body rhythms continue even if the stimuli from the external environment are changed artificially. For example, many rhythms continue if the organism is placed in a lighted area for several days or is moved to a different time zone. Living things appear to have a sort of timer, or **biological clock**, inside the body that keeps their cycles going. An actual time-keeping structure has not been found in any organism, but some time-keeping mechanism is believed to exist.

Studies of leaf movement indicate that biological clocks also exist in plants.

Fiddler crabs have darkened skin during the day. At night their skin becomes lighter. Their color is darkest every fifteen days when the lowest low tide occurs during the day. In one study, crabs were kept in constant darkness. Their skin became dark and light just as it did in their natural environment. In another study fiddler crabs from the Atlantic Ocean were sent in dark containers by plane to the Pacific. In Pacific waters, the crabs' color changed according to the Atlantic tides. Both these experiments support the existence of biological clocks.

Our bodies also have diurnal rhythms. The bodily rhythms of the people living in a given time zone are in tune with local time. When people fly to a new time zone, their biological clocks are still working on the old schedule. The travelers find themselves hungry and sleepy at different times from usual.

Scientists have wondered what happens to people living under constant environmental conditions. In an experiment, individuals remained completely alone in special underground rooms for four weeks. Light and other environmental factors were kept constant, and each subject's activities and bodily functions were monitored. The people in the experiment slept, ate, and worked when they wanted. Under these conditions, each subject maintained an activity cycle that was nearly but not exactly 24 hours.

25-15 The daily color changes of fiddler crabs appear to be controlled by a biological clock.

Day coloration

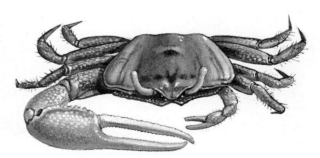

Night coloration

circa = about
diem = day

Circadian Rhythms

When most organisms are kept under constant conditions, they maintain many bodily rhythms that approximate 24 hours. A rhythm that repeats approximately every 24 hours is called a **circadian** (sur-KAY-dee-un) **rhythm**. Human sleep and wakefulness is an example of a circadian rhythm. When an organism returns to its natural environment, environmental clues help set the circadian rhythm at the more exact, 24-hour period.

All biological clocks are reset by stimuli in the environment. For example, people overcome jet lag as their biological clocks are reset by the day-night cycle in their new surroundings. Fiddler crabs transplanted from the Atlantic to the Pacific eventually reset their cycle of color change to Pacific time.

Photoperiodism in Plants

Photoperiod also affects animals. For example, poultry breeders regulate light in their henhouses for maximum egg-laying and growth in chickens.

Photoperiodism is the influence of light duration on an organism. For many years it was thought that the length of time a plant is exposed to light causes flowering. Scientists now know that it is the length of time a plant is exposed to darkness, not light, that affects flowering.

Certain plants flower only if they have a long uninterrupted period of darkness. These **short-day plants** include tobacco, soybeans, and chrysanthemums. Short-day plants will not flower if a flash of light interrupts their period of darkness. **Long-day plants**, such as spinach, flower only if they have a short period of darkness. A flash of light during the night can cause long-day plants to flower. Plant growers sometimes do this to force long-day plants to flower. Many plants are not affected by the length of the dark period and so are called **day-neutral plants**. They flower whether the night is long or short. Dandelions are day-neutral plants.

25–16 Photoperiodism.
a. The poinsettia, a short-day plant, blooms in the winter.
b. The long-day clover flowers in the summer.

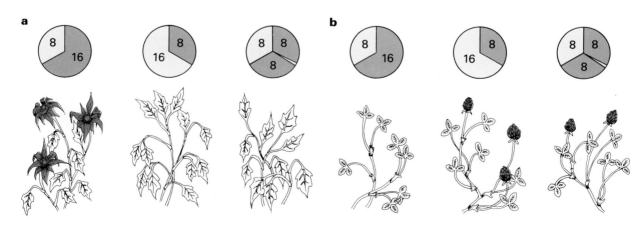

Migration

The spreading out of organisms into new territories is dispersal. Organisms that disperse do not return home as migrating ones do.

Some birds fly south in the winter and return in the spring. Large herds of wildebeests travel over the African plains in search of food from season to season. Schools of gray whales travel from the Arctic Ocean to the warm waters off the coast of California to breed. These journeys are examples of **migration**, which is the movement of organisms from one region to another and back to the original location. Migration is a complex type of instinctive behavior.

Robins, thrushes, warblers, geese, ducks, and other birds can be seen leaving their northern homes each autumn. They make this yearly voyage as food becomes scarce. When they travel south, birds find food for themselves and their young. Some birds travel thousands of kilometres.

Salmon and other fish are guided back to their home streams by smell. The fish detect organic molecules from the home stream, which have been carried to the ocean, and follow them to their source.

Fish are also known to travel great distances to spawn. Salmon, for example, hatch in freshwater streams. Some time after they hatch, they travel to the sea, where they live their adult lives. When they are ready to reproduce, males and females swim upstream to spawn. In the case of the Pacific salmon, the fish die after they spawn. Amazingly, their journey ends in exactly the same stream in which they hatched.

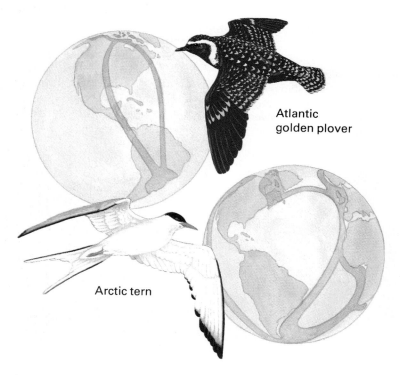

25–17 The Atlantic golden plover follows a southerly route in autumn to its winter range in South America, returning by another route in spring. The arctic tern has the longest journey of any bird species, a migratory route that sweeps past four different continents.

Introduction to Social Behavior

Social behavior in animals involves interaction between two or more individuals. A large number of moths attracted to your outside light in the summer is *not* an example of social behavior. It is just a group of moths. There is no interaction among them. Animals do things together during social behavior. Geese on a lake feed together, swim together, and sleep together.

Migration is also a type of social behavior.

Many species of animals display social behavior; however, it is mostly seen among insects and vertebrates. Some aspects of social behavior are inherited; others are learned. Whereas social behavior in insects is mostly inherited, much social behavior in vertebrates is learned.

In many species social behavior is limited to interactions involving mating and parental care of offspring. In other species social behavior is highly developed. Many insect and vertebrate species, for example, are organized into **societies**, which are populations in which members perform specialized functions.

Social animals often work together in finding food and defending the group. For example, when gulls in a breeding colony join together to attack a predator, they have a better chance of forcing the enemy away than an individual would. On the other hand, the breeding colony as a group is more apt to be noticed by a predator than an individual. It is important to note that organizing into societies involves trade-offs.

25–18 Humans are highly social organisms, as seen in this picture. An old-fashioned barn raising brought together members of many families in the community.

The Role of Communication in Social Behavior

A male tree frog sings from a pond on a spring night. A female tree frog, hearing his song from a distant tree, descends from her hidden position in the tree and approaches the pond in search of the male frog. Communication between individuals is a very important aspect of social behavior. When animals communicate, one individual sends messages and another individual receives the messages. The recipient of the message is able to interpret the sender's signals and respond to them. Several types of communication commonly occur in social behavior.

Communication refers only to signals that carry messages from one living thing to another.

Chemical Communication Many animals use chemicals as their main method of communication. Chemical signals produced by one animal and transmitted to another animal are called **pheromones** (FAYR-uh-mōn). The word *pheromone* resembles the word *hormone*. Both substances are secreted by glands. A hormone affects the individual that produces it. A pheromone affects another individual.

The chemical odor signal produced by female mammals in estrus is an example of a pheromone. The female releases the pheromone in her urine, and the male detects it by smell. The pheromone indicates the reproductive condition of the female to the male. Insects produce pheromones that act as sex attractants. Tiny amounts of these substances have powerful effects. As little as .01 µg of the pheromone produced by one female gypsy moth is enough to attract more than a billion male moths. In certain species of ants, a scout ant that finds a source of food uses its stinger to lay down a pheromone trail. The trail is detected and followed by other ants to the source of the food.

Pheromones help animals recognize members of their own species. They are also used by some species to signal danger.

25–19 Pheromones are widely used by ants. **a.** Ant laying down a pheromone trail. **b.** Ants follow a pheromone trail between their nest and a food source. Although the chemical evaporates, ants following the trail constantly renew it. As the last of the food is being gathered, fewer ants follow the trail.

25-20 Touch communication. Male and female banded coral shrimps touch feelers in a ritual greeting.

Humans also use touch for communication: a hug denotes affection; a nudge calls for attention; and a slap informs the receiver of the sender's anger.

Communication Involving Touch Touch signals are often used during mating. The female stickleback fish cannot lay her eggs unless the male touches her on the tail. A snail does not begin its mating movements until touched by its sexual partner. The mother herring gull delivers food to her offspring when it taps on her beak. Primates communicate by touch when they groom each other. During grooming, one animal uses its fingers to pick through the hair of another animal, cleaning off debris and parasites. Grooming has a calming effect and so reduces aggression between individuals in groups such as baboons, chimpanzees, and gorillas.

Touch signals are important in the development of young. Young primates, such as chimpanzees and baboons, require physical contact with their mothers in order to develop normally. If young primates do not receive proper touch signals from their mothers, they show abnormal behavior as adults. They do not respond normally to a mate or to their own offspring.

Communication Involving Sound and Hearing Communication by sound serves various functions. Voice signals help animals identify members of their species. Among members of species that move about in places where it is hard to see each other, sound communication helps keep the group together. The short squeals of piglike mammals called tapirs keep them "in touch" with each other in the dense rain forests of South America. Some scientists think that the "songs" of the humpback whale help keep groups of whales together during migration.

25-21 Sound communication. A Baird's tapir, a native of the rain forests of South America, communicates with squeals.

Section 25.2 Cyclic Behavior and an Overview of Social Behavior

Humans keep "in touch" by talking, seeing each other, and writing. People from all over the world communicate through newspapers, television, and radio.

Many animals use their voices to signal alarm. The vervet monkey uses three different alarm calls, depending on the kind of predator it sees. When the monkey sees a leopard, its alarm call sends all the monkeys in the area up into the trees. An alarm call for a snake causes monkeys to jump up on their hind legs and check the grass. When a monkey sees an eagle, the alarm call sends monkeys into dense vegetation.

Various animals use their voices for threat signals. A dog's bark and a male gorilla's roar deter intruders. Voice signals also play a role in reproduction. The spring chorus of frog sounds helps bring frogs into breeding condition at the same time.

Communication Involving Sight Many animals detect visual signals from the size, shape, and color of another animal's body. In this way they distinguish males from females and familiar animals from strangers.

Much communication in humans depends on sight. Facial expressions and body position reflect emotions and attitudes. Printed matter, drawings, paintings, photographs, and movies all make use of visual communication.

In addition to the way they look, animals send sight signals by the way they act. They move their bodies, take on certain positions, and change their facial expression. These behaviors communicate messages about danger, food, territory, and threat. Courtship and mating behavior also involve sight signals passed from one sex to the other. In mallard ducks, the female chooses a male as a mating partner. In order to be noticed by the female, the male must perform a series of specific movements. This breeding "dance" and the bright color of the male mallard are visual signals to the female.

25–22 Sight communication. Highly mobile faces of chimpanzees express many emotions, from the excitement conveyed by a pant-hoot (left) to a frustration yawn (right).

25-23 Like most social animals, the elephant relies on several senses for communication. Elephants use the position of their trunks, ears, and tusks, as well as their trumpetlike voices in communicating with members of their own and other species.

Communication May Involve Several Senses Animals detect signals of several types. When the silver-backed male gorilla issues a threat, he transmits sound and sight signals. First he emits loud hoots. Then he rises up on his hind legs and beats his chest. He runs along shaking and tearing at trees. Finally he thumps the ground with one or both palms. These outbursts are both seen and heard. Social animals generally use several senses for communication; however, they may use some senses more than others. Whereas social insects usually communicate with chemicals, vertebrates communicate mostly with sound and sight.

Building a Science Vocabulary

diurnal
biological clock
circadian rhythm
photoperiodism

short-day plant
long-day plant
day-neutral plant
migration

social behavior
society
pheromone

Questions for Review

1. What is a biological clock?
2. What are circadian rhythms?
3. Explain why chrysanthemums do not flower if a flash of light interrupts their period of darkness.
4. What is migration? What is an advantage of migration to birds?
5. What is a pheromone? Explain how pheromones help ants find food.
6. What major senses do animals use in communicating?
7. What two important functions are performed by social grooming?

25.3 SOCIAL BEHAVIOR IN INSECTS AND VERTEBRATES

Social Insects

Insects that form societies include termites, wasps, ants, and bees. A population of insects living together in one place make up a **colony**. In the social insects, most behavior is stereotyped and inherited. An insect's behavior benefits the group and not necessarily the individual. Cooperation among individuals is common; competition is rare. An insect colony works like a kind of superorganism, the parts of which are the individual insects.

The insects within a colony are divided into **castes**. A caste is a group of individuals that are physically different from other members of the colony. The members of a caste are physically adapted to perform certain tasks within the colony.

The males in an insect colony are specialized to produce sperm and fertilize eggs; however, they generally cannot feed themselves. The females in an insect colony are usually divided into two castes. The **queen** belongs to one caste. The queen has a very large abdomen and is larger than the other members of the colony. Her life is devoted to laying eggs. Her daughters belong to the second female caste, the **workers**, which cannot reproduce. Workers protect and provide for the queen. They raise the young and obtain food for the colony. Some workers have detachable stingers. These workers die when they sting an intruder in defense of the colony. The death of an individual for the benefit of the colony is part of the pattern of social behavior in insects.

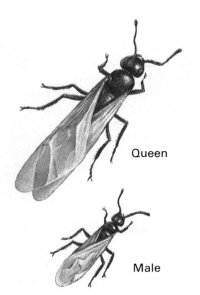

25–24 Insect colonies are organized into three castes, shown here in ants: queen, male, and worker. Members of an insect colony recognize castes but not individuals.

Social Behavior in Honeybees

A colony, or hive, of honeybees is made up of 20 000 to 80 000 bees and one queen. During the queen's life, which lasts five to seven years, she produces as many as 1000 eggs a day. The workers, which develop from fertilized eggs, are smaller than the queen and usually live about six weeks.

During her short lifetime, the worker bee performs many tasks. Her first priority is to feed the queen and other members of the colony, including the developing offspring. The worker's next function is as a housekeeper—or "hivekeeper," in this case. She produces wax, which is made into new cells. The cells are used to store food and to house the eggs laid by the queen. She also cleans existing cells and removes dead bees from the hive. The worker also protects the hive and gathers food.

25–25 Honeybees. **a.** Honeybee castes include the queen, the drone, and the worker. **b.** Bees in a hive attend to the queen in the center, grooming and feeding her.

25–26 Bees feed by sucking honey that is stored in the cells of the hive.

The male caste within a honeybee colony consists of **drones**, which are haploid male bees that develop from unfertilized eggs. A drone's only function is to produce sperm to fertilize eggs produced by the queen.

Life Within the Colony Life in the hive is always active. Workers build new cells, feed members of the hive, tend the developing young, and search for food. The queen lays eggs. Some of the eggs are deposited in special wax cells. The bees developing in the special cells are fed a substance called **royal jelly**. The special diet transforms these bees into queens.

During the spring, the hive becomes overcrowded with workers and drones. The queen bee leaves the existing hive with about half the workers and starts a new colony in a different location. Once the queen leaves the hive, one young queen-to-be emerges in the original hive. She produces a pheromone that signals drones in her own and other hives that she is ready to mate. She leaves the hive and mates with many drones. During mating she receives enough sperm to last her lifetime. She stores the sperm in her abdomen. A drone dies after mating, but the new queen lays unfertilized eggs to produce more drones.

When fall comes and food supplies dwindle, the useless drones are killed by workers or are driven out of the hive and left to starve. Over the winter, the honeybees cluster together in the hive and move their legs, wings, and abdomens. This produces heat, which keeps them warm. They eat honey stored in the hive. When spring returns, the cycle starts over again.

Section 25.3 Social Behavior in Insects and Vertebrates

Our understanding of the bee's remarkable system of communication is based on studies done by the famous Austrian biologist, Karl von Frisch.

Communication When a worker bee finds food, she returns to the hive and communicates information about the food to the other workers. First, she performs a dance that tells the other bees exactly where the food is. When the food is less than 100 metres from the hive, the worker does a **round dance** on a vertical side of the hive. She circles one way and then the other. The faster she dances, the larger the source of food. Other workers soon join in the dance. The worker who started the dance also carried the odor of the food source on her body. The message from the round dance and the odor of the food stimulates other workers to leave the hive to find the food.

When the food is more that 100 m from the hive, the worker does a **waggle dance**. The waggle dance, like the round dance, is performed on a vertical side of the hive. The worker first dances in a straight line while wagging her abdomen. Next, she turns to the left and dances a half circle back to where she started. Then, she dances along the straight line again. This time she turns to the right and dances a half circle back to the beginning. The pattern is repeated over and over. The direction of the straight line in the dance communicates where the food is in relation to the sun. The time the worker takes to dance the straight line signals how far away the food is from the hive. The farther away the food is, the longer it takes for her to dance the straight line. Other workers join in the waggle dance. Once they learn the distance and direction to the food source, they fly directly to the food.

Chemical communication is also important in a honeybee colony. The queen produces a chemical called **queen substance**.

25–27 Where a food source is less than 100 m from the hive, a scout bee will perform a round dance (**a**). When a food source is more than 100 m from the hive, a scout bee will perform a waggle dance. Usually, the waggle dance is performed inside the hive (**b**). Occasionally, a scout bee will perform the waggle dance on a horizontal surface outside the hive (**c**). In this case the straight line of the dance points directly to the food source.

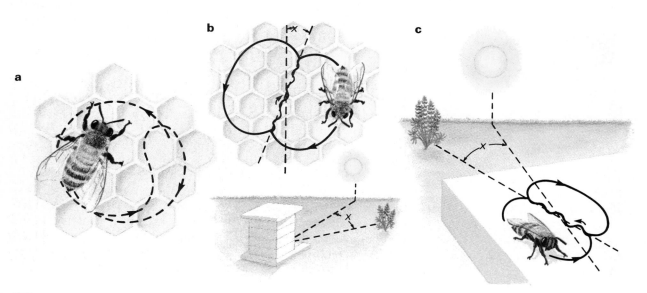

720   Chapter 25 Behavior

Chemical communication is also important in helping bees identify each other's caste.

The workers lick queen substance off the body of the queen bee and transfer it to all the other bees in the hive by exchanging the contents of their stomachs. A pheromone in queen substance prevents workers from reproducing and prevents production of other queens. When queen substance is not produced, the colony members become restless. The absence of queen substance is known throughout the hive within about 30 minutes. By that time, the workers begin preparations for a new queen.

Social Behavior in Vertebrates

Each member of a vertebrate society is adapted to survive as an individual and as a member of the group. Individuals within a vertebrate society form small subgroups, including paired mates, parents and offspring, brothers and sisters, and other relationships. Individuals and subgroups keep their individuality even though they are part of the larger society.

Learning plays an important role in the social behavior of vertebrates. As a result, their social behavior is more flexible than that of social insects. An ant responds to an alarm pheromone in a stereotyped way. It raises its antenna, opens its jaws, walks, and then runs toward the disturbance. A rhesus monkey, however, quickly perceives different aspects of an alarm situation. It learns whether the alarm is due to something within the social group or is due to an outside danger and behaves according to the particular type of alarm.

Cooperation and competition are balanced in vertebrate societies.

Vertebrate societies show more competition and aggression than insect societies. Too much aggression, however, is a threat to species survival. Accordingly, you can observe adaptations among vertebrates that reduce aggression, including rituals, territoriality, and social dominance.

25–28 Aggression in Japanese macaques is expressed between rival males (left) as they compete for leadership of the troop.

Male deer use their antlers in ritual fighting, which is largely ceremonial. A winner is established, yet both contestants survive.

Rituals are also seen during courtship and mating. There they reduce aggression between mating partners.

Rituals Rituals are a form of inherited social behavior found among many species. They are performed in an exaggerated way and are often used to convey a message during a dispute. Rituals help animals assess each others' strength and prevent them from hurting each other. When a dog sends a message of aggression to another animal, it stands with its tail high, legs straight, ears up, and eyes staring at its opponent. Even its hair stands on end. The threatened animal may respond by crouching. By this behavior it submits, and the contest is over. The winner walks away, and neither animal is hurt.

Territoriality A **territory** is an area occupied and defended by one or more animals. The behavior an animal exhibits while occupying and defending its territory is called **territoriality**. The territory of a small male fish known as the pike blenny is a burrow or hole that a worm left behind. A ritual fight occurs when another male pike blenny approaches. A wolf pack occupies an area at least 130 km² in size. Wolves mark their territory with urine. A pheromone in the urine is an odor signal that lets other wolves know to stay out. The howling of a wolf pack reveals the presence of the pack and its size.

Once an animal or group of animals establishes a territory, other animals usually stay away. In this way, territoriality reduces aggression and competition. It also keeps the size of the population in a territory balanced with the available food.

An animal must be strong to establish and keep its territory. Weak animals are driven out of existing territories. Without nesting sites, weak animals have difficulties finding mates and breeding.

25–29 Oryx gazelles lock horns in a ritual display (**a**). A red-winged blackbird indicates its territory by puffing up its feathers and calling from a high perch (**b**).

25–30 Social dominance in wolves. The wolf on the left, with its teeth bared, ears back, and head thrust forward, has established its dominance over the wolf on the right, which is holding its head down and withdrawn into its neck in a submissive posture.

Social dominance reduces aggression and competition. It also ensures that strong parents produce the next generation. These benefits help species survive.

Social Dominance In many vertebrate societies, some individuals are dominant, or more powerful and important than others. In fact, the society is organized into "power" groups. Within such a society, a dominant individual or group of individuals wins ritual as well as real fights. Dominant animals maintain order and protect the group. They eat and are groomed before others and have first selection of mates and nesting sites.

The **pecking order** seen in chickens is an example of social dominance. When a flock of hens is put in a pen, they threaten each other and fight. This behavior establishes a ranking order. One hen dominates all the others: she pecks every other chicken and does not get pecked in return. Another hen pecks every chicken except the first one. This hen is second in the rank. The third hen in the pecking order pecks all but the first two. All the hens in the flock are ranked in this way. The last member of the pecking order gets pecked by everyone and pecks no one. The pecking order determines how food, living space, and mates are obtained. Once established, it eliminates fighting. If a new hen is added to the group, a new pecking order is determined.

The Social Life of Baboons

Baboons are large primates that inhabit the grasslands of Africa. They live in organized social groups called **troops**. Baboon troops range in size between 10 and 200 baboons. In general, a baboon stays in the same troop from the time of its birth until its death. An individual baboon is almost always within a few feet of other troop members.

Unlike most other primates, baboons live on the ground, not in trees. During the day, the baboon troop searches for food, which consists of leaves, roots, stems, fruits, and flowers. Baboons do not normally eat meat or hunt for prey. After searching for food, the baboon troop returns to the nesting site near trees or cliffs to spend the night.

Organization Within a Baboon Troop Adult males weigh about 55 kg and are nearly twice the size of adult females. They also have large canine teeth, which they display during courtship and ritual behavior.

A baboon troop is organized by social dominance. One adult male is the most dominant individual. Other adult males are ranked below him. The most dominant male has first choice of mates and nesting sites.

Other dominant male baboons in the troop stay close to the most dominant male. If fighting breaks out within the troop, the group of dominant males stops the fight by ritual behavior, including staring, yawning, lowering ears, and raising eyebrows. This ritual communicates that fighting is to cease. If the ritual does not work, the male will resort to a harmless bite. The actions of the dominant males maintain peace and order in the troop.

Baboon females are also ranked by social dominance. When a dominant female is in estrus, only the most dominant male mates with her. Reproduction by the dominant members of the band increases the likelihood of strong, healthy offspring.

When a troop moves over open ground, females and babies stay in the center close to the dominant adult males. Less dominant adult males walk on the outside. The females and young are protected from attack by leopards or other predators. A baboon troop defends its territory. Baboons that try to intrude are sent away with ritual calls, showing of teeth, and other threat gestures.

25–31 Baboon troop in Kenya. **a.** The troop segregates into subgroups. **b.** A dominant adult male in the troop threatens a newly arrived male.

25–32 Among the subgroups within the troop are found an adult male and female grooming pair (**a**) and a mother with her young (**b**).

Subgroups Within a Baboon Troop If you watch a baboon troop at rest, you can see small groups of individuals grooming each other. These **grooming clusters** frequently stay together while the troop is walking. A mother with a newborn infant is often the center of a grooming cluster. The mother and infant are groomed by other adult females and young baboons. Other grooming clusters are made up of pairs of adult females or young baboons in play groups. In grooming clusters, adult females usually groom the males and other baboons in the troop.

The mother-child relationship forms a very important subgroup in baboon societies. The infant clings to the mother's fur as soon as it is born, riding along as she moves about. Other members of the troop are attracted to the newborn. Females come over and pay it lots of attention. Even the dominant male sits close by, walks along, and protects the mother and her infant. At first the infant stays with the mother 24 hours a day. As it grows older, it rides on her back and leaves her for periods of time. When not with its mother, a young baboon learns skills and behavior used in adult life from adults and its play group.

Communication Within a Baboon Troop Baboons use sounds and gestures to communicate threat, attack, fear, escape, and other situations. When startled by a snake, a baboon stands up on its hind legs, spreads its arms and issues a shrill warning bark. In a difficult situation, a baboon may gaze off in the distance, scratch its back or shrug its shoulders. This behavior seems to be an attempt on the part of the animal to stall for time. After not seeing each other for a while, friendly baboons grunt, lip-smack, and may even embrace. Communication establishes bonds between members of the troop and keeps them in touch with each other. It also aids in defense, helps keep order, and helps individuals learn their role in the social order.

Applying Information About Social Behavior in Animals

Learning about social behavior in animals helps us understand the behavior of animals with which we interact. In addition, scientists wonder how social behavior in animals relates to humans. Some scientists think that certain elements of vertebrate social behavior, such as territoriality, can be found in human behavior. Unfortunately, it is very difficult to draw conclusions about human behavior. Social behavior in animals largely depends on heredity. Human behavior depends a great

25-33 A human imitation of a whooping crane courtship dance was used to bring a female raised away from other cranes into reproductive readiness. The crane eventually joined in the dance.

deal on self-awareness and conscious choice. It is true that primates and other vertebrates behave in a more flexible way than do insects. But human behavior is still far more flexible. Baboons do not analyze their own behavior or that of other animals. They neither choose the nature of their social organization, nor do they make decisions that change their society. Humans, however, control their social interactions by deciding how they want to act and carrying out their decisions.

Building a Science Vocabulary

colony	royal jelly	territory
caste	round dance	territoriality
queen	waggle dance	pecking order
worker	queen substance	troop
drone	ritual	grooming cluster

Questions for Review

1. What is the function of the queen in a honeybee hive? a drone? a worker?

2. Give an example of communication by pheromones within a colony of honeybees.

3. What is communicated by the round dance? the waggle dance?

4. What are the major differences between insect societies and vertebrate societies?

5. In a vertebrate society, how does territoriality help keep peace from without and social dominance help keep peace from within?

6. State one important way that social behavior aids survival of baboons.

7. Describe two subgroups within the baboon troop.

Perspective: Discovery
A Woman in the Wild

Instead of a fairly comfortable life in a laboratory, British biologist Jane Goodall took to the wilds. As a young woman, she travelled to the Gombe Stream Reserve on the shore of Lake Tanganyika in eastern Africa to study chimpanzees in their natural habitat. Dr. Goodall was alone in this tropical jungle with but a few camp helpers. In her work in the jungle she was able to show that much of what had been observed about chimpanzees in zoos and laboratories was wrong.

Study of the great apes—orangutans, gorillas, and chimpanzees—provides an opportunity to study human-like patterns of behavior. Since the habitats of these primates is increasingly being destroyed by human activity, they are endangered species. Studies like Dr. Goodall's become all the more important.

During the first ten years of her work, Dr. Goodall observed social bonds, which were especially strong between a mother and her offspring. Just as in humans, recognition of family members lasts a lifetime. Jane Goodall even learned how to recognize all of the chimpanzees she observed and gave them names, just as we do.

She also observed that chimpanzees have body language—they stare, kiss, and hold hands, just as we do. And, they communicate with hoots, pants, and howls. She found that they use tools. For example, they could remove the leaves from a twig and insert the twig into a termites' nest like a probe. When the twig is withdrawn, the termites sticking to it are then eaten.

She also found that chimpanzees hunted and ate small monkeys. Previously it was thought they were pure vegetarians. And, desire for food—especially bananas—became the source of severe competition and fighting. Even more violent behavior in defending territories was also observed.

Dr. Goodall's work was summarized in a wonderful book, "In the Shadow of Man" (1971, Houghton Mifflin Co.). Her work continues today in a well-known research field station. The station is a well deserved recognition of a superb scientist, a woman in the wild.

Chapter 25

Summary and Review

Summary

1. An organism's behavior is made up of responses to stimuli in its internal and external environments. Behavior aids the survival of the organism and the species.

2. Plant behavior consists of tropisms, which are growth movements toward or away from a stimulus.

3. Inherited behavior in protists and animals is automatic and predictable. It includes kineses, taxes, reflexes, and instincts. Specific stimuli called releasers are believed to cause the expression of inherited behaviors.

4. Learning is a change in behavior that is based on experience. Learned behavior includes habituation, imprinting, classical conditioning, and operant conditioning.

5. Memory is the process of storing and retrieving learned information. Memory may be short-term or long-term. Reasoning uses memory and learning to solve new problems.

6. Certain behaviors cycle with environmental cycles. Mechanisms called biological clocks cause many cyclic behaviors to persist in animals even when they are placed in constant environmental conditions.

7. In migration, birds, fish, and other animals travel great distances at certain times of the year and later return to their original locations.

8. Social behavior involves interactions among animals of the same species. It occurs during courtship, mating, parental care, feeding in groups, defense, and migration.

9. Animals communicate using chemical signals, touch, sound, and visual signals. Most animals use more than one type of signal when they communicate.

10. Most social behavior in insects is inherited and stereotyped. In insect societies, the individual's behavior benefits the group. Most insect societies are made up of different castes.

11. In a honeybee society, the queen lays eggs, the workers perform the chores of the hive, and the drones fertilize eggs. Honeybees communicate the location of food by performing a dance.

12. In vertebrate societies, the individual is more important than in insect societies. Learning plays a greater role and aggression and competition occur more frequently in vertebrate societies than in insect societies. Rituals, territoriality, and social dominance help reduce aggression and competition.

13. Baboon societies are organized into troops. A baboon troop protects its members, shows territoriality, and is organized according to social dominance.

14. It is difficult to apply information about the social behavior of animals to humans. Human social behavior involves a great deal of self-awareness and conscious choice, factors not found in other animal societies.

Review Questions

Application

1. In a certain species of blind ants, a scout finds a source of food. Even though the workers of this species are blind, they are able to locate the food that the scout found. How are they able to do this without seeing where they are going?

2. Female dogs in estrus produce a pheromone that male dogs detect. How does the pheromone act as both an attractant and a releaser?

3. Thousands of grunion (small fish living off the coast of southern California) spawn in the spring on three to four nights when there is a new full moon and the tides are highest. On those nights, females come up on the beach between waves and deposit their eggs in the sand. The males then fertilize them. State the cause, function, and kind of behavior exhibited by the grunion.

4. Morning glory flowers open in the morning and close later in the day. Design an experiment to determine whether this behavior is controlled by a biological clock.

5. Regardless of the age of the plants, species of soybeans flower only in September. Why does this occur?

Interpretation

1. An animal behaviorist repeatedly observed the following behavior between two chimpanzees named Olly and Mike. "When nervous Olly greets Mike she may hold out her hand toward him, or bow to the ground, crouching submissively with downbent head." What does Olly's behavior communicate to Mike? What is the benefit to Olly of this behavior?

2. If you were watching a chimpanzee society in the forest, you might observe that when a chimpanzee finds food, it utters loud barks. How would you explain this behavior? What comparable behavior can be found in honeybees?

3. What type of symbiosis could be considered social behavior? Give an example.

Extra Research

1. Research the migration pattern of the California gray whale (*Eschrichtius gibbosus*). Draw a map of the areas the whales inhabit, showing the path of whale migration. Find out the functions of their migration.

2. Design and carry out an experiment to determine whether gravity or water is the stronger stimulus on the behavior of plant roots.

3. Research the social organization of army ants, termites, chimpanzees, or dolphins. Describe the society in writing and with drawings.

4. Use classical conditioning to train a goldfish to swim to the top of its bowl when you appear.

5. Test your own memory. Using a dictionary or the glossary of this book, write a list of ten unfamiliar words. Memorize them and see how many words you remember the following day. Then, make up a second list of ten unfamiliar words. For each word, look up the definition and derivation, and use each word in a sentence. Try to associate each word with something that you already know. Test your memory of the second list of ten words the following day. Compare your results from both experiments. What conclusions can you draw? Devise and perform additional experiments to test your long-term memory.

Career Opportunities

Psychologists study human behavior and mental processes. Many work to help people improve their mental health. Others teach, do research, or help solve institutional problems. A master's degree is usually required. For information, write the Canadian Psychological Association, 558 King Edward Ave., Ottawa, Ontario K1N 7N6.

Commercial artists use their knowledge of human behavior to design attractive and effective books, advertisements, displays, brochures, and packages. Some work independently; others work for advertising agencies or private companies. Design school or free-lance experience is required. Write the Graphic Arts Industries Association, 75 Albert Street, #906, Ottawa, Ontario K1P 5E7.

Chapter 26

Health and Disease

Focus *Good health depends on good nutrition, plenty of rest, and the proper functioning of the body's defense mechanisms against disease. Through the study of medicine, we learn much about the nature of many diseases, including how they are caused, cured, and prevented.*

26.1 CAUSES OF DISEASE

When most people get sick, they want to know, "What caused my illness?" and "How can I get better?" These are the essential questions about disease. The causes of disease fall into three main categories: body malfunctions; environmental factors, including tobacco, alcohol, and drugs; and infectious organisms.

Body Malfunctions

The word *malfunction* means failure to work or function. When someone has a body malfunction, a part of the body does not work properly. Body malfunctions come about in several ways.

Damaged Body Parts A part of the body can be injured by an accident, by infection, by tissue breakdown due to aging, and by harmful substances in the environment. An injured body part

26-1 A foundation of good nutrition, rest, and exercise in childhood is important for resistance to disease as an adult.

may suffer severe damage and fail to work properly, resulting in a disease. **Emphysema** (EM-fuh-SEE-muh) is an example of such a disease. People with emphysema have damaged lungs. The damage may be caused by smoking or certain kinds of air pollution. The lung impairment prevents sufferers of emphysema from taking in sufficient oxygen.

Hereditary Disorders A hereditary disorder occurs when someone inherits a faulty gene or does not inherit the correct number of chromosomes. Sickle-cell anemia and hemophilia are hereditary disorders of the transport system. Phenylketonuria is a hereditary disorder of the nervous system, and cystic fibrosis is a hereditary disorder of the respiratory system.

Endocrine Malfunction Disease may result when an endocrine gland does not secrete the proper amount of a hormone. If the pancreas secretes too little insulin, diabetes mellitus results. If the pituitary gland does not secrete the right amount of growth hormone, giantism or dwarfism result.

Immune System Malfunction The **immune system** is the body's chemical defense mechanism. If it is not working properly, disease can result. Allergies are diseases that happen in people whose immune system is not functioning properly. In certain individuals, the immune system attacks the person instead of the disease. Such action on the part of a malfunctioning immune system damages the body. Scientists believe that rheumatoid arthritis, a disease that affects the body's joints, is due to the improper working of the immune system.

26-2 Cancerous lung tissue appears as a dense white mass at the top of this photograph.

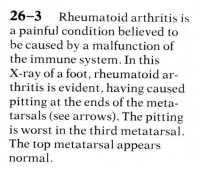

26-3 Rheumatoid arthritis is a painful condition believed to be caused by a malfunction of the immune system. In this X-ray of a foot, rheumatoid arthritis is evident, having caused pitting at the ends of the metatarsals (see arrows). The pitting is worst in the third metatarsal. The top metatarsal appears normal.

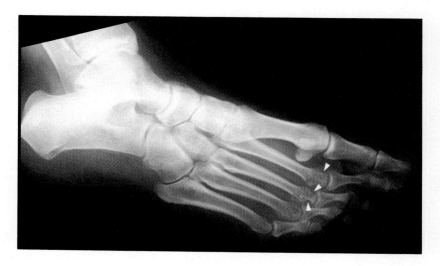

Aging Aging is a natural process and is not itself a disease. Many old people are healthy; however, old people have a greater chance of getting sick than young people. As a person ages, the number of body cells decreases. Body systems do not work as efficiently as they did when the person was younger. There are more cases of hardening of the arteries (atherosclerosis), heart disease, senility, and cancer among the aged than in young people. Age is not the direct cause of these illnesses. Medical researchers are trying to find the link between disease and the aging process in hopes of finding ways to keep more people healthy when they get old.

Senility is a type of intellectual impairment accompanied by memory loss.

Environmental Factors

The environment provides food, air, water, shelter, protection, and other necessities. Yet some environmental factors are harmful and can result in disease.

Pollution Pollution is considered a major health hazard. Burning fuel for vehicles, homes, and industry can pollute the atmosphere. Air pollution is one cause of respiratory illness. Water and soil can be polluted by waste products of industry, sewage, fertilizers, and pesticides. Chemical pollutants in soil and water can harm people directly when they seep into the water supply. They can also reach humans and cause disease through food chains. These pollutants can act as poisons or cause diseases such as cancer. Radioactive substances can be dangerous when released into the atmosphere or accidentally spilled onto land or water. Exposure to large amounts of radiation can result in radiation burns, cancer, and impaired reproductive function.

Pollution control measures seek to limit the amount and kinds of pollutants that reach the environment.

Improper Sanitary Conditions The proper disposal of human wastes is vital to maintaining health. Without proper sewage treatment, water supplies can become contaminated with disease-causing parasites. Disease can become widespread under unsanitary conditions. Typhoid fever is one disease caused by a type of bacteria that is transmitted by contaminated water. Cholera and amebic dysentery are also spread in contaminated drinking water.

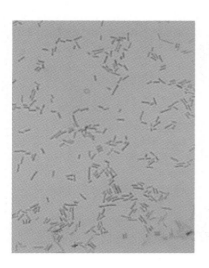

26-4 Micrograph of the bacterium *Salmonella typhosa*, which causes typhoid fever. The disease, which can be fatal, affects the intestines, causing fever and discomfort usually lasting two to three weeks.

Section 26.1 Causes of Disease

Lack of Medical Aid Many diseases can be prevented as well as treated with medical aid. Medical aid involves regular checkups, vaccinations, and information about staying healthy. In communities that have no connection to a health-care facility, women during childbirth and newborn babies are at risk. There is also a greater chance of diseases spreading quickly.

Improper Diet An improper diet often results in malnutrition, which can cause disease. Malnutrition comes from dietary shortages and from dietary excesses. Scurvy, pellagra, and anemia result from nutritional deficiencies. Obesity, hardening of the arteries, and certain types of heart disease are linked to eating too much of the wrong kinds of food.

Stress Stress is a stimulus, such as worry or pain, that upsets the body's equilibrium. Many conditions in modern society seem to cause stress. The human body can overcome a certain amount of stress; however, some people are unable to adapt to the stress in their lives. For them, stress can cause disease or can aggravate another disease that they have. High blood pressure, ulcers, and emotional disorders are examples of diseases that are worsened by stress.

26–5 The traffic jams experienced by drivers can lead to high levels of stress on a regular basis.

Tobacco, Alcohol, and Drugs

Tobacco, alcohol, and drugs are environmental factors that can cause severe illnesses. In each case, a person suffers from a disease, the cause of which is self-administered. These diseases have harmful effects on the body and affect millions of people.

Tobacco Cigarette smoke contains over 1000 poisonous, or toxic, substances, many of which cause cancer in laboratory animals. **Nicotine** (NIK-uh-TEEN), which is usually present in high concentrations in cigarette smoke, affects the nervous system. It has a stimulating effect at first and then a depressing one, setting up a need for habitual smoking.

Cigarette smokers have a 30 percent to 80 percent higher death rate than nonsmokers, depending on how long and how much they smoke. The increased death rate of smokers is due to lung cancer, emphysema and other respiratory ailments, and heart disease. Smokers also have a higher incidence of stomach ulcers and cancer of the esophagus, trachea, bladder, pancreas, and kidneys than nonsmokers. Regardless of the age of the smoker, smoking impairs every aspect of lung function. Furthermore, when a pregnant woman smokes, it can have harmful effects on her unborn child.

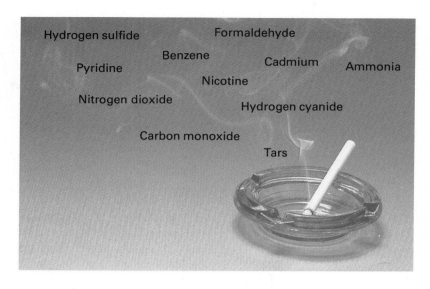

26-6 Cigarette smoke contains many harmful substances, including carcinogens, such as tar, and toxic gases, such as carbon monoxide. When inhaled, these substances can cause serious damage to lungs and other organs.

Smoking is not just a personal problem. It also pollutes the atmosphere and causes nonsmokers to inhale toxic fumes. It has been shown that the breathing of cigarette smoke by nonsmokers is harmful to their health. Smoking creates another social problem as well: lit cigarettes cause fires. Careless smokers can cause destruction of forests, loss of property, injury to people, and loss of life.

Alcohol The alcohol that people drink is ethanol, which is a product of fermentation. Alcohol is a depressant of the nervous system. It affects the parts of the brain that control judgment and memory, usually giving a person a feeling of release. Alcohol also reduces coordination and skill. An increased intake of alcohol may result in confusion and hallucinations and can also bring about mental breakdown.

More than half the deaths resulting from highway accidents in North America are caused by drunk drivers.

Besides its effects on the nervous system, alcohol increases the heart rate and widens the blood vessels near the skin. The increased diameter of blood vessels near the skin gives the drinker a feeling of warmth, while he or she is actually losing body heat. Because of its effects on the transport system, alcohol can be especially dangerous to someone with a heart condition.

Alcohol can also irritate the stomach lining and help cause digestive disorders. Prolonged use of alcohol can result in deterioration, or cirrhosis, of the liver. Alcohol is absorbed directly through the stomach wall into the bloodstream. When it reaches the brain, it has an intoxifying effect. Alcohol supplies calories, but not essential vitamins, minerals, or other nutrients. When it is used in place of food, deficiency diseases result. Alcohol is particularly dangerous when it is used with other drugs.

26-7 Alcohol and narcotics have a variety of effects on the body, as does withdrawal from drug addiction.

Alcohol	Narcotics	Withdrawal
Loss of judgment	Feeling of well-being (euphoria)	Irritability
Loss of self-control	Slurred speech	Depression
Slurred speech	Change in emotions	Anxiety
Change in emotions	Can cause unconsciousness or death	Confusion
Blurred vision		Insomnia
Loss of depth-perception		Tremors
Can cause unconsciousness or death		Physical pain

Alcoholism is a disease in which a person is physically and emotionally dependent (addicted) on alcohol. There are 2 to 9 million alcoholics in the United States alone. The causes of alcoholism are complex, involving heredity as well as environment. Alcoholics suffer from their addiction. Their health becomes impaired, they cannot perform their jobs properly, and their family and social lives usually suffer. If the alcohol supply is suddenly stopped, an alcoholic becomes physically ill, a condition known as **withdrawal sickness**. Withdrawal sickness is severe and hard to live through. For this reason it is very difficult for an alcoholic to stop drinking.

Drugs Drugs are substances used in medicine. When prescribed by a physician, drugs can help a patient regain physical or mental health. Certain drugs are prescribed to help a patient sleep or to reduce pain. Some people administer drugs to themselves for other than medical reasons. These people may become **drug abusers**. There are different reasons why people become drug abusers. Some want relief from their problems; others do it for fun.

Drug abusers use various chemicals, all of which can have harmful physical and mental effects. These chemicals include **barbiturates** (bar-BICH-ur-its), which are used to induce sleep, and **amphetamines** (am-FET-uh-MEENZ), which are used to keep people awake. They include **mind-altering drugs**, such as marijuana and LSD, which are used to change perceptions, thoughts, and feelings. They also include **narcotics**, such as heroin and cocaine, which are used to deaden pain and produce a temporary feeling of pleasure.

Repeated use of many drugs can cause addiction. The addiction may involve physical and emotional dependency. Once dependent on a drug, a user may take an overdose, which can cause death. If an addict's supply of the drug stops, the resulting withdrawal sickness can be extremely severe.

The word addiction *may also be used to describe a person's inability to control consumption of caffeine, nicotine, or alcohol.*

Infectious Organisms

Infectious organisms are what most people call "germs." They are parasites that live, grow, and reproduce inside a host. Under certain conditions, they make the host sick. Infectious organisms that make their host ill are called **pathogens** (PATH-uh-jenz). Sometimes people are host to infectious organisms, but they do not get sick. Many people carry the bacteria that cause strep throat, for example; yet, they usually remain in good health. If something upsets the body balance, such as a chill or lowered resistance, the disease may take hold.

Most infectious diseases can be transmitted from person to person. Such diseases are said to be **communicable** (or contagious). A disease is communicable when the parasite is adapted for dispersal from host to host. Some parasitic bacteria, for example, produce spores that can survive outside the host's body. The disease spreads when the spores are dispersed to a new host. Parasitic bacteria that are not adapted for dispersal remain in the host's body. The diseases they cause are not communicable. Tooth decay is an example of a disease that is infectious but not communicable.

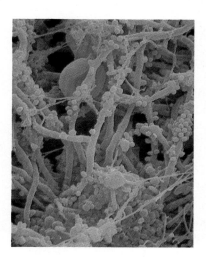

26-8 Tooth plaque bacteria are visible in this micrograph as small beads.

Building a Science Vocabulary

emphysema	drug	narcotic
immune system	drug abuser	pathogen
nicotine	barbiturate	communicable
alcoholism	amphetamine	
withdrawal sickness	mind-altering drug	

Questions for Review

1. State the cause and the body part affected in each of the following diseases.
   a. emphysema   b. cystic fibrosis   c. giantism

2. Name one disease that is caused by each of the following.
   a. air pollution   c. malnutrition
   b. improper sanitary conditions   d. cigarette smoking

3. How do drug abuse and drug addiction differ?

4. When is an infectious disease communicable?

Section 26.1 Causes of Disease

26.2 INFECTIOUS DISEASE AND BODY DEFENSE

The Spread of Communicable Diseases

Communicable diseases may be transmitted directly from person to person or indirectly by means of some living or nonliving thing. Pathogens exit the body of one host and enter a new host's body through a break in the skin or through a body opening. The nose, mouth, reproductive openings, and excretory openings are possible entrances for infectious organisms. The transmission from host to host occurs in several ways.

Direct Physical Contact Some pathogens are transmitted when someone touches a person with the disease. Venereal diseases, such as syphilis and gonorrhea, are transmitted by direct physical contact during sexual intercourse. Viruses that cause cold sores, warts, and type B hepatitis (a liver disease) are also transmitted by direct physical contact.

Type B hepatitis is also called serum hepatitis. It can be transmitted on an infected needle during an injection.

Droplet Infection Certain pathogens are adapted to travel through the air on dust particles or in water droplets. Large numbers of pathogens travel from host to host on droplets expelled during a sneeze or cough. A person sends out about 20 000 droplets during a sneeze. Cold viruses and other pathogens can be picked up by a new host as far away from the source as five metres.

Contaminated Food and Water Some parasites are adapted for dispersal in body wastes. These may get into food or water in places where sanitary conditions are poor. They are transmitted when a new host ingests contaminated food or water. Typhoid fever is transmitted in this way, as is type A hepatitis (infectious hepatitis). Certain parasitic worms, such as tapeworm and trichina worm, live in animals other than humans for part of their life cycle. When these animals are eaten by humans, the pathogenic worms can be transmitted to a new host.

The type A hepatitis virus can also be transmitted on objects contaminated by feces.

Contaminated Articles Some pathogens are transmitted on objects that a host touches. Eating utensils, clothing, combs, facial tissues, and napkins can all be contaminated. The fungus that causes athlete's foot can be transmitted from host to host by a bare floor covered with fungal spores.

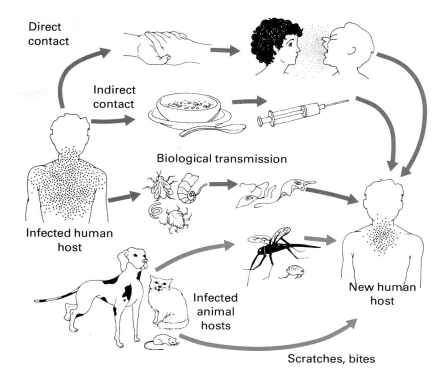

26–9 Transmission of disease from an infected host to a new host may occur by direct contact, indirect contact, or biological transmission.

Insects Insects may also transmit disease organisms. The bacteria that cause bubonic plague, which was responsible for the deaths of many millions of people during the late Middle Ages, are carried by fleas that live on rodents, particularly rats. The pathogen enters a new host when the flea bites a person. Yellow fever, caused by a virus, is transmitted by the bite of the *Aedes* mosquito. Malaria, caused by a protozoan, is transmitted when the *Anopheles* mosquito bites a person.

Epidemics Body defenses and health measures help control the spread of pathogens. Nevertheless, under certain conditions widespread transmission of an infectious disease, known as an **epidemic**, occurs. Millions of people can be affected in an epidemic. During World War I, more than 200 million people suffered from influenza ("the flu"), a disease caused by a virus. About 400 000 people died of it within a period of six months. Epidemics can quickly spread from country to country, and their control depends on worldwide cooperation.

26–10 "Bring out your dead" was a call often heard during the Bubonic plague epidemic that ravaged Europe in the late Middle Ages.

Section 26.2 Infectious Disease and Body Defense

Types of Disease Organisms

Most pathogens are microscopic. Bacteria, viruses, rickettsia, disease-causing fungi, and some worms are too small to be seen by the naked eye. A type of tapeworm, on the other hand, can grow to a length of 15 m in the intestine. Regardless of their size and structure, all pathogens are parasites that damage their hosts.

Bacteria Certain species of bacteria produce spores that are surrounded by a unique wall that is almost impossible to destroy. Bacterial spores can withstand boiling, freezing, and drying. Bacterial spores help disperse the organisms and help them survive for long periods of time even under unfavorable conditions.

Bacteria cause many different diseases in humans, including tuberculosis, typhoid fever, cholera, bacterial pneumonia, scarlet fever, strep throat, tetanus, syphilis, gonorrhea, diphtheria, staphylococcus infection, and bubonic plague. They also cause tooth decay and gum disease.

Many bacteria harm their hosts by producing poisons called **toxins**. The bacteria that cause tetanus produce a toxin that interferes with synapses between nerve cells in the central nervous system. The toxin causes several muscle spasms. Diphtheria bacteria live in the lungs and produce a toxin that is carried in the blood. When the toxin reaches the body's cells, it destroys the cells' ability to produce protein, which can eventually kill the host.

In some diseases, the host's response to the bacteria causes damage. Lung cells produce large amounts of fluid when they are invaded by pneumonic bacteria. As a result, a person with pneumonia has difficulty breathing.

Viruses The protein coat of a virus is adapted to attach to a specific place on a host's cell membrane or cell wall. Cold viruses, for example, attach only to certain areas of mucous membrane cells in the respiratory system. Once attached to the outside of a host cell, the viral nucleic acid core can penetrate and enter the cell. The protein coat remains outside. Under the proper conditions, the viral nucleic acid within a cell uses host materials to replicate and produce proteins, forming many new viruses. Eventually, the new viruses break out of the host cell and invade more cells.

The means by which viruses reproduce and disperse damages host tissue. Pneumonia viruses damage lung tissue. Chicken pox viruses damage skin tissue. Polio viruses sometimes damage nerve tissue. Bacteriophages, which are parasites in

Tuberculosis has claimed the lives of many people, including such famous artists as Yeats, Schubert, Chopin, Kafka, and Robert Lewis Stevenson.

26-11 The botulism bacterium, *Clostridium botulinum*, thrives in anaerobic conditions such as that provided inside a sealed jar or can. *Clostridium* produces spores that can withstand boiling but not acid conditions or high heat under pressure. As a result, acid foods, such as peaches, nectarines, tomatoes, and many other fruits can be canned without a pressure cooker. Most meats and vegetables require cooking under pressure. Botulism poisoning is characterized by convulsions and blindness and is often fatal.

bacterial cells, contribute to certain diseases. The bacteria that cause diphtheria, for example, only produce their harmful toxin when a particular kind of virus occupies their cells. Diseases caused by viruses include the common cold, polio, flu, viral pneumonia, chicken pox, smallpox, measles, and viral hepatitis.

Protozoa Protozoa differ in their means of locomotion. They can have cilia, flagella, pseudopods, or no specialized means of moving about. Only one species of ciliate is parasitic in humans: *Balantidium coli*. It inhabits the human colon and destroys colon tissues. It is transmitted in contaminated food and water.

Flagellated protozoa known as trypanosomes cause **African sleeping sickness**. These parasites are transmitted by tsetse flies. In the tropics, many animals as well as people have trypanosomes in their blood. The tsetse fly picks up the protozoa when it bites a person or animal harboring them. Trypanosomes grow and reproduce inside the fly's intestine and then enter the fly's salivary gland. Trypanosomes are transmitted along with saliva when the fly bites a new victim. African sleeping sickness is characterized in humans by headache, fever, and fatigue. As the parasites live and reproduce in the host's bloodstream, blood cells are destroyed. Unless the disease is treated, the parasites enter the fluid surrounding the brain and spinal cord, where they destroy nerve tissue. The victim becomes unconscious and often dies.

The dysentery ameba moves by pseudopods. It is a human parasite that is transmitted in food or water contaminated with human wastes. Dysentery amebas also live in human colon cells. The parasites feed on colon tissue. A person with amebic dysentery suffers with diarrhea and sores of the colon.

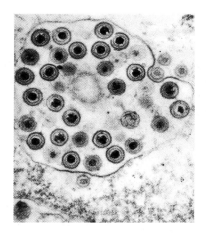

26-12 In this tissue culture *Herpes simplex* viruses are visible as dark spheres within a host cell. The virus invades the body through mucus membranes and lays dormant for long periods in nerve tissue. Periodically, especially under times of stress, the virus comes to the surface of the skin, causing an outbreak of sores at the site of the original infection.

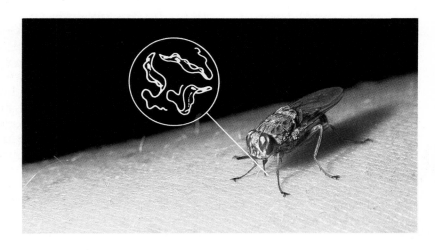

26-13 A tsetse fly on human skin. This fly, common to Africa, transmits trypanosomes, which cause African sleeping sickness, from infected hosts to new hosts.

26-14 Malaria is caused by the protozoan *Plasmodium falciparum*, which completes its life cycle in two host organisms. The female *Anopheles* mosquito is host during the sexual reproductive stages and infects a human during a bite. The parasites reproduce asexually in blood and liver cells and eventually enter the blood, where they may infect an uninfected mosquito that bites the human host.

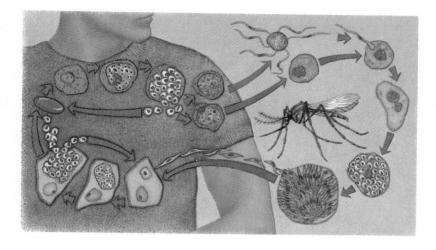

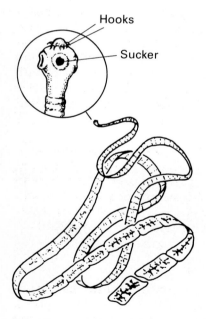

26-15 An adult tapeworm consists of segments that each contain testes and ovaries. A tapeworm attaches itself to the intestinal wall.

Malaria is caused by a protozoan that has no means of locomotion. The malarial parasite produces spores and is transmitted by the bite of the female *Anopheles* mosquito. When malarial parasites first enter the body, they travel in the human bloodstream to the liver. After about two weeks, they reenter the blood. Each parasite then enters a red blood cell, undergoes repeated cell division, and forms spores. The spores break out of the host's red blood cell, destroying the cell. Then the spores enter new red blood cells. The process works in cycles. All the liberated spores attack red blood cells at the same time. With each cycle of spores released, large numbers of host blood cells are destroyed. The host experiences high fever, chills, and sweating. Malaria can be treated with medications. Like other diseases transmitted by insects, however, its control depends on eradication of the insect carriers.

Parasitic Worms Most parasitic worms are flatworms or roundworms. Leeches are the only parasitic segmented worms. Tapeworms and flukes are parasitic flatworms. Certain species of tapeworm live in the human intestine and eat their host's food. The victim loses weight and has little energy. The head of a tapeworm has suckers and hooks that cling to the intestinal wall. Tapeworms have no mouth or digestive system of their own. Digested food from the host merely diffuses into their cells. A tapeworm can produce both eggs and sperm. Body sections containing fertilized eggs break off the tapeworm and leave the host's digestive system with wastes. If a pig or other animal ingests tapeworm sections, the tapeworm continues its life cycle in the new host's body. The tapeworm reinhabits a human host when the animal host is used for food. Medication and meat inspection are used in controlling these parasites.

26-16 The hookworm *Necator americanus* enters its human host while in its larval stage. Larvae travel through the blood to the lungs, move up the trachea, and are swallowed. Once inside the small intestine the larvae mature and lay eggs, which are eliminated with the feces, reinfecting the soil.

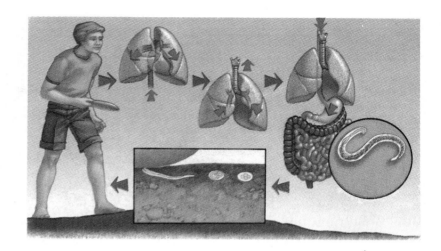

Diseases caused by roundworms include trichinosis and hookworm. A person gets trichinosis by eating improperly cooked pork that contains the parasites. Inside its human host, the parasitic worms reproduce in tremendous numbers and infest muscle tissue. Unlike trichina, hookworm larvae develop in the soil. They enter the body by boring through the skin on the soles of the feet. They travel through various parts of the body. As adult worms, they live in the intestine and ingest blood from the vessels in the intestinal walls. Hookworm eggs leave the host's body with wastes. If feces are deposited in the soil, the hookworm can then be dispersed to a new host. Hookworm is common in areas where the climate is warm, sewage facilities are poor, and people go barefoot. The disease can be controlled with improved sanitation and the wearing of shoes.

Rickettsia Rickettsia (rih-KET-see-uh) are unicellular prokaryotes that are smaller than bacteria and live only as parasites. Rocky Mountain spotted fever is a disease caused by rickettsia. The disease is transmitted by the bite of a wood tick and affects many people in the Western Hemisphere. Typhus, a disease transmitted by the body louse, is also caused by rickettsia. Rickettsia injure their hosts by damaging cells in the blood vessel walls.

Parasitic Fungi Parasitic fungi include those that cause ringworm, athlete's foot, and dandruff. Fungi produce spores, which serve to disperse the organism. Consequently, spores are also the means by which fungal diseases are transmitted. A person can pick up fungal spores on contaminated articles such as combs, towels, or floors. Fungal spores grow and develop in the body, causing discomfort and damage to host tissues.

Body Defenses Against Disease

We are constantly exposed to disease-causing factors. Yet, for the most part, we manage to stay fairly healthy. The body is adapted to defend itself against foreign substances, or **antigens**, such as bacteria and foreign tissues. The body combats antigens in three important ways, known as the first, second, and third lines of defense.

First Line of Defense The first line of defense keeps foreign substances out of the body, preventing them from entering the bloodstream and tissues. The first line of defense works in several ways. The unbroken skin acts as a physical barrier against entering microbes or particles. Cilia in the trachea and bronchial tubes sweep particles out of the respiratory tract. The mucus-coated lining of the respiratory tract, the digestive tract, the urinary system, and the reproductive system trap and remove incoming substances. Saliva contains a substance that kills bacteria. Tears, sweat, and urine flush bacteria and other foreign substances away. Hydrochloric acid in the stomach destroys bacteria. All these adaptations present an effective barrier against foreign substances.

Smoking tends to put the cilia "to sleep" and allows particles that would normally be kept out of the lungs to enter them.

Second Line of Defense If a foreign substance passes the first line of defense and enters the bloodstream or tissues, the body mobilizes its second line of defense: the white blood cells. White blood cells seek out, ingest, and kill bacteria. White blood cells travel in the blood and lymph systems. They also squeeze through capillary and lymph vessel walls and travel in tissues. When bacteria invade the body, the bone marrow is stimulated to produce and release large numbers of white blood cells. Once bacteria are in the bloodstream, they give off substances that

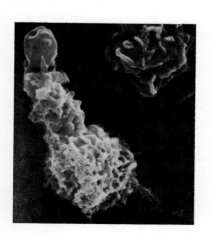
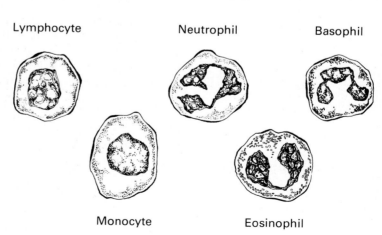

26–17 Leukocytes. **a.** Alveolar macrophage ingesting a yeast cell. **b.** Illustrations of the five major types of leukocytes.

combine with factors in the blood plasma. The product that is formed attracts white blood cells to the infected area. Other factors in the plasma coat the bacteria, further attracting white blood cells to them.

Third Line of Defense: The Immune System The immune system is really two systems in one. One system is a major defense against bacteria. The other system operates in rejecting transplants of foreign tissue, rejecting tumor cells, and destroying viruses, fungi, and bacteria. In general, the immune system fights antigens by the production of **antibodies**. Antibodies are chemical substances produced by the action of certain white blood cells, the **lymphocytes** (LIM-fuh-sīts). Antibodies help prevent disease. Once a disease takes hold, they also help to overcome it. **Immunity**, a body condition in which a person is protected against disease, is usually brought about by the activity of the immune system.

Scientists have developed a theory of immunity that explains how the immune system works. According to the theory, lymphocytes "patrol" the body and "recognize" antigens by fitting into their surfaces. Then they produce antibodies to destroy the invaders. Many different kinds of lymphocytes exist in the body, each of which recognizes a specific antigen by its shape.

Each type of lymphocyte has a different shape, enabling it to attach to a specific antigen. Lymphocyte shape is inherited.

26–18 Immune system function. 1. A foreign body enters tissue and is ingested by a macrophage. 2. The macrophage processes an antigen from the foreign body and attaches it to its surface. 3. A lymphocyte carrying the appropriate antibodies on its surface is stimulated when its antibodies contact the antigens. 4. The lymphocyte divides and produces offspring plasma cells able to synthesize antibody molecules in large numbers. 5. The antibodies are released and combine with the antigens, deactivating the foreign body.

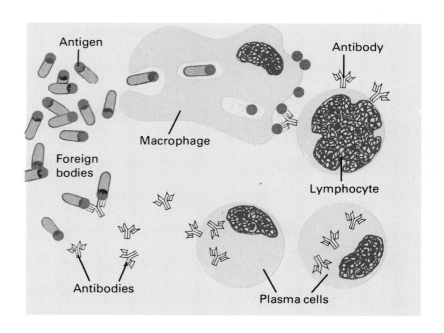

Section 26.2 Infectious Disease and Body Defense

The reaction that occurs between antibody and antigen is called the **immune reaction**. You are not aware of most of the immune reactions that occur in your body; however, sometimes an immune reaction occurs near the skin and you may see redness and swelling. When your body defenses succeed, the inflammation disappears.

As an embryo develops, special cells enter the thymus, liver, and spleen. These cells become differentiated into lymphocytes. Some of the lymphocytes then migrate to the lymph nodes and the bone marrow. Throughout life, lymphocytes continue to form in the bone marrow, lymph nodes, and spleen.

Lymphocytes recognize foreign substances, yet they also recognize "self." According to present theory, the lymphocytes in the developing embryo identify all the substances in the embryos' body as self. It is believed that the embryo's tissues suppress the production of antibodies against themselves. Later in life, antibodies against self are not produced. However, the production of antibodies against nonself substances (antigens) is not suppressed in the embryo. Later in life, antibodies are produced against antigens.

In certain diseases of the immune system, known as autoimmune diseases, antibodies that damage body tissues are produced.

Building a Science Vocabulary

epidemic	antigen
toxin	antibody
African sleeping sickness	lymphocyte
malaria	immunity
rickettsia	immune reaction

Questions for Review

1. State the cause and means of transmission of each of the following infectious diseases.
   a. athlete's foot   c. common cold
   b. malaria   d. trichinosis

2. Why was the flu considered an epidemic disease during World War I?

3. Why are white blood cells considered a second line of body defense? How do they defend the body against disease?

4. What is the function of the immune system?

Perspective: Careers
The Research Laboratory: Who Does What?

When a foreign protein is introduced into the body, the body responds by producing an antibody. The interaction of the antibody with the foreign protein, or antigen, enables the body to resist disease. We call such resistance immunity. The study of immunity is known as immunology.

In experiments in immunology, proteins from blood, urine, and other body fluids can be injected into laboratory animals. These proteins act as antigens, causing the production of antibodies. Although the antigen-antibody reaction can be seen by the eye, it is also analyzed by a wide variety of instruments. Some of these instruments are connected to computers. It is now possible to measure less than one billionth of a gram of an antibody in body fluid.

Like much other research, immunology studies are conducted by a team. The team consists of laboratory assistants, technicians, and technologists who work under the direction of research investigators. The research investigator usually holds a doctoral degree, either an M.D. or a Ph.D., degrees which require study for four or more years beyond college. The research investigator designs the research, supervises the staff that carries out most of the procedures, and analyzes the results of the research. The results are published and studied around the world.

The research technologists have completed four years of college, specializing in a field such as medical technology. They carry out the more complex research procedures. The research technicians have two years of study beyond high school, usually in a community college. They carry out less technical and less complex procedures. The laboratory assistants are high school graduates. They have general responsibility for keeping the laboratory clean and orderly and for taking care of the laboratory animals.

Each member of the immunology team contributes to the operation of the whole laboratory. Each member of the team is dependent on the other members. Only by working together can they achieve their common goals.

26.3 CONQUERING INFECTIOUS DISEASES

Infectious diseases were the leading cause of death one hundred years ago. Since that time, the cause, prevention, and cure of many serious diseases caused by pathogens have been discovered.

The first person to identify microorganisms, or microbes, as a cause of disease was Louis Pasteur. In 1857, Pasteur advanced the **germ theory**, which states that microbes cause disease. Since that time, scientists have built on Pasteur's work, developing preventives and treatments that help protect us from infectious diseases.

Antiseptic Surgery

In 1865, Joseph Lister, a British physician, first applied Pasteur's theory to surgery. Lister was concerned with the large number of people that died after surgery. For example, half the people that underwent amputation at that time died after the operation. Lister tried to destroy microbes in the operating room by using various disinfectants. The death rate after surgery finally decreased when Lister treated his patients' wounds and sprayed the operating area with carbolic acid.

Identification of Pathogens

Curing any infectious disease is easier when the exact cause is known. Robert Koch, a German country doctor, devised a technique for isolating and identifying the microbe responsible for a disease. Koch's process involves four steps:

1. Obtain the pathogen from the diseased host.
2. Isolate the pathogen in pure culture.
3. Inject the pathogen from the pure culture into a healthy animal.
4. If disease occurs, obtain the pathogen from the experimentally infected animal, grow it in pure culture, and compare it with the pathogen from the original culture.

Using this technique, Koch found the bacteria that cause anthrax, a disease of cows, sheep, and people. He also discovered the bacteria that caused tuberculosis. Since that time, hundreds of scientists have used Koch's techniques.

26-19 The operating room in a hospital is kept as sterile as possible to help prevent infections during surgery.

Vaccination

In 1796, Edward Jenner, a British country doctor, discovered a way to protect people from infectious diseases. In eighteenth-century England, one out of three children died of smallpox before the age of three. However, those who survived were immune to smallpox for life. Jenner had no idea what caused smallpox. He observed that milkmaids often got a mild form of the disease from cows that had cowpox. He also noted that milkmaids that had had cowpox rarely got smallpox. These observations gave him the idea for a way to prevent smallpox. Jenner took cowpox pus from the sore on a milkmaid's hand and smeared it into scratches on the arm of a healthy boy. Two months later, the boy was found to be immune to smallpox.

The introduction of antigen into a body with the purpose of causing immunity to a disease is called **vaccination** (VAK-suh-NAY-shun). The antigen that is introduced into the body is called a **vaccine** (vak-SEEN). The word *vaccination* was first used in reference to Jenner's use of cowpox to prevent smallpox; today, it refers to immunization in general.

Vaccines and the Immune System A vaccine contains a safe amount and kind of antigen for a given disease. The antigen stimulates the body to produce the appropriate antibodies. Some vaccines, such as the Salk polio vaccine, are made up of dead disease organisms. Other vaccines, such as those for diphtheria and tetanus, use weakened toxins from a pathogen. Another type of vaccine contains a harmless amount and kind of live pathogens. This type of vaccine is used mainly against viral diseases such as yellow fever, polio (Sabin vaccine), flu, smallpox, measles, and mumps.

Vaccination is also referred to as immunization.

vacca = a cow

A vaccine does not cause the illness; it provides immunity to a disease by stimulating the production of antibodies. The antibodies remain in the blood and destroy the disease organisms if they should enter the body.

26–20 At first, many people feared that the use of cowpox vaccine would cause strange side effects. This attitude gave rise to satirical cartoons, such as this one, published in London in 1802, in which people who have been vaccinated sprout parts of cows.

Section 26.3 Conquering Infectious Diseases

Vaccines help protect populations from epidemics of infectious disease.

Length of Immunity Some antibodies produced during a disease or after administration of a vaccine last a long time, giving years or even a lifetime of immunity. Some antibodies are destroyed after a short time, giving only temporary immunity. In certain cases, a person exposed to a disease is directly injected with antibodies. This treatment, known as **passive immunity**, gives quick and temporary protection against disease. In passive immunity the body does not make its own antibodies. In contrast, the body does make its own antibodies during **active immunity**, which results from having a disease or getting the vaccine against it.

Use of Chemicals Against Infectious Diseases

Throughout history, people have used medicines to cure disease. During the Middle Ages, monks grew herbs to be used as medicines. The first documented case of a chemical cure for a specific disease came in 1619. The wife of the Spanish governor of Peru was cured of malaria by an extract made of the bark of the chincona tree. The bark of this tree contains **quinine** (KWĪ-nīn), which is used as a cure for malaria to this day.

After the discoveries of Pasteur and Koch, scientists searched for chemicals that would kill specific pathogens. A breakthrough came in the early twentieth century when Paul Ehrlich found a medicine that killed syphilis bacteria and thereby cured the disease. In the 1930s a group of chemical compounds called **sulfa drugs** were found to combat bacteria. Bacterial pneumonia was one of the infectious diseases that was helped by sulfa drugs.

The greatest success of chemical treatment for infectious organisms came with the use of **antibiotics** (AN-tih-bī-AHT-iks), which are chemical substances produced by fungi that inhibit the growth of bacteria. The first antibiotic, **penicillin**, was discovered in 1928 by Alexander Fleming, a British microbiologist. However, it was not until 1940 that this antibiotic was produced in purified form and used as a medicine. Since the discovery of penicillin, other antibiotic drugs have been developed, including streptomycin, the tetracyclines, and erythromycin. Antibiotics are used successfully to cure major diseases, such as bacterial pneumonia, tuberculosis, and scarlet fever.

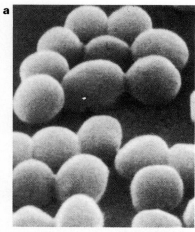
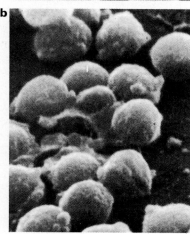

26–21 *Staphylococcus aureus* bacteria, commonly found in wound infections, normally have smooth outer cell walls (**a**). Treatment with penicillin causes the cell walls to break down by interfering with cell wall synthesis (**b**).

26–22 Many antibiotics kill bacteria by inhibiting protein synthesis. The arrows indicate the specific points at which several common antibiotics act.

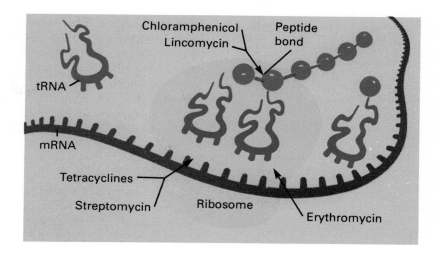

The use of chemical medicines is not without problems. Drugs can produce side effects in certain people. For example, some people are allergic to penicillin. Another problem in the use of antibiotics is the growth of antibiotic-resistant strains of bacteria. The use of an antibiotic creates an environment that favors the survival of resistant strains. Scientists try to overcome this problem by developing new and effective antibiotics. Physicians try to overcome the problem by limiting the use of antibiotics. They hope that by using antibiotics less, resistant strains will not develop as rapidly.

Infectious Disease and Large Populations

During the years between 1900 and 1970, the leading causes of death in North America shifted from infectious to noninfectious diseases. Tuberculosis, once a widespread cause of death, has become rare. Only influenza and pneumonia remain as major causes of death due to infectious disease. Noninfectious diseases, such as heart disease, cancer, and cerebrovascular diseases (strokes), have become the three leading causes of death.

In the same period of time, the average life expectancy at birth rose from 47 years to 71 years. Better sanitary conditions, housing, clothing, and food are partly responsible for the increase in life expectancy. Treatment of many infectious diseases is also responsible.

Research in science and technology has resulted in improved health among individuals and populations as a whole. One approach to the improvement of public health is to study diseases of populations.

Cardiovascular disease claimed over 1 million lives in North America in 1980—more than double the toll taken by cancer and nearly half the total of deaths from all causes.

epidemios = among the people
-logy = study of

Epidemiology The study of diseases of populations is known as **epidemiology** (EP-uh-DEE-mee-AHL-uh-jee). According to epidemiologists at the National Cancer Institute in the United States, more deaths from bladder cancer occur in areas where industrial pollutants are concentrated in the water supply. In addition, they estimate that at least 60 percent of all cancers are caused by factors in the environment. Some of these factors are cancer-causing chemicals called **carcinogens** (kar-SIN-uh-jenz).

Removing carcinogens from the environment helps reduce death from cancer; however, no one really knows how many different carcinogens are in use. More than 1500 different chemical substances are suspected of causing cancer. Some chemicals, such as asbestos, arsenic, and vinyl chloride, have been definitely established as carcinogens. As a result, governments have adopted regulations to control their use.

Modern Technology Aids Diagnosis and Treatment

Many types of machines and tools are available to physicians for use in diagnosing illnesses. Most physicians also make use of medical laboratories to analyze blood and urine samples and to grow cultures. X-rays, radioactive elements, electronic machines, and computers are other tools that aid physicians in their work.

Treatment has also come a long way since the days of Joseph Lister. Today's surgeon uses laser beams, anesthetics, intricate instruments, and artificial body parts. In addition to the use of artificial organs, living tissues and organs are transplanted as treatment for certain diseases. Sometimes technology is used to help the parts of a person's body work better. Pacemakers, artificial kidney machines, and respirators all help keep the body functioning properly.

Chemical treatment of disease, or **chemotherapy** (KEE-mō-THER-uh-pee), is not merely limited to infectious disease. Medicines are also used effectively to control certain types of cancer, some diseases of the transport system, and certain mental illnesses. Scientists have synthesized body substances such as hormones and vitamins to treat chemical deficiencies in the body. Gene-splicing is used to obtain bacteria that produce im-

26–23 The world's first artificial human heart, the Jarvik-7, was successfully implanted in December, 1982 into a 61-year-old man. The heart consists of aluminum valves and two hollow plastic chambers that substitute for the ventricles. Pumping action is provided by a "heart driver" support machine connected to the heart from outside the body.

26-24 Ultrasound scanning equipment is used in diagnosis. The technique produces a "picture" of the internal organs being scanned.

portant body substances, such as insulin and growth hormone. This technique is also being used to provide interferon, an antiviral protein that may have an effect against certain kinds of cancer.

Health Care: A Modern Concept

Research aims to improve health by investigating the cause, diagnosis, cure, and prevention of disease. New instruments and techniques are constantly being developed to aid in these areas. In addition, people need to know how to maintain good health and what to do when they become ill. Through education, individuals and populations can learn about health and disease and contribute to the better health of their families and of the communities in which they live.

Building a Science Vocabulary

germ theory	active immunity	penicillin
vaccination	quinine	epidemiology
vaccine	sulfa drug	carcinogen
passive immunity	antibiotic	chemotherapy

Questions for Review

1. Why did many people of Lister's day die after undergoing surgery?
2. How does the body develop active immunity?
3. What are antibiotics? What problem arises with the extensive use of an antibiotic?
4. What is chemotherapy? Give three examples of diseases that are treated with chemotherapy.
5. How does epidemiology aid in combatting disease?

Chapter 26
Summary and Review

Summary

1. Body malfunctions are one of three main categories of cause of disease. Body malfunctions can be caused by physical damage (emphysema), hereditary disorders (hemophilia), endocrine malfunctions (diabetes mellitus), and immune system malfunctions (allergy).

2. Aging is a natural process that places older people at greater risk of becoming ill.

3. Environmental factors that cause disease include pollution (respiratory illness), improper sanitation (typhoid fever), lack of medical aid (childhood diseases), improper diet (scurvy), and stress (high blood pressure).

4. Tobacco, alcohol, and drugs can cause diseases that have dangerous effects on the user and society.

5. Infectious diseases can be communicable or noncommunicable. Communicable diseases may be transmitted by direct contact (syphilis), droplet infection (common cold), contaminated food and water (tapeworm), contaminated articles (athlete's foot), and insects (bubonic plague).

6. Epidemics involve the widespread transmission of a disease and can affect millions of people (influenza).

7. Organisms that cause disease include bacteria (tuberculosis), viruses (chicken pox), protozoa (amebic dysentery), worms (hookworm), rickettsia (typhus), and fungi (ringworm).

8. The body has three lines of defense against disease: the skin and other features keep antigens outside the body; white blood cells destroy bacteria; and the immune system produces antibodies that counteract specific antigens.

9. Early discoveries in the conquest of infectious disease include Pasteur's germ theory, Lister's use of antiseptics, Koch's method for finding the causative organism of a disease, and Jenner's vaccination against smallpox.

10. Immunity can be temporary or long-lived, active or passive.

11. Modern methods of combatting disease include chemotherapy, epidemiology, and technology aiding in diagnosis and treatment.

Review Questions

Application

1. Explain how the Sabin vaccine provides protection against polio.

2. The organism that causes a disease is often different from the one that transmits it. What are the causative organism and the transmitting organism for each of the following diseases?
   a. malaria
   b. yellow fever
   c. bubonic plague

3. A person stepped on a dirty nail and punctured the skin. In order to avoid tetanus, a doctor injected the person with antibodies. Describe the type of immunity obtained by this procedure.

4. An epidemiological study showed a large number of people suffering with hepatitis B, a viral disease of the liver, in a neighborhood of drug addicts. What could explain the outbreak of this disease?

5. Emphysema and cystic fibrosis both affect the respiratory system. What is the difference in causation of the two diseases?

6. A teenager began smoking to be sociable. After a time, smoking became a habit. Explain this change in terms of the effect of smoking on the body.

Interpretation

1. A researcher found two different types of bacteria (*A* and *B*) in the body of a patient with disease *X*. How could the researcher determine whether either *A* or *B* caused the disease?

2. When a tissue is transplanted from one person to another, drugs are given to suppress the action of the immune system of the recipient. This is done to avoid rejection of the transplanted tissue. What is the danger of this procedure to the recipient?

3. What two measures could help bring an outbreak of yellow fever under control?

Extra Research

1. In the library, research the incidence of malaria in the world today. Write a newspaper story about the current significance of this disease.

2. Look up the story of the discovery of penicillin. Write a series of questions that you could use in an interview with Alexander Fleming if he were alive. Fill in his answers. By the end of the interview the listener should know how penicillin was discovered and tested.

3. Prepare arguments for and against increasing the use of antibiotics by physicians or increasing the money spent on community health care programs.

4. Prepare an encyclopedia of pathogenic organisms. Include a description of the organism and the disease it causes. Also discuss how the disease is transmitted, how widespread it is, and how it is diagnosed, prevented, controlled, and treated.

5. Visit a local hospital and observe medical equipment, laboratories, treatment rooms, and other facilities in use.

6. Write a short dramatic sketch that could be used on a children's television program showing what happens to an antigen when it attacks a body.

7. Use reference materials to discover the role played by disease in historic battles. Write a report entitled "Bugs and Battles."

8. Find out how six famous people of the past died and the age at which they died. State your opinion as to whether they would have been able to live longer had today's modern medical techniques been available to them.

Career Opportunities

Bacteriologists often work with disease-causing and food-spoilage organisms. Their work is important in medicine, agriculture, and food processing. Many do research or supervise manufacturing operations for pharmaceutical companies and food processing plants. Others work in hospitals and universities. Four or more years of university are required. For information write the Canadian Association of Medical Microbiologists, Henderson General Hospital, Concession St., Hamilton, Ontario L8V 1C3.

Health services administrators work for hospitals, nursing homes, social service agencies, and health care organizations. They must have good organizational skills in order to plan for future needs and manage the health care resources and staff. At least four years of university are usually required. Contact the Canadian College of Health Service Executives, 17 York Street, #201, Ottawa, Ontario K1N 5S7.

Unit VIII
Suggested Readings

Adamson, Joy, *Born Free: A Lioness of Two Worlds*, New York: Random House, Inc., 1974. *Story of a lioness who lives in the worlds of the jungle and of humans. The author lived in Kenya and helped introduce the lioness back into the wild.*

De Kruif, Paul, *Microbe Hunters*, New York: Harcourt Brace Jovanovich, Inc., 1959. *An enthralling account of the struggles and successes of the scientists who helped conquer infectious diseases.*

Dubos, Rene, Maya Pines, and the editors of *Life, Health and Disease*, Life Science Library, Time Inc., New York, 1965. *Describes various causes of disease and gives information about specific diseases. Well illustrated.*

Fisher, Allan C. Jr., "Mysteries of Bird Migration," *National Geographic*, August, 1979. *Describes the advantages to birds of migrating, as well as birds' adaptations for migration.*

Goodall, Jane, *In the Shadow of Man*, Boston: Houghton Mifflin Co., 1971. *Fascinating account of the author's intimate study of chimpanzees in East Africa.*

Gould, James L., "Do Honeybees Know What They Are Doing?" *Natural History*, June-July, 1979. *Bees navigate and communicate without ever having done so before; they inherit these behaviors. In the course of their lives they also learn sets of information about the environment, such as color, odor, shape, landmarks, and locations.*

Human Physiology and the Environment in Health and Disease, Readings from *Scientific American*, San Francisco: W. H. Freeman and Co., 1976. *Informative articles discuss the relationships between environment and health, the immune system, and aging.*

Lott, John A., "Heart Attack: Desperate Hours for the Body," *SciQuest*, October, 1980. *Describes what happens in the body during a heart attack, tests for diagnosing heart attacks, and treatment for patients.*

Matthews, Robert W., and Janice R. Matthews, "War of the Yellow Jacket Queens," *Natural History*, October, 1979. *Interesting and beautifully illustrated article about queen wasps. Describes combat between queens from different colonies for possession of a nest.*

Milne, Lorus, and others, *The Secret Life of Animals: Pioneering Discoveries in Animal Behavior*, New York: Elsevier-Dutton Publishing Co., Inc., 1975. *A handsome book loaded with facts about animal senses, survival tactics, migration, mating, predation, etc. Contains excellent photographs supplemented by descriptions of experiments and observations.*

Moore-Ede, Martin C., "What Hath Night to Do With Sleep?" *Natural History*, September, 1982. *Two circadian pacemakers in the human body regulate the rhythm of body temperature and sleep and arousal timetables. Disruption of these rhythms is common in today's world and has its consequences.*

Murchie, Guy, "A Special Parasite: The Rock That Wakes to Cause Trouble," *Science Digest*, September, 1979. *The author characterizes the virus as a gene with a coat on and tells what occurs when a virus's nucleic acid enters cells.*

Nassif, Janet Zhun, *Handbook of Health Careers: A Guide to Employment Opportunities*, New York: Human Sciences, 1980. *An excellent guide to health careers and how to select one.*

Payne, Roger, "Humpbacks: Their Mysterious Songs," *National Geographic*, January, 1979. *Describes and analyzes the beautiful and complex songs of whales.*

The International System of Units and the Metric System

The International System of Units (SI) was established in 1960 in an effort to standardize the metric system and to replace all other former systems of measurement.

The metric system focuses on the number 10 and measures by multiples and divisions of 10. Basic metric units include the metre (length), the litre (liquid volume), and the gram (mass). Other units are formed by adding prefixes. For example, *kilo*metre and *milli*metre.

The Celsius temperature scale assigns 0° to the freezing point of water and 100° to the boiling point of water. The Celsius scale was adopted in 1948 by an international conference on weights and measures.

SI Prefixes

Prefix	Symbol	Meaning	Multipliers		Prefix	Symbol	Meaning	Multipliers	
tera	T	trillion	(10^{12})	1 000 000 000 000	deci	d	tenth	(10^{-1})	0.1
giga	G	billion	(10^{9})	1 000 000 000	centi	c	hundredth	(10^{-2})	0.01
mega	M	million	(10^{6})	1 000 000	milli	m	thousandth	(10^{-3})	0.001
kilo	k	thousand	(10^{3})	1 000	micro	μ	millionth	(10^{-6})	0.000 001
hecto	h	hundred	(10^{2})	100	nano	n	billionth	(10^{-9})	0.000 000 001
deca	da	ten	(10^{1})	10	pico	p	trillionth	(10^{-12})	0.000 000 000 001

Some SI Base Units and Their Symbols

Length	metre	(m)
Mass	gram	(g)
Liquid volume	litre	(L)
Time	second	(s)

Some SI Compound Units and Their Symbols

Area	square metre	(m^2)
Dry volume	cubic metre	(m^3)
Speed	metre per second	(m/s)

The following tables give a range of metric or SI units for common measures and quantities.

Length

1 km	length of a brisk 10-minute walk
1 m	height of a 3-drawer filing cabinet
1 mm	thickness of a dime
1 μm	size of a bacterium

Volume (dry)

50 m^3	average classroom
500 cm^3	softball
1 cm^3	marble

Volume (liquid)

200 L	oil drum
10 L	large pail
150 mL	tea cup

Mass

1000 kg	compact car
250 kg	large motorcycle
10 kg	large turkey
2 kg	desk telephone
450 g	pound of butter
25 g	slice of bread
1 g	medium raisin
20 mg	postage stamp

Temperature: The Celsius Scale

250°	wood burns
200°	paper burns
100°	boiling point of water
37°	normal body temperature
20°	room temperature
0°	freezing point of water
−30°	extremely cold
−118°	alcohol freezes

Temperature conversion: $°F = 9/5°C + 32$
$°C = 5/9(°F - 32)$

Additional Units

calorie (cal): the quantity of heat required to raise the temperature of 1 g of water 1°C.

Calorie (Cal): a kilocalorie, or 1000 calories.

Five-Kingdom Classification of Living Things

Classification of organisms into five kingdoms was originally proposed by Robert Whittaker and recently revived by Lynn Margulis. The system of classification used in this book is based largely on their work. This appendix is designed to give you an overview of all organisms. It does not include some minor groups or any extinct groups.

KINGDOM MONERA

About 3000 species. Prokaryotic; unicellular; cell walls of complex proteins and polysaccharides. Chemosynthetic, autotrophic, or heterotrophic.

Phylum Schizophyta About 1600 species. Some form chains or clusters. Most are heterotrophic. Reproduction asexual (cell division). Bacteria.

Phylum Cyanophyta About 1500 species. Some form chains or filaments. Usually autotrophic. Live in damp soil or rocks or in water. Reproduction asexual (fission). Blue-green algae.

KINGDOM PROTISTA

About 30 000 species. Eukaryotic; unicellular. Autotrophic or heterotrophic.

Phylum Euglenophyta About 800 species. Most are autotrophic; some are parasitic. Aquatic. Move by flagella. *Euglena*.

Phylum Chrysophyta About 8000 species. Autotrophic. Marine. Flagellated, ameboid, or nonmotile. Golden algae and diatoms.

Phylum Pyrrophyta About 1100 species. Autotrophic. Aquatic. Most have two flagella. Golden-brown algae (dinoflagellates).

Phylum Ciliophora About 5000 species. Heterotrophic. Move by beating cilia. *Paramecium*, *Stentor*.

Phylum Mastigophora About 1200 species. Heterotrophic. Move by lashing flagella. *Trypanosoma* (cause of sleeping sickness).

Phylum Sarcodina About 8000 species. Some have shells. Heterotrophic. Move and feed with pseudopods. *Ameba*.

Phylum Sporozoa About 2000 species. Parasitic. No organs for moving. Reproduction asexual (spore formation). *Plasmodium* (cause of malaria).

Phylum Myxomycota About 450 species. Multinucleate. Heterotrophic. Reproduction asexual with alternating plasmodial and sporangial stages. Slime molds.

KINGDOM FUNGI

About 100 000 species. Eukaryotic; most are multicellular; most are multinucleate; cell wall composed mainly of chitin. Heterotrophic.

Phylum Zygomycota About 1500 species. Multicellular. Terrestrial. Reproduction asexual (spores produced in sporangia). Black bread mold.

Phylum Ascomycota About 30 000 species. Most are multicellular. Terrestrial or aquatic. Reproduction asexual (budding) or sexual (ascus formation). *Penicillium*, mildew, morels, truffles, yeasts.

Phylum Basidiomycota About 25 000 species. Multicellular. Terrestrial. Reproduction asexual (spore formation in basidia) or sexual (gametangia). Club fungi, mushrooms.

Phylum Fungi Imperfecti About 25 000 species. Classification for convenience; fungi whose spore-bearing structures are not yet known. Molds used for cheese; fungus of athlete's foot.

KINGDOM PLANTAE

About 285 000 species. Eukaryotic; most are multicellular; cell walls of cellulose. Autotrophic (with a few exceptions).

Phylum Chlorophyta About 7000 species. Some form filaments or colonies. Some move by flagella. Green algae.

Phylum Rhodophyta About 4000 species. Filamentous or branchlike; no conducting tissue or motile cells. Most are marine. Red algae.

Phylum Phaeophyta About 1100 species. Most are stationary. Motile cells have two flagella. Most are marine. Brown algae.

Phylum Bryophyta About 23 500 species. Little or no conducting tissue. Moist habitat. Reproduction by alternation of generations with dominant gametophyte. Mosses, liverworts, hornworts.

Phylum Tracheophyta About 261 000 species. Bodies differentiated into leaves, stems, and roots; well-developed vascular systems conduct water, minerals, and organic substances. Reproduction by alternation of generations with dominant sporophyte and reduced gametophyte; produce seeds; male gametes motile in some species.

 Subphylum Sphenopsida About 30 species. Vascular tissue; leaves and branches in whorled pattern; cones. Horsetails.

 Subphylum Pteropsida About 211 000 species. The most complex tracheophytes; gap in stem's vascular tissue at position of each leaf.

  Class Gymnospermae About 700 species. Needlelike leaves. Seeds in cones. Pine, spruce, and other conifers.

  Class Angiospermae About 200 000 species. Flowering plants bearing seeds enclosed in fruits.

   Subclass Monocotyledonae About 34 000 species. Flower parts usually in threes; leaf veins parallel; one cotyledon; vascular bundles scattered in stem. Lily, grasses.

   Subclass Dicotyledonae About 166 000 species. Flower parts in fours or fives; leaf veins branching; two cotyledons; vascular bundles arranged in ring in stem. Maple, begonia, rose, oak.

KINGDOM ANIMALIA

About 1 000 000 species. Eukaryotic; multicellular; no cell walls. Heterotrophic.

Phylum Porifera About 4200 species. Stiff skeletons, porous bodies: no symmetry; two cell layers; no locomotion or nervous systems. All are aquatic; most are marine. Sponges.

Phylum Coelenterata About 11 000 species. Jellylike bodies; radial symmetry; two cell layers; stinging cells for predation; tentacles; nerve nets; gut with only one opening. Aquatic. Swimming or stationary. *Hydra*, jellyfishes, sea anemones, corals.

Phylum Platyhelminthes About 15 000 species. Flattened bodies; bilateral symmetry; three cell layers; gut with only one opening; no circulatory system or body cavity; flame cells for excretion. Many are parasitic. Aquatic and free-living or internal parasites. Hermaphroditic. Planarians, flukes, tapeworms.

Phylum Aschelminthes About 12 500 species. Diverse group; bilateral symmetry; three cell layers; body covered with cuticle; muscular organ for feeding at anterior end of gut. Parasitic or free-living. Ribbon worms, rotifers, priapulids, gastrotrichs, horsehair worms, roundworms (nematodes).

Phylum Bryozoa About 4000 species. Bilateral symmetry; three cell layers; U-shaped digestive tract; primitive nervous system. Feed with ciliated tentacles. Aquatic. Usually stationary and colonial. Retain larvae in brood pouch. Small "moss" animals.

Phylum Brachiopoda About 260 species. Clamlike paired shells; bilateral symmetry; "arms" with tentacles. Marine. Adults sessile, using stalk for support. Lamp shells.

Phylum Annelida About 8800 species. Segmented worms; bilateral symmetry; most have well-developed body cavity, complete digestive tract, nephridia, circulatory system, ventral nerve cord. Parasitic or free-living. Many are aquatic. Leeches, earthworms.

Phylum Mollusca About 110 000 species. Not segmented; bilateral symmetry; muscular foot, mantle, and head; soft bodies, some with hard shells; three-chambered heart; most have radula for scraping; complex digestive, circulatory, respiratory, and excretory systems. Most are aquatic. Squids, clams, octopuses, snails, slugs.

Phylum Arthropoda About 774 000 species. Segmented, with paired, jointed appendages; hard, jointed exoskeleton; no nephridia; reduced body cavity; complete digestive tract; dorsal "brain," ventral, solid nerve cord and ganglia.

 Class Crustacea About 30 000 species. Appendages on thorax and sometimes on abdomen; five or more pairs of legs; two pairs of antennae, one pair of mandibles, usually two pairs of maxillae. Most are aquatic. Brine shrimps, water fleas, lobsters, crabs, wood lice.

 Class Arachnida About 35 000 species. Usually four or five pairs of legs, first pair of appendages for grasping; no jaws or antennae; fangs or pincers; two body sections; most are air-breathing. Most are terrestrial. Spiders, scorpions, ticks.

 Class Chilopoda About 2000 species. 15-173 segments, each with one pair of appendages; poison glands. Centipedes.

Class Diplopoda About 7000 species. 20-100 segments, each with two pairs of appendages. Decomposers. Millipedes.

Class Insecta About 700 000 species. Most have tracheas; most are winged; one pair of antennae; three pairs of legs; three distinct body divisions (head, thorax, abdomen). Most are terrestrial.

Order Coleoptera Hard front wings, membranous hind wings; chewing mouthparts. Beetles.

Order Diptera One pair of wings; sucking and piercing mouthparts. Flies, mosquitoes.

Order Hemiptera Wings thin from base to tip; mouthparts for sucking. True bugs.

Order Homoptera Wingless or wings lifted above body; sucking mouthparts. Leafhoppers, plant lice, cicadas.

Order Hymenoptera Wingless or with hind wings shorter than front wings; chewing or sucking mouthparts; some with stingers. Bees, ants, wasps.

Order Isoptera Two pairs of wings; chewing mouthparts. Termites.

Order Lepidoptera Two pairs of scaly wings; sucking mouthparts. Butterflies, moths.

Order Odonata Two pairs of long wings; chewing mouthparts. Dragonflies, damselflies.

Order Orthoptera Long, tough front wings covering membranous hind wings; chewing mouthparts. Grasshoppers.

Phylum Echinodermata About 6000 species. Spiny bodies; radial symmetry in adults; three cell layers; well-developed body cavity; water vascular system; tube feet. Marine. Starfishes, sea urchins, sand dollars.

Phylum Chordata About 45 000 species. Bilateral symmetry; three cell layers; paired appendages; have gill slits, notochord, and dorsal, hollow nerve cord at some time in development. Reproduction sexual.

Subphylum Hemichordata About 100 species. Small, wormlike; controversial "notochord." Marine. Tongue worms.

Subphylum Urochordata About 1600 species. Saclike; may form colonies; filter feeders. Marine. Usually stationary as adults. Sea squirts.

Subphylum Cephalochordata About 13 species. Fishlike; no cartilage or bone; notochord throughout life; filter feeders. Marine. Lancelets (*Amphioxus*).

Subphylum Vertebrata About 43 000 species. Notochord replaced in adult by cartilage or bone; brain enclosed in skull. Most move about freely, a few are stationary. Aquatic or terrestrial.

Class Agnatha About 10 species. Jawless fish; eel-like; no limbs, bones, scales, or fins; cartilaginous skeleton; coldblooded. Aquatic. Lampreys, hagfish.

Class Chondrichthyes About 600 species. Cartilaginous fish; scales; paired fins; large gill openings; no air bladder; coldblooded. Most are marine. Sharks, rays.

Class Osteichthyes About 20 000 species. Bony fish; usually have air bladder; a few have lungs; gills with hard cover; paired fins; scales; coldblooded. Aquatic. Sturgeon, trout, perch, lungfishes.

Class Amphibia About 2000 species. Skin usually naked; most have legs; usually larvae have gills, adults have lungs; incomplete double circulation; coldblooded. Aquatic as larvae; aquatic or terrestrial as adults. External fertilization; eggs not protected by shell or membranes. Salamanders, frogs, toads.

Class Reptilia About 5000 species. Dry skin usually covered with scales; four legs (none in snakes); lungs; incomplete double circulation; coldblooded. Terrestrial or aquatic. Internal fertilization; embryo enclosed in egg shell and protective membranes. Lizards, snakes, turtles, alligators.

Class Aves About 8600 species. Skin covered with feathers; forelimbs are wings; lungs; complete double circulation; warmblooded. Terrestrial or aquatic. Internal fertilization; embryo enclosed in egg shell and protective membranes. Pigeons, chickens, finches, woodpeckers, and other birds.

Class Mammalia About 4500 species. Four limbs; lower jaw made of one pair of bones; three bones in middle ear; hair or fur; diaphragm used in respiration; complete double circulation; warmblooded. Terrestrial or aquatic. Internal fertilization; young nourished with milk from mother. Rats, whales, humans, apes, hamsters, cats, horses, kangaroos, bats.

Order Monotremata About 5 species. Only mammals that lay eggs. Platypus.

Order Marsupalia About 250 species. Only mammals in which young are born only partly developed; development completed in pouch on mother's lower belly. Kangaroo, opossum, phalanger.

Order Insectivora About 400 species. Small; have many small teeth; omnivorous; most are nocturnal. Star-nosed mole, shrew.

Order Edentata About 30 species. Small-brained; no teeth or back teeth only; many eat ants and termites. Sloth, armadillo, anteater.

Order Pholidota About 8 species. No teeth; body covered with scales; eat ants and termites. Pangolin.

Order Tubulidentata 1 species. Only a few teeth in adults; toes end in structures that are somewhat hooflike, somewhat clawlike. Aardvark.

Order Chiroptera About 900 species. Relatively large brains; fly using webs of skin between appendages and between fingers. Large-eared bat, little brown bat.

Order Dermoptera 2 species. Glide through air using winglike webs of skin between appendages and tail. Colugo.

Order Carnivora About 280 species. Have large, pointed canine teeth and some sharp molars; claws. Most are carnivorous. Terrestrial or aquatic. Tiger, giant panda, dog, bear, walrus, seal, skunk.

Order Rodentia About 1700 species. Gnaw food using chisellike incisors that grow continually. Diverse habitats; most are terrestrial, some are semiaquatic. Porcupine, rat, prairie dog, vole.

Order Lagomorpha About 60 species. Short tails; four rodentlike incisors in upper jaw. Herbivorous. Rabbit, pika, hare.

Order Hyracoidea 1 species. Have nails in place of hooves; skeleton somewhat like proboscideans'. Herbivorous. Hyrax.

Order Primates About 200 species. Eyes face forward; well-developed incisors, canines, and molars; most have nails on toes. Monkey, human, ape, tarsier.

Order Artiodactyla About 170 species. Complex stomachs for digestion of plant materials; even number of hooves on each leg; many with horns. Herbivorous. Hippopotamus, giraffe, oxen, bison, deer.

Order Perissodactyla About 15 species. Odd number of hooves on each leg; large, grinding teeth. Herbivorous. Tapir, rhinoceros, horse.

Order Proboscidea 2 species. Incisor teeth modified as tusks; temporary, replaceable molars; trunk. Herbivorous. Elephant.

Order Cetacea About 80 species. Horizontal tails; front limbs are flippers, no hind limbs; large heads, no necks; hairless as adults. Aquatic; most are marine. Whale, dolphin.

Order Sirenia 5 species. No hind legs; wide tails aid in swimming; almost hairless. Herbivorous. Aquatic. Manatee, dugong.

Glossary

A simple, phonetic pronunciation is given for words in this book that may be unfamiliar or hard to pronounce. Syllables deserving main stress are printed in capital letters. Syllables deserving minor stress are printed in small capitals. Unstressed syllables are printed in lowercase letters. The key below gives the sound of letters that are commonly used for more than one sound.

Example: Multicellular is pronounced MUL-tuh-SEL-yoo-lur.

Pronunciation key					
a	cat	ew	new	or	for
ah	father	g	grass	ow	now
ar	car	i, ih	him	oy	boy
ay	say	ī	kite	s	so
ayr	air	j	jam	sh	shine
e, eh	hen	ng	sing	th	thick
ee	meet	o	frog	u, uh	sun
eer	deer	ō	hole	z	zebra
er	her	oo	moon	zh	pleasure

abiogenesis (AY-bī-ō-JEN-uh-sis) The idea that, under certain conditions, living things arise from nonliving things.
absorption The movement of substances such as nutrients across a cell membrane and into the cytoplasm of a cell.
actin A protein found in the thinner filaments of muscle fibril.
activation energy The energy that is required to start a chemical reaction.
active site The place on an enzyme into which the substrate fits.
active transport The transport of materials through a cell membrane to a region of higher concentration; requires energy.
adaptation An inherited characteristic that helps an organism survive in its environment.
adenine (AD-un-EEN) A nitrogen base of DNA; pairs with thymine.
adenosine diphosphate (uh-DEN-uh-seen DĪ-FAHS-fayt), **(ADP)** A nucleotide that functions in energy storage and transfer in a cell.
adenosine triphosphate (ATP) A nucleotide that temporarily stores chemical energy used in life processes.
adrenal (uh-DREE-nul) **cortex** The outer part of the adrenal gland; it secretes several steroid hormones.
adrenal gland An endocrine gland on top of each kidney.
adrenal medulla (mih-DUL-uh) The inner part of the adrenal gland.
aerobic (ayr-Ō-bik) Able to thrive where oxygen is present.
afterbirth The placenta, amniotic sac, and blood and other fluids that are expelled following childbirth.
air sac In insects, a tiny balloonlike structure connected to each trachea which functions in breathing.
allantois (uh-LAN-tuh-wis) In birds, some reptiles, and some mammals, a membranous bag inside the eggshell that collects the embryo's wastes; functions in gas exchange.
allele (uh-LEEL) Alternate version of a gene for a trait.
alternation of generations Plant life cycle that alternates between a sexual stage and an asexual stage.
alveolus Air sac in a lung, where gas exchange occurs.
amino (uh-MEE-nō) **acid** Chain of protein units.
ammonia A highly poisonous nitrogenous waste.
ammonification The work done by bacteria decomposing protein and releasing ammonia.
amnion (AM-nee-un) Second membrane to form around the embryo in birds, reptiles and mammals.
amniotic fluid Liquid that fills the amnion, providing a watery environment in which an embryo develops.
anaerobic (AN-ayr-Ō-bik) Able to thrive where there is no oxygen present.
analogous (uh-NAL-uh-gus) **structures** Structures in different organisms that do similar jobs but develop in different ways.
anaphase (AN-uh-FAYZ) Second stage of mitosis.
annual ring An alternating light or dark inner ring of a tree that marks the growth of the tree.
anther One main part of a stamen, which produces the male reproductive cells in a flower.
anthropomorphism (AN-thruh-puh-MŌR-fizm) Attributing human attitudes to other organisms.
antibiotic (AN-tih-bī-AHT-ik) A chemical substance produced by fungi that inhibits the growth of bacteria.
antibody A chemical substance produced by certain white blood cells to combat antigens.
antigen (AN-tih-jen) A foreign substance in the body.
anus (AY-nus) An opening (in the body) at the end of the digestive tract, through which undigested food passes.
anvil Bone in the middle ear that conducts vibrations from the eardrum to the oval window.
aorta (ay-OR-tuh) Largest artery in the body; leads from the heart and branches into smaller and smaller vessels.
apical (AYP-uh-kul) **meristem** An area of rapid cell division found at the tips of stems and roots.
aqueous humor A watery fluid that nourishes the eye and fills the area between the cornea and the lens.
arteriole A tiny artery that feeds blood into capillaries.

artery A thick-walled tube that carries blood away from the heart.
asexual (ay-SEK-shoo-wul) **reproduction** The production of offspring by one parent.
atom The smallest particle of an element.
atrium One of two thin-walled chambers at the top of the heart that receives blood from veins.
auditory canal Part of the outer ear that leads to the tympanic membrane.
auditory nerve A nerve within the ear that carries nerve impulses from the cochlea to the brain.
autonomic nervous system Part of the peripheral nervous system that controls involuntary actions.
autosome (AHT-uh-SOM) Any of the pairs of chromosomes in humans, which are not sex chromosomes.
autotroph (AHT-uh-TROF) An organism that makes its own food by photosynthesis.
axon An extension of the cell body of a neuron, along which an impulse travels.

bacteriophage A virus that attacks bacteria; phage.
bark The tissues outside of the vascular cambium.
basal metabolic energy The minimum amount of energy needed to keep vital internal processes going.
basal metabolic rate The metabolic rate of a resting organism.
behavior The response of an organism to stimuli in its internal or external environment.
bile A fluid used in the digestive process that is secreted by the liver and stored in the gallbladder.
bile salts Found in bile; they break up fats and oils.
binary fission An asexual process whereby one parent cell divides into two offspring cells of equal size.
biogenesis The idea that all living things arise from other living things; life from life.
biological clock A timer inside the body that keeps cycles going.
biology The branch of science that deals with living things.
biome (BI-ōm) A large ecosystem occurring in a major land region. Identified by its climax vegetation.
blastula (BLAS-choo-luh) Stage in an animal embryo's development; a hollow, fluid-filled ball of cells.
bone marrow A tissue found in bones that makes red blood cells for the body's use.
Bowman's capsule The cupped end of a nephron.
brainstem The main part of the brain that attaches to the spinal cord.
brain wave An electrical impulse given off by the brain.
bronchus A small tube that enters each lung and functions in breathing.
bronchiole Microscopic branch of the bronchus.
budding A process of asexual reproduction in which the parent's body produces a miniature version of itself, known as a bud.

calyx (KAY-liks) The outer ring of flower parts, which is composed of individual, leaflike parts called sepals.
cambium (KAM-bee-um) A thin layer of cells that produces secondary growth in plants.
canopy In a rain forest, a kind of ceiling formed by tree branches and leaves.
capillarity The tendency of liquids to cling to the sides of narrow vessels.
capillary (KAP-uh-LAYR-ee) The smallest blood vessel in the transport system; connects an artery with a vein.
carbohydrate (KAR-buh-HI-drayt) An organic compound that contains only carbon, hydrogen, and oxygen.
carcinogen (kar-SIN-uh-jen) A cancer-causing chemical.
carnivore Animal that exists solely on a meat diet.
carrier A female heterozygous for a sex-linked trait, who exhibits the normal phenotype and who may pass the trait to her offspring as a recessive gene.
carrier molecule A molecule in the cell membrane that transports materials passively into and out of the cell.
carrying capacity Term that is used to describe the population size that an ecosystem can support.
cartilage (KART-il-ij) A flexible, elastic tissue that supports parts of the body and lines the surfaces of joints.
caste A physically different group within a colony.
catalyst (KAT-ul-ist) A substance that affects the rate of a chemical reaction.
cell The smallest unit of life which makes up the bodies of all living things.
cell body Part of a neuron that contains the nucleus and organelles.
cell membrane A living, thin, semipermeable membrane that surrounds all cells.
cell plate In cell division in plants, the structure that forms between the two nuclei as the cytoplasm divides.
cellular respiration The process by which cells obtain energy from food.
cellulose (SEL-yoo-lōs) A fibrous or woody material found in the cell walls of plants and fungi. It is a polysaccharide.
cell wall A nonliving, semi-rigid container that surrounds cell membranes in plants, monerans, and fungi.
central nervous system In vertebrates, the brain and spinal cord.
centriole (SEN-tree-ōl) A small, cylindrical structure made up of microtubules; found in animal cells.
centromere (SEN-truh-mir) The point where chromatids are attached to each other.
cerebral cortex The outer part of the cerebrum.
cerebrospinal (suh-REE-brō-SPI-nul) **fluid** A liquid substance that cushions the central nervous system.
cerebellum (SAYR-uh-BEL-um) The main part of the brain located behind and below the cerebrum.
cerebrum (suh-REE-brum) The main part of the brain located behind the forehead.
cervix (SER-viks) The opening of the uterus.
chemical bond The bond that holds atoms together when they combine.
chemical equation A statement in which symbols and formulas are used to describe a chemical reaction.
chemical reaction A process in which chemical bonds between atoms are broken, rearranged, and new bonds are formed, resulting in new compounds that have properties different from those of the original compounds.
chemoreceptor (KEE-mō-rih-SEP-tur) A receptor that detects chemicals.
chitin (KIT-in) A complex carbohydrate that makes up the exoskeleton of most insects.

chlorophyll (KLOR-uh-FIL) A green substance found in the chloroplasts of plants. Functions in photosynthesis.

chloroplast (KLOR-uh-PLAST) A plant cell organelle that has a double membrane and contains chlorophyll.

chorion (KOR-ee-AHN) The first and outermost membrane to form around the embryo in mammals, birds, and reptiles.

choroid (KOR-oyd) The middle layer of the eye, which contains blood vessels and black pigment.

chromatid (KRŌ-muh-tid) The name for each chromosome of a double chromosome following replication.

chromosome (KRŌ-muh-sōm) A threadlike strand of DNA molecules found inside the cell. It is composed of genes.

cilia (SIL-ee-uh) Tiny, hairlike projections covering the surface of some protozoa and some cells in multicellular organisms.

circadian (sur-KAY-dee-un) **rhythm** A rhythm that repeats approximately every 24 hours.

citric acid cycle The second stage of cellular respiration.

cleavage The repeated cell division of a zygote.

climax community The last residents to live in an area; it represents a balanced community.

clone All the organisms produced from one parent during asexual reproduction.

cochlea (KAHK-lee-uh) Coiled structure in the inner ear that picks up vibrations from the oval window.

codon Code words that make up the genetic code; each codon consists of three nucleotide bases.

coldblooded Having a body temperature that changes with the surrounding temperature.

collecting tubule A tubule which connects to the tubule of a nephron in the kidney.

colon (KŌ-lun) Another name for the large intestine.

colony A population of a particular species living together.

commensalism (kuh-MEN-sul-izm) A type of symbiosis in which one of the organisms benefits and the other is not affected.

community A group of different kinds of organisms living and interacting in an area.

compact bone A type of bone tissue found along the length of the long bones in the body.

competition Rivalry between individuals for a specific resource or life necessity.

compound A chemical substance that is made up of two or more elements.

compound eye A hemispherical structure made up of many small regular surfaces called facets.

cone One of the two types of light receptor cells contained in the retina; involved in color vision.

consumer An organism that does not make its own food, depends on other living things for food.

contractile vacuole A vacuole found in the paramecium, which fills up with excess water and pumps the water out.

controlled experiment An experiment that tests one variable factor against a control.

cork The protective outer layer in woody plants.

cork cambium Produces cork in woody plants.

cornea (KOR-nee-uh) The rounded, transparent part of the sclera at the front of the eye.

corolla (kuh-RŌL-uh) A structure found inside the calyx of a flower, made up of petals.

corpus callosum (KOR-pus kuh-LŌ-sum) A bridge of tissue that connects the two hemispheres of the cerebrum.

corpus luteum (KOR-pus LOO-tee-um) A temporary endocrine gland formed when an empty follicle in the human ovary is stimulated by LH; secretes progesterone.

cotyledon (KOT-ul-EED-un) A seed leaf that provides food for the plant embryo after germination.

covalent (kō-VAY-lunt) **bond** A type of bond that forms between atoms that share some of their outer electrons.

crista A shelflike structure formed by folding of the inner membrane of the mitochondrion.

crop A food storage organ.

crossing-over During meiosis, the exchange of parts of two homologous chromatids.

cuticle (KYOOT-ih-kul) A waxy layer covering the surface of a leaf, which is secreted by the leaf epidermis.

cytoplasm (SĪT-uh-plazm) A substance that is found in the space between the nucleus and the cell membrane.

cytoplasmic division The second stage of cell division.

cytosine (SĪT-uh-SEEN) A nitrogen base of DNA; pairs with guanine.

dark reactions The second stage of photosynthesis.

deciduous Kind of tree that loses its leaves in the fall.

decomposer An organism that breaks down the bodies of other dead organisms and the waste products of living organisms into simple chemical substances.

dehydration synthesis A common kind of synthesis in cells, in which a water molecule "splits off" from the reacting molecules at the site where a new bond is formed.

dendrite An extension of a neuron's cell body, which picks up impulses and transmits them to the cell body.

deoxyribonucleic (dee-AHK-see-RĪ-bō-noo-KLEE-ik) **acid (DNA)** Nucleic acid found in all cells; it plays a major role in heredity and in controlling the cell's activities.

diaphragm (DĪ-uh-FRAM) A dome-shaped muscle located at the bottom of the chest cavity.

dicot (DĪ-kot) A plant whose seeds have two cotyledons.

differentiation Specialization, as in formation of sperm.

diffusion The movement of a substance from a place of high concentration to a place of low concentration.

digestion A life function in which food is broken down and made usable for cells.

diploid (DIP-loyd) **number (2n)** The number of chromosomes in a body cell with a full set of homologous chromosomes.

disaccharide Two sugar units, (or simple sugars) bonded together; another name for double sugar.

diurnal (dī-YOOR-nul) Day-night cycle; active during the daytime.

division of labor The specialization of body parts into tissues that perform different functions.

dominant Term applied to an inherited trait which is expressed in the offspring when either one or both genes for the trait are present.

dominant generation The most notable form in the life cycle of plants exhibiting alternation of generations.

double fertilization The process of fertilization by two sperm; occurs only in flowering plants.

double helix The double-spiral shape of the DNA molecule.

drone A haploid male bee that develops from an unfertilized egg. Its only function is to produce sperm.
drug A substance used in medicine; can help a patient regain physical or mental health.
dynamic equilibrium The tendency of a biological system to remain in balance.

ecological succession The process of change from a pioneer community to a climax community.
ecosystem (EE-kō-SIS-tum) All the living and nonliving things in an area that interact and exchange materials.
ectoderm (EK-tuh-DERM) In animal embryonic development, the outer layer of the gastrula.
effector The part of an organism that responds to information transmitted from a receptor. A gland is an example.
egestion (ee-JES-chun) The process of removing undigested wastes from the body.
egg A female reproductive cell; also called egg cell.
electron (ih-LEK-tron) A negatively charged particle of an atom, which surrounds the nucleus of the atom.
electron carrier A biological molecule that transports electrons (e^-), protons (H^+), and energy within a cell.
electron transport The third stage of cellular respiration.
element Any of 100 simple substances that cannot be separated into different kinds of matter.
embryo (EM-bree-ō) A developing zygote; the organism formed before germination, hatching, or birth.
embryo sac A group of seven cells produced from cytoplasmic division; functions as the female gametophyte in flowers.
endocrine gland Hormone-producing gland in animals that releases hormones directly into the blood.
endocytosis (EN-dō-sī-TŌ-sis) The process of bringing large molecules and particles into a cell.
endoderm (EN-duh-DERM) In animal embryonic development, the inner layer of the gastrula.
endoplasmic reticulum (EN-dō-PLAZ-mik ri-TIK-yoo-lum) The elaborately folded system of membranes running through the cytoplasm of most cells.
endoskeleton An internal skeleton.
endosperm Stored food that will be used by the developing embryo during the process of germination in plants.
energy The ability to do work or to cause change.
enzyme (EN-zīm) Protein that functions as a catalyst in biological reactions.
epidermis (ep-uh-DUR-mis) In plants, a thin layer of cells that lines and protects the outside of the plant.
epiglottis (EP-uh-GLOT-is) Flap of tissue in the throat that functions to keep food out of the air passage.
erythrocyte (ih-RITH-ruh-sīT) Red blood cell.
esophagus (ih-SAHF-uh-gus) Part of the digestive tract that allows food to pass from the pharynx to the stomach.
estrus (ES-trus) The female reproductive cycle in most mammals other than humans.
estuary Body of water that forms where a river meets the ocean.
eukaryote (yoo-KAYR-ee-ōt) An organism that has cells containing a true nucleus.
Eustachian (yoo-STAY-shun) **tube** A tube that connects the middle ear to the pharynx.
evolution The changing of species over time.

excretion A process whereby an organism's cells are cleansed of wastes; maintains salt and water balance.
exocytosis The process of release of large molecules by cells.
exoskeleton A skeleton that is found outside the body.
experimental variable A single difference between the experimental organisms and the control organisms.
extensor A muscle that straightens a body part.
extinction The dying out of a species.
eyespot A spot of pigment that detects light and darkness but does not form images or detect movement.

facet (FAS-it) One of the many surfaces of a compound eye; the lens of an ommatidium.
feces (FEE-seez) Waste product of digestion expelled from the intestines.
fermentation An incomplete form of cellular respiration.
fertilization The joining, or fusion, of a sperm with an egg.
fetus (FEET-us) Term used to describe the human embryo after the eighth week of pregnancy.
fibril (FĪ-brul) One of the narrow parts of a fiber in skeletal muscle; composed of filaments.
fibrin (FĪ-brin) Protein threads that form a mesh at the site of damage in a blood vessel wall; aids in blood clotting.
filament In muscle cells, the structural component of fibrils; made of protein. In seed plants, the stalk of the stamen, the male reproduction organ.
filtrate The plasma in a Bowman's capsule, resulting from movement of plasma from the glomerulus.
filtration A kidney function that involves the movement of plasma from the glomerulus into the Bowman's capsule.
flagellum (fluh-JEL-um) A long, hairlike structure at the front of some protozoa; moves by a whiplike action.
flexor A muscle that bends a body part.
flower The specialized reproductive structure of the angiosperm.
follicle An egg sac in the human ovary.
food cavity The place where food is taken during ingestion.
food chain A sequence of organisms through which food passes in a community, beginning with producers.
food web The different food chains in an area which are interrelated to form a kind of feeding network.
fovea (FO-vee-uh) The small depression in the middle of the retina which is the site of the eye's sharpest vision.
fragmentation Asexual reproduction in which small pieces of an organism's body grow into new organisms.
fruit A ripened ovary of a seed plant.

gallbladder A sac that stores bile secreted by the liver.
gamete (GAM-eet) A specialized reproductive cell found in organisms which reproduce sexually.
gametogenesis (guh-MEET-uh-JEN-uh-sis) The process of producing gametes.
gametophyte (guh-MEET-uh-FĪT) The sexual stage of a plant life cycle; a plant that produces gametes.
ganglion A clump or cluster of cell bodies of neurons.
gastric glands Glands in the walls of the stomach.
gastric juice Produced by the gastric glands; contains enzymes that mix with food and promote digestion.

gastrula (GAS-troo-luh) In animal embryonic development, a pouchlike structure formed from the blastula and made up of two layers.
gene The segment of DNA that carries the code for a specific protein.
gene splicing Introducing a short piece of DNA from one organism into the DNA of another.
genetic code The order of the bases in DNA. Made up of codons, or triplets, of nucleotide bases.
genetic recombination The changing of the gene combination on DNA.
genetics (juh-NET-iks) The study of heredity.
genotype (JEE-nō-tīp) Symbols which represent an organism's gene combination.
germination The sprouting of a seed.
gestation (jes-TAY-shun) The period of time between fertilization of the egg and birth of the offspring.
gizzard Muscular portion of the digestive tract in birds which helps grind food; a second stomach.
glial (GLEE-ul) **cell** Type of cell that surrounds a neuron and that appears to nourish and protect the neuron.
glomerulus (glah-MER-yoo-lus) The capillary network inside a Bowman's capsule; functions during filtration.
glucose (GLOO-kōs) The most common simple sugar. The main product of photosynthesis.
glycogen (GLĪ-kuh-jun) Carbohydrates stored in the liver and muscle tissues of humans and other animals.
glycolysis (glō-KOL-uh-sis) The first stage of cellular respiration.
Golgi (GOL-jee) **apparatus** A cell organelle composed of stacks of flattened membrane-bounded sacs.
gonad (GŌ-nad) One of the pair of reproductive organs in an animal; produces haploid gametes.
grafting A type of artificial vegetative propagation in which a stem cut from one plant, the scion, is attached, or grafted, to a rooted growing plant, the stock.
granum (GRAN-um) A structure found within the chloroplast, composed of thylakoid discs stacked in piles.
gray matter Gray areas of the brain and spinal cord that contain masses of nerve cell bodies and glial cells.
guanine (GWAH-neen) A nitrogen base of DNA; pairs with cytosine.
guard cell One of the specialized paired cells in a leaf that open and close the stomates.

habitat The place where an organism lives.
habituation (huh-BICH-oo-WAY-shun) A type of learning by which an animal learns not to respond to stimuli.
hammer Bone in the middle ear that conducts vibrations from the eardrum to the oval window.
haploid (HAP-loyd) **number (n)** Half of the diploid number of chromosomes.
heart valve A valve that exists between the atria and ventricles or between the ventricles and arteries leading away from the heart; prevents back flow of blood.
heartwood The older xylem toward the center of a tree which becomes filled with waste products.
hemoglobin (HEE-muh-GLŌ-bin) Protein in blood that carries oxygen.
herbivore (ERB-bih-vōr) An animal that eats only plants.

heredity The transmission of traits or characteristics from parents to offspring.
heterogamete A gamete that is different from another gamete in size and shape.
heterotroph (HET-ur-uh-TROFS) An organism that does not make its own food; obtains food from other living things.
heterozygous (HET-ur-ō-ZĪ-gus) Describes a genotype in which two genes for a trait are different.
hibernation A deep sleep in which the body systems function much more slowly.
homeostasis (HO-mee-uh-STAY-sis) The maintenance of a balanced internal environment.
homologous chromosomes A pair of chromosomes.
homologous (hō-MOL-uh-gus) **structures** Similar structures that develop in the same way in different organisms.
homozygous (HŌ-mō-ZĪ-gus) Describes a genotype in which genes for a trait are alike.
hormone A chemical carried in the blood which regulates body functions in complex plants and animals.
host The organism that a parasite lives on or in.
hybrid An organism that results from a cross of parents having different expressions of the same trait.
hybrid vigor Term used to describe a hybrid that is sturdier and more resistant to disease than its parents.
hydrolysis (hī-DROL-uh-sis) Biological reaction in which water is taken up at a broken bond; digestion.
hypothalamus (HĪ-pō-THAL-uh-mus) Part of the brain located below the thalamus at the bottom of the cerebrum; acts as an endocrine gland and is the center for the sex drive and certain emotions.
hypothesis (hī-PAHTH-uh-sis) A possible explanation of observations.

immune reaction The reaction that occurs between antibody and antigen.
immunity A body condition in which a person is protected against disease.
imperfect flower A flower that has either male or female reproductive organs—but not both.
imprinting A type of learning in which animals, during their early life, follow the first moving thing they see.
impulse A signal carried between nerve cells.
incomplete dominance The effects of two genes, seeming to blend in the phenotype of the offspring.
infrared radiation A kind of heat wave.
ingestion The process of taking food into the body.
inorganic compound Compounds that do not contain carbon and hydrogen.
instinct A complex stereotyped behavior.
interphase Period during which a cell is not dividing.
intestinal juice A juice secreted by the intestinal glands; contains enzymes that complete the digestion of proteins, carbohydrates, and nucleic acids.
intestine Part of the digestive tube in animals.
invertebrate An animal without vertebrae.
involuntary muscle Smooth muscle that, under most circumstances, cannot be consciously controlled.
ion (Ī-on) An atom that has a charge.
ionic bond A type of bond in which one atom takes electrons from the other atom.

iris The colorful part of the choroid at the front of the eye of most vertebrates.
isogametes (ī-sō-GAM-eet) Gametes that are alike in size and shape.

joint The point where two bones come together.

kidney A major organ of excretion in animals; produces urine.
kinesis Simple form of inherited behavior in which an organism responds to stimuli by moving randomly.
kinetic (kih-NET-ik) **energy** The energy of motion.

labor The process of human childbirth.
lacteal A microscopic lymph vessel in the villus.
larva A small wormlike form of an insect.
larynx (LAYR-anks) The voice box; located in the trachea.
lateral line A series of grooves on a fish's head and sides which contain receptor cells that detect water currents.
lateral meristem An area of rapid cell division that accomplishes secondary growth in plants.
lens A structure in the eye that focuses light.
lenticel (LEN-tih-sel) A group of loosely packed cells on stems and bark that allow gas exchange.
leukocyte (LOO-kuh-SĪT) White blood cell.
life zone A zone of climax plants and animals that varies according to elevation or altitude.
ligament A tough tissue composed mainly of collagen, which attaches bones to each other in movable joints.
light reactions The first stage of photosynthesis.
linkage Two or more genes located on one chromosome.
lipid Fatty substance; includes fats, oils, and waxes.
liver As the largest gland in the body it functions in digestion, excretion, and many metabolic processes.
locus Position in a chromosome.
lymph (LIMF) Tissue fluid absorbed by the vessels of the lymph system; called plasma when it returns to the blood.
lymph node A clump of tissue in the lymph system which filters out bacteria, dead cells, and foreign substances.
lymphocyte (LIM-fuh-SĪT) A white blood cell that produces antibodies.
lysosome (LĪ-so-SOM) A membrane-covered cell organelle that contains digestive enzymes.

Malpighian (mal-PIG-ee-un) **tubule** Excretory organ in the grasshopper.
mammary glands Milk-producing glands in mammals.
mantle A strong membrane that covers the soft body and often the gills of a mollusk and secretes the shell.
mechanoreceptor (muh-KAN-ō-rih-SEP-tur) A receptor that detects a stimulus such as pressure or vibrations.
meiosis (mī-Ō-sis) A special type of cell division that produces haploid cells.
meninges (meh-NIN-jeez) Three layers of tissue that surround and protect the brain and spinal cord.
menstrual (MEN-strul) **cycle** The reproductive cycle of the human female.

meristem (MEHR-uh-STEM) An area of rapid cell division in a plant.
mesoderm (MES-uh-DERM) In animal embryonic development, the middle layer of the gastrula.
messenger RNA (mRNA) One of the three types of RNA found in cells; carries the message for protein synthesis.
metabolic rate The rate at which organisms use energy.
metabolism The sum of all the chemical reactions of an organism.
metamorphosis In insects, stages of development, either complete (egg, larva, pupa, adult) or incomplete (egg, nymph, adult).
metaphase (MET-uh-FAYZ) The third stage of mitosis.
microfilament A solid threadlike structure made of protein and able to create movement by contracting.
microtubule A long, thin, hollow tube found in cilia and flagella that creates movement by contracting.
microvillus Microscopic fold of the surface of the villi in the intestine.
mineral An inorganic substance; important for the formation and functioning of body parts. An essential nutrient.
mitochondrion (MĪT-uh-KON-dree-on) A cell organelle found in eukaryotes; supplies most of the cell's energy.
mitosis (mī-TO-sis) The first stage of cell division.
molecule (MOL-eh-KYOOL) A combination of atoms from one or more elements held together by a covalent bond.
molting The process of shedding an exoskeleton and growing a new one; found in arthropods.
monocot (MON-uh-KOT) A plant whose seeds have one cotyledon. (Corn and lilies are examples.)
monosaccharide (MON-uh-SAK-uh-RĪD) A simple sugar; the basic structural unit of all carbohydrates.
morula (MOR-yoo-luh) In animal embryonic development, a solid ball of cells produced by animal cell division.
motor neuron A neuron that transmits impulses to an effector.
motor unit A motor neuron axon and all the muscle fibers its branches contact.
mucus (MYOO-kus) A thick fluid secreted by the mucous membranes.
muscle fiber One of the fibers that composes skeletal muscle; consists of one cell containing several nuclei.
mutagen A substance that causes mutations.
mutation (myoo-TAY-shun) An accidental change in DNA.
mutualism (MYOO-choo-wul-izm) A symbiotic relationship in which both species benefit from the interaction.
myelin (MI-uh-LIN) A white, fatty material that forms a sheath between the Schwann cells and the axon in some vertebrate neurons.
myosin (MĪ-uh-sin) A protein found in the thicker filaments of fibrils in skeletal muscle.

nectar A sugary solution produced by flowers.
nematocyst (neh-MA-tō-sist) Specialized stingers on the tentacles of hydras and other coelenterates.
nephridium An excretory organ occurring in pairs in earthworms.
nephron A system of tubes that acts as a filter, the working unit of the kidney.
nerve A cable composed of many axons.

nerve net Type of nervous system found in the hydra, in which neurons are scattered throughout the body.
nerve pathway The route traveled by a nerve impulse.
neuron (NER-on) A specialized cell that makes up nervous systems; consists of a cell body, dendrites, and an axon.
neutron (NOO-tron) An uncharged particle found in the nucleus of an atom.
niche (NITCH) What an organism does in an ecosystem.
nicotinamide adenine dinucleotide (NAD$^+$) An electron carrier in cellular respiration.
nicotinamide adenine dinucleotide phosphate (NADP$^+$) An electron carrier in photosynthesis.
nitrate A nitrogen-containing material found in soil.
nitrification The process whereby bacteria change ammonia to nitrates.
nitrite Nitrogen-containing substance produced from the decomposition of ammonia by bacteria.
nitrogen fixation A process in which bacteria convert nitrogen gas to nitrates.
nondisjunction The failure of chromosomes to separate during meiosis.
notochord (NŌT-uh-KORD) A supporting rod of tissue that runs along an animal's back.
nuclear membrane A sac that surrounds the nucleus in all eukaryotic cells.
nucleic acid An organic molecule composed of nucleotides bonded together.
nucleolus (noo-KLEE-uh-lus) A large organelle found in the nuclei of eukaryotic cells; acts in making ribosomes.
nucleotide (NOO-klee-uh-TĪD) Basic structural unit of nucleic acids; consists of a phosphate group, a simple sugar, and a nitrogen base.
nucleus (NOO-klee-us) A structure found in eukaryotic cells that directs the cell's activities and holds the code for heredity. Also, the dense area in the center of an atom.
nymph (NIMF) A stage in incomplete metamorphosis in insects; the juvenile form of an organism.

ommatidium (AHM-uh-TID-ee-um) Unit of a compound eye.
omnivore Animal that eats both plants and animals.
oogenesis (ō-uh-JEN-uh-sis) The process of egg formation.
oral groove A channel located on the side of a paramecium, into which cilia sweep bacteria, protozoa, and other food.
organ A body structure composed of several tissues that work together to perform a certain function.
organelle (or-guh-NEL) A small structure in a cell.
organic compound Compounds that contain carbon and hydrogen; also may contain oxygen and other elements.
organism A living thing.
osmosis (os-MŌ-sis) The diffusion of water across a semipermeable membrane.
oval window A membrane in the inner ear.
ovary (Ō-vuh-ree) A female reproductive organ.
oviduct (Ō-vuh-dukt) A tube through which an egg passes from an ovary to the uterus.
ovulation (ah-vyoo-LAY-shun) A stage of the menstrual cycle in which an egg is released from the ovary.
ovule A part of the ovary of a flower in which female spores are produced.
ovum An egg produced in the ovaries of a female animal.

pacemaker Another name for sinoatrial node.
palisade layer A dense layer of cells containing chloroplasts located beneath the upper epidermis of a leaf.
pancreas A gland that secretes pancreatic juice into the small intestine during digestion; also produces insulin.
pancreatic (PAN-kree-AT-ic) **juice** A digestive juice secreted by the pancreas and used in the small intestine.
parasite An organism that obtains food directly from the live organisms in or on which it lives.
parasympathetic nervous system A division of the autonomic nervous system that controls vital organs.
parathyroid (PAYR-uh-THĪ-royd) **gland** One of four small glands located just behind the thyroid gland.
parthenogenesis (PAR-thuh-nō-JEN-uh-sis) The development of an organism from an unfertilized egg.
passive transport The movement of materials by diffusion and osmosis.
pathogen (PATH-uh-jen) An infectious organism that makes its host ill.
penis (PEE-nis) Part of the male reproductive system in many animals through which semen and urine pass to the outside of the body.
peptide bond A bond between amino acids.
perfect flower A flower that contains both male and female reproductive organs.
peripheral nervous system All the nerves and ganglia that exist outside of the brain and spinal cord.
peristalsis (PAYR-ih-STOL-sis) The wavelike contractions caused by muscles in the wall of the digestive tract.
permafrost Soil that is permanently frozen.
permeable Allows most materials to pass through freely.
petal A flower part which is often brightly colored and may produce special odors and nectar.
pH A measure of how acidic or basic a solution is.
pharynx (FAYR-anks) The throat; located behind the mouth and before the larynx and esophagus.
phenotype The expression of an organism's genotype.
pheromone (FAYR-uh-MŌN) A chemical signal produced by one animal and transmitted to another animal.
phloem (FLŌ-um) System of narrow tubes in plants that transports sugars and other nutrients.
photoperiodism The influence of light duration on an organism. Affects flowering in plants.
photoreceptor A receptor that detects light.
photosynthesis (FŌT-ō-SIN-thuh-sis) Process by which organisms that contain chlorophyll produce sugar from carbon dioxide and water in the presence of light.
phototropism The tendency of stems and leaves of a plant to grow in the direction of a light source (usually the sun).
pioneer community The earliest community of plants, insects, and animals that settle in an area.
pistil The female reproductive organ in a flower.
pituitary (pih-TOO-ih-TAYR-ee) **gland** A pea-size gland at the base of the brain, composed of two lobes.
placenta (pluh-SENT-tuh) In mammals, the organ through which materials pass between embryo and mother during pregnancy.
platelet (PLAYT-let) A cell fragment carried by the plasma, which functions in blood clotting.
polar body A small haploid cell formed during oogenesis.
polarized The state of a neuron membrane in which the outside has a positive charge relative to the inside.

pollen Haploid male spores produced in the stamens of flowers.
pollen grain The male gametophyte in flowering plants.
pollen tube A tube formed after pollination, originating from the second nucleus of a pollen grain.
pollination The transfer of a pollen grain from the anther to the stigma.
polypeptide chain A series of amino acids bonded together; a protein.
polyploid Having more than two sets of chromosomes.
polysaccharide Long-chain carbohydrates bonded together by many simple sugars.
polyunsaturated fat A fat that has fatty acids with more than one double bond.
population All the organisms of a species in an area.
population crash A dramatic reduction in a population's size.
population explosion Uncontrolled population growth.
potential energy Inactive or stored energy.
predation A symbiotic relationship between predators and prey.
predator An organism that kills another living animal and feeds on it.
pregnancy The period of development of the embryo in the uterus.
prey The animal that serves as food for a predator.
primary growth The lengthwise or vertical growth of stems and roots.
producer An organism that makes its own food.
product A compound produced in a chemical reaction.
prokaryote (prō-KAYR-ee-ōt) An organism that lacks a true nucleus.
prophase (PRŌ-fayz) The first stage of mitosis.
proprioceptor (PRO-pree-uh-SEP-tur) A type of receptor that detects body movements and position.
protein (PRŌ-teen) A polypeptide chain. A large molecule consisting of a chain of amino acids bonded together.
protein synthesis The process by which protein is made.
proton (PRŌ-ton) Very small particle found in the nucleus of an atom; has a positive electrical charge.
pseudopod (SOO-duh-POD) A false foot; the extended part of the body of a protozoan, such as the ameba.
puberty (PYOO-ber-tee) A stage in human development during which the production of gametes begins.
pulmonary circulation The circulation of blood from the heart to the lungs and back to the heart.
pupa (PYOO-puh) A stage during complete metamorphosis in insects, in which the insect lives inside a protective covering and obtains nourishment from stored food.
pupil The opening in the center of the iris in an eye.
pylorus (pī-LOR-us) A valvelike muscle in the stomach, which opens to allow partially digested food to move from the stomach into the small intestine.

queen Egg-laying female caste found in social insects.

radioactive decay The process during which radioactive substances change to a nonradioactive form.
ray In woody plants, conducting tissue that transports materials laterally.
reabsorption One of the processes involved in waste removal by the kidneys in which materials pass from the tubule of a nephron back into the transport system.
receptacle The base of a flower.
receptor A kind of sensor that picks up information about an organism's internal or external environment.
recessive Term applied to an inherited trait whose expression is hidden in a heterozygous offspring, but which is expressed in its homologous form.
recombinant DNA DNA with a gene combination that has been changed by genetic recombination.
recombination Another name for crossing-over.
rectum The end of the large intestine.
red blood cell A disc-shaped blood cell containing hemoglobin; involved in oxygen transport; erythrocyte.
reflex An immediate response to a stimulus.
reflex arc A simple nerve pathway consisting of a receptor, a sensory neuron, one or more interneurons, a motor neuron, and an effector.
reinforcer A factor that makes an organism repeat a behavior—food, water, praise, money, and attention.
related organism One of a group of organisms believed to have the same or common ancestors.
releaser A specific stimulus in the environment that causes an organism's inherited response.
respiration The process of gas exchange in organisms.
respiratory center One of two centers within the brain that respond to the level of carbon dioxide in the blood.
reticular activating system (RAS) A complex nerve pathway located in the brainstem which allows a person to be conscious and alert.
retina The inner layer of the eye.
ribonucleic acid (RNA) A nucleic acid that exists in three forms, each having a function in protein synthesis.
ribosomal RNA (rRNA) One of the three types of RNA; the main structural component of ribosomes.
ribosome (RĪ-buh-sōm) Beadlike cellular organelle found in the cytoplasm of cells, located on the rough endoplasmic reticulum; assembles proteins.
ritual A form of inherited social behavior found in animals; often used to convey a message, assess each others' strength, and prevent injury by controlling aggression.
rod A rod-shaped cell in the retina that contains light-sensitive pigments that are responsible for black and white vision.
root hair A fine projection of a root epidermal cell that extends outward into the soil.
royal jelly The substance fed to bees developing into queens in special cells in a hive.
runner A type of stem in plants.
rutting A mating period during which the males of certain animals change in behavior and physical appearance.

sapwood The lighter-colored, conducting wood of a tree.
saturated fat A fat containing fatty acids with only single bonds between carbon atoms.
scavenger An animal that eats animals that it does not kill.
Schwann cell Type of glial cell that forms a covering around the axons of many vertebrate neurons.
scion (SĪ-un) A stem cut from a plant which is to be grafted onto another plant.

sclera (SKLIR-uh) The tough, protective outside layer of a vertebrate eye; the white of the eye.
scrotum (SKROT-um) Sac that contains the testes.
secondary growth Growth which causes plants' stems, branches, and roots to increase in width.
secondary sex characteristic Any of the physical changes associated with the change in hormone levels at puberty.
secretion An active transport process in the kidney in which ions or molecules are removed from blood.
seed A plant embryo that is surrounded by food and wrapped in a protective seed coat.
seed coat The hardened walls of an ovule that forms a protective covering for a seed.
semen (SEE-mun) A fluid produced by male animals that contains sperm and accompanying secretions.
semicircular canal A structure found in the inner ear which contains fluid that moves when the head is moved.
seminiferous (SEM-uh-NIF-er-us) **tubules** Loops of folded tubes inside a testis.
semipermeable membrane A membrane that allows only certain substances to pass through.
sensory neuron A neuron that transmits impulses from a receptor.
sepal (SEE-pul) A leaflike part of a flower.
septum A solid wall of tissue that separates the left and right sides of the heart.
seta A bristle that protrudes from the body wall in an earthworm.
sex chromosomes The X and Y chromosomes in a male and female that determine the sex of an offspring.
sexual reproduction Reproduction that requires the joining of the nuclei of two types of specialized cells.
shell Outside of an egg that protects the embryo.
simple eye An eye that detects movement and forms crude images at close range.
simple sugar The basic structural unit of all carbohydrates; also known as a monosaccharide.
sinoatrial (S-A) node A small mass of tissue on the wall of the right atrium that sends electrical impulses across the heart in rhythmic waves; also known as the pacemaker.
skeletal muscle Muscle involved in moving the bones of the skeleton.
smooth muscle Unstriated muscle composed of spindle-shaped cells interlaced with one another.
sodium pump A mechanism in the cell membrane of a neuron that actively transports sodium ions out of the cell and potassium ions into the cell.
somatic nervous system Part of the peripheral nervous system that transmits impulses for voluntary actions.
sperm Male reproductive cell produced in the testes.
spermatogenesis (spur-MAT-uh-JEN-uh-sis) The process of sperm production.
spinal cord A nerve cord that runs down the center of the back in vertebrates and which is enclosed in vertebrae.
spindle The structure of spindle fibers stretched from centriole to centriole across the entire cell.
spindle fibers Fibers made of microtubules which form between centrioles.
spiracle (SPEER-uh-kul) Hole in the side of a grasshopper's body through which gas exchange occurs.
spongy bone Type of bone tissue found at the ends of the long bones in the body; has many spaces and is very light.
spongy layer A loosely arranged layer of cells in a leaf, found beneath the palisade layer.
spontaneous generation Another name for abiogenesis.
sporangium A specialized reproductive organ found on the underside of fern leaves.
spore A reproductive cell that contains a nucleus and a small amount of cytoplasm, usually having a hard protective covering that resists drying.
spore formation Asexual reproduction by means of spores.
sporophyte (SPOR-uh-FIT) In plants, the asexual stage in their life cycle. A plant resulting from a diploid zygote.
stamen (STAY-mun) Male reproductive organ in a flower.
starch A polysaccharide made up of glucose units.
stigma The top part of a pistil in a flower.
stimulus An environmental factor that has an effect on an organism.
stirrup Bone in the middle ear that conducts vibrations from the eardrum to the oval window.
stock A rooted, growing plant to which a cutting is grafted.
stomach A receptacle for food after it passes through the pharynx and esophagus.
stomate (STO-mayt) One of many small openings in the surface of a leaf through which gas exchange occurs.
stroma (STRO-muh) A fluid or gel-like substance that surrounds grana in a leaf's chloroplast.
structural formula A written formula that shows the arrangement of atoms in a molecule.
style A slender stalk in the pistil of a flower located below the stigma and above the ovary.
substrate A compound upon which an enzyme acts during a chemical reaction.
symbiosis (SIM-bi-O-sis) The relationship of two kinds of organisms that interact closely and repeatedly.
sympathetic nervous system A division of the autonomic nervous system that controls vital organs.
synapse (SI-naps) A microscopic space that exists between the axon of one neuron and the dendrites of another.
synapsis (sih-NAP-sis) Pairing of homologous chromosomes during the process of cell division.
synaptic knob A knob at the tip of end fibers of an axon.
synthesis The building of large molecules from small ones.
systemic circulation The circulation of blood from the heart to the body and back to the heart.

target structure A muscle, gland, or individual cell whose activity is regulated by a hormone.
taste bud A receptor for taste.
taxis A movement directly toward or away from a stimulus.
telophase (TEL-uh-FAYZ) The final stage of mitosis.
template The DNA strand on which mRNA forms.
tendon A connective tissue which attaches muscle to bone.
territoriality The behavior an animal exhibits while occupying and defending its territory.
testis One of the paired gonads of the male reproductive system in animals; produces sperm and male hormones.
tetrad A homologous pair of chromosomes, consisting of four chromatids and two centromeres.
thalamus A structure at the bottom of the cerebrum that transmits impulses between the cerebrum and brainstem below.

thermoreceptor A receptor that detects warmth or cold.
thylakoid (THI-luh-koyd) **disc** A flattened membrane sac found in the chloroplast.
thymine (THI-meen) A nitrogen base of DNA; pairs with adenine.
thyroid (THI-royd) **gland** An H-shaped structure located in the neck; produces thyroxin.
timberline The elevation above which trees do not grow.
tissue A group of similar cells that work together.
tissue fluid The plasma that is pushed out of the artery end of capillaries into the spaces around cells.
trachea (TRAY-kee-uh) Part of a system of tubes that carries air to and from cells in an insect. In humans, a tube located between the pharynx and the bronchus.
trait A characteristic.
transfer RNA (tRNA) One of the three types of RNA found in cells; decodes the information carried by mRNA.
transmitter substance A chemical contained in sacs within the synaptic knob of an axon; transmits nerve impulses.
transpiration (TRAN-spuh-RAY-shun) The process o evaporation of water from the stomates of leaves into the air.
tropism (TRŌ-pizm) The growth movements of certain plant parts toward or away from stimulus.
tubule Long coiled part of a nephron in the kidney.
turgor movement A plant's movement due to changes in water content of various parts of the plant.
twitch The action of a muscle fiber, consisting of a brief contraction followed by relaxation.
tympanic (tim-PAN-ic) **membrane** A thin sheet of cells within the ear that detects and transmits vibrations.

ultraviolet light Part of the light spectrum that is not seen by humans.
umbilical (um-BIL-ih-kul) **cord** Cord that attaches the embryo to the placenta in mammals.
unsaturated fat Fat that has fatty acids with double bonds between atoms.
uracil (YUR-uh-sil) A nitrogen base of RNA; bonds with the DNA nucleotide adenine.
urea (yoo-REE-uh) The substance to which ammonia is converted in many organisms.
ureter (yoo-REE-ter) A tube through which urine leaves each kidney.
urethra (yoo-ree-thruh) Duct through which urine from the bladder leaves the body.
uric acid A nitrogenous waste excreted as a powder or paste by reptiles, insects, and birds.
urinary bladder Structure that receives urine from the ureters and kidneys and stores it until it can be excreted.
urine Liquid produced in the kidneys, which consists of urea, water, and other wastes removed from the blood.
uterus (YOOT-er-us) A structure of the female reproductive system, in which the embryo and fetus develop.

vaccination (VAK-suh-NAY-shun) Immunization; the introduction of antigen into a body.
vaccine (vak-SEEN) The antigen that is introduced into the body during vaccination.
vacuole (VAK-yoo-ōl) A saclike structure in a cell which usually contains solutions of salts and food materials.

vagina (vuh-JI-nuh) Part of the female reproductive system in animals; a channel extending from the uterus to the outside of the body.
variation An inherited difference that exists between members of a species.
vascular (VAS-kyoo-ler) **bundle** A plant stem structure consisting of xylem and phloem.
vascular cambium A type of cambium that produces secondary growth of xylem and phloem in plants.
vascular cylinder A plant root structure containing xylem and phloem surrounded by one or more layers of cells.
vascular tissue Plant tissue made up of hollow tubes, or veins, that support the plant and carry food and water.
vas deferens Part of the male reproductive system in animals; a duct that stores sperm.
vegetative propagation Asexual reproduction in plants, in which a whole new plant is reproduced from a growing portion of one parent plant.
vein In plants, vascular tissue in the spongy layer of a leaf. In animals, vessel that transports blood back to heart.
vena cava One of two large veins that return blood to the heart.
ventral nerve cord A double nerve cord that runs along the entire underside of an earthworm.
ventricle One of two thick-walled chambers at the bottom of the heart which pumps blood into the arteries.
venule A tiny vessel that receives blood leaving capillaries.
vertebra (VER-teh-bruh) One of a series of bones that run down the center of a vertebrate's back.
vertebrate (VER-tuh-BRIT) An animal with a backbone.
vestigial (ves-TIJ-ee-ul) **organ** A structure or body part that has no apparent use.
villus One of the microscopic, fingerlike projections of some of the cells lining the digestive tract.
visible light Part of light spectrum seen by humans.
vitamin An organic substance that forms part of the structure of many enzymes.
vitreous (VIT-ree-us) **humor** A transparent jelly that fills the area between the lens and the retina.
voluntary muscle Skeletal muscle that is under conscious control.

warmblooded The ability to maintain a constant body temperature.
white blood cell A leukocyte; functions in body defense.
white matter White areas of the brain and spinal cord that contain myelin-covered axons.
worker A member of a female caste in social insects, such as bees and ants; usually sterile and cannot reproduce.

xylem (ZĪ-lum) Narrow tubes that carry water and provide support for the plant.

yolk sac Third and last membrane to form around the embryo in birds, reptiles, and mammals; provides food.

zygote (ZĪ-gōt) The diploid cell resulting from the fusion of two haploid gametes during sexual reproduction.

Index

Page references for illustrations and photographs are printed in **boldface** type.

A vitamin, 169, 334
Abdomen, grasshopper, 329, **329**
Abiogenesis, 589-90, **590**, 597, 616
Abscissic acid, 352, **352**, 374
Absorption, 240, 253, 254, 260, 270, **270, 627**
Absorption spectra, **156**
Acetylcholine, 383
Acids, pH, 142
Acromegaly, 363, **363**, 374
Actin, 338, **338**
Activation energy, 138-39, **138-39**, 146
Active immunity, 750, 754
Active principle, 590, 616
Active site, **140**
Active transport, 267, **267**, 292
Adaptations, 25-28, **25, 27**, 615, 663-65, **664**
Addictions, 735-36
Adenine, 437, **437**, 454, **596**
Adenosine, **134**
Adenosine diphosphate (ADP), **134,** 157-59
Adenosine monophosphate (AMP), **134**, 355, 374
Adenosine triphosphate (ATP), 103, **103**, 134, **134**, 157, **157**, 178, 344; in cellular respiration, 206; in citric acid cycle, 190-92, **193**; in digestion, 242; as energy go-between, 182-83, **183**; in food-energy processes, 182-88; formation of, **159**, 160; in glycolysis, 189; in muscle contraction, 339, **339**; nature of, 158; production of, 185; in transport, 267
ADH. *See* Antidiuretic hormone
ADP. *See* Adenosine diphosphate
Adrenal cortex, **366**, 371
Adrenal glands, **361**, 366, **366**, 371
Adrenal medulla, 366, **366**, 370-71
Adrenalin, 286, 306, **366**, 371
Adrenocorticotropic hormone (ACTH), 371
Adult insect, 358-59, **359**
Aerobic organisms, 211, 232
African sleeping sickness, 741
Afterbirth, 513, 526
Aging, 733, 754
Air pressure, breathing and, 230, 232

Air sacs, grasshopper, 221-22, **222**
Alanine, **133**, 449
Albinism, gorilla, **579**
Alcohol, 735-36, **736**, 754
Alcoholism, 736
Algae, **21, 48**, 110, 277
Algal bloom, **48**
All-or-nothing response, 380
Allantois, **520**, 521
Alleles, 570, **570**, 586
Allergy, 751, 754
Alternation of generations, 482-83, **482**, 500
Altitude, breathing and, 230
Alveoli, 224-25, **225-26**, 232
Amaranth, food resource, **657**
Ameba, 11, 51, **52**, 104, **105**, 109, **213**, 226, **236**, 240, **241**, 322, **322**
Amebic dysentery, 733, 741, 754
Ameboid movement, 322
Amino acids, 132, **132**, 163, 166, 296-98, **296**, 308-09, **308**, 449
Amino group, 132, **132**, 296-97
Amnion, 509, **509, 512**, 520, 526
Amniotic fluid, 509, **509, 512**, 520, 526
Ammonia, 124, **142**, 296, 297-98, **297**, 308, **308**, 312, 595, **596**, 629, **629**, 735
Ammonification, 629
AMP. *See* Adenosine monophosphate
Amphetamines, 399, 736
Amphibians, 76-77, **76-77**, 81
Amphioxus, **75**, 82
Amylase, 239, **239**, 251, 252
Anaerobic, 184, 206
Anaerobic bacteria, 186
Anal pore, 243, **243**
Analogous structure, 40, **40**
Anaphase, 441-43, **442-43**, 456
Anesthetics, 399
Aneurism, 399, 402
Angiosperms, 59, **60, 61**, 62, 84, 485, **485**
Animal(s), 42-44, **42, 43**, 84; hormone function in, 355-60, 374; locomotion in, 323-25; pollination by, 491; structures in the cells of, 110
Animal kingdom, 63-83, **63**
Animalcules, 90
Annelids, 68-69, **69**, 82, 84
Annuals, plants, 497, 500
Annual rings, 497, **497**, 500
Anole lizard, 319-20, **320, 472**
Anopheles mosquito, 742
Ant(s), 72, **718**, 721

Anther, 486, **486**
Anthocyanins, 155, **155**, 178
Anthrax, 748
Anthropomorphism, 700
Antibiotics, 613-14, 616, 749-51, **749-50**
Antibodies, 280, **280**, 571, 745, 747, 754
Antidiuretic hormone (ADH), 312-13, 316, 350, **350**, 371
Antigens, 280, **280**, 282, 571, 744, 747, 754
Antigen-antibody reactions, 571, 745
Antiseptic surgery, 748, **748**
Anus, 241, 245-49, **245, 247, 249**, 260
Anvil, in the ear, 415, **415**
Aorta, 281, **281, 284**, 285
Apes, 80
Apical meristems, 494, **495**
Appendix, human, **249**
Aqueous humor, 423, **423**
Aquatic ecosystems, 685-692, 694
Arachnids, 72, **72**
Archeopteryx, **603**
Arginine, 350, 449
Aristotle, **5**, 42
Arteriole, **281**
Arteriosclerosis, 131, 288, **288**
Arteries, 281-82, **282, 284**, 292
Arthropods, 71-73, **71**, 82, 84
Artificial cloning, 460, 474
Artificial pollination, 490
Artificial selection, 550, **551**
Asbestos, 145
Ascorbic acid, 169
Asexual reproduction, 17, **17**, 457-58, **457**, 461, 463, 474, 477-81, **479**, 500
Asparagine, **350**, 449
Aspartic acid, 449
Astigmatism, 426
Athlete's foot, 54, 738, 743, 754
Atmosphere, 595, **597**
Atoms, 118-20, **119**, 146
ATP. *See* Adenosine triphosphate
Atrium, in human heart, 283-84, **283-84**, 292
Auditory canal, 414, **415**
Auditory nerve, 413-15, **415**
Auto-immune diseases, 746
Autonomic nervous system, 396-97, **397-98**, 402
Autosomes, 560
Autotrophs, 595-97, 616
Auxins, 351-52, **351-52**, 354, 374
Avery, O. T., 577
Axons, 378, **378-380**, 388, 390, **390**, 402; giant, in squid, **385**

772

B-complex vitamins, 166, 168-70
Baboons, 723-26
Bacilli, **47**
Backbone, in nucleotides, 134
Bacterium/Bacteria, **42-43**, 46, 47, **50**; antibiotic resistant, 751; basic types of, 47, **47**; as decomposers, 54; disease-causing, 754; diseases caused by, 740; intestinal, 254, **254**; in nitrogen cycle, 629, **629**; oil-digesting, 555; Pasteur's experiment, 593, **593**; tooth-plaque, **737**; in a vacuole, **105**
Bacteriophages, **49**, 577; disease caused by, 740; Griffith's experiment, **577**
Balance, 660; detection of, 430; perception of, 416-17, **416**
Balanced ecosystem, 640-41, 649-54, 660
Ball and socket joints, **335**
Banding, 584, **584**, 586
Barbiturates, 736
Bark, of plants, 496
Barriers to dispersal, 666, **666**, 694; separating related organisms, 606, 616; types of, **606**
Basal metabolic energy, 173
Basal metabolic rate, 200, 206
Bases, pH, 142
Bats, **39**, 80, **81**, 413-14, **413-14**, 430
Beadle, George W., 578, **579**
Bean plant, 62, **62**, 215-18, 232, 270-73, 292, 303-04, 316, 326, **326**, 344
Bear, **43**, 80, **523**, **662-63**, **682**
Beaver, 663, **688**
Bee(s), 6, 407-08, **720**
Beetles, 72, **73**, 277
Behavior, 698-729. *See also* Inherited behavior; Learned behavior; Social behavior
Beriberi, 168
Biceps, **336**, **341**
Biennials, 497, 500
Bile, 251
Bile ducts, **309**, **398**
bile salts, 251-52, **251**
Binary fission, 458, **458**, 474
Binomial nomenclature, 34, 84
Biochemical pathways, 183-84
Biogenesis, 590, 592-93
Biological clocks, 710-711, **710**, 728
Biological history, fossil clues and, 601-04
Biologists, 6, 113, 277
Biology, 3-8, **7**, 28, 113
Biomes, 673-84, **673**, **683-84**, 694
Birds, 31-32, **32**, 76, 79, **79**, 81-82, 277, 299, 633
Birth, 512-14, **513**, 521
Birth rate, 652, **652**, 660

Blastula, **508**, 509, 526
Blind spot, 425, **426**
Blood: clotting of, 280, **280**, 292; clumping in, **571**; donation of, 572, **572**; in gas exchange, 226-27, **226**, 232; path in humans, 284-85, **284**; pH, **142**; transfusions, 570-72, **572** types of, 570-72
Blood cells, 278-80, **279**, 281
Blood pressure, **286**, 289, 292, 734, 754
Blood supply, 286-87, **287**
Blood vessels, 281-83, **281-83**; in bone, 334; in digestion, **14**; diseases of, 288-89; earthworm, 274-75, **275**; grasshopper, 275-76, **275**; to liver, 308-09, **308-09**; pulmonary, **284**
Blue-green algae, **42-43**, 46, 48, **48**, **50**
Body chemistry, 605, 616
Body fluids, disease defense, 744
Body language, chimpanzees, 727
Body malfunctions, 730-33, 754
Body movement, perception of, 416-17
Body parts, artificial, 752, **752**
Body position, detection of, 416-17, **417**
Body processes, hormones in, 370, **370**
Body structures, fossil and living species, 604, 616
Bohr diagram, **123**
Bond: chemical, 122-24; covalent, 122-23; differences in, **129**; double and single, 124; formation and breakdown of, 143-44; ionic, 123-24
Bone(s), 332-33, **332**, 344; composition and structure, 334, **334**
Bone marrow, 279, **279**
Bony fish, 76, 81
Botany, 4
Botulism, **740**
Boveri, Theodore, 559, 575
Bowman's capsule, 310-11, **310**, 312
Brain, 379, **379**, 392-93; in central nervous system, 391, **391**; in earthworm, 386; eye connections, **424**; fish and primate, 411, **411**; grasshopper, 387, **387**; hemispheres, 394, **394**; human embryo, **510**; lobes, 394; optic nerve connections, **424**; parts of, 393-96, **393**, **402**; in perception, 410-11, **410**; senses and, 430; ventral surface, **397**
Brain waves, 392-93, **392**
Brainstem, 393, **393**, 396, 402
Bread mold, **43**, 578

Breathing, 211-12; in grasshopper, 221-22, **221-22**; in humans, 223-25, **223-25**, 232; rate of, 229, **229**, 232. *See also* Respiration
Breeding experiments in, 607
Breeding periods, 517, 526
Breeding studies, 41
Breeding tests, 84
Breeding true, 532
Bridges, Calvin, 564-65
Bronchial tree, **225**, 232
Bronchioles, **224**, 225
Bronchus/Bronchi, **224**, 225
Bronowski, Jacob, 559
Brown, Robert, 95-96
Bryophyllum, budding, **458**
Bryophytes, 55, 57, **57**
Bud, 458-59, 474; hydra, **66**; in vegetative propagation, 500
Budding, 458-59, **458**, 474
Bulbs, 478
Burning, in carbon cycle, 628, **628**
Butterflies, **5**, 72, 277

C vitamin, 166, 169, 334
Caffeine, 399
Calcium, 119, 171, 339, 356-57, **357**
Calorie (Cal), 172
Calorimeter, 170, **172**
Calvin, Melvin, 151, 162, **162**
Calvin cycle, **162-63**
Calyx, 486
Cambium, 495-96
Camera-type eyes, 430
Cancer, **445**, 732, 752
Canopy, forest, 679, **679**
Canis familiaris, 24, **26**, 34, **34**, 36, **37**
Capillarity, 271-72, **271-72**, 292
Capillary(-ies)—earthworm, 220, 232, 274-75, **275**, 292
Capillary(-ies)—human, 281-82, **281-82**, 292; blood supply and, **287**; in digestion, 253, 260; lymph system and, **289**; in respiration, **224**, 232; transport through, 263
Capillary beds, **285**
Capillary tube, **278**
Carbohydrates, 128-29, 146, 178; breakdown, **197**; digstion, 239, **239**; essential/nonessential, 166; functions and sources, 168; long-chain, **128**
Carbon, 118, 119, **120**, 637
Carbon cycle, 628, **628**, 638
Carbon dating, 601, **601**
Carbon dioxide, 124, **152**, 595, **596**; in carbon cycle, 628, **628**; excretion, 298, 303; in the environment, 637; in gas exchange, 211-30, 232
Carbon monoxide, **596**, 735
Carcinogens, 752

773

Cardiac muscle, 337, **337**, 339, 342, 344
Caribou, 682, **682**, 698-99
Carnivores, 622, **622**, 638
Carotenoids, 155, **155**, 178
Carpals, **332**
Carrier(s) of sex-linked traits, 563-64, **563**
Carrier molecules, in transport, 267, **267**
Carrying capacity, 650, **650**, 656-57
Carson, Rachel, 6
Cartilage, 76, 324, 333-35, **333**, 344
Cartilaginous fish, 76, 81
Castes, insect, 718-19, **718-19**, 728
Castings, earthworm, 246
Cat, 75, **406**, **530-31**
CAT scanner, **399**
Catalyst, 139
Cell/Cells, 28, 86-207; basic life unit, 88-115; boundaries of, 100-01; catalysts in, 139-40; centrifuging and chromatography, 146; chemical components of, 126-35; chemical reactions in, 136-44; early observation of, 89-94; energy and, 146; eukaryotic, *see* Eukaryotes; as factory, 263; functions of, **99**; gas exchange in, 213, **213**; internal skeleton, **108**; investigation of, 95-99; light detection by, 427; molecular machinery of, 117-47; in osmosis, **266**; prokaryotic, *see* Prokaryotes; reproduction of, 435-39; size, 94, 114; specialization of, 98-99, **98-99**; structure of, 11, **11**; structures in, 109-10, **111**, 114; swollen and shrunken, **266**; transport into and out of, 263, 267-68, **267-68**; waste products of, 14-15, **14**, 296-98, 316; water balance and, 303, **303**
Cell body, 378, **378**, **383**, 402
Cell digestion, **105**
Cell division, 352, 441, 443, 454
Cell membrane, 11, 100-01, **100-01**, 107-10, **107**, **108**, **111**, 114
Cell plate, 444, **444**
Cell structure, 110-11, **111**, 114
Cell theory, 96, 114
Cell types, **46**
Cell wall, 101, **101**, 109-10, **111**, 114
Cell wastes, 14-15, **14**, 316
Cellular movement, 320-22
Cellular organization, 11-12
Cellular respiration, 181-83, 206, 211-12; biochemical pathways, 183-84; fermentation and, 196; mechanism of, 189-96; metabolism and, 197-203; nutrients and, **198**; photosynthesis and, 186-88, **187-88**, 206; and salt/water balance, 300, **300**; stages of, 184-86, 206; summary of, **193**
Cellular waste products, 14-15, **14**, 296-98, 316
Cellulose, 100, **100**, 110
Central nervous system, 378, 391, **391**, **397**, 402
Centrifuge, 126, **126**, 146, **278**
Centrioles, 108-10, **108**, **111**, 114
Centromere, 442-43
Cerebellum, 393, **393**, 395, **395**, 402
Cerebral cortex, 393, **393**
Cerebrospinal fluid, 391, **391**
Cerebrum, 393-95, **393**, 402
Cervix, 367, **505**, 506, **506**
Chance: heredity and, 539-40; laws of, 539, **539**, 556
Change over time, 588-617; clues that suggest, 605-07; in ecosystems, 668-69; in species, 26-27; theories explaining, 608-14. *See also* Evolution; Heredity
Characteristics of living things, 10
Chargaff, Erwin, 453
Chase, Martha, 577, **577**
Chemical(s): cycling of, 24, 28; detection of, 430; from food, **14**; hazardous, 145
Chemical bonds. *See* Bond
Chemical components of matter, 117-25
Chemical compounds. *See* Compounds
Chemical control, 348-75
Chemical cycles, 627-31, 638
Chemical energy, **136**
Chemical feedback mechanisms, 356-57, **356-57**, 374
Chemical reactions. *See* Reactions
Chemical signals: communication by, 714, **714**, 728; in life processes, 14, 728
Chemical symbols, 118-19
Chemoreceptors, 406-07, **407**, 410, 427-30
Chemotherapy, 750-52, 754
Chick, development in egg, **520**
Chicken pox, 754
Childhood diseases, 754
Chimpanzee, **700**, **708**, **716**, 727
Chitin, 324
Chlorine, 119, **124**, 171, 380, **380**
Chlorophyll, 48, 104, 155-56, **156-57**, 160, 178
Chloroplasts, 104, **104**, 109-10, **111**, 114, 153, **153-54**, 178
Cholera, 733
Cholesterol, 288, **288**
Cholinesterase, 383
Chordates, **43**, 74-82, 84
Chorion, **509**, 520, **520**
Choroid, 423, **423**
Chromatids, 442-43, **442**
Chromatograph, 126, 127
Chromatography, 127, **127**, 146
Chromosomal disorders, 564-66
Chromosomal genetics, 559-66
Chromosomal mutations, 568
Chromosome(s), 102-03, **103**, 110, 114, 440, 474; of *E. coli*, **440**; function, 109; genes and, 559, 567-74; in heredity, 559-87; homologous, 440, 562; length, **94**; of male and female fruit flies, **560**; radiation damage, **568**; in sex determination, 560-62
Cigarette smoke, contents, 734, **735**
Cilia, 51, **52**, 107, 108; disease defense, 744; in locomotion, 107, 327, **327**, 344; movement by, 321, **321**, 344; in respiratory system, 224-25, **225**; of *Stentor*, **108**
Circadian rhythms, 710-11
Circulatory system, **5**, 214, **214**, 284-85, **284-85**
Citric acid cycle, 184, **185**, 190-91, **191**, 206
Class, 35, **35**, 37, 84
Classical conditioning, 705, **705**, 728
Classification of living things, 10, 28, **37**, 38-44; modern plan, 84; perspective on, 45; scientific names and, 31-37, 84; systems of, **42-43**, 44
Clavicle, **332**, 336
Cleavage of cell, 508, **508**, 526
Climate, 660; as barrier to dispersal, 666, 694; in biomes, 674, **674**, 677, **677-78**, 680, **680**, 681, **681**; population growth, 645
Climax community: forest, 669, **669**, 671-72, 694
Climax organisms: in coniferous forest, **680**, 681; in deciduous forest, 677-78, **678**; in desert, 674-75; **675**; elevation and, 683-84; in tropical rain forest, **679**; in tundra, 681-83, **682**
Cloaca, frog, 300-01, **301**
Clone/Cloning, 460, 463, 474
Clotting of blood, 280, **280**, 292
Clumping, **571**, 598
Cocci, **47**
Cochlea, 415, **415**, 430
Codons, 580
Coelenterates, 65, **65-66**, 82, 84
Cohesion, in water, 272
Cohesion-tension theory, 272, **272**, 292
Colchicine, 569
Coldblooded animals, 76, 206; metabolic rates, 202-03

Cold receptors, 417-18, **417**
Cold sores, 738
Collagen, 334-35
Collecting tubule, in kidney, 310, **310**
Colon, 254
Colony of social insects, 718, **718**
Colorblindness, inheritance, **564**
Color vision, 425, **425**
Commensalism, 633, **633**, 638
Common names, 33
Communicable diseases, 737-39, 754
Communication: in animals, 725-28; in honeybees, 720-21, **720**, 728; multiple signals in, 717, 728; in social behavior, 714-21
Community(ies), 20-23, **20**, 28
Compact bone, 334, **334**
Companion cells, **273**
Competition: adaptations that reduce, 636, **636**; as barrier to dispersal, 666, 694; between organisms, 23; as symbiosis, 635, **635**, 638
Complete metamorphosis, 359, **359**, 374
Compounds, 121, 146; biological, 128-34; chemical, 121-23; classes of, 128-34; formation of, 122-24; isolation of, 127
Compound eyes, 421-22, **421**, 430
Computerized axial tomography (CAT), **399**
Condensation, 627, **627**
Conditioned response, 705
Cones, in the eye, 424-25, **424-25**, 430
Coniferous forest(s), **680**, 683
Coniferous forest biome, 673, **673**, 680, **680-81**, 694
Consumer(s), 22, 28, 84, 622, **622**, 623, 638; aquatic, **21**; in energy pyramid, 625-26; food getting, 235-38, 260; forest, **21-22**
Consumer protists, 51-52
Contaminated articles, disease carriers, 738, 754
Contamination, 738
Contractile tissues, 328, **328**
Contractile vacuole, 304, **304**, 316
Contraction of muscle, 338-39, **338-39**
Controlled experiment, 114
Cork, 495-96, **496-97**
Cork cambium, 495-96, **496-97**
Cork cells, **91**, 114
Corn, **22**, 62, **175**, 490
Corn smut, **54**
Cornea, 423, **423**
Corolla, 486
Corpus callosum, 393, **393**, 394
Corpus luteum, 368, **368**, 470
Corpus luteum stage, in menstrual cycle, 368, 369, **369**

Correns, Carl, 575
Cortisol, 366, **366**
Cortisone, 499
Cotyledons, 61, **61**, 492, **492**
Courtship behavior(s), 518, **518**, 526, 728
Covalent bonds/bonding, 122-23, **123**, 146
Cowpox, 749, **749**
Cranial nerves, **397**
Cranium, **332**
Crayfish, **622**
Crepuscular animal, 709, **709**
Crick, Francis, 439, 453, **453**, 575-76, 584
Crista, Cristae, 185, **185**
Crop: in earthworm, 245, **245**; in grasshopper, 246, **246-47**
Cross(-es): dihybrid, **546**; of peas (round × wrinkled), **533**; predicted outcomes, 541-44, **544**; probability and, 539-44; rye × wheat, **550**
Crossing-over, 467, **467**, 567-68, **567**
Cross-pollination, 490, **490**
Crowding, population growth and, 645, **645**, 660
Crustaceans, 71, **71**
Cuticle, leaf, 153, **153**
Cuttings, in plant reproduction, 479, 500
Cyclic AMP, 355, **355**, 374
Cyclic behavior, 709-13, **709**, 728
Cysteine, **350**, 449
Cytokinins, 352, **352**, 374
Cytoplasm, 95, 109-10, 114; organelles in, 103-09
Cytoplasmic division, 441, 444, **444**, 458-59
Cytoplasmic streaming, 273, **273**
Cytosine, 437, **437**, 454, 596

D vitamin, 166, 169, 170, 334
Dance, honeybee, 720, **720**, 728
Dandruff, 743
Dark reactions, 157, **157-58**, 162, **162-63**, 178
Darwin, Charles, 575, 608-10, **609**, 616
Data, collection of, 97
Day-neutral plants, 711
Death rate, 652, **652**, 660
Decay, molds, 54
Decibel scale, **415**
Deciduous forest, 677, **677**, 683
Deciduous forest biome, 673, **673**, 677-78, **677-78**, 694
Deciduous trees, 677
Decomposers, 22, **22**, 28, 47, 84, 622-23, 638; in the carbon cycle, **628**; in energy pyramid, 626; food

getting, 237; in the nitrogen cycle, 629, **629**; protists, 53
Deer, 20, **644**, 678
Defense hormones, 353, **353**, 374; social behavior and, 728
Dehydration, 295-96, **296**
Dehydration synthesis, 143, **143**, 146
Dendrites, 378, **378**, 402
Deoxyribonucleic acid. *See* DNA
Deoxyribose, 447, **447**, 596
Depth perception, 425-26
Desert(s), 674-75, **674**
Desert biome, 673-75, **673-74**, 694
Detection, receptor function, 406
Development: adaptations for, 519-522; early stages, 508-11, 526; of embryos, 519-22; of frog, **519**; in egg, **520**; narsupial, **522**; nonplacental, 521, 526; in plants, 494, 497; in uterus, 521; vertebrate, 508-15, 519-24, 526
De Vries, Hugo, 575
Diabetes mellitus, 365-66, 374, 754
Diagnosis, 752, 754
Diamond, as carbon, **118**
Diaphragm, action in breathing, 223, **223**, 232
Diatoms, **43**, 50-51, **51**
Diatomaceous earth, 51
Dichotomous key, 31-32
Dicots, 61-62, **61-62**
Didinium, **268**, 458
Diet(s), 174, 177, 178, 734, 754
Differentiation, 450, 474
Diffusion, 264-66, **264**, 292, 304, 316
Digestion, 14-15, **14**, 238, 245, **245**, 251-53, 260; chemical, 238-39, **239**, 252, 260; extracellular, 240, **240**, 260; intracellular, 240, **240**, 260; mechanical, 238-39, **238**, 260. *See also* Food processing
Digestive cavity, 244, **244**, 260
Digestive glands, 248, **249**, 260
Digestive systems, **249**, 255-56, 260
Digestive tract, 236, 245, 248, **249**, 260
Digitalis, 499
Dihybrid cross, 545-46
Dinoflagellate, **51**
Dinosaurs, 78, 615
Dipeptide, **133**
Diploid cells, 482, 500
Diploid number, 464, 474
Disaccharides, 128
Disease, 730-55; organisms causing, 740-43, 754
Dispersal of organisms, 663-72, **664-65**, 694; barriers to, 666-67, 694
Diurnal animals, 709, **709**
Diurnal cycle, 709, **709**
Division of labor, 98, 114

775

DNA, 134, 435-39, **435**, **439-40**; genes and, 576-84; **586**; geometric representation, **437**; models of, **434-35**, 436-39, **436**, 454; model research, 576-78, 586; molecules, **117**, 435-55; new organisms and, 555; nucleotides, 437, **437**; reading of, 448, **448**; replication, 435, 438-39, **438**, 454; research, 453; RNA and, 447-48, **447**
DNA polymerase, 439
Dog(s), **27**, **37**, **75**, **264**, 522
Dolphin, 307, **307**, 706
Dominance, law of, 534-35, 556
Dominant, 534
Dominant generation, 483-84, 487-88, 500
Donkeys, **41**
Dopamine, 383
Double fertilization, 488, **488**, 500
Double helix, 437, 454
Double sugar, 128
Down, Langdon, 565
Down's syndrome, 565-66, **566**
Dragonfly, 72, **73**, 359
Drones, 719, **719**, 728
Droplet infection, 738, 754
Drosophila (fruit fly), **94**, 560-68, **560**, **565**, **567**, **569**
Drugs, 399, 736, **736**, 754; abusers of, 736
Duckbill platypus, 80, **80**
Dwarfs, 361-62, **361**
Dye, diffusion, **265**
Dynamic equilibrium, in ecosystem, 649, 660

E vitamin, 166, 169
Ears, **406-07**, 414-15, **414-15**
Earth: formation of, 594, **594**; life origin on, 593-95; photo of, **594**
Earthquake prediction, 405-06
Earthworm(s), 68, **69**, 82, **471**; digestion in, 260; excretion in, 305, **305**, 316; food processing in, 245-46, **245-46**; hearts, 275, **275**, 292; movement and locomotion, 328-29, **328**, 344; nervous control in, 386-87, 402; respiration in, 219-20, **219-20**, 232; transport in, 274-75, **275**, 292
Echinoderms, 74, **74**, 82, 84
Echolocation, 414, 430
Eckstein, Gustav, 235
Ecological succession, 668-72, 694
Ecologist, **6**
Ecology, 20
Ecosystem, 24-26, 28; balanced, 649-54, 660; carrying capacity, 650, **650**; changes over time, 668-69; chemical cycles in, 627-31;

energy in, 624, 638; food in, 621-26, 638; geography of, 662-95; human factor in, 692, **692**; interactions in, 620-39; latitude and elevation, **683**; organisms in, 28; populations in, 640-61; water cycle in, **24**
Ectoderm, 509, 526
Effectors, 378, **379**, 389, **389**
Egestion, 241, **241**, 254, 260
Egg(s), 469, 474; chromosomes in, 561; development in, **520**; human, 367-69, **368-69**, **470**, 505-06, **505**, **507**, 526; insect, 358-59, **359**; plant, 482-84, **482**, 488, 500; protein source, **167**; quantity released, 516, 526; queen laying, **719**; reptile, **78**; uric acid storage, 297, **297**
Egg-laying mammals, 80, **80**
Ehrlich, Paul, 750
Elastin, 335
Electric fish, **409**
Electrical energy, **136**
Electricity, sense organs for, **409**
Electron(s), 92, **92-93**, 119, 120, 136, **157**, 160-61
Electron carriers, 159, 164, **192**
Electron microscope, 49, 92-93, **92**
Electron transport, 184, **185**, 191-92, **192**, 206
Elements, 117-19, **117**, 146
Elephant, 522, **717**
Elevation, life zones, 683-84, **683**, 694
Elodea, cells of, 101
Embryo(s), 492, 500, **508**, 509, **509**, 519-22, 526
Embryo sac, 487-88, **487-88**
Embryology, 605, **605**, 616
Embryonic induction, 510, **510**
Emigration rate, 653, 660
Emphysema, 732, 754
End chain, in mRNA, 449
End plates, in phloem, **273**
Endangered species, **603**, 615
Endocrine glands, 355-56, 361-74, **361**
Endocrine malfunction, 732, 754
Endocrine system, 361-71, **361**, 374
Endocytosis, 268, **268**, 292
Endoderm, 509, 526
Endoplastic reticulum, 106-11, **106-07**
Endoskeleton, 324-25, **325**
Endosperm, **61**, 488, 494
Energy, 136-38, **136**, 146, 172, 180-81, **188**, **200**, 624-25, **624**, 638
Energy pyramids, 625-26, **625**, 638
Entamoeba coli, E. histolytica, 254
Environment, 16-28, **21**, 145, 230, **230**; heredity and, 553-54, **554**, 556; respiration and, 230, **230**

Enzymes, 104, 139-43, **139**, 146; in chemical digestion, 238-40, **240**, 252; genes and, 578-79, 585; in lysosomes, 104-05; pH and, 142-43, **142**, 146; temperature and, 141-43, **141**
Epidemics, 739, **739**, 754
Epidemiology, 750, 754
Epidermis, 153, **153**, 421, 495
Epididymus, **504**
Epiglottis, 248, **249**
Epilepsy, 393
Equilibrium of populations, 649, 659, 660
ER. See Endoplasmic reticulum
Erythrocytes, 278-79, **279**, 281, 292, 309
Erythromycin, 750
Escherichia coli, 254
Esophagus, **13**, 245, **245**, 248-49, 260
Estrogen, 367-69, **369**, 371, 506
Estrus, 518-19
Estuaries, 685, 690-91, **691**, 694
Ethanol, 193, 194-95, 206
Ethylene, **123**, 353, 374
Euglena, 42-43, **42-43**, 50, 231
Eukaryotes/Eukaryotic cells, **46**, 47, **50**, 84, 102-10, 114
Eustachian tube, 414, **415**
Evaporation, 627, **627**
Evolution, 589-98, 602-16, **604**
Excretion/Excretory systems: in complex organisms, 301-02, 316; in earthworm, 305, **306**, 316; in grasshopper, 305-06, **306**; in humans, 307-14, **307-10**, 316; in hydra, 304, 316; nephridia in, 316; in paramecium, 304, **304**, 316; in plants, 303-04, 316; steps in, 301, 316
Exercise, muscle action and, 340-41
Exhaling, 223
Exocytosis, 268, **268**, 292
Exoskeleton, **325**, 329-30, 344
Experiment, controlled, 96
Extensor muscles, 336, **336**
External fertilization, 471-72, **472**, 474, 519
Extinction of species, 603; adaptation vs., 615
Eye(s), 378, **406-07**, 420-26, **421**, 430; camera type, 422, **422**; dim/bright adjustment, 425; human, **397-98**, 423-24, **423-24**, 510; insect, 421-22, **421**; as receptors, 377, **378**; spider, 420, **421**
Eyespots, 420, **420**, 430

Facets, of compound eyes, 421, **421**
Fallopian tube, **505**
Family, 35, **35**, 37, **37**, 84

Farsightedness, 426, **426**
Fast foods, 177
Fast pain, 418
Fats, 130-31, **130-31**, **242**, 257
Fatty acids, **130**, 131, 166, 290
Feces, 254
Felis domesticus, 34, **34**
Female chromosomes, 560
Female reproductive system, 505-06, **505**
Femur, **332**
Fermentation, **186**, 193-96, **193**, 206
Ferns, 10, **42-43**, 55, 58, **58**, **354**, 459, 484, **484**, 500
Fern relatives, **43**, 55, 58, **58**
Fertilization, 471-72, 474; human, **505**, 506, 526; plant, **488**, 492
Fertilizers, in nitrogen cycle, 630, 638
Fetus, 511-12, **511**
Fiber, plant, 499
Fibrils, 338-39, **338-39**
Fibrin, 280, **280**
Fibula, **332**
Fiddler crab, color changes, 710, **710**
Filament: flower, 486, **486**; muscle, 338-39, **338**
Filial generations (F_1, F_2, F_3), 533
Filter feeding, 237, **237**
Filtrate, 311, **312**
Filtration, kidney, 311, **312**, 316
Finches, Darwin's, **611**
Fish, 75-76, **76**, 81-82, 132, **132**
Flagellum(a), 52, 107-08; movement by, 321, **321**, 344
Flatworms, 67, **67**, 82
Fleas, 72, **73**
Fleming, Alexander, 750
Flemming, Walther, 441
Flexor muscles, 336, **336**
Flies, 590-92, 613-14, 616. *See also* Drosophila
Flower(s), **118**, 486-87, 500
Flowering hormones, 353
Flowering plants, 490-98
Fluorocarbons, 145
Folic acid/Folacin, 109, 254
Follicle(s), 368, **470**, 505-06, 526
Follicle stage, in menstrual cycle, 368-69, **369**
Follicle stimulating hormone (FSH), 368-71, **369**
Food: acquisition of, 236-38; breakdown process, 181-88; contaminated, 738, 754; digestion, 14-15, **14**; in the ecosystem, 621-26, 638; energy and, 170-73, 178, 180-207; getting and using, 13-15, 14, 28, 235-42; needs for, 150, 175-76; quality of, 177; source of, 150; world needs for, 175-76
Food cavity, **236**

Food chains, 19, 21, **21**, 28, 622-25, **623**, 638; strontium-90 in, 631, **631**
Food crops, world production, **478**
Food processing, 234-61; in representative organisms, 243-47. *See also* Digestion
Food production, 148-64, 175, 178-79
Food supply, 660; population growth and, 644, **644**
Food vacuoles, 236, **236**, 243, **243**, 260
Food wastes, **14**, 15, 241
Food webs, 21, 28, 622-23, **623**, 638, 685-86
Forest climax community, 669, **669**, 671-72
Formaldehyde, **735**
Formulas: chemical, 121; structural, 123, **123**, **129**, **596**
Fossils, 588-89, 599-605, **599**, 616, **628**
Fovea, 423, **423**
Fox, Sydney, 598
Fragmentation, 460, **460**, 474, 478, 500
Franklin, Rosalind, 453
Fraternal twins, 514, **514**, 526
Fresh water: water/salt balance in, 300, 316
Freshwater algae, 55, 56
Frisch, Karl von, 6
Frog(s), **43**, 77, 82, **461**; development, **508**, **519**; excretory system, 300-01, **301**
Frontal lobe, 394
Fructose, **128**
Fruit(s), 60, **60**, 492, **492**, 499
Fruit flies. *See* Drosophila
Fruit-ripening hormones, 353
Fungal kingdom, 53-54
Fungi, 42, 43-44, **43**, 50, 54, **54**, 84, 110, 114, **114**, 743, 754

Gallbladder, **249**, 251
Gamete(s), 57, 464, 474
Gamete cells, 469, 474
Gamete release, 517-18, 526
Gametogenesis, 469, 474
Gametophyte, 482, **483**, 500
Gametophyte generation, 483-84, 488
Ganglion/Ganglia, 379, **379**, 385
Gas exchange, 210-33; in aerobic organisms, 211; blood role in, 226-27, **226**; in cells, 213, **213**; in grasshopper, 276; in the lungs, 225-27; in plants, 216-18, **217-18**. *See also* Respiration
Gastric glands, 248, **249**, **250**, 260
Gastric juice, **142**, 250, 252
Gastrin, 256, **256**, 371

Gastropod, **70**
Gastrula, 509, 526
Gazelles, **722**
Gene(s), 465, 474, 559-87; chromosomes and, 559, 567-74; DNA and, 576-84, 586; enzymes and, 578-79, 585; how they work, 580; mutations and, 581-82
Gene mapping/maps, 567, **567**, 584, 586
Gene mutations, 568, 581, 586
Gene splicing, 555, 583, 752-53
Generations, terms for, 533
Genetic code, 449, 580, 584
Genetic counseling, 549
Genetic counselors, 549, 556
Genetic engineering, 583
Genetic recombination, 582-83, **583**, 586
Genetic research, modern, 582-84
Genetics, 531-57, 559-66
Genotypes, 534, 556
Genus, 35-37, **35**, **37**, 82, 84
Geography, as barrier to dispersal, 694
Geological periods, life and, 602
Geotropism, 701, **701**
Germ theory of disease, 748, 754
Germination, 493-94, **494**, 500
Gestation, 507-12, 522, 526
Giant axons, **385**
Giant cell body, **385**
Gibberellins, 352, **352**, 374
Gill(s), 70, **71**, 214
Gill slits, 75, **75**
Gizzards, 239, **245**, 246
Gland cells, 13
Gland tissue, 13
Glial cells, **378**, 379
Gliding joints, **335**
Glomerulus, **310**, 311
Glucagon, 257, 365-66, **365**, 370-71
Glucagon diseases, 365-66, 374
Glucose, **128**, **152**, **157**, 162-64, **162**, **164**, 178, 182-83, **183**, 206, 257
Glutamic acid, 449
Glutamine, **350**, 449
Glycerol, 130, **130**
Glycine, **133**, **350**, 449
Glycogen, 129, 257
Glycolysis, 184, **185**, 189-90, **189**, 206
Goiter, 361, 374
Golgi apparatus, 106-11, **107**, **111**
Gonads, 503-05, 526
Gonadotropin, 356
Gonorrhea, 738
Goodall, Jane, 727
Grafting, 479, **479**, 500
Grains, **60**, 168
Grand Canyon, **500**, **666**
Granum/Grana, **154**, 178
Grass, **60**, 62

Grasshopper(s), 72, **73**, 82, 385; body structure, 329-30, **329-30**, 344; excretion in, 305-06, **306**; food processing in, 246-49, 260; gas exchange in, 276; head of, **246**, 329, **329**; hearing in, 414, 430; heart, 275-76, **275**, 292; hormone controls, 358, **358**; locomotion in, 329-30, **329-30**, 344; nervous control in, 387, **387**, 402; respiration and breathing, 221-22, **221-22**, 232; transport in, 275-76, **275**, 292
Grassland(s), 675-76, **675-76**
Grassland biomes, 673, **673**, 675-76, 676, 694
Gravity, plant tropism and, 701
Gray matter, 391, 393, **396**
Great Lakes, 686
Green algae, 55
Greenhouse effect, 637
Griffith, Fred, 576; experiment, **577**
Grooming, 725, **725**
Growth: human, after birth, 514-15, 514; plant, 494-97
Growth hormones, 351-52, **354**, 370-71; human, 361-62, **363**; synthetic, 354
Growth movements, 322, 326, **326**, 344
Guanine, 437, **437**, 454, **596**
Guano, **297**
Guard cells, **153**, **215**
Gulls, 79, **299, 523, 703**, 715
Gurdon, J. B., 460, **461**, 474
Gymnosperms, 62, 84, **85**, **485**

Habitat, 667, **667**, 694
Habituation, 278, 704
Hairs in human ear, 415-17, **415-17**, 430
Hammer, in human ear, 415, **415**
Haploid cells, in plant reproduction, 482, 487, 500
Haploid number, 464-65, 474
Harvey, William, 5
Health, 730-55, 730-31, 753
Hearing, 410, 414-15, 430
Heart. *See name of organism*
Heartwood, 497, **497**
Heat, 410; metabolism and, 200-01; perception of, 417-18, 430; receptors for, 417-18, **417**
Heat energy, **136**
Helmont, J. B. van, 150
Hemoglobin, 227, **227**, 232, 278-79, **279**, 309
Hemophilia, 563-64, **564**, 724-25
Hepatitis, 738
Herbivores, 622, **622**, 638
Hereditary disorders, 732, 754

Heredity, 436, 531-57; chance and, 539-40; environment and, 554, 555-56; genes in, 559-87
Herpes simplex virus, **741**
Hershey, Alfred D., 577, **577**
Herpetologists, 277
Heterogametes, 469, **469**, 474
Heterotrophs, 595-98, 616
Heterozygotes, 534, 556
Hibernation, 202-03, **202**
Hinge joints, **335**
Histadine, 449
Holdfast, **56**
Homeostasis, 228, 290-91, 316
Homo sapiens, 34, **34**, 37, 45. *See also* Human
Homologous structures, **39**
Homozygous, 534, 556
Honeybee(s), **491**, 718-21, **719**
Hooke, Robert, 91-92, **91**, 95, 114
Hookworm, 68, 82, **743**, 754
Hormonal control, in humans, 361-70, 374
Hormones, 255-56, 341-42, 344, 349-60, **350, 352**, 358, **358**, 374
Horse(s), **41**, **664**; evolution, 615
Horsefly, eyes, 421
Horseshoe crab, 72, **72**, 615
Hosts, 23, 634, **635**
Housefly, metamorphosis, 359
Human(s), **37**, **39**, 80, 82; breathing and respiration, 223-30, 232; carbon in, **118**; as ecosystem factor, 692; excretion in, 307-14, **307-10, 312-13**, 316; fertilization in, 506, 526; food processing in, 248-58, 260; gestation period, 522; hearing in, 414-15, 430; hormonal control in, 361-70, 374; movement and locomotion in, 331-42, 344; nervous control, 391-400; parental care, **436, 524**; sight in, 423-26; smell in, 427-28, **427**; social behavior, 728; as social organisms, 715; transport in, 278-91
Human heart, 283-84, **283-84**; artificial, **752**; beating of, 283-86, **283, 286**, 292; blood vessels of, 288, **288**; chambers of, 283, **283**, 292; diseases of, 278, 288-89; nerves for, **398**; regulation of, 285-86; valves, 284
Human population: growth, 655-58, **655**, 660; problems of, 659
Human reproduction, 504-07, 526
Humerus, **332**, 336
Hummingbird, **40**
Hummingbird moth, **40**
Hybrids, 41, 533, **533**, 536, **536**, 556
Hybrid vigor, 550, **550**
Hydra, **43**, 66, **66**, 82; digestive

cavity, **244**; excretion in, 304, 316; food processing in, 244, **244**; movement and locomotion, 327-28, **328**, 344; nervous control in, 385-86, **385**, 402; photosynthesis in, **157**; respiration in, 219, **219**, 232; transport in, 274, **274**, 292
Hydra littoralis, 66, **66**
Hydrocarbon chains, 131
Hydrochloric acid, 250, **744**
Hydrogen, 119-20, **119, 162**, 595, **596**
Hydrogen cyanide, **735**
Hydrogen sulfide, **735**
Hydrolysis, 144, **144**, 239
Hydrotropism, 701
Hypertension, **286**, 289, 292, 734, 754
Hyperthyroidism, 362
Hypoglycemia, 365-366
Hypothalamus, 361, 363-64, 368, 370, 374, **393**, 395
Hypothermia, 203-04
Hypothesis, 96, 97, 114
Hypothyroidism, 362

Identical twins, 514, **514**, 526
Ilium, **332**
Immigration rate, 653, 660
Immune reaction, **745**, 746
Immune system, 732, 745, **745**, 754
Immunity, 745, 747, 750, 754
Immunology, 745-47
Imperfect flower, 486
Imprinting, 704, **704**, 728
Inbreeding, 551, **551**
Incomplete dominance, 546-548, **547-48**, 556
Incomplete metamorphosis, 358, 359, 374
Independent assortment, 536-38, **537**
Infection, routes of, **739**
Influenza, 739, 754; virus, **49**
Infrared light, 408, 430
Ingestion, 236, 260
Ingenhousz, Jan, 151
Inhaling, **223**
Inheritance, of pairs of genes, 541-46
Inherited behavior, 700, **700**, 702-03, 728
Inner ear, 414, **415**, 430
Inorganic compounds, 121-22, 146
Insects, **63**, 72, **73**, 754; development, hormone control of, 358-60, 374; as disease communicators, 739; eyes and vision, 421, **422**, 430; fossil, **599**; pollination by, 491; social behavior, 718-21, 728; specialists on, 277
Instincts, 703, **703**, 728
Insulin, 132, **132**, 257, 365-66, **365**, 370, 371, 374

Insulin-making bacteria, 555
Intelligence, 707-08
Interactions, **20**, 20-28, 632-36, 646
Interferon, 752-53
Interneurons, 379, **379**, 402
Intermediate fibers, in cells, **108**
Internal fertilization, 472, **472**, 474, 516, 526
Interphase, 441, **441, 443**, 454
Intestinal bacteria, 254, **254**
Intestinal folding, 241, **241**
Intestinal glands, 248, 260, 371
Intestinal juice, 251-252
Intestines, 245-47, **245, 247**, 253-54, 260, 308, **308, 361, 398**
Invertebrates, **43**, 63, **63**; nervous system, 385-87, 402
Involuntary muscles, 337
Iodine, 119, 171, 362
Ions, 120-21, **121**, 279, 379, **379**, 402
Ionic bonds, 123-24, **124**, 146
Ionic compounds, 146
Iris—human eye, 423, **423**
Iron, 119, 171
Ischium, **332**
Isogametes, 469, **469**, 474
Isoleucine, 166, 449

J-shaped growth pattern, 647-48, **648**, 660
Jawless fish, 76, 81
Jellyfish, **66**, 82, 237, **237-38**; medusa, 65
Jenner, Edward, 749, 754
Joints, 333, 335, **335**
Jungle ecosystem, **623**
Juvenile hormone, 358-60, **358**

K vitamin, 166, 169, 254
Kaibab deer, 643
Kangaroo rat, 300, **675**
Kanuri, lactose intolerance, 259
Kelp, **43**, 56, **56**
Kepone, 145
Kidneys, **285**, 294-95, 308, **308**, 310-14, **310, 361, 398**
Kilojoule (kJ), 172, 178
Kinesis(es), 702, **702**, 728
Kinetic energy, 136, 146
Kingdom, 35, **35, 37**, 42-44, 46-84
Knee-jerk reflex, 396
Koch, Robert, 748, 750, 754
Krebs, Hans, 190
Krebs cycle. *See* Citric acid cycle

Labor, 512-13, 526
Lacteal vessel, 290
Lactic acid, 193, **195**, 206, 340, **340**
Lactase, 259

Lactose, 259
Lake(s), 669, 685-87, **687**, 694
Lake communities, 670
Lake-to-forest succession, 669-72, **670-71**, 694
Lamarck, J. B., 608-09, **608**, 616
Large intestine, 13, 248, **249**, 254, 260, **285**
Larva, insect, 359-60, **359**
Larynx, 224, 232, **397, 398**
Lateral line, 407, 409, **409**
Lateral meristems, **495**
Latitude, 683, 694
Leaf/Leaves, 153, **153**, 178, 216-17, **216-17**, 326, 344, 478, 480, 499
Leaf hairs, **153**
Learned behavior, 700, **700**
Learning, 704-06, 721, 728
Leeches, 68, **69**, 82, 742
Leeuwenhoek, A. van, 89-91, **90**, 92, 114, 320, 592
Legumes, **629**, 630, **630**
Lens: of human eye, 423, **423, 510**; of simple eye, 421; in vertebrate eyes, 422
Lenticels, 215, 232
Lethal mutations, 568
Leucine, 166, 449
Leukocytes. *See* White blood cells
Levan, Albert, 560
Lichens, 54, **54**, 237, 668, **668**
Life: beginning, 593-95; origin theories, 589-98; world of, 1-85
Life functions, 13-18, **15**, 28
Life histories, in classification, 40-41
Life processes, controlling, 13, 15-17, **15**, 28
Life span, 17, **444**, 454, 497
Life structures, 11-12
Life zones, 670-71, 683-84, **683-84**, 687, 694
Ligaments, 335, **335**
Light, **156**, 160, 351, 407-08, 420-22, 430, 701
Light energy, **136**, 152
Light reactions, 157-61, **158, 161, 163**, 178
Light receptions, 424-25
Light-sensitive cells, 421-22
Light waves, **408**
Lightning, **629**, 638
Linkage, 548, **548**, 586
Linnaeus, Carolus, 34, **34**
Linoleic acid, **131**, 166
Lipase, 239, 251, **251**, 252
Lipids, 100, **100**, 114, 128, **130**, 146, 166, 178; **197**; molecules, 100
Lister, Joseph, 748, 752, 754
Liver, 13, 248, **249**, 251; amino acids and, 308-309, **308**; blood vessels and, 308-09, **308-09**; in circulatory system, 285; in digestion, 257,

257; erythrocytes and, 309; in excretion, 308-09, **308-09**, 316; hemoglobin and, 309; intestines and, 308, **308**; kidneys and, 308, **308**; nerves for, **398**; pyruvic acid and, 308, **308**; red blood cells and, 309; tissue, **309**; urea and, 308, **308**
Liver fluke, 67, **67**, 82
Liverworts, **43**, 55, 57, **57**
Living things, 2-29; basic functions, 28; classification of, 30-85; kinds of, **33**; order of appearance on earth, 601, 602
Lizards, 319, 320, **320, 477, 675**
Lock-and-key theory, 140-41, **140**
Locomotion, 319-45; in a cell, 107; in representative organisms, 326-30
Locus, 570
Long-day plants, 711, **711**
Long-lived immunity, 750, 754
Long-term memory, 706, **707**, 728
Lorenz, Konrad L., 704, **704**
Loudness, **415**
Lungs: air path to, 224-25, 232; cancerous tissue, **732**; in circulatory system, 285; excretion and, 307, **307**, 316; gas exchange in, 225-27; nerves for, **398**; in respiration, 232; surface area, 225, **225**
Luteinizing hormone, 368-70, **369**
Lymph, **289**, 290
Lymph nodes, **289**, 290
Lymph system, 289-90, **289**
Lymph vessel, 252, **260**
Lymphocytes, 745-46
Lysine, 166, 449
Lysosomes, 104-05, **105**, 107, 109-10, 111, 114

McCarty, Maclyn, 577
Macleod, Colin, 577
Maggots, Redi's experiment, **591**
Magnesium, 119, 171
Maize, 22, 62, **175**
Malaria, 742, **742**
Malarial parasite, 742, **742**
Male chromosomes, 560
Male reproductive system, 504-05, **504**, 526
Malfunctions, body, 730-33, 754
Malnutrition, 165, **174**, 178
Malpighi, Marcello, 5
Malpighian tubules, 305-06, **306**, 316
Malt sugar, 128
Maltase, 128, **129**, 239, **239, 439**
Maltose, **143, 144**, 239, **239**
Mammals, **63**, 76, 80-81, **80-81**, 277
Mammary glands, 80
Mandible, **332**

Mangold, Hilde, 510
Mantle, of mollusk, 70, **71**
Marrow, **334**
Marsh, 670, **690**
Marsupials, 80, **80**, 520, 522, **522**, 606, **606**
Mating etc., 516-17, **517**, 526, **700**, 728
Matrix: of bone and cartilage, 333-34, **333-34**, 344
Matter, components of, 117-25
Mature lake community, 670
Maxilla, **332**
Mayer, J. R. von, 151
Memory, 706, **706**, 728, 750, 754
Mechanical energy, **136**
Mechanoreceptors, 406-07, **407**
Medical aid, lack of, 734, 754
Medicine, 499
Mediterranean fruit fly, **404-05**
Medulla, in brain, **393**, 396
Medusa, **65**, **66**
Meiosis, 465-68, **465-66**, 474, 482, **482**, 500
Membrane, in cell, **107**, 265
Membrane pockets, in muscles, 339
Memory, 706, **707**, 728
Mendel, Gregor, **532**, 608
Mendel's Laws, **531-57**; exceptions to, 546-48
Meninges, 391, **391**
Meningitis, 399, 402
Menopause, 504
Menstrual cycle, 367-69, **369**, 374, 506, 526
Menstruation, 368-69, **369**, 506
Meristems, 494-95, **495**, 500
Mesoderm, 509, 526
Messenger RNA (mRNA), 250, 448, **448-49**, 454, **750**
Metabolic rates, 198-199, 203, 206
Metabolism, 197-203, **197**
Metacarpals, **332**
Metatarsals, **332**
Metaphase, 441-43, **442-43**, 454
Methane, 595, **596**
Methionine, 166, 449
Metric units, 94, **94**
Microfilaments, 107, **108**, 109, 114
Micrograph, 92
Micrographia, Hooke, 91
Microscopes, 89-93, **90-93**, 114; kinds of, 83-93, 114
Microtubules, 107, **108**, 109-11, **111**, 114; triplets, **108**
Microvillus(-i), 253, **253**
Midbrain, 393, **393**, 396
Middle ear, 414, **415**, 430
Miescher, Frederick, 576
Migration, **640-41**, 712, **712**, 728
Milk: digestion problems, 259; human, 513
Miller, Stanley, 597, **597**

Mind-altering drugs, 736
Minerals, **164**, 178, 260, 270; essential, 166; in nutrition, 170-71
Minimal medium, 578
Mitochondrion(-ia), 103, **103**, 110, 111, 114, 185, **185**, 339, **339**
Mitosis, 441-43, **441**, **443**, 454, 458, 463, 482, 500; meiosis compared with, **465**
Mixed nerves, 379
Models, space filling, **122**
Molds, decay, 459
Molecular compounds, 146
Molecular formula, **123**
Molecules, 117-47, **122**, 263-67, **267**, 292, 435-39, **598**
Mollusks, 70-71, **71**, 82, 84, 237
Molting, 324, **324**, 358, **358**
Monerans/Moneran kingdom, **42-44**, 43, 46-48, **50**, 84, 110, 114
Monkeys, 80
Monocots, 61-62, **61**
Monohybrid cross, 541
Monosaccharides, 128
Monotremata, 80, **80**
Morgan, Thomas Hunt, 560, 562-64, 567
Morula, 508
Mosquitoes, 72, 236, **236**, **237**
Mosses, 42, 43, 55, 57, 459, 483, **483**
Moth(s), **40**, **43**, **73**, 413-14, **414**, 427, **427**, 612-13, **612**
Motor nerves, 379, **379**
Motor neurons, 379, **379**, 389, **389**, 390, **390**, 402
Motor unit, 390, **390**
Mount St. Helens, 646
Mouse, **22**, **94**, 522
Mouths, 13, 245, **245-47**, 248, **249**, 260, **407**
Movement, 319-45, 420, 430
mRNA, 250, 448, **448-49**, 454, **750**
Mucus, 219-20, **219-20**, 250, 744
Mules, 41
Mueller, Herman J., 568
Multicellular algae, **42**, **43**, **56**, 57
Multicellular animals, 42
Multicellular organisms, 12, **12**, 28, 213-14, **214**, 232, 240
Multiple alleles, 570, 586
Multiple-gene inheritance, 573, **573**
Multiple sclerosis, 399, 402
Muscle(s), 323, 336-42, 344; action and exercise, 340-42; atrophy, 341; cardiac, 337; cells, 12, 193, **193**, 195, 213-14, **213**; development, **341**; earthworm, 328-29, **328**, 344; electrical stimulation, **340**; fatigue, 340; fibers, 336-37, **337**, **389**, 390, **390**, 416; fish sense organ, **407**; movement, 323, **323**; receptor, **416**; skeletal, 336, **336**;

338-39, **338-39**, 344, 390, 402; skeletons and, **325**, 344; smooth, 337, **337**, 344; tendon connection, **335**; voluntary, 336
Mushrooms, **22**, **43**, **53**, 54, **54**, **459**, **469**
Mutagens, 568-69
Mutations, 461, **461**, 568-69, **568-69**, 581-82, **582**
Mutualism, 632, **633**, 638
Myelin, 388, **388**
Myosin, 338-39, **338**

NAD$^+$, 190
NADH, 184-85, 190, 206
NADP$^+$, **158**, **159**, 161
NADPH, 157, **157-58**, 159-60, **161**, 178
Narcotics, 736
Nasal passages, 224, **224**
Natural disasters, 646, 660
Natural selection, 610-14, 616
Navel, 510
Nearsightedness, 426, **426**
Nectar, bee collecting, **491**
Nematocysts, 244, **244**
Nematodes, 68, **68**, 82, 84
Nephridium(-ia), 305, **305**, 316
Nephrons, 311-12, **310-12**
Nerve(s), 255-56, 341-42, **341**, 344, 379
Nerve cells, **95**
Nerve impulses, 378, 379, 381-84, **383**, 410, **410**, 430
Nerve net, in hydra, 385, **385**, 390
Nerve pathways, 389, 402
Nerve tissue, disorders of, 402
Nerve tube, in chordates, 75
Nervous control, 377-403
Nervous system, 16, 387-84, 397, **397**, 399, 402
Neurons, 378-82, **378**, **380**, **385**, 388, **388**, 402, 416, **416**
Neurospora, 578, **578**
Neutrons, 119, 120
Niacin, 169
Niche, 667, **667**, 694
Nicotine, 399, 734, **734**
Nicotinamide, **159**
Nicotinamide adenine dinucleotide (NAD), 190
Nicotinamide adenine dinucleotide phosphate (NADPH), 157, **157-58**, 159-60, 161, 178
Nictitating membrane, **605**
Nitrates, 629-30, **629**, 638
Nitrification, 629, **629**
Nitrites, 629, **629**
Nitrogen, **119**, **596**
Nitrogen base, in nucleotides, 134, **134**

Nitrogen base pairs in DNA, 437-38, 454
Nitrogen cycle, 629-30, **629**, 638
Nitrogen fixation, 630, **630**, 638
Nitrogen-fixing bacteria, 555, **629**, 630, **630**
Nocturnal animals, 707, **709**, 709
Noncommunicable diseases, 754
Nondisjunction, 564-66, **564**
Nonliving things, 24, 28
Nonmovable joints, **335**
Noradrenalin, 366, **366**, 371, 383
Nose, **406**
Nostrils, catfish, **407**
Notochord, 75, **75**
Nuclear membrane, 46, 47, **83**, 102, **102**, **107**, 109-10, **111**, 114
Nuclear pores, 102, **102**, **111**
Nucleases, 239, 251, 252
Nucleic acids, 102, 128, 134, **134**, 146
Nucleolus, 102, **102**, 103, 109-10, **111**, 114
Nucleotides, 134, 146, **437**, 454
Nucleus, 11, 95, 96, 102-03, **102**, 109, **111**, 114, 378, **388**
Nudibranch, **70**
Nutrients, 163, 178, 238-40, 242, **242**, 257, 260; absorption of, 240, 257, **257**, 260; in cellular respiration, **198**; energy from, 197-98, 206; essential/nonessential, 165-66, 178; nutrition and, 165-170; use and storage, 242, **242**, 257, 260
Nutrition, 165-179
Nymph, 358-60, **359**

Oak, **23**, 43, 62
Oak gall, **23**
Observation, methods of, 89-94
Occipital lobe, 394
Oceans, 595, 666, 689-90, **689-90**, 694
Octopus, **16**, **70**, 82, **378**
Offspring, 13, 17-18, 523-24, 526, 541-44
Oils, 130, 163, 555
Oleic acid, **124**
Ommatidium(-a), 421-22, **421**
Omnivores, 622, **622**, 638
One-gene-one-enzyme hypothesis, 578
Oogenesis, 470, **470**, 474
Oparin, A. I., 595-97
Open transport system, 275-76, **275**, 292
Operant conditioning, 705-06, **705**, 728
Operating room, antiseptic environment, 748, **748**
Opsins, 425
Optic cups, **510**
Optic nerve, 423-24, **423**, 424

Oral groove, 243, **243**
Order, 35, **35**, **37**, 84
Organ(s), **12**, **13**, 28; artificial, 752, **752**
Organ systems, 12, **13**, 28
Organelles, 96, 99, 103-10, 114
Organic acid group, 132, **133**
Organic compounds, 112-22, 146
Organic debris, 671, 694
Organic molecules, 595
Organisms, 9
Origin of Species, Darwin, 610
Osculum, 64
Osmosis, 266, **266**, 292
Outbreeding, 552
Outer ear, 414, **415**, 430
Outer electrons, **120**
Oval window, 415, **415**
Ovary(ies), 467, 470, 486, **486**, 505-06, **505**, 526
Oviduct, 367-68, **505**, 367-68
Ovulation, 367-68, **368**, **470**, 506
Ovulation stage, in menstrual cycle, 368-69, **369**
Ovule, **486**, 487-88, 500
Ovum/Ova, **94**, 470, **470**, **505**, **507**. *See also* Eggs
Oxygen, 119; in cellular respiration, 184; diffusion of, 264; in digestion, **14**; electron transport and, 192; in food energy process, 186; in gas exchange, 211-230, 232; in metabolism, 199, **199**; in photosynthesis, **152**; plant sources of, 499; producers and, 178
Oxygenated blood, 227
Oxyhemoglobin, 227, 278-79
Oxytocin, 362, 371
Ozone layer, 596-97, **597**

Pacemaker, heart, 285-86, **286**
Pacemaker cells, 342, 344
Pain, 407, **407**, **410**, 417-18, 430
Palisade layer, 153, **153**
Pancreas, 13, 248, **249**, 251, 260, **361**, 365-66, **365**, 371, 398
Pancreatic juice, in human digestion, 251, 252
Pancreatic secretions, 256
Paper chromatography, 127, **127**
Paramecium caudatum, 43, 52, **52**, 105, **268**; excretion in, 304, **304**, 316; as food, **236**; food processing in, 243; movement and locomotion in, 327, **327**, 344; reproduction, **17**; respiration in, 219, **219**, 232; transport in, 273, **273**, 292; vacuoles in, 104
Paraplegics, 396
Parasites, **23**, 49, 67, **67**, 634, **635**
Parasite-host interaction, 646, **646**

Parasitic fungi, 743, 754
Parasitic wasp, **23**
Parasitic worms. *See* Leeches; Hookworms, Tapeworms
Parasitism, 634, **635**, 638
Parasympathetic nerves, 398, **398**
Parasympathetic nervous system, 397-98, **397-98**, 402
Parathyroid glands, 357, **357**, **361-62**, 364, 371
Parathyroid hormone, 357, **357**, 371
Parent(s), **436**, 471, 533
Parental care, 523-34, **523-24**, 526, 728
Parietal lobe, of brain, 394
Parthenogenesis, 472, **473**, 474
Passive immunity, 750, 754
Passive transport, 267, 292
Past, information about the, 599-607, 616
Pasteur, Louis, 592-93, 616, 748, 750, 754
Patella, **332**
Pathogens, 737
Pauling, Linus, 453, 575
Pavlov, Ivan, 705
Pea(s), 148-49; **532**; Mendel's research on, 532-40
Pearls, 70
Pecking order, 723
Penfield, Wilder, 393
Penicillin, 57, **749**, **750**, 751
Penis, **504**, 505
Peppered moth, 612-13, **612**
Peptides/Peptide bonds, **133**
Perception, 406, **406**, 410-12, **411**
Perennials, plant, 497, 500
Perfect flower, 486, **486**
Peripheral nervous system, 391, **391**, 397-98, **397**, 402
Peristalsis, 245, 260
Permafrost, 681
Permeable membrane, 100, 114, 265, **265**
Perspiration, 308
Petals, 486, **486**
pH, 142-43, **142**, 146
Phage (bacteriophage), 577
Phalanges, 332
Pharynx: earthworm, 245, **245**; human, 224, 248, **249**
Phaseolus vulgaris, 62, **62**
Phenotypes, 534, **534**, 556
Phenylalanine, 166, **350**, 449, 579
Phenylketonuria (PKU), 579
Pheromones, 714, **714**
Phloem, 272-73, **273**, 495-497
Phosphate, **596**
Phosphate group, 134, **134**, 437, **437**
Phosphorus, 119, 171
Phosphorus cycle, 630-31
Photoperiodism, 711, **711**

Photoreceptors, 406-07, **407, 410, 424-25**, 430
Photosynthesis, 13-15, **48**, 150-56, **152**, 178; in the carbon cycle, **628**, 638; cellular respiration and, 186-88, **187-88**, 206, 217; in chloroplasts, 104; equation for, 152; experiment on, 150-51, **150**; leaf and, 153; major events of, **158**; mechanism of, 157-64; origin of, 595; producers in, 164; summary of, **163**; water splitting in, **161**
Phototropism, 701, **701**
Phylum/Phyla, 35, **35**, 37, 82, 84
Physical conditions: in water environments, 686-87, 689-90, **691**
Physical contact, disease communication via, 738, 754
Phytoplankton, **625**, 685-86, **685**, 694
Pigments, plant, 155, **155**, 178
Pines, **43**, 59, 62
Pioneer(s), plant, 668
Pioneer communities, 668, **668**, 670, 694
Pistil, 486-87, **486**
Pitch, musical, **409**
Pituitary gland, **361**, 362-64, **363-64, 369**, 370-71, **393**, 394
Placenta, 369, 509-10, **509**, 521, 526
Placental birth, **521**
Placental mammals, 80, **80**
Plague, **739**, 754
Planaria, 67, 82, 427, **702**
Plane of polarization, **408**
Plankton, 685-86, **685**, 688, 694
Plant(s), **42-44**, 55-62, 84; behavior, 701, **701**, 728; cell reproduction and hormones, 352; cell structure, 110, 114; development in, 494-97; dispersal, 663; excretion in, 303-04, 316; as food source, 499; growth in, 351-52, 374, 494-97, 500; hormones, 351-53, 374; life span, 497; movements in, 222, 326, **326**, 344; petroleum from, 205; primary growth, 494-95, **495**; reproduction in, 477-94; respiration in, 215-18, 232; secondary growth, 495-96, 500; sexual reproduction in, 482-89; structure, 43; structures in cells of, 110; transport in, 270-73, 292; uses of, 499; water/salt balance, 299
Plant breeders/breeding, 549-50, 555
Plant hormones, synthetic, 354, **354**
Plant kingdom, 55-62
Plant pigments, 155, **155**, 178
Plant stem, 89, **95**
Plasma, 278-79, **278**, 282, 290, 292
Plasma proteins, 278

Platelets, 280, **280**, 281, 292
Platyhelminthes, 67, **67**, 82, 84
Pneumonia, 740
Poisons, 309, 399
Polar bodies, 470, **470**, 474
Polarization, 380-81, **380**, 402, **408**
Polarized light, 407-08, **408**, 430
Pollen, 487, **487**, 500
Pollen grains, 95, 487, **487-88**, 500
Pollen tube, 488, **488**
Pollination, 488, **488**, 490-92, **490**; artificial, 490, **490, 532**
Pollution, 6, 733, 754
Polyp, **66**
Polypeptides, 596, **596, 598**
Polypeptide chains, **132, 133**, 280, **280**
Polyploidy, 569, **569**
Polyribosome, 446
Polysaccharides, **129**, 146
Polyunsaturated fats, 131, **131**
Pond, 686-87, **687**
Pons, in human brainstem, **393**, 396
Population(s), 640-41, 649-54, 656, 660; infectious diseases and, 751-52
Population balance, 649-54, 656-60
Population crash, **648**
Population density, 651-54, **651-52, 654**, 658
Population explosions, 643
Population growth, 655-58, **656**; carrying capacity and, 650; characteristics of, 640-48; doubling time, **643**; limiting factors, 644-46; mathematics of, 642, 646-48, **646-48**; patterns of, 647-48
Poriferans, 64, **64**, 82, 84
Potassium, 119, 171
Potassium-argon dating, 601
Potassium ions, 380-82, **380-81**
Potential energy, 136-37, 146
Pouched mammals, 80, **80**, 522, **522**, 606, **606**
PPB, 145
Prairie(s), 675, **676**
Precipitation, 627, **627**
Predation, 634, **634**, 638
Predator(s), 622, 634, **634**, 638
Predator-prey cycle, 646
Prediction, with Mendel's laws, 541-44
Pregnancy, 369, **369**, 506
Pressure, 417-18, 430
Priestly, Joseph, 151
Primary consumers, 623, **623**
Primary growth in plants, 494-95, **495**
Primary phloem, 495
Primary xylem, 495
Primitive streak, 510, **510**

Probability, Mendel's laws and, 539-44
Producer(s), 22, 28, 84, 621; 638; aquatic, **21**; in energy pyramid, **625**; food and, 178; food getting, 235; forest, **21, 22**; in photosynthesis, 164
Producer protists, 50-51, **51**
Product, 140, **140**
Progesterone, 367, 369, **369**, 371
Prokaryotes/Prokaryotic cells, **46, 50**, 84, 103, 106, 108-10, 114
Prolactin (PRL), 370-71
Proline, **350**, 449
Prophase, 441-43, **442-43**, 454
Proprioceptors, 407, **407**, 416-17
Prostate gland, **504**
Proteases, 239, 250, 252, **252**
Protein(s): 128, 132-33, 178, 446; breakdown, **197**; in cell membrane; 100, 114; decomposition of, 630, 638; digestion of, 252, **252**; essential/nonessential, 166-67; formation on mRNA, 450; functions and sources, 167; from glucose, 163; plasma, 278; twisting and folding of, 133, **133**
Protein hormones, 355, **355**, 374
Protein ion, 380, **380**
Protein molecules, **94, 100**
Protein synthesis, 446-52, **450-51**, 454
Protists, **42-44**, 50-84, **50**; inherited behavior in, 702-03, 728; structures in cells of, 110
Protons, 119-20, **119**
Protozoa, **42, 43**, 50, **50**, 51, **52, 90**, 741-42, 754
Pseudopods, 51, 236, **236**, 322, **322**
Puberty, 504, 515, 526
Pubis, **332**
Public health, advances in, 751-53
Pulmonary blood vessels, **284**
Pulmonary circulation, 284, **284**
Pulse, 263; rate, 283
Pure-breeding plants, 532-33
Pus, 279
Punnett, R. D., 540
Punnett square, 540-46, **540**
Pupa, 359, **359**
Pupil, of the eye, 423, **423**
Pylorus, **250**
Pyridine, **735**
Pyruvic acid, **189**, 190, **191**, 206, 296, 308, **308**

Queen, in insect colony, 718, **718, 719**, 728
Queen substance, 720-21
Quercus species, 34, **34**, 36, **36, 37**
Quinine, 750

R groups, 133, **133**
Radiant energy, **136**
Radiation risks, 568, **568**
Radioactive decay, 601, **601**
Radius, human, **332, 336**
Rays—in plant tissue, 496, **496**
Reabsorption, 311-12, **311**, 316
Reactions. *See* Chemical reactions
Reasoning, 707-08, **707**, 728
Receptors, 378, **379**, 402, 405-12, 430; action of, 410-11; kinds of, 406-07; light, 424-25; in reflex arc, 389, **389**; in semicircular canals, 416, **416**; in skin, **417**, 430; in tendon, **416**; types of, 430
Recessive (genetics), 534-35, **535**
Recombinant DNA, 582. *See also* Genetic recombination
Rectum; grasshopper, 246, **247**; human, **249**
Red algae, 55, 56
Red blood cells, 93, 278-79, **279**, 281; 292; function, **99**; in gas transport, 226-27, 232; liver and, 309; sickled and normal, **581**; size, **94**
Red marrow, **334**
Redi, Francesco, 590-92, 616
Redwood trees, **59**
Reflex(es), 396, 703, **703**, 728
Reflex arc, 389, **389**, 402
Regulation/Regulators, 649-50; of digestive system, 255-56, **255-56**; of excretory system, 312-13, **313**; of respiratory system, 227-30, **228-29**; of transport system, 285-87, **287**
Reinforcer, **706**
Releasers, 703, **703**, 728
Renal blood vessels, 310, 311
Replication of DNA, 435, 438-39, **438**, 454
Reproduction, **17-18**, 28, **435**, 454, 456-75; cellular basis of, 440-45; human, 504-07, 526; of molecules and cells, 434-55; in plants, 477-94; types of, 457-62; in vertebrates, 503-07, 516-18
Reproductive hormones, 367-68, 374
Reptiles, 76, 78, 81, 277
Reserpine, 499
Respiration, 211-12, **212**; in the carbon cycle, **628**; cellular, *see* Cellular respiration; circulatory system and, **214**; in the earthworm, 219, **219**; environment and, 230, **230**; in the grasshopper, 221-22, **221-22**; homeostasis and, **228**; in the hydra, 219, **219**; in the paramecium, 219, **219**, 232; photosynthesis and, 217; in plant, 215, 232; in representative organisms, 215-22; in the water

cycle, 627. *See also* Breathing: Gas exchange
Respiratory centers, 229, **229**, 232
Respiratory illnesses, 734, 754
Respiratory systems, 214, 222, 224, 227-230
Responses to stimuli, 16-17, **16**
Resting state in neuron and axon, 380, **380, 402**
Reticular activating system, 396
Retina, 430; in camera-type eyes, **422**, 423, **423**; structure of, 424-25, **424-25**
Retinene, 425, **425**
Rheumatoid arthritis, 732, **732**
Rhizomes, 478, 480
Rhodopsin, 425, **425**
Rib(s), **223, 332**
Rib muscles, **223**, 223
Riboflavin, 169
Ribonucleic acid. *See* RNA
Ribose, **159**, **447**, 591
Ribosomal RNA (rRNA), 448, 454
Ribosomes, 103, 106, **106**, 109, **110**, **111**, 114, **446**, 448, **751**
Rickets, 168
Rickettsia, 743, 754
Ringworm, 54, 743, 754
Ritual(s), 722, **722**, 728
RNA, 134, 446-48, **447**, 454
Rocky Mountain spotted fever, 743
Rods, in human eye, 424-25, **424**, 430
Romalea microptera, 73
Root(s), 218, 270, **270**, 326, 344, 352, 470, 480, 499
Root cells, 110
Root hairs, 270, **270**
Root pressure, 271-72, **272**, 292
Rough ER, **111**
Round dance, honeybee, 720, **720**
Roundworms, **68**, 82
Royal jelly, 719
Royal Society of England, 90, 91
rRNA, 448, 454
Runners, in plant, 478, 480, 500
Run-off in water, **627**
Rutting, **517**, 518-19

S-shaped growth pattern, 647, **647**, 650, 660
Sabine vaccine, 575
Sacrum, **332**
Saliva, 246, 252
Salivary glands: grasshopper, **247**; human, 248, **249**, 255, **255**, 398
Salk vaccine, 575
Salmonella typhi, **733**
Salt, in ocean water, 689
Salt balance, 299-301, 316
Salt water, salt in, 299-300, **299**, 316, 689

Sanitary conditions, in disease, 733, 754
Sapwood, 497, **497**
Saturated fats, 131, **131**
Saussure, Nicolas T. de, 151
Sawtooth curve, **649**
Scapula, **332, 336**
Scavengers, 622, **622**, 638
Schleiden, Matthias, 96, 114, 119, 575
Schull, George M., 549-50
Schwann, Theodore, 96, 114, 575
Schwann cells, 388, **388**
Scientific method, 96-97, 114
Scientific names, 31-37, 84
Scion, 479, **479**
Sclera, 423, **423**
Scrotum, 504-05
Scurvy, 734, 754
Second signal, hormone, 355, **355** 374
Secondary growth, 495-96, **496**, 50
Secondary phloem, 496, **496**
Secondary sex characteristics, 367, 374, 504, 526
Secondary xylem, 496, **496**
Secretin, 256, 355, 371
Secretion. 311-12, **312**, 316
Sediment, 600-01
Sedimentary rock, **500**, 600-01, 616
Seed(s), 492-93, **492-93**, 499, **664**, 665
Seed coat, **61**, 492, **492**, 500
Seed plants, **42-43**, 55, 59-62; life cycle, 485-89, **485**
Seed types, 485
Seedless vascular plants, 58
Segregation, law of, 536-37, **536**, 556
Self-pollination, 490, **490**
Semen, 505
Semicircular canals, 415-16, **415-16**, 430
Seminal vesicles, **504**
Seminiferous tubules, **504**, 505, 526
Semipermeable membrane, 100, 114
Senses, 404-31
Sensory nerves, 379, **410**
Sensory neurons, 379, **379**, 389, **389** 402; in reflex arc, 389, **389**
Sepals, **486**
Septum, in heart, 283, 284
Seratonin, 383
Serine, 449
Seta/Setae, 328-29, **328**, 344
Sex chromosomes, 560, 562
Sex determination: *Drosophila*, 565; human, 560-61
Sex-linked inheritance, 561, **563**
Sex-linked traits, 562-64, 562-64
Sex organs, nerves for, 398
Sexual reproduction, 18, 457, 464-74; in plants, 482-89
Short-day plants, 711, **711**
Short-term memory, 706, **707**, 728

783

Sickle-cell anemia, 581-82, 586, 732
Sickled red blood cell, **581**
Sight, 410, 420-26, **426**, 716-17
Sign languages, chimpanzee learning, **700**
Simple eyes, 420, **421**, 430
Sinoatrial node, 285-86, **285**, 292
Size, of biological objects, **94**
Skeletal muscles, 336, **336**, 338-39, **338-39**, 344, 390, 402
Skeletal tissues, 333-35, **333**
Skeletons, 323-25, 325, 331-35, **332**, 344; bony, 324, **325**; cartilaginous, 324; exoskeleton, **325**, 329-30, 344; fossil, **599**
Skin, **95**, 308, **308**, 316, **407, 417**, 430, 744, 754
Skinner, B. F., 705
Skull, 391, **391**
Sleep movements, in bean plant, 326, **326**, 344
Sleeping sickness, 741, **741**
Sliding filament theory, 338, **338**, 344
Slime molds, **42-43, 50,** 53
Slow pain, 418
Small intestine, 13, 248, **249**, 251-53, 260
Smallpox, 749, **749**, 754
Smell(s), 410-11, **410-11**, 426-28, **427**, 430; diffusion of, **264**
Smoking, 734-35
Smooth ER, **111**
Smooth muscle, 337, **337**, 344
Snake, **23, 43,** 78, **78**, 82, 203, 235, **325**, 408, **408, 427, 623,** 679
Social behavior, 713-28, **713, 726**; human and animal, 725-27; insect, 718-21, 728; vertebrate, 721-26
Social bonds, chimpanzees, 727
Social dominance, 723-25, **723**, 728
Social insects, 718-21. See also Ants; Bees
Sodium, 119, **124**, 171
Sodium ions, in axons, 380-81, **380-81**
Sodium pump, 381-82, **381**
Soil(s): pH, **142**; in the various biomes, 674, 676, **678**, 680-81
Somatic nervous system, **397**
Sound, 409, **409**, 413-15, **415**; communication and, 715-16, **715**, 728
Sound waves, 413-14, **413**, 430
Space filling models, **122-23**
Specialization, 98-99, **98-99**, 114, 245, **245**
Species, 9, 28, 35-36, **35, 37,** 84; change in, 26-27; clues to change, 606, 616; extinction of, 603; fossil-living relationship, 603-05

Spectra, absorption, **156**
Spemann, Hans, 510
Sperm, 469, 474; chromosomes in, 561; human, 504-505, **504, 507**; plant, 482-84, **482-83**, 488, 500; quantity released, 516, 526
Spermatogenesis, 470, **470**, 474
Sphygmomanometer, **286**
Spices, 499
Spiders, **72,** 82, **523,** 663, **703**
Spinal cord, 63, 379, **379,** 389, **389, 391, 393,** 396, **396**
Spinal nerves, 391
Spindle, 442-43
Spindle fibers, 442, **442**
Spiracles, 221-22, **221-22,** 232
Spirilla, **47**
Spleen, human, **285**
Sponge(s), 64, **64,** 82
Spongy bone, 334, **334**
Spongy layer in leaf, 153, **153**
Spontaneous generation. See Abiogenesis
Sporangium(-a), 484, **484,** 500
Spores, **47,** 459, 474; formation of, 459, 474
Sporophyte, 482, 482-83, 487, 500
Squash, 60, **558-59**
Stains, 93, **93**
Stamens, 486-87, **486**
Staphylococcus, **749**
Starch, **128,** 129, 163
Starfish, 74, **74,** 82, **460,** 622
Stearic acid, **131**
Stem(s), 218, **218,** 326, 478, 480, 499
Stem cells, **11**
Stentor, **52, 108**
Stereotyped behavior, 702-03, **702,** 728
Sternum, **332**
Steroids, 350, 356, 371, 374
Stickleback fish, 700, **700**
Stimulus(i), 16-17, **16**
Stigma, 486, **486**
Stipe, **56**
Stirrup, 415, **415**
Stock, 479, **479,** 500
Stomach, 12, **13,** 246, **247,** 248, 249, 255-56, **256,** 260, **285,** 361, 399
Stomach pouches, grasshopper, 246, **247**
Stomach wall (glands), 371
Stomates, 152, **215,** 217, **217,** 232
Stratton, Charles, 361-62, **361**
Stratton, Lavinia, 361-62, **361**
Strawberry(ies), reproduction, **18,** 479
Streams, 685, 687-88, **688,** 694
Streptococcus, **47**
Streptomycin, 750
Stress, 734, **734,** 754
Stroke, 399, 402

Stroma, 154, **154,** 178
Strontium-90, 631, **631**
Structure, of living things, 9-18, 84
Strychnine, 399
Style, 486, **486**
Substrate, **140,** 351
Sucrose, 128
Sugars, 128-29. See also Fructose; Glucose; Lactose; Maltose; Simple sugars
Sulfa drugs, 750
Sulfur, 119, 171; cycling, 630-31
Sun, **152, 157, 163,** 624, **624,** 638
Surgery, 752
Survival, 28, 412, 523-24, 526, 610-611
Sutton, Walter S., 559, 575
Swallowing, 248, **249**
Swamplands Act, 693
Swan-neck flask, 592-93
Sweat glands, 308, **308**
Symbiosis, **23,** 28, 632-36, 638
Symbiotic relationships, 632-36
Symbols, chemical, 118-19
Sympathetic nervous system, 387-98, **397-98,** 402
Synapse, 382, **382,** 402
Synapsis, 466
Synaptic knob, 382-83, **382**, 402
Synthesis, 143
Syphilis, 738, 754
Systemic circulation, 284

Tadpole, 77
Tapeworms, 67, **67,** 82, 237, 740, **742,** 754
Target cells, of hormones, 355, 364
Target structures, 349-51
Tarsals, **332**
Taste, **410,** 426, 428-30, **428-29**
Taste buds, 428-29, **428-29**
Tatum, Edward L., 578-79
Taxis/Taxes, 702, 728
Taxonomic categories, 35-36, 84
Taxonomic tests, 39-41
Taxonomists, 84
Taxonomy, 34
Technology: medical-surgical, 752-54; population and, 656-58
Telophase, 441-43, **442-43,** 454
Temperature, 125, 141-43, **141,** 146, **410,** 417-18
Template, in DNA, 448
Temporal lobe, 394
Temporary immunity, 750, 754
Tendons, 335, **335, 416**
Tendrils, 326
Tension, in water, 272
Territoriality, 722, **722,** 728
Tertiary consumer, **623**
Test cross, 552-53

Testis/Testes, **361**, 367, 371, 374, 470, 504, **504**, 526
Testosterone, 367, 371, 504
Tetanus, 399
Tetany, 364, 374
Tetracyclines, 750
Tetrad, 366-67
Thalamus, **393**, 395
Theory, 97
Thermoreceptors, 406-08, **407-08**, **410**, 417-18
Thiamin, 169
Thorax, grasshopper, 329, **329**
Threonine, 166, 449
Thylakoid discs, **154**, 178
Thylakoid membrane, 156
Thymine, 437, **437**, **447**, 454, **596**
Thyroid gland, 361-64, **361-64**, 371, 374
Thyroid stimulating hormone (TSH), 364, **364**, 370, 371
Thyrotropic releasing factor (TRF), 364, **364**
Thyroxin, 361, 363, 364, **364**, 370, 374
Tibia, human, **332**
Tick(s), **23**, **72**
Tijio, Joe Tin, 560
Timberline, 684, **684**
Tissue cells, 12-13, **12**, 28
Tissue fluid, 282, **289**, 290, **307**
Tissue spaces, 292
Tobacco, 734-35
Tongue, **406**
Tool use, chimpanzee, **708**, 727
Tooth-plaque bacteria, **737**
Touch, **410**, 417, 430, **715**, 715, 728
Toxins, bacteria-formed, 740
Trachea: dolphin, **307**; grasshopper, 221-22, **221-22**; human, **224**, 232, **232**, 249, 397-98
Tracheophytes, 55, 58-62
Traits, hereditary, 533, **535**
Transfer RNA (tRNA), 448-50, **449**, **451**, **750**
Transmitter substances, 382-83, **383**, 402
Transpiration, 272, **272**, 627
Transport, 262-93; into and out of cells, 267-68, **267-68**; characteristics of, 263-69; closed, 292; in earthworm, 274-75, **275**; function of, 263, 292; in grasshopper, 275-76, **275**, 292; in homeostasis, 290-91; in humans, 278-91; in hydra, 274, **274**, 292; models of, 267, **267**; in paramecium, 273, **273**, 292; in plants, 270-73, 292; in representative organisms, 270-73
Transport systems, 269, 274-75, **275**, 292, 350, 362, 374
Trees: in coniferous forests, **680**;

energy and growth of, 137, **137**; trunk section, **496**
Triceps, **336**
Trichinella spiralis, **68**
Trichinosis, **68**, 743
Triplet code, **449**, 454
TRIS, carcinogen, 145
tRNA, 448-50, **449**, **451**, **750**
Troops, baboon, 723-25, **724**, 728
Tropical rain forest, 678, 683, **683**
Tropical rain forest biome, 673, **673**, 678-79, **678-79**, 694
Tropisms, 701, **701**, 728
Trypanosoma, **52**
Tryptophan, 166, 449
Tschermak von Seysenegg, Erich, 575
Tsetse fly, 741, **741**
Tubers, in reproduction, 478
Tuberculosis, 748, **754**
Tubeworms, 68, **69**
Tubules, 305-06, **306**, 310, **310**, 316
Tumors, 399, 402
Tundra, 681, 683, **683**
Tundra biome, 673, 673, 681-83, **681-82**, 694
Tuning fork, **413**
Turgor movement, 322, **322**, 344
Twins, 3-4, **4**, 514, 526
Twitches, 340-41, **340**
2, 4-D, **354**
Tympanic membrane, 414, **414**
Typhoid bacteria, 109
Typhoid fever, 733, **733**, 754
Typhus, 743, 754
Tyrosine, **350**, 449

Ulcer, stomach, 250, **250**
Ulna, **332**, **336**
Ultracentrifuge, 127
Ultrasound, 409, 413-14, 430; scanner, **753**
Ultraviolet, 407, **407**, 597
Umbilical cord, **509**, 510
Unicellular algae, **42-43**, 50
Unicellular organisms, 12, **12**, 28, **213**, 232, 240
Universal donor, 571
Universal recipient, 571
Unpolarized light, **408**
Unsaturated fats, 131, **131**
Uracil, **447**, 454, **596**
Urea, 297-98, **297**, 301, 308, **308**, 316
Ureter, 310, **310**
Urethra, **310**
Urey, Harold C., 597, **597**
Uric acid, 297-98, **297-98**
Urinary bladder, 310, **310**, **504**
Urines, 299-301, **300**, 310, **310**, 316
Use and disuse, Lamarck theory, 608-609, 616

Uterus, 367-69, **369**, **398**, **505**, 506, **509**, **514**, 526

Vaccination, 749-50, **749**, 754
Vaccines, 749, **749**
Vacuoles, 104, **105**, 109-110, **111**, 114, **153**, 236, **236**, 243, **243**, 260, 304, **304**
Vagina, **367**, **505**, 506, **506**
Valine, 166, 449
Valves: heart, 284; in veins, 282-83, **283**
Variations: 26-28, **26**, **27**, 461-62, 474; in natural selection, 610-11, 616; within same litter, 530-31, **558-59**
Varicose veins, 283
Varieties, 36; developed by breeding, **607**
Vas deferens, **504**, 505
Vascular bundles, 273, **273**
Vascular cambium, 495-96, **496-97**
Vascular cylinders, **273**
Vascular tissue, 55, **55**
Vegetative propagation, 460, 478-81, 500
Vegetative reproduction, 463
Veins: human, 281-84, **281-83**, 292; leaf, **153**, **216**
Vena cava, 282-83, **284**, 285
Venereal disease, 738
Ventral nerve cords, 386, **386-87**, 390
Ventricles, in heart, 283-84, **283**, 292
Venule, **281-82**
Venus' flytrap, 322, **322**
Vertebrae, 63
Vertebral column, **332**
Vertebrates, **43**, 63, **63**, 84; adaptations in, 516-24; as chordates, 76; classes of, 76, 81; 508-15, 519-24; reproduction in, 503-07, 516-18; social behavior in, 721-26
Vertebrate eyes, **422**, 422, 430
Vertebrate nervous system, 388-90
Vertebrate neurons, 388, **388**
Vestigial organs, 605, **605**, 616
Vibrations: detection of, 430
Victoria, Queen, 563, 564
Villus/Villi, **241**, 251, **253**, 260
Viral DNA, 577
Virgil, 589-90
Virus(es), 49, **49**, 94
Virus diseases, 49, 738, 740-41
Visible light, 407, **407**
Vision, 407-08, 411, **411**, 421, **422**, 424-25, **426**
Visual signals, communication and, 728
Vitamins, 163, 166, 168-70, 178, 254, 260, 334
Vitreous humor, **423**, 424

Waggle dance, honeybee, 720, **720**
Wallace, Alfred Russel, 575, 608-10, 60?, 616
Warmblooded animals, 79, 202-203, 206
Wart(s), 738
Waste products, 660; of cells, 296-98, 310; from embryo, 521; nitrogenous, 296-98, 316; population growth, 645
Water: absorption by roots, 270; in the carbon cycle, 628; in diet, 171, 254, 260; diffusion of molecules through, 265, **265**; excretion of, 300-01; intake in organisms, 300-01; pH of pure, **142**; in photosynthesis, **152**; polluted, **6**; in primitive atmosphere, 595; splitting in photosynthesis, **161**, 171; structural formula, **596**; transport upward in plants, 271-72, **272**

Water balance, 299-301, **300, 303**, 316
Water contamination, 738, 754
Water cycle, **24**, 627, **627**, 638
Water molecule, **94, 123**
Water table, **627**
Water vacuole, **105, 111**
Watson, James, 277, 439, 453, **453**, 575-76, 584
Waxes, 130
Weismann, August, 609
Whales, **39**, 40, **41**, 80, **211**, 625, 715
Wheat rust, 54
White blood cells, **99**, 279-81, **279**, 292, 754; disease defense, 744, **744**; migration, **320**; types of, **744**
White matter, 391. 393, **396**
Whitefish cell, 441-42
Wilkins, Maurice, 453
Wilting, water balance and, 303
Wind pollination, 491, **491**
Wings, 330, **330**
Withdrawal from alcohol, 736

Wolves, 722, **723**
Wood, growth of, 496-97, **496-97**, 500
Workers, insect, 718-19, **718-19**, 728
Worms, 709, 742-43, 754

X-ray crystallography, 453
Xylem, 270-73, **270-71**, 292, 495-97

Yeast, 53-54, **54**, 186, 193-95, 194, 206
Yellow marrow, **334**
Yolk feeding, 519-21, 526
Yolk sac, 509, 520, 526
Yucca, 353

Zoology, 4
Zooplankton, 685-86, **685**, 694
Zygotes, 464, 474

Acknowledgments

Photographs

Unit I

1–1 C. Lockwood/Animals, Animals; 1–2 Wayland Lee*/Addison-Wesley Publishing Company; 1–5 Stephen Frisch & Associates*; 1–6a Elaine Wicks/Taurus Photos; 1–6b Dr. W. Aubrey Crich; 1–6c J. Menschenfreund/Taurus Photos; 1–7 © 1980 Barrie Rokeach; 1–8 Rosemary Scott/Taurus Photos; 1–10 B. J. W./Tom Stack & Associates; 1–11 Frank S. Balthis; 1–12 E. J. Cable/Tom Stack & Associates; 1–13a E. J. Cable/Tom Stack & Associates; 1–13b Manfred Kage/Peter Arnold Inc.; 1–14b Tom Stack/Tom Stack & Associates; 1–20a Jack Wilburn/Animals, Animals; 1–20b Phil & Loretta Hermann/Tom Stack & Associates; 1–21b Charles Marden Fitch/Taurus Photos; 1–21c Wayland Lee*/Addison-Wesley Publishing Company; 1–22a Dr. W. Aubrey Crich; 1–22b R. Mitchell/Tom Stack & Associates; 1–22c Grant Heilman Photography; 1–23a Chris Newbert/Tom Stack & Associates; 1–23b Curt Weinhold/Tom Stack & Associates; 1–23c Charles Palek/Animals, Animals; 1–26 Tom Brakefield/Taurus Photos; 1–27 George H. Harrison/Grant Heilman Photography; 1–28 Ryan DeMarr/Tom Stack & Associates; 1–29a George K. Bryce/Animals, Animals; 1–29b Dr. W. Aubrey Crich; 1–31 K. Preston-Matham/Animals, Animals; 1–32a Grant Heilman Photography; 1–32b Zig Leszczynski/Animals, Animals; 1–32c Brian Parker/Tom Stack & Associates; 1–32d Wolfgang Bayer/Bruce Coleman Inc.; 1–33 Timothy O'Keefe/Tom Stack & Associates; 1–34 Leonard L. Rue III/Bruce Coleman Inc. Art in-text (p. 28) Breck P. Kent/Animals, Animals; 2–1 Gary Milburn/Tom Stack & Associates; 2–4 David Overcash/Bruce Coleman, Inc.; 2–11a Grant Heilman Photography; 2–11b Ed Robinson/Tom Stack & Associates; 2–11c Howard Hall/Tom Stack & Associates; 2–13a Oxford Scientific Films/Animals, Animals; 2–13b D. R. Specker/Animals, Animals; 2–14a Jeff Foott/Bruce Coleman Inc.; 2–14b Ed Robinson/Tom Stack & Associates; 2–15a State Historical Society of Colorado; 2–15b Santa Fe Railway; 2–15c Addison-Wesley Files; 2–16 Oxford Scientific Films/Animals, Animals; 2–18a S. C. Holt, University of Massachusetts/BPS; 2–18b R. Rodewald, University of Virginia/BPS; 2–19a Z. Skobe, Forsythe Dental Center/BPS; 2–19b Z. Skobe, Forsyth Dental Center/BPS; 2–19c S. C. Holt, University of Massachusetts/BPS; 2–20a Manfred Kage/Peter Arnold Inc.; 2–20b E. R. Degginger/Animals, Animals; 2–21 Brian Parker/Tom Stack & Associates; 2–23a A. Murphy, Viral Pathology Branch, Centers for Disease Control, Atlanta, Georgia; 2–23b T. J. Beveridge, University of Guelph/BPS; 2–23c S. Dales, University of Western Ontario; 2–25a Oxford Scientific Films/Animals, Animals; 2–25b Manfred Kage/Peter Arnold Inc.; 2–27 Runk-Schoenberger/Grant Heilman Photography; 2–28 Manfred Kage/Peter Arnold Inc.; 2–29 Robert Mitchell/Earth Scenes; 2–31a D. T. Maunder, Microbiology Section, Continental Can Co.; 2–31b Stephen Krasemann/Peter Arnold Inc.; 2–31c Margaret Brandow/Tom Stack & Associates; 2–31d Harry Ellis/Tom Stack & Associates; 2–31e W. H. Hodge/Peter Arnold Inc.; 2–31f Alice Kessler/Tom Stack & Associates; 2–33 J. Robert Waaland, University of Washington/BPS; 2–35 Tom Bean/Tom Stack & Associates; 2–36a Stephen J. Krasemann/Peter Arnold Inc.; 2–36b W. H. Hodge/Peter Arnold Inc.; 2–36c W. H. Hodge/Peter Arnold Inc.; 2–36d W. H. Hodge/Peter Arnold Inc.; 2–36e Donald Orenstein/Earth Scenes; 2–37 Clyde H. Smith/Peter Arnold Inc.; 2–39a B. Wallace/Tom Stack & Associates; 2–39b Arthur Phaneuf Jr./Earth Scenes; 2–39c Robert W. Mitchell/Earth Scenes; 2–39d Oxford Scientific Films/Earth Scenes; 2–39e Tom Stack/Tom Stack & Associates; 2–42a R. Head/Earth Scenes; 2–42b Grant Heilman Photography; 2–45a Tom Stack/Tom Stack & Associates; 2–45b Tom Stack/Tom Stack & Associates; 2–46a Ed Robinson/Tom Stack & Associates; 2–46b A. Kerstitch/Tom Stack & Associates; 2–46c Lewis Trusty/Animals, Animals; 2–46d Tom Stack/Tom Stack & Associates; 2–48 Oxford Scientific Films/Animals, Animals; 2–49a Runk-Schoenberger/Grant Heilman Photography; 2–49b Runk-Schoenberger/Grant Heilman Photography; 2–49c Robert Mitchell/Tom Stack & Associates; 2–50a David Wacker/Tom Stack & Associates; 2–50b Centers for Disease Control, Atlanta, Georgia; 2–51a Peter Arnold/Peter Arnold Inc.; 2–51b Brian Parker/Tom Stack & Associates; 2–51c Oxford Scientific Films/Animals, Animals; 2–52 John Shaw/Bruce Coleman Inc.; 2–53a A. Kerstitch/Tom Stack & Associates; 2–53b Zig Leszczynski/Animals, Animals; 2–53c Raymond Mendez/Animals, Animals; 2–53d Richard Humbert, Stanford University 2–53e Brian Parker/Tom Stack & Associates; 2–55 Timothy O'Keefe/Tom Stack & Associates; 2–56a M. Chappell/Animals, Animals; 2–56b Stephen Krasemann/Peter Arnold Inc.; 2–56c Zig Leszczynski/Animals, Animals; 2–56d Tom Stack & Associates; 2–56e R. Jordan/Animals, Animals; 2–56f Robert W. Mitchell/Animals, Animals; 2–58 R. Head/Animals, Animals; 2–62a Fred Whitehead/Animals, Animals; 2–62b Dave Spier/Tom Stack & Associates; 2–62c Steve Martin/Tom Stack & Associates; 2–63a R. Andrew Odum/Peter Arnold Inc.; 2–63b Amil Myshin/Animals, Animals; 2–64a Dr. W. Aubrey Crich; 2–64b D. Specker/Animals, Animals; 2–65 Zig Leszczynski/Animals, Animals; 2–66a Zig Leszczynski/Animals, Animals; 2–66b Zig Leszczynski/Animals, Animals; 2–67a Richard Kolar/Animals, Animals; 2–67b Brian Milne/Animals, Animals; 2–67c Henry Fox/Animals, Animals; 2–67d G. Kooyman/Animals, Animals; 2–68a Steve Martin/Tom Stack & Associates; 2–68b Grant Heilman Photography; 2–68c Hans & Judy Beste/Tom Stack & Associates; 2–69a Mitchell Robert/Tom Stack & Associates; 2–69b F. Mitchell/Tom Stack & Associates.

Unit II

3–1 E. Cable/Tom Stack & Associates; 3–2 *Antony Van Leeuwenhoek and His "Little Animals"*/Dover Book; 3–4 The Bettman Archive Inc.; 3–5 Courtesy, Brundy Library; 3–6a Stephen Frisch*; 3–6b E. Degginger/Earth Scenes; 3–7a Manfred Kage/Peter Arnold Inc.; 3–7b W. Rosenberg, Iona College, BPS; 3–7c W. Rosenberg, Iona College/BPS; 3–8 Manfred Kage/Peter Arnold Inc.; 3–10a E. Degginger/Earth Scenes; 3–10b Manfred Kage/Peter Arnold Inc.; 3–10c E. Cable/Tom Stack & Associates; 3–10d E. Cable/Tom Stack & Associates; 3–12 Michael Gadomski/Earth Scenes; 3–13 Manfred Kage/Peter Arnold Inc.; 3–15a E. Cable/Tom Stack & Associates; 3–15b E. Cable/Tom Stack & Associates; 3–16 J. R. Waaland, University of Washington/BPS; 3–17 R. Rodewald, University of Virginia/BPS; 3–18 Manfred Kage/Peter Arnold Inc.; 3–19 R. Rodewald, University of Virginia/BPS; 3–20 E. H. Newcomb, University of Wisconsin, Madison/BPS; 3–21a T. J. Beveridge, University of Guelph/BPS; 3–21b E. H. Newcomb, University of Wisconsin, Madison/BPS; 3–23 R. Rodewald, University of Virginia/BPS; 3–24 S. C. Holt, University of Massachusetts/BPS; 3–27 J. J. Paulin, University of Georgia/BPS; 3–28 W. Rosenberg, Iona College/BPS; Perspective 3 (p. 113) Howard Hall/Tom Stack & Associates; 4–1 Nelson Max/Lawrence Livermore National Laboratory; 4–2b Wayland Lee*/Addison-Wesley Publishing Company; 4–2c © 1982 DeBeers Consolidated Mines Ltd./Diamond Information Center; 4–2d Andree Abecassis*; 4–2e Keith Murakami/Tom Stack & Associates; 4–6 Wayland Lee*/Addison-Wesley Publishing Company; 4–7d Wayland Lee*/Addison-Wesley Publishing Company; 4–10 Wayland Lee*/Addison-Wesley Publishing Company; 4–11 Stephen Frisch*; 4–12 (left) E. Degginger/Earth Scenes; 4–12 (right) Andree Abecassis*; 4–15 Andree Abecassis*; 4–19 Andree Abecassis*; 4–22 Andree Abecassis*; 4–31 (left) Brian Parker/Tom Stack & Associates; 4–31 (right) Andree Abecassis*; 5–1 Robert Carr/Bruce Coleman Inc.; 5–2 (left) Charles Harbutt/Archive Pictures Inc.; 5–2 (right) © Baron Wolman; 5–6b E. H. Newcomb, University of Wisconsin, Madison/BPS; 5–6c E. H. Newcomb, University of Wisconsin, Madison/BPS; 5–7a W. H. Hodge/Peter Arnold Inc.; 5–7b Em Ahart/Tom Stack & Associates; 5–9 G. I. Barnard, Oxford Scientific Films/Animals, Animals; 5–10 Tom Stack/Tom Stack & Associates; 5–16 John McDermott; 5–19 Andree Abecassis*; 5–20 Andree Abecassis*; 5–21 Andree Abecassis*; 5–22 Mike & Carol Werner/Tom Stack & Associates; 5–23 Andree Abecassis*; 25 (left) Andree Abecassis*; 5–25 (right) Andree Abecassis*; 5–27a Jessica Ehlers/Bruce Coleman Inc.; 5–27b Dr. Nigel Smith/Earth Scenes. 6–1 John McDermott; 6–2 M. P. Kahl/Bruce Coleman Inc.; 6–7 Alice Taylor/University of California, Berkeley; 6–15 (left) Dr. W. Aubrey Crich; 6–15 (right) Andree Abecassis*; 6–16 Tom Stack/Tom Stack & Associates; 6–19 Andree Abecassis*; 6–20 Stephen Frisch*; 6–21 Andree Abecassis*; 6–22 (left) B. Crader/Tom Stack & Associates; 6–22 (right) Tom Walker/Tom Stack & Associates; 6–23 C. Summers/Tom Stack & Associates; 6–24 Zig Leszczynski/Animals, Animals.

Unit III

7–1 Bruce Wellman/Tom Stack & Associates; 7–4 Runk-Schoenberger/Grant Heilman Photography; 7–5 E. J. Cable/Tom Stack & Associates; 7–6 E. J. Cable/Tom Stack & Associates; 7–7 Jack Wilburn/Earth Scenes; 7–8 E. J. Cable/Tom Stack & Associates; 7–10 Peter Arnold/Peter Arnold Inc.; 7–11 Frank S. Balthis; 7–14 Hans Pfletschinger/Peter Arnold Inc.; 7–17 Bob & Miriam Francis/Tom Stack & Associates; 7–18 Tom Stack/Tom Stack & Associates; 7–26 M. Murayama, Bethesda, Maryland/BPS; 7–28 Tom Stack/Tom Stack & Associates; 7–31 Galen Rowell/Peter Arnold Inc.; Perspective 7 (p. 231) Andree Abecassis*. 8–1 Charles G. Summers/Tom Stack & Associates; 8–2 Hans Pfletschinger/Peter Arnold Inc.; 8–3 Kwang W. Jeaon, University of Tennessee/BPS; 8–4a Ron Dillow/Tom Stack & Associates; 8–4b Hans Pfletschinger/Peter Arnold Inc.; 8–4c Kenneth Read/Tom Stack & Associates; 8–5 Walter Fendrich/Animals, Animals; 8–9 Tom Stack/Tom Stack & Associates; 8–11 Manfred Kage/Peter Arnold Inc.; 8–13 E. Degginger/Animals, Animals; 8–21 Andree Abecassis*; 8–25 Evelyn Tronca/Tom Stack & Associates; 8–26 Centers for Disease Control, Atlanta, Georgia; 8–29 Manfred Kage/Peter Arnold Inc. 9–1 Thomas Eisner/Cornell University; 9–3 Andree Abecassis*; 9–4 Wayland Lee*/Addison-Wesley Publishing Company; 9–11 Gary W. Grimes & S. W. Hernault/Taurus Photos; 9–12 Gary W. Grimes/Taurus Photos; 9-13 Wayland Lee*/Addison-Wesley Publishing Company; 9–14 E. Cable/Tom Stack & Associates; 9–15 Wayland Lee*/Addison-Wesley Publishing Company; 9–18 E. Cable/Tom Stack & Associates; 9–23 Andree Abecassis*; 9–24 R. Kolberg/Taurus Photos; 9–25 Grant Heilman Photography; 9–35 Stephen Frisch*; 9–38 American Heart Association. 10–5 (left) Lenard Lee Rue/Animals, Animals; 10–5 (right) W. H. Hodge/Peter Arnold Inc.; 10–8 Frank S. Balthis. 11–1 J. Serafin/Peter Arnold Inc.; 11–2 Kim Taylor/Bruce Coleman Inc.; 11–3 Lynn M. Stone/Bruce Coleman Inc.; 11–4 Manfred Kage/Peter Arnold Inc.; 11–7 Robert W. Mitchell/Tom Stack & Associates; 11–9 Oxford Scientific Films/Animals, Animals; 11–10 Runk-Schoenberger/Grant Heilman Photography; 11–12 Leonard Lee Rue III/Bruce Coleman Inc.; 11–21 Cindy Charles; 11–23 Wayland Lee*/Addison-Wesley Publishing Company; 11–27 Tom Stack/Tom Stack & Associates; 11–29a William Patterson/Tom Stack & Associates; 11–29b Manfred Kage/Peter Arnold Inc.; 11–33 Cindy Charles; 11–34 Barbara F. Reese/Marine Biological Laboratory, Woods Hole, Massachusetts.

Unit IV

12–1 Dave Lissy/Focus on Sports; 12–6 Runk-Schoenberger/Grant Heilman Photography; 12–8 Ronald F. Thomas/Taurus Photos; 12–9 E. Degginger/Earth Scenes; 12–16 Culver Pictures; 12–20 A. I. Mendeloff & D. E. Smith (eds.), CLINICAL PATHOLOGICAL CONFERENCE, American Journal of Medicine 20:133, 1956; 12–28 Jerry Wachter/Focus On Sports; Perspective 12 (p. 373) Stephen Frisch*; 13–1 Howard Sochurek/Woodfin Camp & Associates; 13–2 Zig Leszczynski/Animals, Animals; 13–3 Manfred Kage/Peter Arnold Inc.; 13–5a Manfred Kage/Peter Arnold Inc.; 13–5b Manfred Kage/Peter Arnold Inc.; 13–9a Edwin R. Lewis/University of California, Berkeley; 13–24 Tom Stack/Tom Stack & Associates; 13–29 Stephen Frisch & Associates*; 14–1 David Scharf/Peter Arnold Inc.; 14–2 Michael & Barbara Reed/Animals, Animals; 14–6 Tom Stack/Tom Stack & Associates; 14–16b J. E. Hawkins, Kresge Hearing Research Institute, University of Michigan; 14–16c J. E. Hawkins, Kresge Hearing Research Institute, University of Michigan; 14–20 E. Degginger/Earth Scenes; 14–21 Dr. W. Aubrey Crich; 14–22 Hans Pfletschinger/Peter Arnold Inc.; 14–27 Edwin R. Lewis/University of California, Berkeley; 14–31 (left) Dr. W. Aubrey Crich; 14–31 (center) G. Mathew Brady/Tom Stack & Associates; 14–31 (right) Hans Pfletschinger/Peter Arnold Inc.

Unit V

15–1 Nelson Max/Lawrence Livermore National Laboratory; 15–2 Andree Abecassis*; 15–7 Jack D. Griffith; 15–8a Jack D. Griffith; 15–8b Runk-Schoenberger/Grant Heilman Photography; 15–12 Runk-Schoenberger/Grant Heilman Photography; 15–14 Manfred Kage/Peter Arnold Inc.; 15–15 Omikron/Taurus Photos; 15–20 David Fritts/Animals, Animals; Perspective 15 (p. 453) The Bettman Archive Inc. 16–1 Jeff Foott/Bruce Coleman Inc.; 16–2 G. Antipa, S. F. State University/BPS; 16–5 Runk-Schoenberger/Grant Heilman Photography; 16–15 Hans Pfletschinger/Peter Arnold Inc.; 16–16a Hans Pfletschinger/Peter Arnold Inc.; 16–16b J. H. Robinson/Animals, Animals; 16–17 Oxford Scientific Films/Animals, Animals; 17–1 © John Elk III 1978; 17–4b Andree Abecassis*; 17–5 Culver Pictures; 17–7 Grant Heilman Photography; 17–8a George Bryce/Earth Scenes; 17–8b Bill Tronca/Tom Stack & Associates; 17–11a S. Krasemann/Peter Arnold Inc.; 17–11b Barry L. Runk/Grant Heilman Photography; 17–12 David Scharf/Peter Arnold Inc.; 17–15 Grant Heilman Photography; 17–17 Robert W. Mitchell/Earth Scenes; 17–18 C. A. Morgan/Peter Arnold Inc.; 17–21 (left) Ralph Reinhold/Animals, Animals; 17–21 (right) Runk-Schoenberger/Grant Heilman Photography; 17–22 (top) Steve Raye/Taurus Photos; 17–22 (bottom) Runk-Schoenberger/Grant Heilman Photography; 17–26 Runk-Schoenberger/Grant Heilman Photography; 17–26 Manfred Kage/Peter Arnold Inc.; 18–1 Phil & Loretta Hermann/Tom Stack & Associates; 18–3 © Lennart Nillsson from *Behold Man*, Little, Brown & Co., Boston; 18–4a © Lennart Nilsson from *A Child Is Born*, Delacorte Press, New York; 18–5 © Lennart Nilsson from *A Child Is Born*, Delacorte Press, New York; 18–9 © Lennart Nilsson from *Behold Man*, Little, Brown & Co., Boston; 18–14a Roy Richardson/Animals, Animals; 18–14b C. Perkins/Animals, Animals; 18–15 Ted Schiffman/Peter Arnold Inc.; 18–16 G. Kooyman/Animals, Animals; 18–18 Runk-Schoenberger/Grant Heilman Photography; 18–19 Oxford Scientific Films/Animals, Animals; 18–20a Breck P. Kent/Animals, Animals; 18–20b Bob McNerling/Taurus Photos; 18–21a J. R. MacGregor/Peter Arnold Inc.; 18–21b Henry Ausloos/Animals, Animals; 18–21c David Fritts/Animals, Animals; 18–22 Erika Stone/Peter Arnold Inc.; Perspective 18 (p. 525) Andree Abecassis*.

Unit VI

19–1 Grant Heilman Photography; 19–8 Wayland Lee*/Addison-Wesley Publishing Company; 19–16a M. Austerman/Animals, Animals; 19–16b Grant Heilman Photography; 19–16c U.S.D.A.; 19–17a Margot Conte/Animals, Animals; 19–17b E. R. Degginger/Animals, Animals; 20–1 Joan Randle/Tom Stack & Associates; 20–3a Runk-Schoenberger/Grant Heilman Photography; 20–3b A. C. Holmes/BPS; 20–8 Andree Abecassis*; 20–11 Grant Heilman Photography; 20–12 Judy Bodycote-Sheldon Wolff/U.C.S.F.; 20–14 Edward A. Robinson/Tom Stack & Associates; 20–16a Runk-Schoenberger/Grant Heilman Photography; 20–16b Runk-Schoenberger/Grant Heilman Photography; 20–17 Andree Abecassis*; 20–18 Wayland Lee*/Addison-Wesley Publishing Company; 20–21 Martin M. Rotker/Taurus Photos; 20–23 Tom McHugh/Photo Researchers, Inc.; 20–24 R. C. Leif/BPS; 20–27 Martin M. Rotker/Taurus Photos; 21–1 L. Bolzoni/Bruce Coleman Inc.; 21–5 NASA; 21–9 Sidney W. Fox/University of Miami; 21–10a Runk-Schoenberger/Grant Heilman Photography; 21–10b Steve Allen/Peter Arnold Inc.; 21–10c W. B. Saunders, Bryn Mawr College/BPS; 21–11 Bjorn Bolstad/Peter Arnold Inc.; 21–13 (left) Steve Martin/Tom Stack & Associates; 21–13 (right) S. Asad/Peter Arnold Inc.; 21–24 Kim Taylor/Bruce Coleman Inc.; Perspective 21 (p. 615) Andre Abecassis*.

Unit VII

22–1 Peter Ward/Bruce Coleman Inc.; 22–2a Darrell Ward/Tom Stack & Associates; 22–2b C. Lockwood/Bruce Coleman Inc.; 22–2c Zig Leszczynski/Animals, Animals; 22–11 Oxford Scientific Films/Earth Scenes; 22–13 Andree Abecassis*; 22–14 Leonard L. Rue III/Bruce Coleman Inc.; 22–15 Norman Tomalin/Bruce Coleman Inc.; 22–16 W. Perry Conway/Grant Heilman Photography; 22–17 Breck P. Kent/Animals, Animals; 22–18 Tom Stack/Tom Stack & Associates; 23–1 Phil &

Loretta Hermann/Tom Stack & Associates; **23–4** Audrey Goldsmith/Department of Forestry and Resource Management/University of California, Berkeley; **23–5** Australian Information Service Photograph; **23–6** C. Lockwood/Animals, Animals; **23–7** Kevin Schafer/Tom Stack & Associates; **23–11** Craig Aurness/West Light; **23–13a** Mickey Gibson/Animals, Animals; **23–13b** M. P. Kahl, Jr./Bruce Coleman Inc.; **23–14a** Frank S. Balthis; **23–14b** W. R. Wright/Taurus Photos; **23–16** Caron Pepper/Tom Stack & Associates; **23–19** M. Kent*; **23–20** J. Messerschmidt/Bruce Coleman Inc.; **24–1** Adrian Davies/Bruce Coleman Inc.; **24–2a** Mark Newman/Animals, Animals; **24–2b** John Gerlach/Earth Scenes; **24–2c** Richard Kolar/Earth Scenes; **24–4** Tom Bean/Tom Stack & Associates; **24–6** Rod Planck/Tom Stack & Associates; **24–11** © John Elk III; **24–12a** Phil & Loretta Hermann/Tom Stack & Associates; **24–12b** Oxford Scientific Films/Animals, Animals; **24–12c** R. Andrew Odum/Peter Arnold Inc.; **24–14** Breck P. Kent/Earth Scenes; **24–15a** Charles G. Summers/Bruce Coleman Inc.; **24–15b** Brian Milne/Animals, Animals; **24–17** Gerhard Gscheidle/Peter Arnold Inc.; **24–18a** Tom Stack/Tom Stack & Associates; **24–18b** Rod Planck/Tom Stack & Associates; **24–18c** L. West/Bruce Coleman Inc.; **24–20** Oxford Scientific Films/Earth Scenes; **24–21a** Wolfgang Bayer/Bruce Coleman Inc.; **24–21b** F. Head/Earth Scenes; **24–21c** G. Ziesler/Peter Arnold Inc.; **24–23** Steve Firebaugh/Bruce Coleman Inc.; **24–24a** Marty Stouffer/Animals, Animals; **24–24b** W. Perry Conway/Grant Heilman Photography; **24–24c** Martin W. Grosnick/Bruce Coleman Inc.; **24–26** Caron Pepper/Tom Stack & Associates; **24–27a** John R. Lewis/Tom Stack & Associates; **24–27b** Stephen Kraseman/Peter Arnold Inc.; **24–29** Runk-Schoenberger/Grant Heilman Photography; **24–30** Grant Heilman Photography; **24–31a** Jeff March/Tom Stack & Associates; **24–31b** Breck P. Kent/Animals, Animals; **24–31c** Richard Langer/Taurus Photos; **24–32** Bill Ross/West Light; **24–33a** Jeff Foott/Bruce Coleman Inc.; **24–33b** Harry Engels/Animals, Animals; **24–33c** F. Mitchell/Tom Stack & Associates; **24–34** Tom Stack/Tom Stack & Associates; **24–35a** Steve Earley/Animals, Animals; **24–35b** Stephen Kraseman/Peter Arnold Inc.; **24–35c** George Harrison/Grant Heilman Photography; **24–36** Frank S. Balthis; **24–37** Zig Leszczynski/Animals, Animals; **24–38** Dale Johnson/Tom Stack & Associates.

Unit VIII

25–1 Warren Garst/Tom Stack & Associates; **25–3** H. S. Terrace/Animals, Animals; **25–5a** Hans Pfletschinger/Peter Arnold Inc.; **25–5b** Tom Stack/Tom Stack & Associates; **25–6** Rod Planck/Tom Stack & Associates; **25–8** Thomas McAvoy/Life Magazine © 1955 Time Inc.; **25–10** Andree Abecassis*; **25–13** Warren & Genny Garst/Tom Stack & Associates; **25–14a** Steven Fuller/Animals, Animals; **25–14b** Fred Whitehead/Animals, Animals; **25–14c** Gary Milburn/Tom Stack & Associates; **25–18** The Massilon Museum/Massillon, Ohio; **25–20** Zig Leszczynski/Animals, Animals; **25–21** Patti Murray/Animals, Animals; **25–22** (left) M. Austerman/Animals, Animals; **25–22** (right) Phil & Loretta Hermann/Tom Stack & Associates; **25–23** Animals, Animals; **25–25** Grant Heilman Photography; **25–26** Hans Pfletschinger/Peter Arnold Inc.; **25–28** Zig Leszczynski/Animals, Animals; **25–29a** Clem Haagner/Bruce Coleman Inc.; **25–29b** Patti Murray/Animals, Animals; **25–31a** Julian Hoffman/Animals, Animals; **25–31b** T. W. Ransom/BPS; **25–32a** T. W. Ransom/BPS; **25–31b** D. Fawcett/Animals, Animals; **Perspective 25** (p. 727) Malcolm Kirk/Peter Arnold Inc. **26–1** Brian Parker/Tom Stack & Associates; **26–2** American Cancer Society; **26–3** Courtesy, Dr. Dennis Mazur; **26–4** Runk-Schoenberger/Grant Heilman Photography; **26–5** Mathew Brady/Tom Stack & Associates; **26–6** Wayland Lee*/Addison-Wesley Publishing Company; **26–8** Manfred Kage/Peter Arnold Inc.; **26–10** Culver Pictures; **26–12** A. K. Harrison, CDC, Atlanta/BPS; **26–13** F. S. Mitchell/Tom Stack & Associates; **26–17** John Hadley, Battelle Pacific Northwest Labs/BPS; **Perspective 26** (p. 747) Andree Abecassis*; **26–19** Andree Abecassis*; **26–20** Culver Pictures; **26–21** Dr. David Greenwood/*Science*, Vol. 163, pp. 1076-1077, 1969; **26–23** Tim Kelly/Black Star; **26–24** Andree Abecassis*.

Cover   Dwight R. Kuhn
Title page   Steven Fuller/Animals, Animals
Unit Openers   **I**—Tom Stack/Tom Stack & Associates; **II**—E. Cable/Tom Stack & Associates; **III**—Dave Davidson/Tom Stack & Associates; **IV**—Steve Firebaugh/Bruce Coleman Inc.; **V**—Langrida-McCoy/Rainbow; **VI**—Karen Donelson/Tom Stack & Associates; **VII**—Frank S. Balthis; **VIII**—Lysbeth Corsi/Tom Stack & Associates.

Special thanks to University of California, San Francisco for assistance with photographs on pages 92, 126, 286, 373 and 399.

*Photographs provided expressly for the publisher

Illustration Credits

4–26 From *The Structure and Action of Proteins* by R. E. Dickerson and I. Geis., Benjamin/Cummings, Menlo Park, CA, 1969. Copyright by Dickerson and Geis. **7–16** From *Animals Without Backbones* 2nd edition by R. Buchsbaum. Copyright © 1948 by The University of Chicago Press. **7–27** from *The Structure and Action of Proteins* by R. E. Dickerson and I. Geis., Benjamin/Cummings, Menlo Park, CA 1969. Copyright by Dickerson and Geis. Perspective 10 (p. 315) Cover illustration by Carol Donner. Copyright January, 1979 by Scientific American, Inc. All rights reserved. **11–24** From *Human Anatomy and Physiology* by A. Spence and E. Mason. Benjamin/Cummings, Menlo Park, CA 1979. **13–25** From Helena Curtis: *Biology* 3rd edition. Worth Publishers, New York 1979. **13–26** From *Human Anatomy and Physiology* by A. Spence and E. Mason. Benjamin/Cummings, Menlo Park, CA 1979. **14–5** From *Physics* by Arthur Beiser. Benjamin/Cummings, Menlo Park, CA 1973. **14–17** From *Human Anatomy and Physiology* by A. Spence and E. Mason. Benjamin/Cummings, Menlo Park, CA 1979. **14–27a** and **14–33b** From *A View of Life* by S. Luria, S. Gould, and S. Singer, Benjamin/Cummings, Menlo Park, CA 1981. **21–19** Adapted from Figure 18–7 from *Life: An Introduction to Biology*, 2nd edition by George Gaylord Simpson and William S. Beck. By permission of Harcourt Brace Jovanovich, Inc., 1965. **25–11** Copyright © 1982 by The New York Times Company. Adapted and reprinted by permission.

Text Credits

Page 235: From p. 199 of *The Body Has A Head* by Gustav Eckstein. Copyright © 1969, 1970 by Gustav Eckstein. Reprinted by permission of Harper & Row, Publishers, Inc. Page 455: Copyright © 1972 by Payson R. Stevens. Reprinted by permission. Page 559: From p. 599 of *The Ascent of Man* by J. Bronowski. Copyright © 1973 by J. Bronowski. Reprinted by permission of Little, Brown and Company.